AF595356

Evolutionary Process for Engineering Optimization

Evolutionary Process for Engineering Optimization

Editors

Amir H. Gandomi
Laith Abualigah

MDPI • Basel • Beijing • Wuhan • Barcelona • Belgrade • Manchester • Tokyo • Cluj • Tianjin

Editors
Amir H. Gandomi
University of Technology
Sydney
Australia

Laith Abualigah
Al-Ahliyya Amman University
Jordan
Middle East University
Jordan
Amman Arab University
Jordan
Universiti Sains Malaysia
Malaysia

Editorial Office
MDPI
St. Alban-Anlage 66
4052 Basel, Switzerland

This is a reprint of articles from the Special Issue published online in the open access journal *Processes* (ISSN 2227-9717) (available at: https://www.mdpi.com/journal/processes/special_issues/Evolutionary_Process).

For citation purposes, cite each article independently as indicated on the article page online and as indicated below:

LastName, A.A.; LastName, B.B.; LastName, C.C. Article Title. *Journal Name* **Year**, *Volume Number*, Page Range.

ISBN 978-3-0365-4771-8 (Hbk)
ISBN 978-3-0365-4772-5 (PDF)

Contents

About the Editors

Amir H. Gandomi

Amir H. Gandomi is among the world's most-cited researchers for his work in the fields of global optimization and big data analytics, in particular, using machine learning and evolutionary computations. He is an ARC DECRA Fellow at the Faculty of Engineering and Information Technology at UTS, where he is a Professor of Data Science. He has received multiple prestigious awards for his research excellence and impact, such as the 2022 Walter L. Huber Prize, which is known as the highest-level mid-career research award in all areas of civil engineering. Amir has published more than 300 journal papers and nine books, which, collectively, have more than 30,000 citations (with an H-index of 79), and he has been named one of the world's most influential scientific minds and highly-cited researchers by the influential Clarivate Analytics for five consecutive years, to 2021. He is also ranked 17th among more than 12,000 researchers in the online computer science bibliography, Genetic Programming bibliography, and ranks first in Australia. Based on Analysis Mendeley data by John Ioannidis et al. (Stanford), looking at researchers' impact, Amir is ranked 16,091 among all researchers for career-long impact. He ranked 237 in AI and Image Processing for career-long impact. He ranked 1,279 among all researchers in 2020; and he ranked 59 in AI and Image Processing in 2020. He has served as associate editor, editor, and guest editor in several prestigious journals such as AE of IEEE TBD and IEEE IoTJ. He regularly delivers keynote addresses at major conferences. Prior to joining UTS, Amir was an Assistant Professor at the School of Business at Stevens Institute of Technology in the US. He was also a distinguished research fellow in the BEACON Center, Michigan State University, where biologists, computer scientists, and engineers together study evolution and apply their knowledge to real-world problems.

Laith Abualigah

Laith Abualigah is an Assistant Professor at the Computer Science Department, Amman Arab University, Jordan. He is also a distinguished researcher at the School of Computer Science, Universiti Sains Malaysia, Malaysia. He received his first degree from Al-Albayt University, Computer Information System, Jordan, in 2011. He earned a Master's degree from Al-Albayt University, Computer Science, Jordan, in 2014. He received a Ph.D. degree from the School of Computer Science in Universiti Sains Malaysia (USM), Malaysia, in 2018. According to the report published by Stanford University in 2020, Abualigah is one of the 2% influential scholars, which depicts the 100,000 top scientists in the world. Abualigah has published more than 220 journal papers and books, which collectively have been cited more than 7000 times (H-index = 38). His main research interests focus on Arithmetic Optimization Algorithm (AOA), Bio-inspired Computing, Nature-inspired Computing, Swarm Intelligence, Artificial Intelligence, Meta-heuristic Modeling, and Optimization Algorithms, Evolutionary Computations, Information Retrieval, Text clustering, Feature Selection, Combinatorial Problems, Optimization, Advanced Machine Learning, Big data, and Natural Language Processing. Abualigah currently serves as an associate editor of the Journal of Cluster Computing (Springer), the Journal of Soft Computing (Springer), and Journal of King Saud University - Computer and Information Sciences (Elsevier).

Preface to "Evolutionary Process for Engineering Optimization"

Various real-world engineering applications, such as engineering design, industrial manufacturing systems, and water distribution networks, are complex problems. Evolutionary computation is a hot topic of interest amongst researchers in various disciplines of engineering and science. Evolutionary computation is a group of optimization algorithms used for solving global optimization problems, which is inspired by biological evolution. It includes various signal and population-based methods with a meta-heuristic or stochastic optimization part.

In recent years, evolutionary computation methods have been successfully utilized to address complex real-world problems. The literature is abundant with several other approaches that share the same goal: to find a new optimal solution with satisfactory quality by alternating research strategies. Many theoretical and experimental studies have proved significant evolutionary computation properties. The most famous evolutionary computation methods are the genetic algorithm (GA), evolution strategy (ES), differential evolution (DE), particle swarm optimization (PSO), bacterial foraging optimization (BFO), ant colony optimization (ACO), and the memetic algorithm (MA). However, with the fast growth of complex systems, optimization problems become much larger and complicated. The common issues facing evolutionary algorithms are the dimension of objective functions, decision variables, or constraints.

In light of the expanding interest for new innovative methods of solving real-world and engineering optimization problems, this Special Issue intends to promote high-quality research outputs in the latest progress and improvement of evolutionary algorithms and engineering applications and offers recent advanced studies in the field to serve researchers and practitioners. The main interest is on interdisciplinary research on the evolutionary algorithm, using modern computational intelligence theories, methods, and practices. We invite researchers to submit their original contributions addressing particular challenging aspects in evolutionary computation from both theoretical and applied viewpoints.

Amir H. Gandomi and Laith Abualigah
Editors

Article

Material Generation Algorithm: A Novel Metaheuristic Algorithm for Optimization of Engineering Problems

Siamak Talatahari [1,2], Mahdi Azizi [1] and Amir H. Gandomi [3,*]

[1] Department of Civil Engineering, University of Tabriz, Tabriz 5166616471, Iran; siamak.talat@gmail.com (S.T.); mehdi.azizi875@gmail.com (M.A.)
[2] Engineering Faculty, Near East University, North Cyprus, Mersin 10, Turkey
[3] Faculty of Engineering & Information Technology, University of Technology Sydney, Ultimo, Sydney, NSW 2007, Australia
* Correspondence: gandomi@uts.edu.au

Abstract: A new algorithm, Material Generation Algorithm (MGA), was developed and applied for the optimum design of engineering problems. Some advanced and basic aspects of material chemistry, specifically the configuration of chemical compounds and chemical reactions in producing new materials, are determined as inspirational concepts of the MGA. For numerical investigations purposes, 10 constrained optimization problems in different dimensions of 10, 30, 50, and 100, which have been benchmarked by the Competitions on Evolutionary Computation (CEC), are selected as test examples while 15 of the well-known engineering design problems are also determined to evaluate the overall performance of the proposed method. The best results of different classical and new metaheuristic optimization algorithms in dealing with the selected problems were taken from the recent literature for comparison with MGA. Additionally, the statistical values of the MGA algorithm, consisting of the mean, worst, and standard deviation, were calculated and compared to the results of other metaheuristic algorithms. Overall, this work demonstrates that the proposed MGA is able provide very competitive, and even outstanding, results and mostly outperforms other metaheuristics.

Keywords: material generation algorithm; constrained problems; metaheuristic algorithm; optimization; engineering design problem

Citation: Talatahari, S.; Azizi, M.; Gandomi, A.H. Material Generation Algorithm: A Novel Metaheuristic Algorithm for Optimization of Engineering Problems. *Processes* **2021**, *9*, 859. https://doi.org/10.3390/pr9050859

Academic Editors: Luis Puigjaner and Ján Piteľ

Received: 30 March 2021
Accepted: 10 May 2021
Published: 13 May 2021

1. Introduction

Optimization techniques have been proposed for the optimum design of different problems of everyday life in order to increase the efficiency of systems and human resources. Most of the design problems in nature are complex, with multiple design variables and constraints that classical optimization algorithms, such as gradient-based algorithms, cannot handle. As a solution, numerous artificial intelligence experts have introduced new algorithms with better performance in different fields. Regarding the recent developments in technology, new optimization methods offering higher efficiency, greater accuracy, and increased speed rate are required to deal with difficult optimization problems.

Based on the mentioned concerns about the capabilities of optimization algorithms, a "metaheuristic" approach has been proposed by optimization experts [1] for solving different optimization problems. 'Metaheuristic' refers to specific solution techniques, where higher-level strategies are implemented into the main searching process of the optimization algorithms to provide a powerful searching method with specific capabilities, including the avoidance of entrapment in local optimal solutions. The history of developing different metaheuristic approaches as solutions in different optimization fields can be classified into five different time periods. A brief summary of these historical time periods is presented in Table 1.

Table 1. Summary of historical time periods for the evolution of metaheuristics [2].

Duration	Period	Achievement
Pre-1940	Pre-Theoretical	Limited applications without formal presentation.
1940–1980	Early	Introduction of heuristics approaches.
1980–2000	Method-Centric	Proposal and improvement of metaheuristics algorithms for different applications.
2000–Present	Framework-Centric	Utilization of metaheuristic frameworks in different fields.
Future	Scientific or Future	Future development and design of metaheuristics as a matter of science rather than a matter of art.

With the evolution of numerous metaheuristic algorithms, four different types could be distinguished in terms of their main concepts and inspirations. The first category includes "evolutionary algorithms," such as the Memetic Algorithm (MA) [3], Genetic Algorithm (GA) [4], Genetic Programming (GP) [5], Differential Evolution (DE) [6], Evolution Strategies (ES) [7], and the Biogeography-Based Optimizer (BBO) [8], that have been proposed based on the biological reproduction and evolution. The second category contains swarm intelligence-based optimization algorithms, which are based on the cooperative behavior of self-organized and decentralized artificial or natural systems. Some well-known methods of this category are Particle Swarm Optimization (PSO) [9], Ant Colony Optimization (ACO) [10], Artificial Bee Colony (ABC) [11], Cat Swarm Optimization (CSA) [12], Firefly Algorithm (FA) [13], and Krill Herd (KH) algorithm [14]. The third category consists of algorithms that are motivated by physical laws, such as Simulated Annealing (SA) [15], Harmony Search (HS) [16], Big-Bang Big-Crunch (BBBC) [17], Gravitational Search Algorithm (GSA) [18], Charged System Search (CSS) algorithm [19], Artificial Chemical Reaction Optimization Algorithm (ACROA) [20], Colliding Bodies Optimization (CBO) [21], Chaos Game Optimization (CGO) [22,23], and Atomic Orbital Search (AOS) [24] algorithm. Finally, metaheuristic approaches inspired by the lifestyle of animals or humans are classified in the fourth category, which includes Imperialistic Competitive Algorithm (ICA) [25], Cuckoo Search Algorithm (CSA) [26]. In addition to these metaheuristic algorithms, other difficult challenges have been solved by upgrading, developing, and hybridizing standard algorithms [27–36].

In this paper, a novel metaheuristic algorithm called the Material Generation Algorithm (MGA) is proposed as an alternative approach for solving optimization problems. The main concept of this novel algorithm is based on the principles of chemistry, regarding the production of new materials according to the configurations of chemical compounds and reactions. To evaluate the performance of MGA, we tested it on 15 well-known engineering design problems and 10 constrained mathematical problems in different dimensions (10, 30, 50, and 100), which have been benchmarked by the Competitions on Evolutionary Computation (CEC) and presented in detail by Wu et al. [37] at CEC 2017. The utilized references include the results of CEC 2017, Tvrdík and Poláková [38], Polakova [39], and Zamuda [40]. The Friedman Test [41] is also conducted as a well-known statistical test in order to have a fair judgment about the performance of the MGA.

In recent decades there has be a great challenge for the algorithm developers to develop new solution methods which could have better performance than the previous methods in dealing with complex real-world problems. Due to the massive emergence of novel metaheuristic algorithms in the past few decades, this aspect has been addressed by Sorensen [42] as a tsunami of methods which will have advantages and also disadvantages in the soft computing fields in the future. However, this issue can be justified by discovering other aspects of proposing novel algorithms which is based on the source of inspirational concept of a novel algorithm which should be reasonable enough to be justified alongside a well-developed mathematical model as two of the most important principles of metaheuristic algorithms. Regarding the fact that when a novel algorithm is proposed, it is evaluated by some of the benchmark test problems which has been solved by multiple methods in order to demonstrate its capability as an independent algorithm among the other methods while this kind of proposing a testing the algorithms is not the only aim of this area. The proposed novel algorithm can be of a great help in the situations that the other alternatives cannot reach to a reasonable response in dealing with a considered problem so there should be other alternatives in order to have a good chance to provide a well-designed plan for the industry and even human-related actions in the everyday life. A brief outline of this work is as follows:

Section 2 discusses the inspirational concept and mathematical model of the MGA optimization algorithm. In Section 3, the problem statements, including the selected mathematical and engineering optimization problems utilized to test the proposed MGA as a novel metaheuristic algorithm, are presented. In Sections 4 and 5, the numerical results of the MGA algorithm and other alternative metaheuristic methods in dealing with the considered mathematical and engineering optimization problems are presented. In Section 6, the key findings of this research work are concluded, future research directions are suggested.

2. Material Generation Algorithm

In this section, the inspiration of MGA as a novel metaheuristic algorithm and the mathematical model of this algorithm are presented.

2.1. Inspiration

A material is a mixture of multiple substances comprised of the stuffs of the universe with volume and mass. The material generation process concerns the capability of different substances to merge with each other in order to generate new materials with higher functionality and improved energy levels. Elements are the basic building blocks of the materials, which cannot be broken into parts or even changed into other elements. Materials are engineered on an atomic, nano-, micro-, or macro-scale in order to control the specific properties and improve the performance of a material. Uniquely-generated materials are classified based on their general properties and specific characteristics and according to physical and chemical changes that influence a material's behavior.

Material chemistry is one of the most important disciplines in the material research field. Material engineers study the configuration of materials in order to improve the specific characteristics of materials, developing new ones that are more sustainable and also superior to the previous ones. Chemical changes in materials are achieved by reacting and combining various chemicals. In general, the chemical properties are altered by the transferring or sharing of electrons between atoms of different materials, specifically, chemical bonds formed between materials result in such modifications. In this work, three main concepts of material chemistry (compounds, reactions, and stability) were considered to formulate a metaheuristic optimization algorithm.

2.1.1. Chemical Compound

Most chemical elements in the universe are created through combinations with other elements. With that being, a few chemical elements exist freely in nature. Compounds are formed by combining multiple chemicals via chemical bonds, or the transferring or sharing of electrons, which result in one of the following:

- Ionic compounds are created when electrons are transferred from the atoms of one element to those of another.
- Covalent compounds form when electrons are shared between atoms of different elements.

In addition, ionic compounds contain multiple ions that are held together by the electrostatic force called ionic bonding. Although these compounds are neutral in nature, they consist of some negatively- and positively-charged ions, called anions and cations, respectively. The evaporation, precipitation, or freezing of the constituent ions are the main factors in the process of producing ionic compounds. When an atom or a small group of atoms starts to lose or gain electrons, an ionic compound forms according to the ionic bonding and charged particles. As an example, the formation of sodium chloride, also known as table salt, is depicted in Figure 1. In the process of electron transformation, a sodium (neutral) becomes a sodium cation (Na^+) when it loses one electron. In addition, Cl becomes a chloride anion (Cl^-) when it gains an electron. Thus, table salt is a solid aggregation of Na^+ and Cl^- ions, which attract each other due to opposite charges.

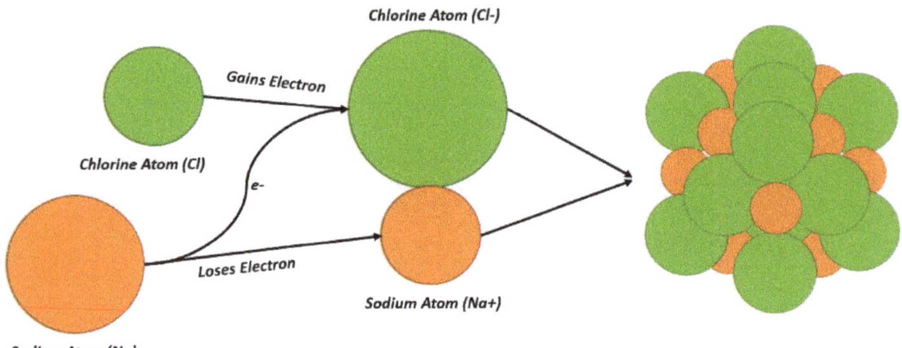

Figure 1. The formation of an ionic compound, NaCl.

Covalent compounds form when an atom of a chemical element shares an electron with another element's atom, which usually occurs between nonmetal elements and results in an electrically neutral atom. Figure 2 displays the formation of a covalent compound that leads to the hydrogen atom. As an example, assuming that two hydrogen atoms begin approaching each other, the nucleus of one atom strongly attracts the electron of the other one. A covalent bond is achieved when a specific distance between the nuclei is reached, and the electrons are equally shared. The net repulsion between nuclei is ignored due to the greater net attraction.

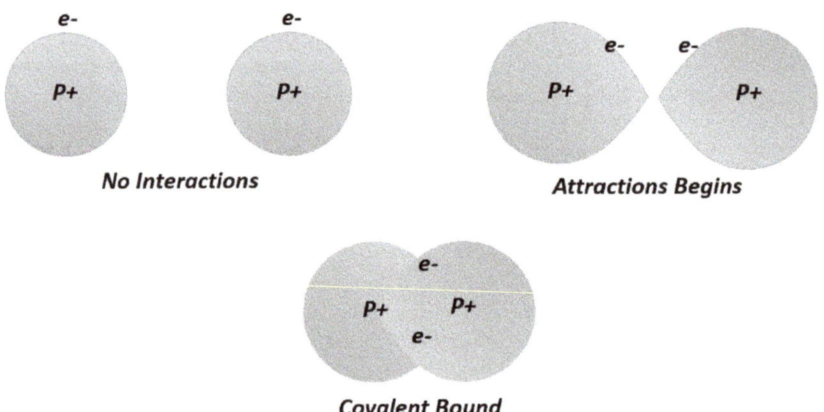

Figure 2. The formation of a covalent compound by means of two hydrogen atoms.

2.1.2. Chemical Reaction

Chemical reactions are the process of transforming one material into another while the chemical equations are used to represent chemical reactions, where the resulting products will have different properties than the starting materials (reactants/reagents), and intermediate materials (in some particular cases).

An example of a chemical reaction is depicted in Figure 3, in which the magnesium wire (Mg) and oxygen gas (O_2) yield powdery magnesium oxide (MgO). As presented in the left bulb, a fine magnesium filament is surrounded by oxygen before the reaction occurs. As the reaction proceeds, the white colored powdery magnesium oxide coats the bulb's inner surface, which is demonstrated in the right bulb. In this reaction, heat and light are also produced as intermediate materials but are not concerned in this description. The chemical equation of the presented chemical reaction is as follows:

$$2\text{Mg (s)} + \text{O}_2\text{ (g)} \overset{\text{Electricity}}{\rightarrow} 2\text{MgO (s)}$$

where s and g stand for solid and gas, respectively.

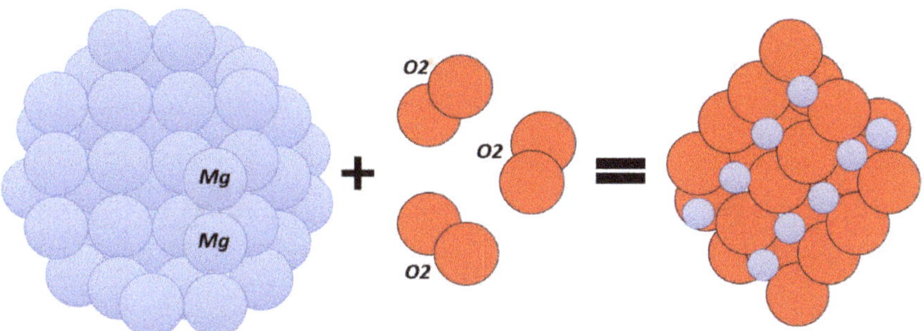

Figure 3. The formation of a chemical reaction.

2.1.3. Chemical Stability

Stability is one of the more important properties of materials in real-world applications. When generating new materials with different characteristics, it is important to consider the stability of the chemical compounds and reactions in different situations. In terms of chemical stability, chemicals have the tendency to resist changes, such as decomposition, due to internal factors and external influences such as heat, air, light, and pressure. Chemical stability is the resistance of a material to change in the presence of other chemicals. A stable chemical product refers to one that has not been specifically reactive in the environment and retains its properties over a specific period of time. Comparatively, unstable chemical materials easily decompose, corrode, polymerize, explode, or burn under certain conditions.

When producing new chemical materials, the processes of transferring or sharing electrons within the initial materials will occur in such a way that the end product will be stable and applicable during a specific period of time.

2.2. Mathematical Model

In order to conduct an optimum design procedure, an optimization algorithm is developed in this section based on the mentioned principles of material chemistry. The basic concepts of the chemical compounds, reactions, and stability are utilized in order to develop and formulate a well-defined mathematical model for the new algorithm. Considering that many natural evolution algorithms establish a predefined population of solution candidates that are evolved through random alterations and selection, MGA determines a number of materials (Mat) comprised of multiple periodic table elements ($PTEs$). In this algorithm, a number of materials is considered as the solution candidates (Mat_n), which are comprised of some elements represented as decision variables (PTE_i^j). The mathematical presentation of these two aspects is as follows:

$$Mat = \begin{bmatrix} Mat_1 \\ Mat_2 \\ \vdots \\ Mat_i \\ \vdots \\ Mat_n \end{bmatrix} = \begin{bmatrix} PTE_1^1 & PTE_1^2 & \cdots & PTE_1^j & \cdots & PTE_1^d \\ PTE_2^1 & PTE_2^2 & \cdots & PTE_2^j & \cdots & PTE_2^d \\ \vdots & \vdots & & \vdots & \ddots & \vdots \\ PTE_i^1 & PTE_i^2 & \cdots & PTE_i^j & \cdots & PTE_i^d \\ \vdots & \vdots & & \vdots & \ddots & \vdots \\ PTE_n^1 & PTE_n^2 & \cdots & PTE_n^j & \cdots & PTE_n^d \end{bmatrix}, \begin{cases} i = 1,2,\ldots,n. \\ j = 1,2,\ldots,d. \end{cases} \tag{1}$$

where d is the number of elements (decision variables) in each material (solution candidates); and n is the number of materials considered to be the solution candidates.

In the first stage of the optimization process, PTE_i^j is determined randomly while the decision variables bounds are defined based on the considered problem. The initial positions of $PTEs$ are determined randomly in the search space as follows:

$$PTE_i^j(0) = PTE_{i,min}^j + Unif(0,1).\left(PTE_{i,max}^j - PTE_{i,min}^j\right), \begin{cases} i = 1,2,\ldots,n. \\ j = 1,2,\ldots,d. \end{cases} \tag{2}$$

where $PTE_i^j(0)$ determines the initial value of the jth element in the ith material; $PTE_{i,min}^j$ and $PTE_{i,max}^j$ are the minimum allowable and maximum allowable values for the jth decision variable of the ith solution candidate, respectively; and $Unif(0,1)$ is a random number in the interval of [0, 1].

2.2.1. Modeling Chemical Compound

To mathematically model the chemical compounds, all $PTEs$ are assumed to be in the ground state, which can be externally excited by the magnetic fields, absorption of energy from photons or light and interactions with different colliding bodies or particles regarding ions or other individual electrons. Due to the different stabilities of elements, they have a tendency to lose, gain, or even share electrons with other $PTEs$, resulting in ionic or covalent compounds. To model the ionic and covalent compounds, d random $PTEs$ are selected using the initial Mat (Equation (1)). For the selected $PTEs$, the processes of losing, gaining, or sharing electrons are modeled through the probability theory. To fulfill this aim, a continuous probability distribution is utilized for each PTE to configure a chemical compound, which is considered as a new PTE, as follows:

$$PTE_{new}^k = PTE_{r_1}^{r_2} \pm e^-, \; k = 1,2,\ldots,d. \tag{3}$$

where r_1 and r_2 are uniformly distributed random integers in the intervals of [1, n] and [1, d], respectively; $PTE_{r_1}^{r_2}$ is a randomly selected PTE from the Mat; e^- is the probabilistic component for modeling the process of losing, gaining or sharing electrons represented with normal Gaussian distribution in the mathematical model; and PTE_{new}^k is the new material.

The newly-created $PTEs$ are utilized for producing a new material (Mat_{new_1}), which is then added to the initial material list (Mat) as a new solution candidate:

$$Mat_{new_1} = \begin{bmatrix} PTE_{new}^1 \, PTE_{new}^2 & \cdots & PTE_{new}^k & \cdots & PTE_{new}^d \end{bmatrix}, \; k = 1,2,\ldots,d. \tag{4}$$

Then, the overall solution candidates are combined and presented as follows:

$$Mat = \begin{bmatrix} Mat_1 \\ Mat_2 \\ \vdots \\ Mat_i \\ \vdots \\ Mat_n \\ Mat_{new_1} \end{bmatrix} = \begin{bmatrix} PTE_1^1 & PTE_1^2 & \cdots & PTE_1^j & \cdots & PTE_1^d \\ PTE_2^1 & PTE_2^2 & \cdots & PTE_2^j & \cdots & PTE_2^d \\ \vdots & \vdots & & \vdots & \ddots & \vdots \\ PTE_i^1 & PTE_i^2 & \cdots & PTE_i^j & \cdots & PTE_i^d \\ \vdots & \vdots & & \vdots & \ddots & \vdots \\ PTE_n^1 & PTE_n^2 & \cdots & PTE_n^j & \cdots & PTE_n^d \\ PTE_{new}^1 & PTE_{new}^2 & \cdots & PTE_{new}^k & \cdots & PTE_{new}^d \end{bmatrix}, \begin{cases} i = 1,2,\ldots,n. \\ j = 1,2,\ldots,d. \\ k = 1,2,\ldots,d. \end{cases} \quad (5)$$

A schematic presentation of the described process for the configuration of new materials based on the concept of chemical compounds (ionic and covalent) is depicted in Figure 4.

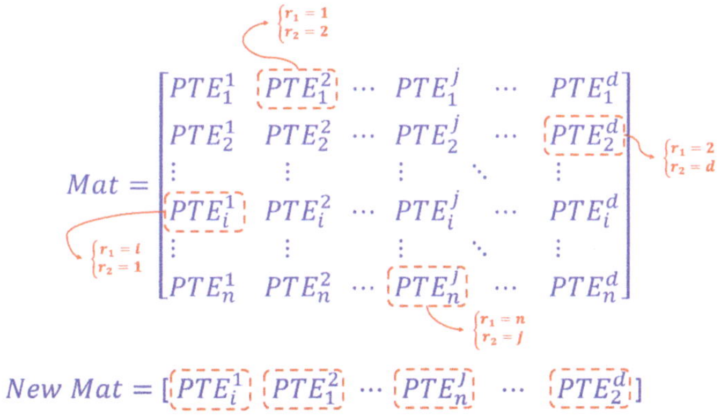

$$Mat = \begin{bmatrix} PTE_1^1 & PTE_1^2 & \cdots & PTE_1^j & \cdots & PTE_1^d \\ PTE_2^1 & PTE_2^2 & \cdots & PTE_2^j & \cdots & PTE_2^d \\ \vdots & \vdots & & \vdots & \ddots & \vdots \\ PTE_i^1 & PTE_i^2 & \cdots & PTE_i^j & \cdots & PTE_i^d \\ \vdots & \vdots & & \vdots & \ddots & \vdots \\ PTE_n^1 & PTE_n^2 & \cdots & PTE_n^j & \cdots & PTE_n^d \end{bmatrix}$$

$$\begin{cases} r_1 = 1 \\ r_2 = 2 \end{cases} \quad \begin{cases} r_1 = 2 \\ r_2 = d \end{cases} \quad \begin{cases} r_1 = i \\ r_2 = 1 \end{cases} \quad \begin{cases} r_1 = n \\ r_2 = j \end{cases}$$

$$New\ Mat = [PTE_i^1 \; PTE_1^2 \; \cdots \; PTE_n^j \; \cdots \; PTE_2^d]$$

Figure 4. The schematic presentation of the random periodic table elements (PTE) selection and creating new materials.

The probabilistic approach for determining e^- is modeled through normal Gaussian distribution, which is important in statistics and often used in the natural and social sciences to represent real-valued random variables with unknown distributions. The probability of selecting a new element (PTE_{new}^k) regarding the randomly selected initial element $(PTE_{r_1}^{r_2})$ is presented as follows:

$$f\left(PTE_{new}^k \middle| \mu, \sigma^2\right) = \frac{1}{\sqrt{2\pi\sigma^2}}.e^{\frac{-(x-\mu)^2}{2\sigma^2}}, \quad k = 1,2,\ldots,d. \quad (6)$$

where μ is the mean, median or expectation of the distribution correspond to the selected random PTE $(PTE_{r_1}^{r_2})$; σ is the standard deviation, which is set to unity in this paper; σ^2 is the variance; and e is the natural base or Naperian base of the natural logarithm.

2.2.2. Modeling Chemical Reaction

Chemical reactions are sort of production process in which different chemical changes are determined in order to produce different products with modified properties even different from the initial reactants. In order to mathematically model the process of producing new materials by the chemical reaction concept, an integer random number (l) is determined regarding the number of materials of the initial Mat are considered for participating in a chemical reaction. Then, l integer random numbers (mj) are generated to determine the positions of the selected materials in the initial Mat so, the new solutions are linear combinations of the other solutions. For each material, a participation factor (p) is also calculated since different materials would participate in the reactions with different amounts. A schematic presentation of the described process is depicted in Figure 5, and the mathematical presentation is as follows:

$$Mat_{new_2} = \frac{\sum_{m=1}^{l}(p_m.Mat_{mj})}{\sum_{m=1}^{l}(p_{mj})}, \; j = 1,2,\ldots,l. \tag{7}$$

where Mat_m is the mth randomly selected material from the initial Mat; p_m is the normal Gaussian distribution for the mth material participation factor; and Mat_{new_2} is the new material produced by the chemical reaction concept.

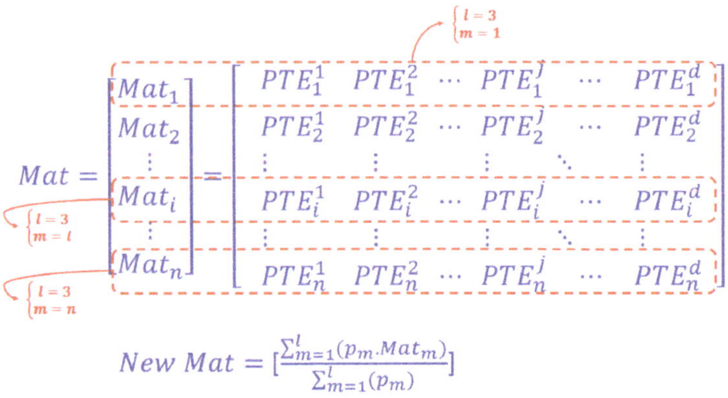

Figure 5. The schematic view of the random material selection for creating new materials.

2.2.3. Modeling Chemical Stability

As previously described, the principle of material stability concerns the tendency of natural systems to seek local and general equilibria at all structural levels. Material stability is mathematically represented by determining the quality of the solutions as *Mat*. Materials with the highest stability levels alongside the ones with lowest stability levels are equivalent to the best and worst fitness values of all solution candidates in the optimization runs.

Considering the chemical compound and chemical reaction configuration approaches, the overall solution candidates are combined as follows:

$$Mat = \begin{bmatrix} Mat_1 \\ Mat_2 \\ \vdots \\ Mat_i \\ \vdots \\ Mat_n \\ Mat_{new1} \\ Mat_{new2} \end{bmatrix}, \; i = 1,2,\ldots,n. \tag{8}$$

Moreover, the stability levels of the initial material and newly0produced materials should be considered in order to decide whether or not the new materials should be included in the overall material list (*Mat*) corresponding to the solution candidates. The quality of new solution candidates is then compared to the initial ones, whereby the new materials should be substituted by initial materials with worst fitness values corresponding to worst stability levels.

For boundary violation control, a flag is determined in order to control the violating solution candidates while a maximum number of iteration or objective function evaluation can be considered as stopping criteria. The flowchart of the MGA algorithm is presented in Figure 6.

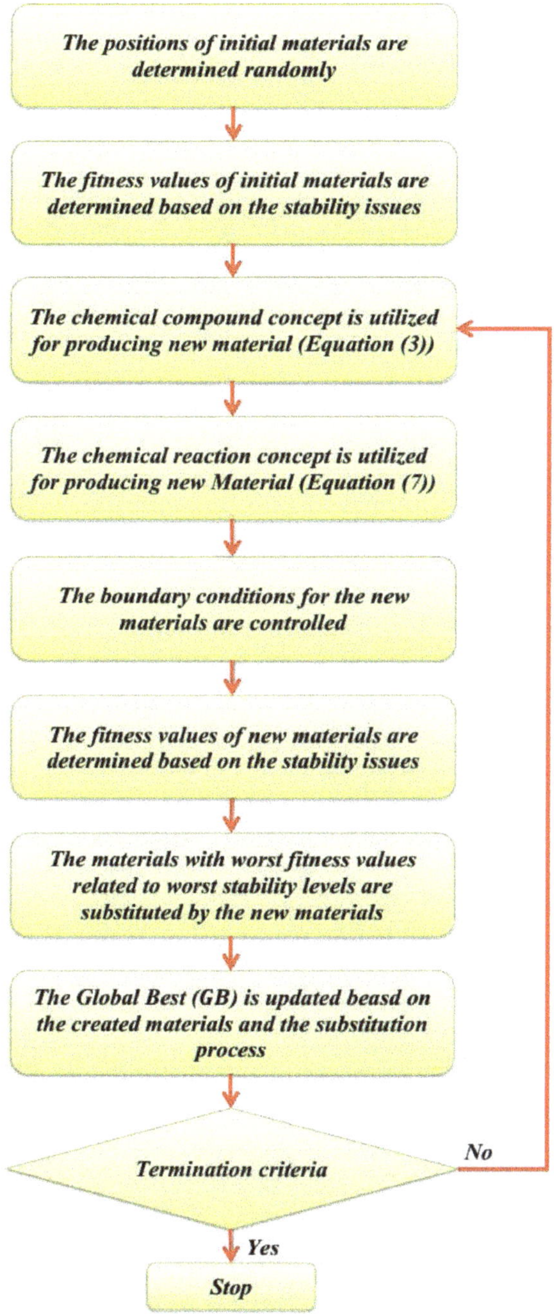

Figure 6. Flowchart of the Material Generation Algorithm (MGA).

3. Problem Statement

In this section, a brief description of the considered design examples is presented. Regarding the fact that these examples are categorized as constrained optimization problems, the general formulations of these kinds of optimization problems are presented as follows:

$$f(\overline{x}),\ \ \overline{x} = x_1,\ x_2,\ \ldots,\ x_n \tag{9}$$

$$g_i(\overline{x}) \leq 0,\ \ i = 1,\ 2,\ \ldots,\ n \tag{10}$$

$$h_j(\overline{x}) = 0,\ \ j = 1,\ 2,\ \ldots,\ m \tag{11}$$

where $f(\overline{x})$ is considered as the objective function of the optimization problem that can be considered to be maximized or minimized; $g_i(\overline{x})$ and $h_j(\overline{x})$ are the ith and jth inequality and equality constraint, respectively; $\overline{x}$ is the position vector related to the optimization variables; and n and m are the total number of inequality and equality constraints, respectively.

In most cases, the equality constraints can be transformed into inequality constraints by considering the following:

$$\left|h_j(\overline{x})\right| - \varepsilon \leq 0,\ \ j = 1,\ 2,\ \ldots,\ m \tag{12}$$

where ε is a predefined small positive number, which is typically near to zero. In this work, ε was set to 0.0001.

3.1. Mathematically-Constrained Problems

The mathematical problems of the CEC 2017 benchmark suite are presented in Table 2, while the specific details and mathematical formulations were presented in detail by Wu et al. [39]. In order to evaluate the results of the proposed MGA, the statistical results of different state-of-the-art metaheuristic algorithms regarding the considered constrained problems were derived of the recent literature [38–40].

Table 2. Brief description of the Competitions on Evolutionary Computation (CEC) 2017 mathematical constrained problems [37].

No.	Type	D	H	G	Bounds
C_1	Non Separable	10, 30, 50 and 100	0	1	$-100 \leq x_i \leq 100$
C_2	Non Separable	10, 30, 50 and 100	0	1	$-100 \leq x_i \leq 100$
C_3	Non Separable	10, 30, 50 and 100	1	1	$-100 \leq x_i \leq 100$
C_4	Separable	10, 30, 50 and 100	0	2	$-10 \leq x_i \leq 10$
C_5	Non Separable	10, 30, 50 and 100	0	2	$-10 \leq x_i \leq 10$
C_6	Separable	10, 30, 50 and 100	6	0	$-20 \leq x_i \leq 20$
C_7	Separable	10, 30, 50 and 100	2	0	$-50 \leq x_i \leq 50$
C_8	Separable	10, 30, 50 and 100	2	0	$-100 \leq x_i \leq 100$
C_9	Separable	10, 30, 50 and 100	2	0	$-10 \leq x_i \leq 10$
C_{10}	Separable	10, 30, 50 and 100	2	0	$-100 \leq x_i \leq 100$

D: Dimensions; G: Number of inequality constraints; H: Number of equality constraints.

3.2. Engineering Design Problems

The second type of constrained problems included 15 well-known engineering problems, which have been solved by different optimization algorithms. A brief description of these design examples is presented in Table 3, and the specific details of each example are provided in the following subsections. These examples have also been benchmarked by Kumar et al. [43] regarding the CEC 2020 engineering design scheme.

Table 3. Description of the constrained engineering design problems.

No.	Name	D	G	H	Formulation
F_1	Speed Reducer	7	11	0	[44]
F_2	Tension/Compression Spring	3	4	0	[45]
F_3	Pressure Vessel	4	4	0	[45]
F_4	Welded Beam	4	7	0	[45]
F_5	Three-Bar Truss	2	3	0	[46]
F_6	Multiple Disk Clutch Brake	5	8	0	[47]
F_7	Planetary Gear Train	9	10	1	[48]
F_8	Step-Cone Pulley	5	8	3	[49]
F_9	Hydrostatic Thrust Bearing	4	7	0	[50]
F_{10}	Ten-Bar Truss	10	3	0	[51]
F_{11}	Rolling Element Bearing	10	9	0	[52]
F_{12}	Gear Train	4	1	1	[53]
F_{13}	Steel I-Shaped Beam	4	2	0	[46]
F_{14}	Piston Lever	4	4	0	[46]
F_{15}	Cantilever Beam	5	1	0	[46]

D: Dimensions; G: Number of inequality constraints; H: Number of equality constraints; Min: Feasible Solutions.

4. Numerical Results of Mathematical Problems

The numerical results based on the CEC 2017 benchmark problems by means of the MGA and other alternatives in dealing with the described constrained problems with different dimensions of 10, 30, 50, and 100 are presented in this section. For comparison, a total of 25 optimization runs was performed, including a maximum number of function evaluations (20,000 $\times$ D), where D is the problem dimension. These results are presented in Tables 4–7 for different dimensions, in which (c) is the number of violated constraints consisting of the number of violations by more than 1, 0.01, and 0.0001; ($\overline{v}$) is the mean violation at the median solution; (SR) is the feasibility rate defined as the ratio of feasible runs to total runs; and ($\overline{vio}$) is the mean constraint violation values of all optimization runs.

Based on the obtained results of MGA in dealing with the mathematical constrained problems of CEC 2017 with a dimension of 10, MGA was superior to the other metaheuristics in most of the cases. Considering the functions with dimensions of 30, MGA outranks two of the alternative metaheuristics while in comparing to the third one, the results of MGA are so competitive. In dealing with functions of 50 and 100 dimensions, the results of MGA are comparable to the others.

Regarding the fact that the considered problems of the CEC 2017 benchmark suite are all the latest problems in the evolutionary computation field with higher levels of complexity and difficulties while there are few approaches that can provide acceptable results in dealing with these problems. In this regard, the reported results by MGA are marginal because there are not any better results for the considered problems in the literature so the MGA calculated the latest reported results which demonstrate the capability of this algorithm in competing with other methods.

In order to have a better perspective on the performance of different metaheuristic algorithms in dealing with the CEC 2017 benchmark problems, the box plots which are derived of the analysis of the variance (ANOVA), which were conducted for the normalized values of the reported bests, means, standard deviations (Std), and worsts for different dimensions of 10, 30, 50, and 100 in Figures 7–10. It can be concluded that the MGA has competitive performance in dealing with these problems.

Table 4. Statistical results of different approaches for mathematical problems of CEC 2017 with 10 dimensions.

Reference	Result	Function									
		C_1	C_2	C_3	C_4	C_5	C_6	C_7	C_8	C_9	C_{10}
Zamuda [38]	Best	0	0	62700	13.573	0	332.30	−178.02	−0.00135	−0.00498	−0.00051
	Median	0	0	2.260×10^5	13.573	0	1750.6	−26.778	−0.00135	−0.00498	−0.00051
	c	0, 0, 0	0, 0, 0	0, 0, 0	0, 0, 0	0, 0, 0	0, 4, 2	0, 0, 0	0, 0, 0	0, 0, 0	0, 0, 0
	$\overline{v}$	0	0	0	0	0	3.83×10^{-2}	0	0	0	0
	Mean	0	0	3.259×10^5	14.418	0	808.36	−34	0	0	0
	Worst	0	0	1.089×10^6	15.919	0	1819.7	−7	0	0	0
	Std	0	0	2.575×10^5	1.1495	0	545.03	57	0	0	0
	SR	100	100	100	100	100	0	80	100	100	100
	$\overline{\text{vio}}$	0	0	0	0	0	3.766×10^{-2}	3.189×10^{-5}	0	0	0
Polakova [39]	Best	0	0	3533.77	13.5728	0	348.977	−101.211	−0.00135	−0.00498	−0.00051
	Median	0	0	21,144.4	13.5853	0	1368.85	12.7815	−0.00135	−0.00498	−0.00051
	c	0, 0, 0	0, 0, 0	0, 0, 0	0, 0, 0	0, 0, 0	0, 4, 2	0, 0, 0	0, 0, 0	0, 0, 0	0, 0, 0
	$\overline{v}$	0	0	0	0	0	0.029702	0	0	0	0
	Mean	0	0	31,548.2	13.6147	0	648.899	3.74362	−0.00135	−0.00497	−0.00051
	Worst	0	0	118,005	13.8018	0	1260.3	105.62	−0.00135	−0.00485	−0.00051
	Std	0	0	37,019.6	0.061549	0	283.706	69.5716	2.21×10^{-19}	2.44×10^{-5}	1.11×10^{-19}
	SR	100	100	92	100	100	0	88	100	100	100
	$\overline{\text{vio}}$	0	0	6.67×10^{-6}	0	0	0.032309	2.11×10^{-5}	0	0	0
Tvrdík and Poláková [40]	Best	0	0	6341.810292	15.919244	0	103.288465	−148.219878	−0.001348	−0.004975	−0.000510
	Median	0	0	40,103.1993	35.818324	0	307.643490	−65.209283	−0.001348	−0.004975	−0.000510
	c	0, 0, 0	0, 0, 0	0, 0, 1	0, 0, 0	0, 0, 0	0, 0, 5	0, 0, 2	0, 0, 2	0, 0, 1	0, 0, 1
	$\overline{v}$	0	0	0.000103	0	0	0	0	0	0	0
	Mean	0	0	110,008	38.738	0.956779	549.617	−48.7352	−0.001348	0.125471	−0.00051
	Worst	0	0	548,034.199888	55.717399	3.986579	2058.812018	102.366112	−0.001348	3.256178	−0.000510
	Std	0	0	1.5587×10^5	8.948×10^5	1.737×10	4.866×10^2	6.826×10^1	6.639×10^{-19}	6.522×10^{-1}	0.0000
	SR	100	100	44	100	100	96	68	100	100	100
	$\overline{\text{vio}}$	0	0	0.00063352	0	0	0.0053656	0.00309144	1.456×10^{-5}	4×10^{-6}	3.96×10^{-6}
Present Study (MGA)	Best	0	0	5731.729	15.91932	0.048494	177.1936	−204.799	−0.00103	−0.00497	−0.00048
	Median	0	0	9655.116	18.9044315	1.554645	189.7318	−99.5936	0.000667	−0.00497	−0.00034
	c	0, 0, 0	0, 0, 0	0, 0, 0	0, 0, 0	0, 0, 0	0, 4, 2	0, 0, 0	0, 0, 0	0, 0, 0	0, 0, 0
	$\overline{v}$	0	0	0	0	0	0.070346	0	0	0	0

Table 4. *Cont.*

Reference	Result	Function									
		C_1	C_2	C_3	C_4	C_5	C_6	C_7	C_8	C_9	C_{10}
	Mean	0	0	22,532.64	18.57982	1.867275	245.6745	−86.6422	0.001115	0.04604	−0.0003
	Worst	0	0	116,693.6	27.8597111	4.016201	1231.20125	8.612641	0.008756	0.57474405	8.06×10^{-5}
	Std	0	0	35,636.84	4.235729	1.393692	308.6091	68.43618	0.002994	0.152843	0.000167
	SR	100	100	100	100	100	11	100	89	100	100
	$\overline{\text{vio}}$	0	0	0	0	0	0.059761	0	1.11×10^{-5}	0	0

Table 5. Statistical results of different approaches for mathematical problems of CEC 2017 with 30 dimensions.

Reference	Result	Function									
		C_1	C_2	C_3	C_4	C_5	C_6	C_7	C_8	C_9	C_{10}
Zamuda [38]	Best	0	0	2.76×10^6	13.573	0	4095.8	−234.05	-2.82×10^{-4}	−0.00267	−0.000103
	Median	0	0	6.58×10^6	13.573	0	4374.9	−80.772	-2.70×10^{-4}	−0.00267	-9.91×10^{-5}
	c	0, 0, 0	0, 0, 0	0, 0, 0	0, 0, 0	0, 0, 0	0, 4, 2	0, 0, 0	0, 0, 0	0, 0, 0	0, 0, 0
	$\overline{v}$	0	0	0	0	0	2.55×10^{-2}	0	0	0	0
	Mean	0	0	6.70×10^6	13.854	0	5526.4	−81.088	-2.63×10^{-4}	-2.67×10^{-3}	-9.78×10^{-5}
	Worst	0	0	1.17×10^7	15.919	0	5018.0	−36.510	-2.12×10^{-4}	-2.67×10^{-3}	-8.96×10^{-5}
	Std	0	0	2.25×10^6	0.7782	0	759.06	90.929	2.04×10^{-5}	0.00×10^0	3.69×10^{-6}
	SR	100	100	100	100	100	0	96	100	100	100
	$\overline{\text{vio}}$	0	0	0	0	0	2.57×10^{-2}	4.06×10^{-6}	0	0	0
Polakova [39]	Best	0	0	39,059.8	13.5728	0	3121.78	−245.715	−0.00028	−0.00267	−0.0001
	Median	0	0	20,5874	13.5728	0	5802.76	−134.373	−0.00028	−0.00267	−0.0001
	c	0, 0, 0	0, 0, 0	0, 0, 0	0, 0, 0	0, 0, 0	0, 4, 2	0, 0, 0	0, 0, 0	0, 0, 0	0, 0, 0
	$\overline{v}$	0	0	0	0	0	0.012659	0	0	0	0
	Mean	3.87×10^{-30}	5.26×10^{-30}	355,118	13.5728	0	4071.08	−109.428	−0.00028	−0.00267	−0.0001
	Worst	2.08×10^{-29}	3.34×10^{-29}	2.18×10^6	13.5728	0	2405.82	81.6284	−0.00028	−0.00267	−0.0001
	Std	6.10×10^{-30}	8.39×10^{-30}	446,751	5.44×10^{-15}	0	981.519	88.7374	0	1.33×10^{-18}	0
	SR	100	100	100	100	100	0	96	100	100	100
	$\overline{\text{vio}}$	0	0	0	0	0	0.015016	8.20×10^{-6}	0	0	0

Table 5. *Cont.*

Reference	Result	Function									
		C_1	C_2	C_3	C_4	C_5	C_6	C_7	C_8	C_9	C_{10}
Tvrdík and Poláková [40]	Best	0	0	217,854.405028	64.671883	0	1976.35821	−330.786337	−0.000284	−0.002666	−0.000103
	Median	0	0	736,404.82	113.424634	0	3827.58828	−32.589365	−0.000284	−0.002666	−0.000103
	c	0, 0, 0	0, 0, 0	0, 0, 1	0, 0, 0	0, 0, 0	0, 0, 4	0, 0, 2	0, 0, 2	0, 0, 1	0, 0, 1
	$\overline{v}$	0	0	0.001441	0	0	0	0.000067	0	0	0
	Mean	0	0	1.299×10^6	115.734	0.797325	3745.32	−24.1162	−0.000284	0.0233628	−0.000103
	Worst	0	0	5,082,420.837959	159.192594	3.986624	5065.298248	185.582813	−0.000284	0.648053	−0.000103
	Std	0	0	1.195×10^6	2.201×10^1	1.627×10	8.431×10^2	1.154×10^2	1.659×10^{-19}	1.301×10^{-1}	4.149×10^{-20}
	SR	100	100	32	100	100	100	52	100	96	100
	$\overline{\text{vio}}$	0	0	0.0242756	0	0	1.164×10^{-5}	0.0035614	0	1.0709×10^6	6.6×10^{-6}
Present Study (MGA)	Best	0	0	101,125.9	72.90983	0	1369.466	−214.361	1.311075	0.000266	0.342705
	Median	0	0	3,769,626.34	106.7165	0	1582.655	−212.331	2.173907	0.574744	0.627412
	c	0, 0, 0	0, 0, 0	0, 0, 0	0, 0, 0	0, 0, 0	0, 1, 4	0, 0, 0	2, 0, 0	0, 0, 0	2, 0, 0
	$\overline{v}$	0	0	0	0	0	0.002259	0	6.590862	0	2.124071
	Mean	0	0	497,341.7	103.0124	0	1639.729	−229.997	2.013486	0.903313	0.587699
	Worst	0	0	1,083,246	196.971781	0	2375.443	−52.3606	3.901132	4.706474	0.887197
	Std	0	0	417,284.9	22.09312	0	459.9857	106.7847	0.979232	1.476608	0.206912
	SR	100	100	100	100	100	77	100	0	100	0
	$\overline{\text{vio}}$	0	0	0	0	0	0.010403	0	3.814903	0	2.547508

Table 6. Statistical results of different approaches for mathematical problems of CEC 2017 with 50 dimensions.

Reference	Result	Function									
		C_1	C_2	C_3	C_4	C_5	C_6	C_7	C_8	C_9	C_{10}
Zamuda [38]	Best	0	0	7.80×10^6	13.573	0	8775	−347.6	1.40×10^{-4}	3.25×10^{-5}	−347.6
	Median	0	0	2.65×10^7	13.573	0	10,224	−134.7	2.87×10^{-4}	8.66×10^{-5}	−134.7
	c	0, 0, 0	0, 0, 0	0, 0, 0	0, 0, 0	0, 0, 0	0,1,5	0, 0, 0	0, 0, 0	0, 0, 0	0, 0, 0
	$\overline{v}$	0	0	0	0	0	1.38×10^{-2}	0	0	0	0
	Mean	1.49×10^{-8}	0	2.65×10^7	13.988	0	8601	−154.0	2.86×10^{-4}	9.12×10^{-5}	−154.0
	Worst	1.00×10^{-7}	5.94×10^{-8}	4.25×10^7	16.914	0	9202	39.3	4.85×10^{-4}	2.28×10^{-4}	39.3
	Std	1.95×10^{-8}	1.17×10^{-8}	8.66×10^6	0.9868	0	1217	106.3	8.44×10^{-5}	3.91×10^{-5}	106.3

Table 6. *Cont.*

Reference	Result	Function									
		C_1	C_2	C_3	C_4	C_5	C_6	C_7	C_8	C_9	C_{10}
	SR	100	100	100	100	100	0	100	100	100	100
	vio	0	0	0	0	0	1.52×10^{-2}	0	0	0	0
Polakova [39]	Best	8.68×10^{-30}	2.50×10^{-29}	286,730	13.5728	0	6708.83	−0.00013	−0.00204	-4.83×10^{-5}	−0.00013
	Median	7.73×10^{-29}	1.02×10^{-28}	633,683	13.5728	1.30×10^{-28}	8636.68	−0.00013	−0.00204	-4.83×10^{-5}	−0.00013
	c	0, 0, 0	0, 0, 0	0, 0, 0	0, 0, 0	0, 0, 0	0,2,4	0, 0, 0	0, 0, 0	0, 0, 0	0, 0, 0
	$\overline{v}$	0	0	0	0	0	0.011381	0	0	0	0
	Mean	7.79×10^{-29}	9.79×10^{-29}	894521	13.5728	1.68×10^{-28}	7514.8	−0.00013	−0.00204	-4.83×10^{-5}	−0.00013
	Worst	1.42×10^{-28}	1.78×10^{-28}	3.87×10^{6}	13.5728	6.40×10^{-28}	6637.22	−0.00013	−0.00204	-4.83×10^{-5}	−0.00013
	Std	3.08×10^{-29}	4.60×10^{-29}	740490	5.44×10^{-15}	1.59×10^{-28}	1417.76	2.77×10^{-20}	1.33×10^{-18}	0	2.77×10^{-20}
	SR	100	100	100	100	100	0	100	100	100	100
	vio	0	0	0	0	0	0.011693	0	0	0	0
Tvrdík and Poláková [40]	Best	0	0	460,407.836	145.263065	0	3486.644298	−340.22487	0.000601	−0.002037	−0.00004
	Median	0	0	4,381,259.215675	181.081674	0	6041.018996	−85.989214	0.000965	−0.002037	−0.00004
	c	0, 0, 0	0, 0, 0	0, 0, 1	0, 0, 0	0, 0, 0	0, 0, 4	0, 0, 2	0, 0, 0	0, 0, 1	0, 0, 0
	$\overline{v}$	0	0	0.000050	0	0	0	0.000075	0	0	0
	Mean	0	0	6.6413×10^{6}	187.37	0.31893	6364.72	−68.1059	0.0009928	0.0810008	-4.284×10^{-5}
	Worst	0	0	27,234,258.49277	0244.758532	3.986624	9005.415965	163.958553	0.001558	1.138593	−0.00001
	Std	0	0	5.9790×10^{6}	2.5905×10^{1}	1.1038×10^{0}	1.6322×10^{3}	1.3458×10^{2}	2.4328×10^{-4}	2.3626×10^{-1}	6.101×10^{-6}
	SR	100	100	48	100	100	100	56	100	84	100
	vio	0	0	0.0694317	0	0	1.02×10^{-5}	0.00180008	3.48×10^{-6}	1.4708×10^{7}	0
Present Study (MGA)	Best	7.73×10^{-6}	3.40×10^{-7}	67,5040.5	214.0131	183.3693	2104.094	−287.24	6.111175	16.76229	11.98384
	Median	6.16×10^{-6}	2.87×10^{-5}	1686140	231.3667	264.3097	2453.639	−121.342	10.1111743	19.39083	19.33752
	c	0, 0, 0	0, 0, 0	0, 0, 0	0, 0, 0	0, 0, 0	0, 0, 0	0, 0, 0	2, 0, 0	1, 0, 0	2, 0, 0
	$\overline{v}$	0	0	0	0	0	0	0	77.75325	1.398507	2716.256
	Mean	2.29×10^{-5}	9.39×10^{-5}	3,733,884	236.5529	295.9803	2601.021	−100.093	7.705596	18.78359	23.86641
	Worst	8.70×10^{-5}	0.000571	65,265,448.5	309.4055	429.773726	4601.16137	152.382091	10.65774	19.76821	54.4833554
	Std	3.41×10^{-5}	0.000167	4,614,030	32.99758	125.1962	591.8259	105.2148	1.261924	0.963696	11.58393
	SR	100	100	100	100	100	76	100	0	0	0
	vio	0	0	0	0	0	0.057435	0	75.19797	1.255203	3025.091

Table 7. Statistical results of different approaches for mathematical problems of CEC 2017 with 100 dimensions.

Reference	Result	Function									
		C_1	C_2	C_3	C_4	C_5	C_6	C_7	C_8	C_9	C_{10}
Zamuda [38]	Best	2.434	1.072	9.39×10^{7}	13.573	0	15,440	−530.12	1.22×10^{-3}	3.51×10^{-4}	−530.12
	Median	6.211	2.318	2.27×10^{8}	13.573	0	15,595	−324.99	1.44×10^{-3}	4.13×10^{-4}	−324.99
	c	0, 0, 0	0, 0, 0	0, 0, 0	0, 0, 0	0, 0, 0	0, 4, 2	0, 0, 0	0, 0, 0	0, 0, 0	0, 0, 0
	$\overline{v}$	0	0	0	0	0	1.18×10^{-2}	0	0	0	0
	Mean	7	3	2.25×10^{8}	14.028	0	15,533	−335.49	1.48×10^{-3}	4.25×10^{-4}	−335.49
	Worst	16.527	6.765	4.21×10^{8}	16.914	0	14,830	−110.18	1.78×10^{-3}	5.53×10^{-4}	−110.18
	Std	3.190	1.397	9.21×10^{7}	1.083	0	1604	122.40	1.77×10^{-4}	4.92×10^{-5}	122.40
	SR	100	100	100	100	100	0	100	100	100	100
	$\overline{vio}$	0	0	0	0	0	1.19×10^{-2}	0	0	0	0
Polakova [39]	Best	1.30×10^{-26}	1.31×10^{-26}	1.34×10^{6}	13.5728	4.34×10^{-7}	17,164.3	-4.83×10^{-5}	−0.00143	-1.72×10^{-5}	-4.83×10^{-5}
	Median	4.50×10^{-26}	4.59×10^{-26}	2.47×10^{6}	13.5728	4.90×10^{-6}	15,803.2	-4.82×10^{-5}	−0.00143	-1.72×10^{-5}	-4.82×10^{-5}
	c	0, 0, 0	0, 0, 0	0, 0, 0	0, 0, 0	0, 0, 0	0,1,5	0, 0, 0	0, 0, 0	0, 0, 0	0, 0, 0
	$\overline{v}$	0	0	0	0	0	0.009677	0	0	0	0
	Mean	1.03×10^{-25}	8.47×10^{-26}	2.73×10^{6}	13.7132	3.28×10^{-5}	15,562.2	-4.81×10^{-5}	−0.00143	-1.72×10^{-5}	-4.81×10^{-5}
	Worst	6.51×10^{-25}	6.38×10^{-25}	4.79×10^{6}	15.5748	0.000416	16,718.9	-4.77×10^{-5}	−0.00143	-1.71×10^{-5}	-4.77×10^{-5}
	Std	1.68×10^{-25}	1.25×10^{-25}	965,593	0.462703	9.25×10^{-5}	1595.41	1.33×10^{-7}	2.21×10^{-19}	1.29×10^{-8}	1.33×10^{-7}
	SR	100	100	100	100	100	0	100	100	100	100
	$\overline{vio}$	0	0	0	0	0	0.009805	0	0	0	0
Tvrdík and Poláková [40]	Best	0.080255	0.072938	1,684,503.31	329.329439	0	10,950.2096	−481.32898	0.013288	0	0.000365
	Median	0.432564	0.184568	9,938,948.89	408.925707	0.011586	15,506.5581	−278.65043	0.027209	0.000217	0.000501
	c	0, 0, 0	0, 0, 0	0, 0, 1	0, 0, 0	0, 0, 0	0, 0, 2	0, 0, 2	0, 0, 2	0, 0, 0	0, 0, 0
	$\overline{v}$	0	0	0.002547	0	0	0	0.000198	0.000832	0	0
	Mean	0.977746	0.366104	1.51413×10^{7}	413.582	0.818836	15,222.9	−193.458	0.0415975	0.522499	0.00051308
	Worst	11.315168	3.620979	60,598,481.7	469.617973	4.066555	18,535.3302	376.526002	0.087460	5.348516	0.000684
	Std	2.1781×10	6.9971×10^{-1}	1.3449×10^{7}	3.6721×10^{1}	1.512×10	1.7824×10^{3}	2.0127×10^{2}	2.4668×10^{-2}	1.1223×10	7.3482×10^{-5}
	SR	100	100	16	100	100	100	40	0	96	100
	$\overline{vio}$	0	0	0.0242065	0	0	5.6×10^{-6}	0.00436664	0.0012406	1.564×10^{-5}	1.268×10^{-5}

Table 7. *Cont.*

Reference	Result	Function									
		C_1	C_2	C_3	C_4	C_5	C_6	C_7	C_8	C_9	C_{10}
Present Study (MGA)	Best	79.17725	82.46377	1,939,226	1035.546	149,057.2	4524.706	−24.7894	11.02866	16.5913	47.71097
	Median	226.9738	216.2455	5,229,008	1115.146	162,884.7	4982.677	68.55985	11.59449	18.9102539	53.43286
	c	0, 0, 0	0, 0, 0	0, 0, 0	0, 0, 0	0, 0, 0	0, 0, 4	2, 0, 0	2, 0, 0	1, 0, 0	2, 0, 0
	$\overline{v}$	0	0	0	0	0	0.002105	1162.713	1530.176	698.7951	102,643
	Mean	242.9825	256.1113	7,766,430	1107.661	167,178.1	5646.481	88.40124	11.62023	18.23359	53.97361
	Worst	400.539192	445.345609	21,948,580	1186.64915	221,456	7418.342	402.6898	13.20224	19.2152074	60.58393
	Std	125.7168	167.4498	6,395,089	61.20722	26,483.47	1233.678	177.8638	1.155593	0.934306	4.846915
	SR	100	100	100	100	100	0	0	0	0	0
	$\overline{vio}$	0	0	0	0	0	0.305535	1485.207	1221.619	773.7698	105,498

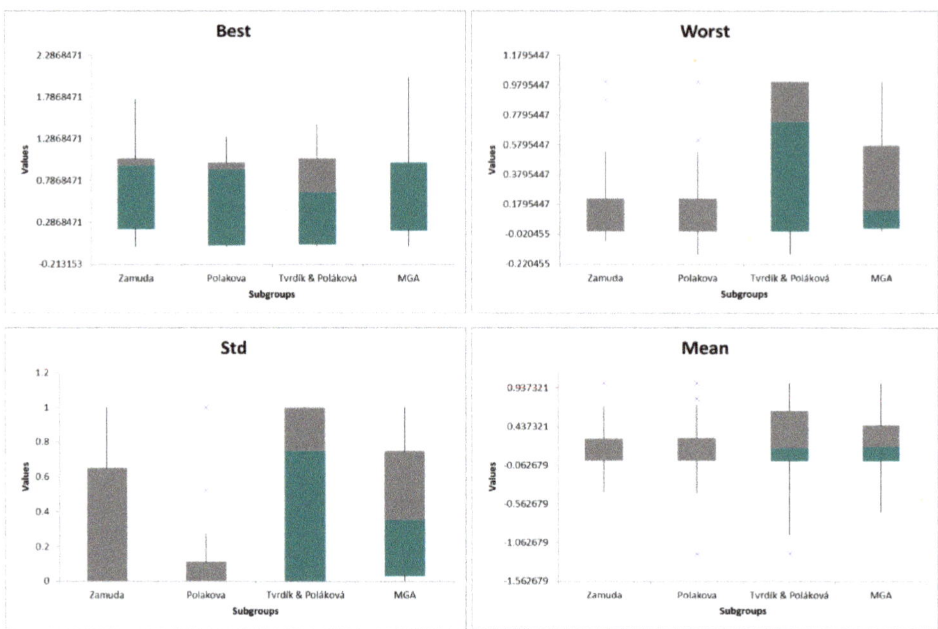

Figure 7. Box plots of the Analysis of Variance (ANOVA) for the 10-dimensional problems.

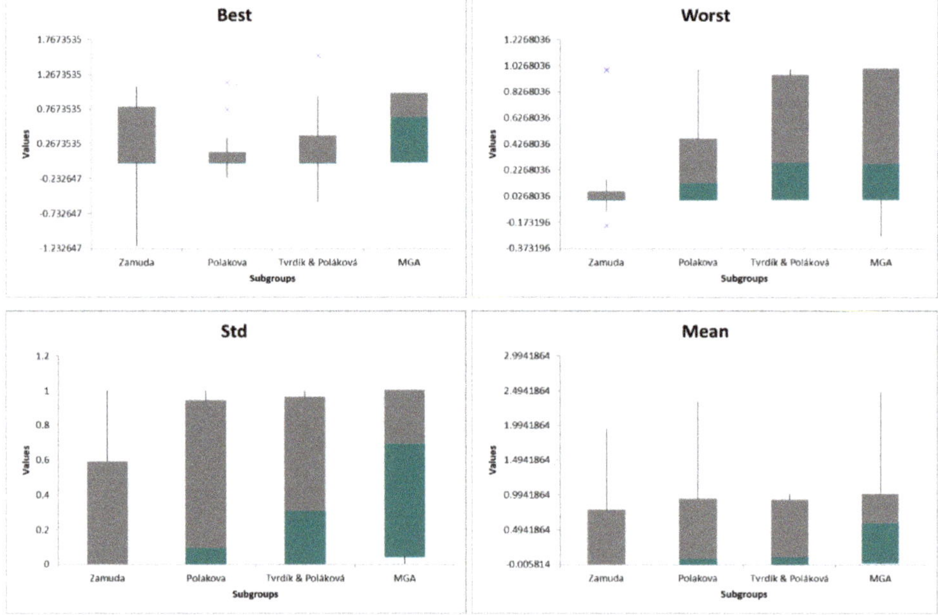

Figure 8. Box plots of the ANOVA for the 30-dimensional problems.

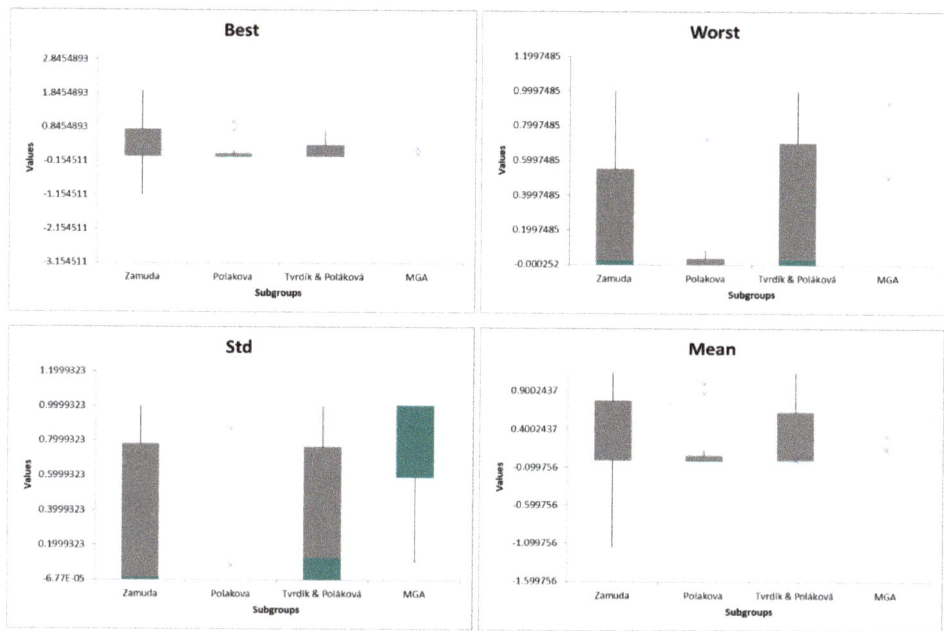

Figure 9. Box plots of the ANOVA for the 50-dimensional problems.

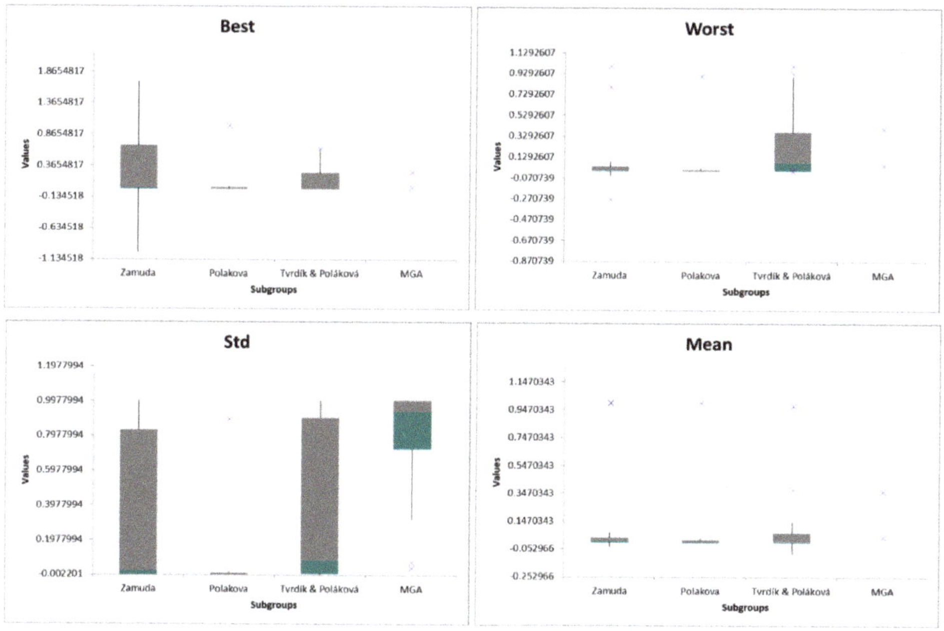

Figure 10. Box plots of the ANOVA for the 100-dimensional problems.

Based on the provided results for the MGA and other state-of-the-art approaches in the evolutionary computation field, the AGA is capable of competing with these excellent algorithms while in some cases even MGA outperforms the others. this performance in dealing with CEC 2017 lead to the fact that MGA's mathematical model is well-established model in which the global and local search are conducted with no need to any parameters to be tuned. In other words, this algorithm does not need any internal parameters to be defined prior to the optimization process which makes this algorithm a best choice in dealing with complex problems in which there are not any information about the complexity level of the problem. Additionally, the MGA generates only two new solution candidates in each iteration which makes the algorithm to require less computational efforts for optimization purposes. Hence, these aspects can be of great importance when the MGA is compared to the other metaheuristic algorithms in the evolutionary computation field. In other words, MGA is a parameter free optimization approach with less computational cost, which makes this algorithm different form the other approaches, while the inspirational concept of this algorithm is also unique.

5. Numerical Results of Engineering Problems

The numerical results of MGA considering the previously-described engineering design problems are presented in this section. In this regard, the results of other metaheuristics in dealing with these design examples were taken from the literature in order to make fair judgments.

The comparative results of the speed reducer design engineering problem, including the obtained design (decision) variables related to the best optimum configuration determined by different methods, are presented in Table A1. In addition, the statistical results, such as the best, mean, and worst fitness values alongside the standard deviation, are presented in Table 8. The results of different metaheuristics show that the best results of MGA are better than the best results of the other approaches in dealing with this design example. The MGA is also capable of providing better statistical results, including mean and standard deviation. The Friedman statistical test results are also presented in Table A2 for comparative purposes.

Table 8. Statistical results of different approaches for the speed reducer design problem.

Approaches	Best	Mean	Worst	Std-Dev
Montes et al. [54]	3025.005	3088.7778	3078.5918	NA
Akhtar et al. [55]	3008.08	3012.1200	3028.2800	NA
Gandomi et al. [46]	3000.9810	3007.1997	3.0090	4.9634
Zhang et al. [56]	2994.471066	2994.471066	2994.471066	3.58×10^{-12}
Present Study (MGA)	2994.438869	2994.47065	2996.558237	4.72×10^{-16}

Considering the spring design problem, the best and statistical results of different metaheuristics, including the obtained design variables related to the best optimum design, are presented in Tables 9 and A3, respectively. It should be mentioned that MGA is capable of obtaining very competitive results for this constrained engineering design problem. It also should be mentioned that MGA yields better statistical results in terms of the mean, worst fitness values alongside the standard deviation than the results of other metaheuristics. The Friedman statistical test results are also presented in Table A4 for comparative purposes.

Tables 10 and A5 present the final and statistical results obtained by the different methods for the pressure vessel engineering design problem, respectively. From these tables, the best result of the MGA method is better than the results of the other approaches. By comparing the statistical results, it is obvious that MGA has better performance in statistical analysis, especially the mean, and worst fitness values alongside the standard deviation. The Friedman statistical test results are also presented in Table A6 for comparative purposes.

Table 9. Statistical results of different approaches for the tension or compression spring design problem.

Approaches	Best	Mean	Worst	Std-Dev
Coello [57]	0.01270478	0.01276920	0.01282208	3.9390×10^{-5}
Ray and Liew [58]	0.0126692	0.0129227	0.0167172	5.1985×10^{-5}
Han et al. [59]	0.01266534	0.01268592	0.01272968	2.1672×10^{-5}
Gandomi et al. [45]	0.01266522	0.01350052	0.0168954	0.001420272
Present Study (MGA)	0.01266523	0.01266558	0.01266723	5.65×10^{-7}

Table 10. Statistical results of different approaches for the pressure vessel design problem.

Approaches	Best	Mean	Worst	Std-Dev
He and Wang [60]	6061.0777	6147.1332	6363.8041	86.4500
Coelho [61]	6059.7208	6440.3786	7544.4925	448.4711
Mezura-Montes and Coello [62]	6059.7456	6850.004948	7332.879883	426
Coello and Montes [63]	6059.9463	6177.2532668	6469.32201	130.9
Present Study (MGA)	6059.714350	6059.694923	6273.765974	0.028912058

The results of the welded beam design problem in Tables 11 and A7 show that MGA is capable of converging to better results than the other approaches. Although the maximum difference between the best results of MGA and the other approaches is only about 4%, MGA is capable of providing better statistical results, including the mean, worst fitness values alongside standard deviation. The Friedman statistical test results are also presented in Table A8 for comparative purposes.

Table 11. Statistical results of different approaches for the welded beam design problem.

Approaches	Best	Mean	Worst	Std-Dev
Huang et al. [64]	1.733461	1.768158	1.824105	0.022194
Eskandar et al. [65]	1.724856	1.726427	1.744697	4.29×10^{-3}
Guedria [66]	1.724852	1.724853	1.724862	2.02×10^{-6}
Han et al. [59]	1.6956397	1.7160908	1.7530472	1.83×10^{-2}
Present Study (MGA)	1.672966512	1.678791422	1.687172363	4.4147×10^{-3}

In Table A9, the final design of different methods and MGA for the three-bar truss design problem, including the obtained design variables, are presented. Table 12 displays the statistical results. Considering the results reported by previous researchers, it is clear that MGA yields very competitive results for this engineering design problem. MGA determined the best optimum value that has been reported thus far, according to the literature, for the considered design example. It also should be noted that the statistical results, including the mean and standard deviation, for the MGA are much better than the results of other approaches. The Friedman statistical test results are also presented in Table A10 for comparative purposes.

The results of the multiple disk clutch brake design problem solved by MGA and other approaches are summarized in Table A11 [47,50,65,68]. The statistical results are presented in Table 13. Accordingly, MGA is capable of calculating very impressive results compared to the other metaheuristics. The maximum and minimum differences between the results of MGA and other metaheuristics are about 49% and 24%, which demonstrates the capability of this algorithm in dealing with multiple disk clutch brake design problem. In addition, the statistical results, including the mean and worst fitness values, demonstrate that MGA can yield extremely better results than the other approaches. The Friedman statistical test results are also presented in Table A12 for comparative purposes.

Table 12. Statistical results of different approaches for the three-bar truss design problem.

Approaches	Best	Mean	Worst	Std-Dev
Gandomi et al. [46]	263.97156	264.0669	NA	0.00009
Ray and Liew [58]	263.8958466	263.9033	263.9033	1.26×10^{-2}
Zhang et al. [56]	263.8958434	263.8958436	263.8958498	9.72×10^{-7}
Grag [67]	263.8958433	263.8958437	263.8958459	5.34×10^{-7}
Present Study (MGA)	263.8958433	263.8958436	263.8959632	2.05×10^{-14}

Table 13. Statistical results of different approaches for the multiple disk clutch brake design problem.

Approaches	Best	Mean	Worst	Std-Dev
Eskandar et al. [65]	0.313656	0.313656	0.313656	1.69×10^{-16}
Rao et al. [50]	0.313657	0.3271662	0.392071	0.67
Ferreira et al. [47]	0.313656	0.313656	0.313656	1.13×10^{-16}
Present Study (MGA)	0.235242467	0.235244323	0.235252239	2.42×10^{-6}

The final results of different metaheuristics in dealing with the planetary gear train design problem, one of the most important and well-established constrained optimization problems, are presented in Tables 14 and A13. By comparing the best results of MGA with other approaches, it can be concluded that MGA can yield outstanding results. Although MGA is also capable of providing better statistical results for the mean and worst fitness values alongside standard deviation results cannot be compared since they have yet to be reported in the literature. The Friedman statistical test results are also presented in Table A14 for comparative purposes.

Table 14. Statistical results of different approaches for the planetary gear train design problem.

Approaches	Best	Mean	Worst	Std-Dev
Rao and Savsani [69] (PSO)	0.53	0.5361934	NA	NA
Rao and Savsani [69] (ABC)	0.525769	0.5272922	NA	NA
Zhang et al. [70]	0.525589	0.525589	NA	NA
Savsani and Savsani [48]	0.525588	0.53063	NA	NA
Present Study (MGA)	0.52325	0.5300526	0.5370588	0.0082564

For the step-cone pulley engineering design problem, the final results of different metaheuristics are presented in Table A15, and the statistical results are provided in Table 15. By comparing the best results, it can be concluded that MGA can yield very impressive results for this constrained engineering problem. The maximum difference between the mean results of MGA and other approaches is about 31%. The Friedman statistical test results are also presented in Table A16 for comparative purposes.

Table 15. Statistical results of different approaches for the step-cone pulley design problem.

Approaches	Best	Mean	Worst	Std-Dev
TLBO [50]	16.63451	24.0113577	74.022951	0.34
WOA [44]	16.6345213	20.93829477	24.8488259	3.3498
WCA [44]	16.63450849	17.53037682	18.83302997	0.9229
MBA [44]	16.6345078	16.702535	18.3237145	0.2627
Present Study (MGA)	16.18595608	16.35528922	16.98647762	0.14824361

TLBO: Teaching-Learning-Based Optimization.

In Table 16, the comparative results of different metaheuristics in dealing with the hydrostatic thrust bearing design problem, including the obtained design and its related best optimum configuration, are presented. Table 17 displays the statistical results. It can be concluded that MGA is capable of converging to better results than the other approaches. The maximum difference between the best results of MGA is about 29%, where MGA yielded better statistical results for the mean, worst fitness values alongside the standard deviation than the other approaches. The Friedman statistical test results are also presented in Table 18 for comparative purposes.

Table 16. Comparison of the best solutions for the hydrostatic thrust bearing design problem.

	Siddall [71]	Deb and Goyal [72]	Coello [73]	Rao et al. [50]	Present Study (MGA)
Best	2288.2268	2161.4215	1950.2860	1625.44276	1623.980938
R	7.155	6.778	6.271	5.9557805026	5.963241516
R_0	6.689	6.234	12.901	5.3890130519	5.395907989
μ	8.321×10^{-6}	6.096×10^{-6}	5.605×10^{-6}	0.0000053586	5.38×10^{-6}
Q	9.168	3.809	2.938	2.2696559728	2.282242505
$g_1(x)$	−11,086.7430	−8329.7681	−2126.86734	−0.0001374735	−144.9586796
$g_2(x)$	−402.4493	−177.3527	−68.0396	−0.0000010103	−1.194802021
$g_3(x)$	−35.057196	−10.684543	−3.705191	−0.0000000210	−0.372450027
$g_4(x)$	−0.001542	−0.000652	−0.000559	−0.0003243625	−0.00032915
$g_5(x)$	−0.466000	−0.544000	−0.666000	−0.5667674507	−0.567333527
$g_6(x)$	−0.000144	−0.000717	−0.000805	−0.0009963614	−0.000996355
$g_7(x)$	−563.644401	−83.618221	−849.718683	−0.0000090762	−4.144258876

Table 17. Statistical results of different approaches for the hydrostatic thrust bearing design problem.

Approaches	Best	Mean	Worst	Std-Dev
Şahin et al. [74]	1625.46467	1627.744198	1650.698747	3.815546973
Rao and Waghmare [75]	1625.44271	1796.89367	2104.3776	0.21
Rao et al. [50]	1625.44276	1797.70798	2096.8012	0.19
Present Study (MGA)	1621.246175	1739.156729	1992.961305	0.11

Table 18. Friedman statistical test results for the hydrostatic thrust bearing design problem.

Rankings	Algorithms	Mean of Ranks
1	Present Study (MGA)	1.5
2	Şahin et al. [74]	2.5
3	Rao and Waghmare [75]	3
4	Rao et al. [50]	3
Chi-sq.	3.6000	
Prob > Chi-sq.	0.3080	

The optimum results of different metaheuristics in dealing with the ten-bar truss design problem are presented in Tables 19 and 20. By comparing the best results, it can be concluded that MGA is capable of outperforming other metaheuristics approaches. Until now, the best value obtained for this example was 529.25, which has been overcome by MGA with 529.12. This indicates the capability of MGA to provide remarkable results for some complex constrained design problems.

Table 19. Comparison of the best solutions for the ten-bar truss design problem.

	Yu et al. [51]	Lamberti and Pappalettere [76]	Baghlani and Makiabadi [77]	Kaveh and Zolghadr [78]	Present Study (MGA)
Best	544.7	534.57	530.76	529.25	529.1204229
A_1	36.380	35.148	35.494	39.569	36.76416
A_2	12.941	13.169	14.777	16.740	16.29897
A_3	35.764	37.69	36.203	34.361	37.94378
A_4	18.314	19.556	15.387	12.994	16.51087
A_5	3.002	1.087	0.6451	0.645	0.659
A_6	5.433	4.844	4.5896	4.802	4.57489
A_7	20.989	18.314	23.211	26.182	22.94023
A_8	24.14	27.415	24.561	21.260	22.63185
A_9	9.753	12.562	12.482	11.766	10.87892
A_{10}	18.102	12.106	12.324	11.392	11.53643

Table 20. Statistical results of the different method for the ten-bar truss bearing design problem.

Approaches	Best	Mean	Worst	Std-Dev
Present Study (MGA)	529.1204229	534.6843574	548.0179132	26.33651675

The results of different methods for the rolling element bearing design problem are presented in Tables 21 and 22. It is clear that the best result of the MGA in this case is better than those of other approaches in the literature. Regarding the fact that this problem is a maximization optimization problem, MGA is also capable of providing remarkable statistical results.

Table 21. Comparison of the best solutions for the rolling element bearing design problem.

	TLBO [50]	Present Study (MGA) *
Best	81,859.74	83,912.87983
D_m	21.42559	125.0002787
D_b	125.7191	21.87451192
Z	11	10.77706583
f_i	0.515	0.515000822
f_0	0.515	0.515002993
K_{Dmin}	0.424266	0.405908353
K_{Dmax}	0.633948	0.65558802
ε	0.3	0.300004155
e	0.068858	0.077544926
ζ	0.799498	0.6

* This problem is a maximization problem.

Table 22. Statistical results of different approaches for the rolling element bearing design problem.

Approaches	Best	Mean	Worst	Std-Dev
TLBO [50]	81,859.74	81,438.987	80,807.8551	0.66
Present Study (MGA)	83,912.87983	83,892.25647	83,711.21317	23.65841

Table A17 [46,79–81] and Table 23 display the comparative and statistical optimization results of multiple optimization algorithms and MGA in dealing with the gear train design problem. It is obvious that MGA outranks the other optimization algorithms, Specifically, MGA obtained a perfect best of zero, which has not been obtained by other metaheuristics, confirming the capability of MGA to yield the lowest possible value in this case. The Friedman statistical test results are also presented in Table A18 for comparative purposes.

Table 23. Statistical results of different approaches for the gear train design problem.

Approaches	Best	Mean	Worst	Std-Dev
Gandomi et al. [46]	2.7009×10^{-12}	1.9841×10^{-9}	2.3576×10^{-9}	3.5546×10^{-9}
Loh and Papalambros [79]	2.7×10^{-12}	2.7×10^{-12}	2.7×10^{-12}	2.2122×10^{-28}
Wang et al. [82] (CPKH)	2.22×10^{-16}	2.22×10^{-16}	8.5×10^{-9}	7.96×10^{-22}
Wang et al. [82] (ABC)	2.92×10^{-15}	3.18×10^{-15}	8.5×10^{-9}	9.81×10^{-10}
Present Study (MGA)	1.06×10^{-19}	7.69×10^{-14}	7.62×10^{-13}	1.78×10^{-13}

CPKH: Chaotic Particle Swarm Krill Herd.

Considering the steel I-shaped beam as one of the most well-formulated design problems, the final and statistical optimization results of multiple metaheuristics are presented in Tables 24 and 25, respectively. By comparing these optimum results, MGA outranked all other well-known algorithms that have been reported recently.

Table 24. Comparison of the best solutions for the steel I-shaped beam design problem.

	ARSM [83]	I-ARSM [83]	MATLAB [83]	CS [46]	Present Study (MGA)
Best	0.0157	0.131	0.0131	0.0130747	0.013074119
h	80	79.99	80	80	79.9999992
b	37.05	48.42	50	50	49.9999985
t_w	1.71	0.9	0.9	0.9	0.9
t_f	2.31	2.4	2.32	2.3216715	2.321792333

ARSM: Adaptive Response Surface Method; I-ARMS: Improved Adaptive Response Surface Method; MATLAB: Matrix Laboratory Optimization Approach.

Table 25. Statistical results of different approaches for the steel I-shaped beam design problem.

Approaches	Best	Mean	Worst	Std-Dev
CS [46]	0.0130747	0.0132165	0.01353646	0.0001345
Present Study (MGA)	0.013074119	0.013074141	0.013074291	3.86×10^{-8}

The final results of different metaheuristics for the piston lever design problem, a frequently occurring optimization problem, are presented in Table A19. The statistical results, including the best, mean, and worst fitness values alongside standard deviation, are presented in Table 26 for comparative purposes. Based on the results, MGA is capable of providing better statistical (mean, worst, and standard deviation of the results) and greatly outranked the other algorithms in terms of the best results. The Friedman statistical test results are also presented in Table A20 for comparative purposes.

Table 26. Statistical results of different approaches for the piston lever design problem.

Approaches	Best	Mean	Worst	Std-Dev
HPSO [46]	162	187	197	13.4
GA [46]	161	185	216	18.2
DE [46]	159	187	199	14.2
CSA [46]	8.4271	40.2319	168.5920	59.0552
Present Study (MGA)	8.413406652	32.4688925	167.4732134	29.96370439

HPSO: Hybrid Particle Swarm Optimization.

Considering the cantilever beam engineering design problem, the optimization results of the different optimization algorithms are all presented in Tables 27 and 28. By comparing the best results of these methods, it can be concluded that MGA is capable of achieving better results. According to the literature, recently-developed algorithms can yield 1.34, at best, for this example. Herein, we found that MGA is capable of providing even better result (1.33997) by conducting a better searching procedure. The statistical results of other optimization algorithms are not reported in the literature; thus, the remarkable results of MGA are beneficial for future works.

Table 27. Comparison of the best solutions for the cantilever beam design problem.

	MMA [46]	GCA-I [46]	GCA-II [46]	CSA [46]	Present Study (MGA)
Best	1.34	1.34	1.34	1.33999	1.339975661
x_1	6.01	6.01	6.01	6.0089	6.011660964
x_2	5.3	5.3	5.3	5.3049	5.315676194
x_3	4.49	4.49	4.49	4.5023	4.510681877
x_4	3.49	3.49	3.49	3.5077	3.485698713
x_5	2.15	2.15	2.15	2.1504	2.150251174

MMA: Method of Moving Asymptotes; GCA: Generalized Convex Approximation.

Table 28. Statistical results of the MGA method for the cantilever beam design problem.

Approaches	Best	Mean	Worst	Std-Dev
Present Study (MGA)	1.339975661	1.340052681	1.340201166	6.99×10^{-5}

By comparing the p-values of the Friedman statistical test which are presented in the table of results by Chi-sq., it is concluded that for the piston lever design example, the lowest p-values is determined which demonstrates the fact that for this example, there are noticeable difference between the results of different approaches. However, the p-values of other examples are also near a mean of 9 which represents the stability of the conducted optimization runs and the statistical tests (Figure 11).

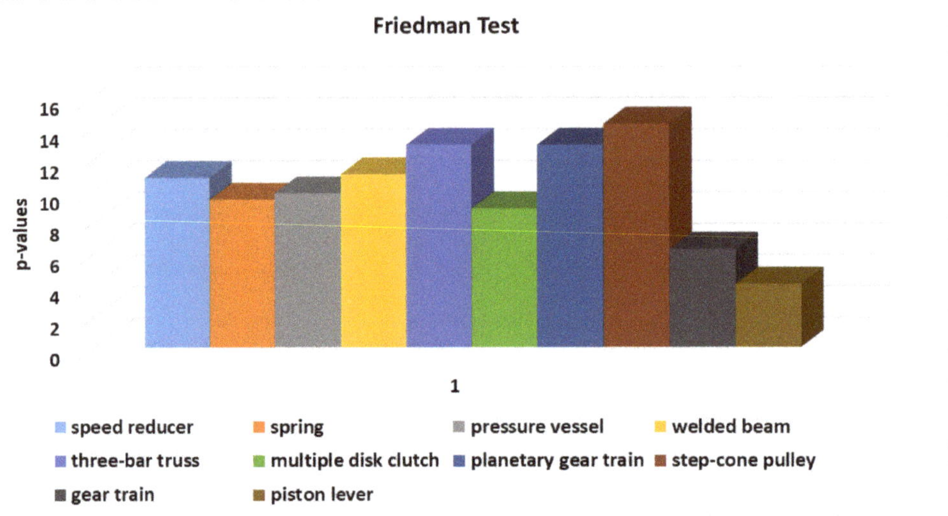

Figure 11. Comparison of Friedman's p-vales for different design examples.

6. Conclusions

In this paper, the Material Generation Algorithm (MGA) is presented as a new metaheuristic for different applications and various optimization problems. In this regard, 25 constrained design problems were considered to evaluate MGA, including 10 mathematically-constrained problems presented by the Competitions on Evolutionary Computation (CEC 2017) and 15 well-known engineering design problems. For comparative purposes, the best results of different metaheuristic algorithms, such as state-of-the-art metaheuristics from CEC 2017, were selected for comparative purposes. Considering the results of MGA in dealing with the mathematical problems, it should be noted that this algorithm is capable of providing very competitive results in different dimensions. In addition, MGA yielded very impressive results in all of constrained engineering design problems compared to the previously reported algorithms. Specifically, the highest difference of about 24% between the best results of MGA and the best results reported thus far in the literature was found for the multiple disk clutch brake engineering design problem. For the three-bar truss design problem, MGA can provide very competitive results and, importantly, nearly the best results reported thus far. For the tension or compression spring, pressure vessel and rolling element bearing problems, the best results were higher for MGA than the best reported results.

While the proposed MGA has been proven to be a powerful method, different applications of this method are suggested for future research. It should be mentioned that the capability of this optimization approach can be controlled in dealing with some complex real-world and even computationally-expensive optimization problems. In addition, some other challenges, such as improving the general formulation of this method and hybridizing with other approaches, should be investigated properly.

Author Contributions: Conceptualization, S.T.; methodology, S.T. and M.A.; software, S.T. and M.A.; Validation, M.A. and S.T.; formal analysis, M.A.; investigation, S.T., M.A. and A.H.G.; resources, S.T., M.A. and A.H.G.; data curation, A.H.G.; writing—original draft preparation, M.A.; writing—review and editing, S.T. and A.H.G.; visualization, M.A.; supervision, S.T.; project administration, S.T.; funding acquisition, S.T., M.A. and A.H.G. All authors have read and agreed to the published version of the manuscript.

Funding: The APC was funded by University of Technology Sydney Internal Fund for A.H. Gandomi.

Institutional Review Board Statement: Not applicable.

Informed Consent Statement: Not applicable.

Data Availability Statement: The Matlab implementation of MGA is accessible at: www.mathworks.com/matlabcentral/fileexchange/92065-material-generation-algorithm-mga (accessed on 3 May 2021).

Acknowledgments: This research was partially supported by the University of Tabriz, grant number 1615 and the APC was funded by University of Technology Sydney Internal Fund for A.H. Gandomi.

Conflicts of Interest: The authors declare no conflict of interest.

Appendix A

Table A1. Comparison of the best solutions for the speed reducer design problem.

	Montes et al. [54]	Akhtar et al. [55]	Gandomi et al. [46]	Zhang et al. [56]	Present Study (MGA)
Best	3025.005	3008.08	3000.9810	2994.471066	2994.438869
b	3.506163	3.506122	3.5015	3.5	3.500007956
m	0.700831	0.700006	0.7000	0.7	0.700000656
z	17	17	17.0000	17	17.00000081
l_1	7.460181	7.549126	7.6050	7.3	7.300541927

Table A1. *Cont.*

	Montes et al. [54]	Akhtar et al. [55]	Gandomi et al. [46]	Zhang et al. [56]	Present Study (MGA)
l_2	7.962143	7.85933	7.8181	7.7153199115	7.715357693
d_1	3.3629	3.365576	3.3520	3.3502146661	3.350542391
d_2	5.3090	5.289773	5.2875	5.2866544650	5.28665793
$g_1(x)$	−0.0777	−0.0755	−0.0743	−0.0739152	−2.155122277
$g_2(x)$	−0.2013	−0.1994	−0.1983	−0.1979985	−98.13710222
$g_3(x)$	−0.4741	−0.4562	−0.4349	−0.9999967	−1.924273761
$g_4(x)$	−0.8971	−0.8994	−0.9008	−0.9999995	−18.30969834
$g_5(x)$	−0.0110	−0.0132	−0.0011	−0.6668526	−0.000437152
$g_6(x)$	−0.0125	−0.0017	−0.0004	−0.0000000	−0.001666474
$g_7(x)$	−0.7022	−0.7025	−0.7025	−0.7025000	−28.09998829
$g_8(x)$	−0.0006	−0.0017	−0.0004	−0.0000000	-6.68×10^{-6}
$g_9(x)$	−0.5831	−0.5826	−0.5832	−0.5833333	−6.999993318
$g_{10}(x)$	−0.0691	−0.0796	−0.0890	−0.0513257	−0.374728341
$g_{11}(x)$	−0.0279	−0.0179	−0.0130	−0.0000000	-3.40×10^{-05}

Table A2. Friedman statistical test results for the speed reducer design problem.

Rankings	Algorithms	Mean of Ranks
1	Present Study (MGA)	1.50
2	Zhang et al. [56]	2.00
3	Gandomi et al. [46]	3.00
4	Akhtar et al. [55]	3.87
5	Montes et al. [54]	4.62
Chi-sq.	10.7848	
Prob > Chi-sq.	0.0291	

Table A3. Comparison of the best solutions for the tension or compression spring design problem.

	Coello [57]	Ray and Liew [58]	Han et al. [59]	Gandomi et al. [45]	Present Study (MGA)
Best	0.01270478	0.0126692	0.01266534	0.01266522	0.01266523
d	0.051480	0.052160	0.0516800	0.05169	0.051689061
D	0.351661	0.368159	0.3565001	0.35673	0.35671774
N	11.632201	10.648442	11.3018335	11.2885	11.28896576
$g_1(x)$	−0.003337	-7.45×10^{-9}	-6.218×10^{-6}	0	0
$g_2(x)$	−0.000110	-3.68×10^{-9}	-1.691×10^{-6}	0	0
$g_3(x)$	−4.026318	−4.075805	−4.0533150	−4.0538	−4.05378563
$g_4(x)$	−0.731239	−0.719787	−0.7278799	−0.7277	−0.7277288

Table A4. Friedman statistical test results for the tension or compression spring design problem.

Rankings	Algorithms	Mean of Ranks
1	Present Study (MGA)	1.25
2	Han et al. [59]	2.25
3	Coello [57]	3.50
4	Ray and Liew [58]	4.00
5	Gandomi et al. [45]	4.00
Chi-sq.	9.4000	
Prob > Chi-sq.	0.0518	

Table A5. Comparison of the best solutions for the pressure vessel design problem.

	He and Wang [60]	Coelho [61]	Mezura-Montes and Coello [62]	Coello and Montes [63]	Present Study (MGA)
Best	6061.0777	6059.7208	6059.7456	6059.9463	6059.714350
T_s	0.8125	0.8125	0.8125	0.8125	0.8125
T_h	0.4375	0.4375	0.4375	0.4375	0.4375
R	42.0913	42.0984	42.098087	42.097398	42.0984
L	176.7465	176.6372	176.640518	176.654050	176.6366
$g_1(x)$	-1.37×10^{-6}	-8.79×10^{-7}	-6.92×10^{-6}	-2.02×10^{-5}	0
$g_2(x)$	-3.59×10^{-4}	-3.58×10^{-2}	−0.03588	−0.03589	−0.0359
$g_3(x)$	−118.7687	−0.2179	2.903372	−24.8998	0
$g_4(x)$	−63.2535	−63.3628	−63.3595	−63.346	−63.3634

Table A6. Friedman statistical test results for the pressure vessel design problem.

Rankings	Algorithms	Mean of Ranks
1	Present Study (MGA)	1
2	He and Wang [60]	2.75
3	Coello and Montes [63]	3.25
4	Coelho [61]	4.00
5	Mezura-Montes and Coello [62]	4.00
Chi-sq.	9.8000	
Prob > Chi-sq.	0.0439	

Table A7. Comparison of the best solutions for the welded beam design problem.

	Huang et al. [64]	Eskandar et al. [65]	Guedria [66]	Han et al. [59]	Present Study (MGA)
Best	1.733461	1.724856	1.724852	1.6956397	1.672966512
h	0.203137	0.205728	0.205730	0.20532536	0.198957505
l	3.542998	3.470522	3.470489	3.26035648	3.341955765
t	9.033498	9.036620	9.036624	9.03664424	9.187291977
b	0.206179	0.205729	0.205730	0.20572991	0.199190532
$g_1(x)$	−44.57856	−0.034128	-1.05×10^{-10}	−0.10520197	−20.76244473
$g_2(x)$	−44.66353	-3.49×10^{-5}	-6.91×10^{-10}	−0.17417862	−23.09392302
$g_3(x)$	−0.003042	-1.19×10^{-6}	-7.66×10^{-15}	−4.04330102	−0.000233027
$g_4(x)$	−3.423726	−3.432980	−3.432984	−3.45179021	−3.469028817
$g_5(x)$	−0.078137	−0.080728	−0.080730	−0.08032536	−0.073957505
$g_6(x)$	−0.235557	−0.235540	−0.235540	−0.22831066	−0.05415088
$g_7(x)$	−38.02826	−0.013503	-5.80×10^{-10}	−0.03397937	−30.47032014

Table A8. Friedman statistical test results for the welded beam design problem.

Rankings	Algorithms	Mean of Ranks
1	Present Study (MGA)	1.50
2	Guedria [66]	2.25
3	Han et al. [59]	3.25
4	Eskandar et al. [65]	3

Table A8. *Cont.*

Rankings	Algorithms	Mean of Ranks
5	Huang et al. [64]	5
Chi-sq.	11.0000	
Prob > Chi-sq.	0.0266	

Table A9. Comparison of the best solutions for the three-bar truss design problem.

	Gandomi et al. [46]	Ray and Liew [58]	Zhang et al. [56]	Grag [67]	Present Study (MGA)
Best	263.97156	263.8958466	263.8958434	263.8958433	263.8958433
A_1	0.78867	0.7886210370	0.7886751359	0.788676171219	0.788675136
A_2	0.40902	0.4084013340	0.4082482868	0.408245358456	0.408248288
$g_1(x)$	−0.00029	-8.275×10^{-9}	-2.104×10^{-11}	-1.587×10^{-13}	0
$g_2(x)$	−0.00029	−1.46392765	−1.46410161	−1.4641049	−1.464101618
$g_3(x)$	−0.73176	−0.536072358	−0.5358983	−0.535895	−0.535898382

Table A10. Friedman statistical test results for the three-bar truss design problem.

Rankings	Algorithms	Mean of Ranks
1	Present Study (MGA)	1.75
2	Grag [67]	1.87
3	Zhang et al. [56]	2.37
4	Ray and Liew [58]	4.25
5	Gandomi et al. [46]	4.75
Chi-sq.	12.8700	
Prob > Chi-sq.	0.0119	

Table A11. Comparison of the best solutions for the multiple disk clutch brake design problem.

	Deb and Srinivasan [68]	Eskandar et al. [65]	Rao et al. [50]	Ferreira et al. [47]	Present Study (MGA)
Best	0.4704	0.313656	0.313656611	0.313656	0.235242467
r_1	70	70	70	70	70.00000008
r_0	90	90	90	90	90.0000003
t	1.5	1	1	1	1.000000013
F	1000	910	810	830	865.6907633
Z	3	3	3	3	2.00000004
$g_1(x)$	0	0	0	0	-2.18×10^{-7}
$g_2(x)$	−22	−24	−24	−24	−25.4999999
$g_3(x)$	−0.9005	−0.909480	−0.91942781	−0.917438	−0.913888149
$g_4(x)$	−9.7906	−9.809429	−9830.371094	−9.826183	−9.985383395
$g_5(x)$	−7.8947	−7.894696	−7894.69659	−7.894697	−9.830260243
$g_6(x)$	−3.3527	−2.231421	−0.702013203	−0.173855	−14.98276443
$g_7(x)$	−60.6250	−49.768749	−37706.25	−40.118750	−83479.16052
$g_8(x)$	−11.6473	−12.768578	−14.2979868	−14.826145	−0.017235569

Table A12. Friedman statistical test results for the multiple disk clutch brake design problem.

Rankings	Algorithms	Mean of Ranks
1	Present Study (MGA)	1.5
2	Ferreira et al. [47]	2.12
3	Eskandar et al. [65]	2.37
4	Rao et al. [50]	4.00
Chi-sq.	8.8378	
Prob > Chi-sq.	0.0315	

Table A13. Comparison of the best solutions for the planetary gear train design problem.

	Savsani and Savsani [48]	Present Study (MGA)
Best	0.525588	0.52325
N_1	34	40
N_2	25	21
N_3	33	14
N_4	32	19
N_5	23	17
N_6	116	69
P	4	3
m_1	2.5	2
m_2	1.75	3

Table A14. Friedman statistical test results for the planetary gear train design problem.

Rankings	Algorithms	Mean of Ranks
1	Present Study (MGA)	2.50
2	Zhang et al. [70]	2.50
3	Savsani and Savsani [48]	3.00
4	Rao and Savsani [69] (ABC)	3.00
5	Rao and Savsani [69] (PSO)	4.00
Chi-sq.	12.8700	
Prob > Chi-sq.	0.0119	

Table A15. Comparison of the best solutions for the step-cone pulley design problem.

	TLBO [50]	WOA [44]	WCA [44]	MBA [44]	Present Study (MGA)
Best	16.63451	16.6345213	16.63450849	16.6345078	16.18595608
d_1	40	40	40	40	38.53034981
d_2	54.7643	54.764326	54.764300	54.764300	53.04151483
d_3	73.01318	54.764326	54.764300	54.764300	70.67294075
d_4	73.01318	54.764326	54.764300	88.428419	84.71470998
w	73.01318	85.986297	54.764300	85.986242	90

WOA: Whale Optimization Algorithm; WCA: Water Cycle Algorithm; MBA: Mine Blast Algorithm.

Table A16. Friedman statistical test results for the step-cone pulley design problem.

Rankings	Algorithms	Mean of Ranks
1	Present Study (MGA)	1.00
2	MBA [44]	2.00
3	WCA [44]	3.25
4	TLBO [50]	4.25
5	WOA [44]	4.50
Chi-sq.	14.2000	
Prob > Chi-sq.	0.0067	

Table A17. Comparison of the best solutions for the gear train design problem.

	Gandomi et al. [46]	Loh and Papalambros [79]	Kannan and Kramer [80]	Sandgren [81]	Present Study (MGA)
Best	2.701×10^{-12}	2.7×10^{-12}	2.146×10^{-8}	5.712×10^{-6}	1.06×10^{-19}
z_d	19	19	13	18	27.32076302
z_b	16	16	15	22	13.75530503
z_a	43	43	33	45	48.25305913
z_f	49	49	41	60	53.98015133

Table A18. Friedman statistical test results for the gear train design problem.

Rankings	Algorithms	Mean of Ranks
1	Present Study (MGA)	2.00
2	Wang et al. [82] (CPKH)	2.37
3	Loh and Papalambros [79]	2.75
4	Wang et al. [82] (ABC)	3.37
5	Gandomi et al. [46]	4.50
Chi-sq.	6.2278	
Prob > Chi-sq.	0.1828	

Table A19. Comparison of the best solutions for the piston lever design problem.

	CSA [46]	Present Study (MGA)
Best	8.4271	8.413406652
H	0.05	0.05
B	2.043	2.041637535
X	120	120
D	4.0851	4.083080224

Table A20. Friedman statistical test results for the piston lever design problem.

Rankings	Algorithms	Mean of Ranks
1	Present Study (MGA)	1.75
2	CSA [46]	2.75
3	HPSO [46]	3.75
4	GA [46]	3.75
5	DE [46]	3.75
Chi-sq.	4.0000	
Prob > Chi-sq.	0.4060	

References

1. Glover, F. Future paths for integer programming and links to artificial intelligence. *Comput. Oper. Res.* **1986**, *13*, 533–549. [CrossRef]
2. Sörensen, K.; Sevaux, M.; Glover, F. A History of Metaheuristics; Handbook of heuristics. *arXiv* **2017**, arXiv:1704.00853.
3. Moscato, P. On Evolution, Search, Optimization, Genetic Algorithms and Martial Arts: Towards Memetic Algorithms. *Caltech Concurr. Comput. Program C3P Rep.* **1989**, *826*, 1989.
4. Holland, J.H. *Adaptation in Natural and Artificial Systems: An Introductory Analysis with Applications to Biology, Control, and Artificial Intelligence*; MIT Press: Cambridge, MA, USA, 1992.
5. Koza, J.R.; Koza, J.R. *Genetic Programming: On the Programming of Computers by Means of Natural Selection*; MIT Press: Cambridge, MA, USA, 1992.
6. Storn, R.; Price, K. Differential evolution—A simple and efficient heuristic for global optimization over continuous spaces. *J. Glob. Optim.* **1997**, *11*, 341–359. [CrossRef]
7. Beyer, H.G.; Schwefel, H.P. Evolution strategies—A comprehensive introduction. *Nat. Comput.* **2002**, *1*, 3–52. [CrossRef]
8. Simon, D. Biogeography-based optimization. *IEEE Trans. Evol. Comput.* **2008**, *12*, 702–713. [CrossRef]
9. Eberhart, R.; Kennedy, J. A New Optimizer Using Particle Swarm Theory. In Proceedings of the Sixth International Symposium on Micro Machine and Human Science, Nagoya, Japan, 4–6 October 1995; pp. 39–43.
10. Dorigo, M.; Maniezzo, V.; Colorni, A. Ant system: Optimization by a colony of cooperating agents. *IEEE Trans. Syst. Man Cybern. Part B Cybern.* **1996**, *26*, 29–41. [CrossRef] [PubMed]
11. Basturk, B. An artificial bee colony (ABC) algorithm for numeric function optimization. In Proceedings of the IEEE Swarm Intelligence Symposium, Indianapolis, IN, USA, 12 May 2006.
12. Chu, S.C.; Tsai, P.W.; Pan, J.S. Cat Swarm Optimization. In *Pacific Rim International Conference on Artificial Intelligence*; Springer: Berlin/Heidelberg, Germany, 2006; pp. 854–858.
13. Yang, X.S. *Nature-Inspired Metaheuristic Algorithms*; Luniver Press: Bristol, UK, 2010.
14. Gandomi, A.H.; Alavi, A.H. Krill herd: A new bio-inspired optimization algorithm. *Commun. Nonlinear Sci. Numer. Simul.* **2012**, *17*, 4831–4845. [CrossRef]
15. Kirkpatrick, S.; Gelatt, C.D.; Vecchi, M.P. Optimization by simulated annealing. *Science* **1983**, *220*, 671–680. [CrossRef]
16. Geem, Z.W.; Kim, J.H.; Loganathan, G.V. A new heuristic optimization algorithm: Harmony search. *Simulation* **2001**, *76*, 60–68. [CrossRef]
17. Erol, O.K.; Eksin, I. A new optimization method: Big bang–big crunch. *Adv. Eng. Softw.* **2006**, *37*, 106–111. [CrossRef]
18. Rashedi, E.; Nezamabadi-Pour, H.; Saryazdi, S. GSA: A gravitational search algorithm. *Inf. Sci.* **2009**, *179*, 2232–2248. [CrossRef]
19. Kaveh, A.; Talatahari, S. A novel heuristic optimization method: Charged system search. *Acta Mech.* **2010**, *213*, 267–289. [CrossRef]
20. Alatas, B. ACROA: Artificial chemical reaction optimization algorithm for global optimization. *Expert Syst. Appl.* **2011**, *38*, 13170–13180. [CrossRef]
21. Kaveh, A.; Mahdavi, V.R. Colliding bodies optimization: A novel meta-heuristic method. *Comput. Struct.* **2014**, *139*, 18–27. [CrossRef]
22. Talatahari, S.; Azizi, M. Chaos Game Optimization: A novel metaheuristic algorithm. *Artif. Intell. Rev.* **2020**, *22*, 1–88. [CrossRef]
23. Talatahari, S.; Azizi, M. Optimization of Constrained Mathematical and Engineering Design Problems Using Chaos Game Optimization. *Comput. Ind. Eng.* **2020**, *145*, 106560. [CrossRef]
24. Azizi, M. Atomic orbital search: A novel metaheuristic algorithm. *Appl. Math. Model.* **2021**, *93*, 657–683. [CrossRef]
25. Atashpaz-Gargari, E.; Lucas, C. Imperialist competitive algorithm: An algorithm for optimization inspired by imperialistic competition. In Proceedings of the 2007 IEEE Congress on Evolutionary Computation, Singapore, 25–28 September 2007; pp. 4661–4667.
26. Yang, X.S.; Deb, S. Cuckoo search via Lévy flights. In Proceedings of the 2009 World Congress on Nature & Biologically Inspired Computing (NaBIC), Coimbatore, India, 9–11 December 2009; pp. 210–214.
27. Talatahari, S.; Azizi, M. Tribe-charged system search for global optimization. *Appl. Math. Model.* **2021**, *93*, 115–133. [CrossRef]
28. Carbas, S. Design optimization of steel frames using an enhanced firefly algorithm. *Eng. Optim.* **2016**, *48*, 2007–2025. [CrossRef]
29. Hasançebi, O.; Çarbaş, S.; Saka, M.P. Improving the performance of simulated annealing in structural optimization. *Struct. Multidiscip. Optim.* **2010**, *41*, 189–203. [CrossRef]
30. Azad, S.K. Design optimization of real-size steel frames using monitored convergence curve. *Struct. Multidiscip. Optim.* **2021**, *63*, 267–288. [CrossRef]
31. Akış, T.; Azad, S.K. Structural Design Optimization of Multi-layer Spherical Pressure Vessels: A Metaheuristic Approach. *Iran. J. Sci. Technol. Trans. Mech. Eng.* **2019**, *43*, 75–90. [CrossRef]
32. Tubishat, M.; Idris, N.; Shuib, L.; Abushariah, M.A.; Mirjalili, S. Improved Salp Swarm Algorithm based on opposition based learning and novel local search algorithm for feature selection. *Expert Syst. Appl.* **2020**, *145*, 113–122. [CrossRef]
33. Mokeddem, D.; Mirjalili, S. Improved Whale Optimization Algorithm applied to design PID plus second-order derivative controller for automatic voltage regulator system. *J. Chin. Inst. Eng.* **2020**, *43*, 541–552. [CrossRef]
34. Kaveh, A.; Hosseini, S.M.; Zaerreza, A. Improved Shuffled Jaya algorithm for sizing optimization of skeletal structures with discrete variables. In *Structures*; Elsevier: Amsterdam, The Netherlands, 2021; pp. 107–128.

35. Ebrahimi, B.; Tavana, M.; Toloo, M.; Charles, V. A novel mixed binary linear DEA model for ranking decision-making units with preference information. *Comput. Ind. Eng.* **2020**, *149*, 106720. [CrossRef]
36. Azizi, M.; Ghasemi, S.A.; Ejlali, R.E.; Talatahari, S. Optimization of Fuzzy Controller for Nonlinear Buildings with Improved Charged System Search. *Struct. Eng. Mech.* **2020**, *76*, 781.
37. Wu, G.; Mallipeddi, R.; Suganthan, P.N. *Problem Definitions and Evaluation Criteria for the CEC 2017 Competition on Constrained Real-Parameter Optimization*; Technical Report; National University of Defense Technology: Changsha, China; Kyungpook National University: Daegu, Korea; Nanyang Technological University: Singapore, 2017.
38. Tvrdík, J.; Poláková, R. Simple framework for constrained problems with application of L-SHADE44 and IDE. In Proceedings of the 2017 IEEE Congress on Evolutionary Computation (CEC), Donostia, Spain, 5 June 2017; pp. 1436–1443.
39. Polakova, R. L-SHADE with competing strategies applied to constrained optimization. In Proceedings of the 2017 IEEE Congress on Evolutionary Computation (CEC), Donostia, Spain, 5 June 2017; pp. 1683–1689.
40. Zamuda, A. Adaptive constraint handling and success history differential evolution for CEC 2017 constrained real-parameter optimization. In Proceedings of the 2017 IEEE Congress on Evolutionary Computation (CEC), Donostia, Spain, 5 June 2017; pp. 2443–2450.
41. Derrac, J.; García, S.; Molina, D.; Herrera, F. A practical tutorial on the use of nonparametric statistical tests as a methodology for comparing evolutionary and swarm intelligence algorithms. *Swarm Evol. Comput.* **2011**, *1*, 3–18. [CrossRef]
42. Sörensen, K. Metaheuristics—The metaphor exposed. *Int. Trans. Oper. Res.* **2015**, *22*, 3–18. [CrossRef]
43. Kumar, A.; Wu, G.; Ali, M.Z.; Mallipeddi, R.; Suganthan, P.N.; Das, S. A test-suite of non-convex constrained optimization problems from the real-world and some baseline results. *Swarm Evol. Comput.* **2020**, *56*, 100693. [CrossRef]
44. Yildiz, A.R.; Abderazek, H.; Mirjalili, S. A Comparative Study of Recent Non-traditional Methods for Mechanical Design Optimization. *Arch. Comput. Methods Eng.* **2020**, *27*, 1031–1048. [CrossRef]
45. Gandomi, A.H.; Yang, X.S.; Alavi, A.H.; Talatahari, S. Bat algorithm for constrained optimization tasks. *Neural Comput. Appl.* **2013**, *22*, 1239–1255. [CrossRef]
46. Gandomi, A.H.; Yang, X.S.; Alavi, A.H. Cuckoo search algorithm: A metaheuristic approach to solve structural optimization problems. *Eng. Comput.* **2013**, *29*, 17–35. [CrossRef]
47. Ferreira, M.P.; Rocha, M.L.; Neto, A.J.; Sacco, W.F. A constrained ITGO heuristic applied to engineering optimization. *Expert Syst. Appl.* **2018**, *110*, 106–124. [CrossRef]
48. Savsani, P.; Savsani, V. Passing vehicle search (PVS): A novel metaheuristic algorithm. *Appl. Math. Model.* **2016**, *40*, 3951–3978. [CrossRef]
49. Rao, S.S. *Engineering Optimization: Theory and Practice*; John Wiley & Sons: Hoboken, NJ, USA, 2019.
50. Rao, R.V.; Savsani, V.J.; Vakharia, D.P. Teaching–learning-based optimization: A novel method for constrained mechanical design optimization problems. *Comput. Aided Des.* **2011**, *43*, 303–315. [CrossRef]
51. Yu, Z.; Xu, T.; Cheng, P.; Zuo, W.; Liu, X.; Yoshino, T. Optimal Design of Truss Structures with Frequency Constraints Using Interior Point Trust Region Method. *Proc. Rom. Acad. Ser.* **2014**, *15*, 165–173.
52. Gupta, S.; Tiwari, R.; Nair, S.B. Multi-objective design optimisation of rolling bearings using genetic algorithms. *Mech. Mach. Theory* **2007**, *42*, 1418–1443. [CrossRef]
53. Zelinka, I.; Lampinen, J. Mechanical Engineering Problem Optimization by SOMA. In *New Optimization Techniques in Engineering*; Springer: Berlin/Heidelberg, Germany, 2004; pp. 633–653.
54. Mezura-Montes, E.; Coello, C.C.; Landa-Becerra, R. Engineering Optimization Using Simple Evolutionary Algorithm. In Proceedings of the 15th IEEE International Conference on Tools with Artificial Intelligence, Sacramento, CA, USA, 5 November 2003; pp. 149–156.
55. Akhtar, S.; Tai, K.; Ray, T. A socio-behavioural simulation model for engineering design optimization. *Eng. Optim.* **2002**, *34*, 341–354. [CrossRef]
56. Zhang, M.; Luo, W.; Wang, X. Differential evolution with dynamic stochastic selection for constrained optimization. *Inf. Sci.* **2008**, *178*, 3043–3074. [CrossRef]
57. Coello, C.A. Use of a self-adaptive penalty approach for engineering optimization problems. *Comput. Ind.* **2000**, *41*, 113–127. [CrossRef]
58. Ray, T.; Liew, K.M. Society and civilization: An optimization algorithm based on the simulation of social behavior. *IEEE Trans. Evol. Comput.* **2003**, *7*, 386–396. [CrossRef]
59. Han, J.; Yang, C.; Zhou, X.; Gui, W. A two-stage state transition algorithm for constrained engineering optimization problems. *Int. J. Control Autom. Syst.* **2018**, *16*, 522–534. [CrossRef]
60. He, Q.; Wang, L. An effective co-evolutionary particle swarm optimization for constrained engineering design problems. *Eng. Appl. Artif. Intell.* **2007**, *20*, 89–99. [CrossRef]
61. Dos Santos Coelho, L. Gaussian quantum-behaved particle swarm optimization approaches for constrained engineering design problems. *Expert Syst. Appl.* **2010**, *37*, 1676–1683. [CrossRef]
62. Zahara, E.; Kao, Y.T. Hybrid Nelder–Mead simplex search and particle swarm optimization for constrained engineering design problems. *Expert Syst. Appl.* **2009**, *36*, 3880–3886. [CrossRef]
63. Sadollah, A.; Bahreininejad, A.; Eskandar, H.; Hamdi, M. Mine blast algorithm: A new population based algorithm for solving constrained engineering optimization problems. *Appl. Soft Comput.* **2013**, *13*, 2592–2612. [CrossRef]

64. Huang, F.Z.; Wang, L.; He, Q. An effective co-evolutionary differential evolution for constrained optimization. *Appl. Math. Comput.* **2007**, *186*, 340–356. [CrossRef]
65. Eskandar, H.; Sadollah, A.; Bahreininejad, A.; Hamdi, M. Water cycle algorithm—A novel metaheuristic optimization method for solving constrained engineering optimization problems. *Comput. Struct.* **2012**, *110*, 151–166. [CrossRef]
66. Guedria, N.B. Improved accelerated PSO algorithm for mechanical engineering optimization problems. *Appl. Soft Comput.* **2016**, *40*, 455–467. [CrossRef]
67. Garg, H. A hybrid GSA-GA algorithm for constrained optimization problems. *Inf. Sci.* **2019**, *478*, 499–523. [CrossRef]
68. Deb, K.; Srinivasan, A. Innovization: Innovating Design Principles through Optimization. In Proceedings of the 8th Annual Conference on Genetic and Evolutionary Computation, Seattle, WA, USA, 8 July 2006; pp. 1629–1636.
69. Rao, R.V.; Savsani, V.J. *Mechanical Design Optimization Using Advanced Optimization Techniques*; Springer Science & Business Media: London, UK, 2012.
70. Zhang, J.; Xiao, M.; Gao, L.; Pan, Q. Queuing search algorithm: A novel metaheuristic algorithm for solving engineering optimization problems. *Appl. Math. Model.* **2018**, *63*, 464–490. [CrossRef]
71. Siddall, J.N. *Optimal Engineering Design: Principles and Applications*; CRC Press: London, UK, 1982.
72. Deb, K.; Goyal, M. Optimizing Engineering Designs Using a Combined Genetic Search. *InICGA* **1997**, 521–528.
73. Coello, C.A. The Use of a Multiobjective Optimization Technique to Handle Constraints. In *Proceedings of the Second International Symposium on Artificial Intelligence (Adaptive Systems)*; Institute of Cybernetics, Mathematics and Physics, Ministry of Science Technology and Environment: La Habana, Cuba, 1999; pp. 251–256.
74. Şahin, İ.; Dörterler, M.; Gokce, H. Optimization of Hydrostatic Thrust Bearing Using Enhanced Grey Wolf Optimizer. *Mechanics* **2019**, *25*, 480–486. [CrossRef]
75. Rao, R.V.; Waghmare, G.G. A new optimization algorithm for solving complex constrained design optimization problems. *Eng. Optim.* **2017**, *49*, 60–83. [CrossRef]
76. Lamberti, L.; Pappalettere, C. Move limits definition in structural optimization with sequential linear programming. Part I: Optimization algorithm. *Comput. Struct.* **2003**, *81*, 197–213. [CrossRef]
77. Baghlani, A.; Makiabadi, M.H. Teaching-learning-based optimization algorithm for shape and size optimization of truss structures with dynamic frequency constraints. *Iran. J. Sci. Technol. Trans. Civ. Eng.* **2013**, *37*, 409.
78. Kaveh, A.; Zolghadr, A. Shape and size optimization of truss structures with frequency constraints using enhanced charged system search algorithm. *Asian J. Civ. Eng. Build. Hous.* **2011**, *12*, 487–509.
79. Loh, H.T.; Papalambros, P.Y. Computational implementation and tests of a sequential linearization algorithm for mixed-discrete nonlinear design optimization. *J. Mech. Des.* **1990**, *5213*, 11–12. [CrossRef]
80. Kannan, B.K.; Kramer, S.N. An augmented Lagrange multiplier based method for mixed integer discrete continuous optimization and its applications to mechanical design. *J. Mech. Des.* **1994**, *116*, 405–411. [CrossRef]
81. Sandgren, E. Nonlinear integer and discrete programming in mechanical design. *J. Mech. Des.* **1988**, *112*, 223–229. [CrossRef]
82. Wang, G.G.; Hossein Gandomi, A.; Hossein Alavi, A. A chaotic particle-swarm krill herd algorithm for global numerical optimization. *Kybernetes* **2013**, *42*, 962–978. [CrossRef]
83. Wang, G.G. Adaptive response surface method using inherited latin hypercube design points. *J. Mech. Des.* **2003**, *125*, 210–220. [CrossRef]

Article

Improved NSGA-III with Second-Order Difference Random Strategy for Dynamic Multi-Objective Optimization

Haijuan Zhang [1], Gai-Ge Wang [1,2,3,4,*], Junyu Dong [1] and Amir H. Gandomi [5]

1 Department of Computer Science and Technology, Ocean University of China, Qingdao 266100, China; zhanghaijuan@stu.ouc.edu.cn (H.Z.); dongjunyu@ouc.edu.cn (J.D.)
2 College of Information Technology, Jilin Agricultural University, Changchun 130118, China
3 Key Laboratory of Symbolic Computation and Knowledge Engineering of Ministry of Education, Jilin University, Changchun 130012, China
4 Guangxi Key Laboratory of Hybrid Computation and IC Design Analysis, Guangxi University for Nationalities, Nanning 530006, China
5 Faculty of Engineering & Information Technology, University of Technology, Sydney, NSW 2007, Australia; gandomi@uts.edu.au
* Correspondence: wgg@ouc.edu.cn

Abstract: Most real-world problems that have two or three objectives are dynamic, and the environment of the problems may change as time goes on. For the purpose of solving dynamic multi-objective problems better, two proposed strategies (second-order difference strategy and random strategy) were incorporated with NSGA-III, namely SDNSGA-III. When the environment changes in SDNSGA-III, the second-order difference strategy and random strategy are first used to improve the individuals in the next generation population, then NSGA-III is employed to optimize the individuals to obtain optimal solutions. Our experiments were conducted with two primary objectives. The first was to test the values of the metrics mean inverted generational distance (*MIGD*), mean generational distance (*MGD*), and mean hyper volume (*MHV*) on the test functions (Fun1 to Fun6) via the proposed algorithm and the four state-of-the-art algorithms. The second aim was to compare the metrics' value of NSGA-III with single strategy and SDNSGA-III, proving the efficiency of the two strategies in SDNSGA-III. The comparative data obtained from the experiments demonstrate that SDNSGA-III has good convergence and diversity compared with four other evolutionary algorithms. What is more, the efficiency of second-order difference strategy and random strategy was also analyzed in this paper.

Keywords: dynamic; multi-objective; NSGA-III; evolutionary algorithm; prediction strategy

Citation: Zhang, H.; Wang, G.-G.; Dong, J.; Gandomi, A.H. Improved NSGA-III with Second-Order Difference Random Strategy for Dynamic Multi-Objective Optimization. *Processes* **2021**, *9*, 911. https://doi.org/10.3390/pr9060911

Academic Editor: Gade Pandu Rangaiah

Received: 8 May 2021
Accepted: 19 May 2021
Published: 21 May 2021

1. Introduction

Most optimization problems in the real world are multi-objective, which include different constraints. For multi-objective optimization, many classical evolutionary algorithms (EAs) have proven to be powerful tools to obtain optimal solutions, like the multi-objective evolutionary algorithm based on decomposition (MOEA/D) [1], fast, elitist multi-objective nondominated sorting genetic algorithm (NSGA-II) [2] and other meta-heuristic algorithms [3–8]. Because most practical problems faced in the real world are not limited to static multiple objectives, environmental changes may lead to different optimal solutions under different time windows. Such issues that have changing objectives and constraints are called dynamic multi-objective optimization problems (DMOPs).

Compared with static multi-objective problems (SMOPs), DMOPs are much closer to practical problems and more complex, which brings some challenges to the optimization process for the following reasons. First, changes in the number of goals or constraints increase the uncertainty of the problem. For SMOPs, convergence and diversity are both important for obtaining optimal solutions, which is also the same for DMOPs. Differently, algorithms for DMOPs have rapid convergence while performing worse for maintaining the diversity of a population, which leads to local, not global optimal solutions. Besides, the Pareto front and optimal solutions tracked by the algorithm also change over time. Second, two parameters, i.e., frequency and severity, are introduced because of the dynamic changes. The change in their values has a great influence on the algorithm's ability to track the optimal solution, which adds much complexity to the problem in a larger part. Therefore, the purpose of the algorithm for DMOPs is to track the changing Pareto front and maintain rapid convergence and diversity when the environment changes.

Over the past few years, dynamic optimization has received increasing attention, and many such techniques have been proposed for solving DMOPs. These techniques can be classified into two categories [9]: (1) memory-based approaches and (2) prediction-based approaches.

In the memory-based method, the initial population is generated by memorizing the historical optimal solution of the previous generation, coping with the changes in the problem. Properly storing the optimal solutions obtained in the past, and restarting these solutions for optimization can greatly improve the efficiency and searchability of the solution as the environmental changes during the optimization process. For instance, Luo et al. [10] combined the species strategy and memory method into species-based particle swarm optimization (SPSO) and enhanced the performance of SPSO. Nakano et al. [11] improved the artificial bee colony algorithm (IABC) by introducing memory and detection schemes for higher dimensional DMOPs. However, several drawbacks in the memory-based approach include large memory consumption and incorrect predictions to solve time-varying nonlinear problems, or more frequent environment.

On the other hand, prediction-based methods aim to predict the changing Pareto front through a certain prediction model, thereby forming a prediction solution to respond to the next change of the environment. In other words, memory-based methods are more suitable for periodic or repeated environmental changes, while various types of changes can be handled through prediction-based techniques. Therefore, many improved algorithms incorporated with different prediction models have been developed. Gong et al. proposed a multidirectional approach [12] based on multi-time optimal solution prediction and a multi-model method [13] based on prediction for DMOPs. The representative individuals generated by the clustering of optimized solutions predict the change of the Pareto front and, finally, generate the predicted solutions. According to the relationship between the type of change and the response strategy, the prediction model corresponding to the type of problem change is selected to generate a predicted solution. Both methods have proved to be superior to other algorithms in performance. Wu et al. [14] added the idea of the knee point into the multi-objective artificial flora algorithm (MOAF), improving the diversity and distribution. In addition, maintaining diversity in the population is also critical for DMOPs. Zhou et al. [15] improved the population diversity through different information obtained from various populations before and after a change occurs, whereby the information is used for guiding the next environmental change. Considering the difficulties in identifying the proportion of diversity introduction methods, [16] proposed an adaptive diversity introduction (ADI) method, wherein the ratio of the diversity introduction can be adjusted adaptively.

For DMOPs, the dynamic version of NSGA-II [2] had been proposed. Similar to NSGA-II, NSGA-III [17] is also a classical evolutionary algorithm for many objectives. Although NSGA-III can be used for solving DMOPs directly, the performance of NSGA-III on DMOPs cannot be guaranteed. Therefore, the motivation of this work was to design dynamic strategies combined with NSGA-III and improve its performance when dealing with DMOPs.

This study mainly focused on the DMOPs that have a fixed number of objectives and decision variables and introduces second-order difference strategy and random strategy to be applied in NSGA-III, combined with the good convergence and diversity of NSGA-III. The general process of the strategy is that when the environment changes are detected, a second-order difference strategy is first used to predict the next centroid based on every two solutions, then new individuals are produced by randomly searching around the predicted solution. Second, in the experimental part, the comparisons on six benchmarks show the performance of the proposed algorithm, and the two parameters (change in frequency and severity) are changed. Subsequently, three different dynamic metrics are evaluated on the proposed algorithm. In addition, the experiment also verifies the efficiency between the second-order difference strategy and the random disturbance strategy in the algorithm.

The rest of this paper is organized as follows. Section 2 presents the background and related work of DMOPs, which we focused on. The proposed strategies are introduced in Section 3, and the experimental settings and results are described in Section 4. At last, Section 5 concludes the results obtained from experiments and discusses future work.

2. Background and Related Work

2.1. Problem Formulation

There are many formulations for dynamic multi-objective optimization problems. Without loss of generality, we considered a minimum DMOP in this study [18]:

$$\begin{cases} \min F(x,t) = (f_1(x,t), f_2(x,t), \cdots, f_M(x,t))^T \\ s.t. x \in \Omega = \prod_{i=1}^{n} [a_i, b_i] \end{cases} \tag{1}$$

where $x = (x_1, x_2, x_3, \ldots, x_n)^T$ is an n-dimensional decision vector, and Ω is the decision space. $F(x, t)$: $R^n \times T \rightarrow R^M$ denotes an M-dimensional objective space, which contains M objectives that have some environmental changes during the evolutionary process. What's more, we suppose that all the objective functions are continuous. There are three definitions of dynamic Pareto as follows.

Definition 1. ***(Dynamic Pareto Dominance** [19]**):*** *In a dynamic environment, a decision vector x_1 Pareto dominates another decision vector x_2 at the time window t, expressed as $x_1 \succ x_2$, if and only if*

$$\begin{cases} \forall i = 1, \ldots, M, f_i(x_1,t) \le f_i(x_2,t) \\ \exists i = 1, \ldots, M, f_i(x_1,t) < f_i(x_2,t) \end{cases} \tag{2}$$

Definition 2. ***(Dynamic Pareto-optimal Set (DPS)** [12]**):*** *For a fixed time window t and a decision vector $x^* \in \Omega$, when there is no other decision vector $x \in \Omega$ such that x dominates x^*, the decision vector x^* is said to be nondominated. The dynamic Pareto-optimal set (DPS) is the set of all non-dominated solutions in decision space, which can be represented by:*

$$DPS(t) = \{x^* \in \Omega | \exists x \in \Omega, x \succ x^*\} \tag{3}$$

Definition 3. ***(Dynamic Pareto-optimal Front (DPF)** [12]**):*** *Similar to the static Pareto-front, the dynamic Pareto-optimal front (DPF) is the set of the corresponding objective values of the DPS:*

$$DPF(t) = \{F(x^*,t) | x^* \in DPS\} \tag{4}$$

Based on the above definitions, the following statement can be made. Since an environmental change can lead to changes of *DPS* or *DPF*, an effective dynamic multi-objective optimization evolutionary algorithm (DMOEA) is expected to track the moving *DPF* in time and still maintain diversity and convergence.

2.2. Evolutionary Algorithms for DMOPs

Due to their wide application in real-world problems, they have acquired rapidly increasing attention in recent years. Because of the changes of *DPS* and *DPF*, DMOEAs pose a higher requirement of efficiency compared with static MOEAs. A well-designed DMOEA should perform well in maintaining both convergence and diversity of the population. Many improved classic evolutionary algorithms can be used in the process of solving DMOPs, such as multi-objective evolutionary algorithm based on decomposition and a first-order difference model (MOEA/D-FD) [20], the dynamic non-dominated sorting genetic algorithm (DNSGA-II) [21], a novel dynamic multi-objective evolutionary algorithm with a cell-based rank and density calculation strategy (DMOEA) [22], and dynamic constrained optimization differential evolution (DyCODE) [23].

As mentioned in Section 1, memory and prediction techniques are general methods for DMOPs and various studies have proposed prediction-based approaches. From the perspective of evolutionary process, prediction-based methods, can generally be divided into two classes: (1) population-based prediction and (2) individual-based prediction. In population-based prediction, the whole population is optimized with a single prediction model while the moving location of each individual is predicted in individual-based prediction methods. Teodoro et al. [24] proposed a method of plane separation for a whole population to solve DMOPs with incorporated references. In terms of computational cost, many algorithms focus on distribution but ignore the excessive computational burden. Moreover, a knee-guided prediction evolutionary algorithm (KPEA) [25] has been proposed to achieve a lower computational cost. Specifically, KPEA generates the non-dominated solutions around the knee points and boundary regions, and then relocates the location of the knee point and boundary regions. Zhou et al. [26] introduced two strategies for application in the process of population re-initialization. The two strategies are based on individuals in the population, and some individuals are predicted according to the information obtained from past environments. A feed-forward prediction approach, which combines a forecasting technique to place an anticipatory group of individuals, was introduced by Hatzakis et al. [27]. Taking the relevance between previous environments and new environment into account, Liu et al. [28] stored the solutions in the past environments into two different archives, so that the stored solutions will provide the information to find the optimal solutions when the next change occurs. Many DMOPs in real-world applications have various characteristics, such as the interval characteristic, which is one of the common applications in DMOPs. To solve interval DMOPs (DI-MOPs), Gong et al. [29] proposed a novel co-evolutionary model in terms of interval similarity. The interval parameters are set, and two different response strategies for change are applied to track the *DPF*. Wang et al. [30] presented a predictive method based on a grey prediction model, which divided the population into some clusters, then the centroid of each cluster was used to build the model. In addition, individuals from different clusters were selected to detect the environmental change.

Herein, we used various standards to measure the performance of a DMOEA in the evolutionary process, which is largely influenced by the speed of convergence and diversity. DMOEAs with rapid convergence track the moving *DPF* rapidly, while DMOEAs with good diversity pursue the effective distribution of the optimal solutions. Keeping a good balance between population diversity and convergence is critical to the performance of DMOEAs. For DMOPs, Chang et al. [31] combined query-based learning with particle swarm optimization (QBLDPSO) to improve the diversity of a population. Instead of maintaining the diversity of particles passively like typical PSO, QBLDPSO actively employs quantum parameter adaptation to achieve population diversity. In order to accelerate the convergence, Liang et al. [32] classified the decision variables into different groups according to different optimization stages, like change detection stage, optimization stage in each time window. The different crossover operators are employed to keep a balance of convergence and diversity.

There are many application scenarios for dynamic optimization algorithms. For example, Wang et al. [33] improved PSO with a mixed-integer programming model and four match-up strategies to manage dynamic scheduling problems with random job arrivals. Luna et al. [34] presented a novel restart method that can react to the changes in the traffic demands, improving the energy efficiency in the fifth generation (5G) of cellular networks.

A novel chance-constrained dynamic optimization method, proposed by Zhou et al. [35], was applied to solve real industrial process issues. In the study of humanoid robots, losing balance and falling down are common problems. Chang et al. [36] applied a DMOEA to the falling down process of robots and was able to reduce damage to the robots. A dynamic multi-objective approach can also be applied to heterogeneous systems. For instance, [37] introduced a multi-objective dynamic load balancing (DLB) approach combined with a genetic heuristic engine, to enhance performance and code portability for such systems.

From the review above, we can identify three essential parts that should be included in an ideal DMOEA, which are change detection, change reaction, and multi-objective optimization. Herein, Algorithm 1 was prepared as the major frame of DMOEA. In the algorithm, after initializing the current generation, some strategies are used for change response when an environmental change occurs. The initialized population will be updated with these useful strategies, and the time window T is going to increase by one, which means the next environmental change. In the next step, the i-th multi-objective problems are optimized using a multi-objective evolutionary algorithm (MOEA) for one generation. The static MOEA employs the updated population as the initial population. At last, if the stop condition is not met, the process is repeated, otherwise the optimization of the next generation will be conducted.

Algorithm 1 The main frame of DMOEA

Input: the number of generations, g;
the time window, T;
Output: optimal solution x^* at every time step;
Initialize population POP_0;
while stop criterion is not met **do**
  if change is detected **then**
    Update the population using some strategies: reuse memory, tune parame-ters, or predict solutions;
    $T = T + 1$;
  end if
  Optimize T-th population with an MOEA for one generation and get optimal solution x^*;
end while
$g = g+1$;
return x^*.

3. NSGA-III and Proposed Strategies

3.1. NSGA-III

There are many classical algorithms for multi-objective or many-objective problems. NSGA-III [17] is one of the classical multi-objective evolutionary algorithms (MOEAs) for static many-objective problems. NSGA-III adds the idea of reference points as an enhancement of NSGA-II [2], which is mainly used for dealing with multi-objective problems. The major framework of NSGA-III is similar to the original NSGA-II algorithm, except for the differences between their environmental selection in the critical layer. The NSGA-II method applies congestion and comparison operation to select and maintain diversity. Comparatively, NSGA-III employs well-distributed reference points to maintain the diversity of a population. Most MOEAs are more effective in solving problems with fewer objectives. When the number of objectives is greater than or equal to four, many methods have reduced selection pressure due to increased dimensions, and the effect is not ideal. Different non-dominated fronts are classified using the method of fast non-dominated sorting. In the selection stage, the congestion distance method is changed to the reference points method, because the former does not perform well for balancing the diversity and convergence in NSGA-II. It is also limited to solving various unconstrained problems, such as normalization, scaling, convexity, concavity, and convergence to the PF plane. In addition, the dynamic version of NSGA-II has been proposed for DMOPs, which was used for comparison in our experiments and NSGA-III was chosen as the evolutionary algorithm combined with the two proposed strategies.

The whole detailed process of NSGA-III is given in Algorithm 2. Firstly, reference points are defined and the initial population is obtained through generating a set of uniformly distributed points randomly. Secondly, genetic operators are used to generate the offspring, and then non-dominated sorting is conducted. The parent population P_t and the generated offspring Q_t are combined as a new population R_t. The new population is placed in a non-dominant order, which means that the solutions in R_t are divided into different non-dominant levels. The N individuals that make up the next population will be selected from the mixed population. If the individuals in the first $l-1$ non-dominated layer are exactly equal to N, then these individuals directly form the next generation population P_{t+1}. Otherwise, the remaining solutions of P_{t+1} are selected from F_l according to the selection mechanism. Then, normalizing the objectives and each solution will be associated with one reference point. The closest K individuals associated with the reference points are selected. Finally, the best N solutions in the combined population are selected, and the next population P_{t+1} is obtained.

Algorithm 2 The procedure of NSGA-III

Input: parent population, P_t;
the archive population, S_t;
i-th non-dominated front, F_i;
Output: the next population, P_{t+1};
Define a set of reference points z^*;
Generate offspring Q_t through cross and mutation using GA operator;
Non-dominated sorting ($R_t = P_t \cup Q_t$);
while $|S_t| < N$ **do**
  $S_t = S_t \cup F_i$;
  $i = i + 1$;
end while
$F_l = F_i$;
if $|S_t| = N$ **then**
  $P_{t+1} = S_t$;
else
  $P_{t+1} = F_1 \cup F_2 \cup \ldots \cup F_{l-1}$;
end if
Normalize objectives and associate each solution with one reference point;
Calculate the number of the associated solutions;
Choose $K = N - |P_{t+1}|$ solutions one by one from F_l;
return P_{t+1}.

3.2. Change Detection

As mentioned above, one of the critical steps for DMOPs is change detection. Before using the change response strategy, change detection is necessary for dynamic multi-objective evolutionary algorithms (DMOEAs). The main process of environmental change detection is described in Algorithm 3, wherein we define a label *flag* indicating whether a change occurs. The initial value of *flag* is set to zero when the change detection starts. Then, individuals randomly selected from the initial population are evaluated for change detection. Herein, 20% of randomly selected individuals are used in our proposed algorithm, and their objectives are stored into P_s, which are then re-evaluated. By investigating whether there is a difference in the objective values between two generations, the results about whether a change occurs can be shown obviously. If there is a difference between the two generations, the value of a *flag* is set to one, and it can be said that an environmental change occurs and the change detection process ends.

Algorithm 3 Change detection

Input: the initial population, P_T;
the current number of iterations, g;
the number of objective functions, F;
the individual in population, p;
stored individuals from past environment, P_s;
the current time window, t;
Output: the sign indicating whether a change occurs, *flag*;
flag = 0;
Select randomly individuals from population P_T, and store individuals into P_s;
for every $p \in P_s$, $i \in F$ **do**
  Caculate $f_i(p, t)$;
  $f_i(P_s) = f_i(p, t)$;
end for
$g = g + 1$;
for every $p \in P_s$ **do**
  Caculate $f_i(p, t)$;
  if $f_i(p, t) \neq f_i(P_s)$ **then**
    flag = 1;
    $t = t + 1$;
    break;
  end if
end for
return *flag*.

3.3. Second-Order Difference and Random Strategies

We propose a new strategy named the second-order difference strategy for DMOPs in this paper. This strategy is based on the first-order difference strategy, which used the centroid of the decision space to describe the trajectory of moving solutions over time. Let C_T be the centroid of the *DPS* and P_T be the obtained *DPS* at the time window T, then C_T can be calculated by the following formula:

$$C_T = \frac{1}{|P_T|} \sum_{x \in P_T} x \quad (5)$$

where C_{T+1} represents the center of the decision space at the next time window, $T + 1$, which can be obtained by Formula (6):

$$C_{T+1} = C_T + \overrightarrow{C_T - C_{T-1}} \quad (6)$$

where $|P_T|$ is the cardinality of P_T; and x is a solution of decision space in P_T. The initial individuals of the population are provided through the prediction model at each generation. As we can see, the first-order difference model employs the centroid of the decision space, and predicts the next centroid for the next time window. Under the dynamic environment, objective functions change over time, but there is a certain relationship between the two objectives before and after the change.

Therefore, we can predict the next solution distribution using the information about the optimal solution before the change. In our proposed second-order difference strategy, the centroid of the objective space is also taken into account. The difference from the first-order difference model is that we consider both objective values and solutions in the decision space. The solutions in *DPS* and the corresponding objectives in *DPF* are mapped to a new two-dimensional mapping space (MS). The x-coordinate of this new plane is the set of *DPS*, and the y-coordinate is the average of objective values. Our proposed strategy is built-in two-dimensional space, and the centroid is the center of the two-dimensional plane related to the objectives. The mapping relation of the strategy is as Figure 1. Therefore, we can define a new $C_T{}'$, which can be computed by:

$$C_T{}' = \frac{1}{\max(|PS_T|, |PF_T|)} \sum_{i=0}^{i=|PS_T|} Euclidean[(x_i, y_i), (x_{i+1}, y_{i+1})] \tag{7}$$

where n_t and $|PF_T|$ are the cardinality of *DPS* and *DPF*, respectively; (x_i, y_i) represents the vector in two-dimensional space, consisting of the optimal solution and the corresponding objective function value; and Euclidean() means to obtain the Euclidean distance between two points in the plane. Similar to the first-order difference model, the location of the centroid at the next time window $T + 1$ is predicted using the formulation as follows:

$$C_{T+1}{}' = C_T{}' + C_T{}' - \overrightarrow{C_{T-1}{}'} \tag{8}$$

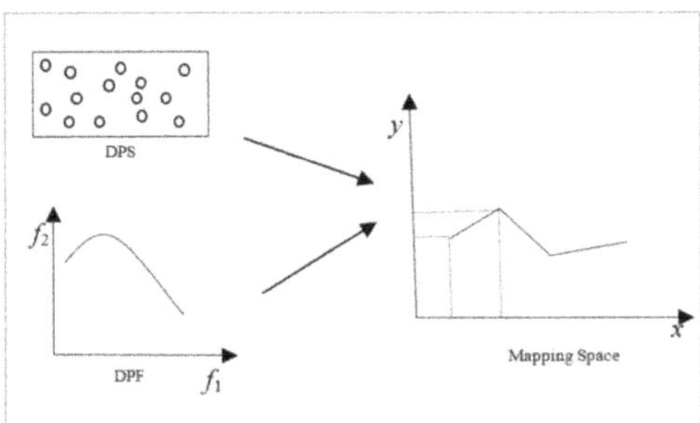

Figure 1. The mapping relation of the proposed strategy.

After determining the centroid location of the next time window, random strategy is then applied to the Algorithm 1 presented in this paper. The random strategy is the random perturbation around the obtained centroid position, so as to obtain a series of solutions similar to the last change. The radius of random strategy is set to 0.1, which means that we will take uniformly random points on the plane centered at the centroid and with the radius. The performance of this strategy can be shown through our next comparative experiments. The whole process of our proposed strategy combined with NSGA-III is described in Algorithm 4.

Some similarities and differences exist between our proposed strategy and the first-order difference model, where the former strategy is based on the latter. Both strategies use Formula (8) to obtain the next centroid, this is why they are called "difference strategies", and the difference between them is the way the centroid is generated. Obviously, the first-order difference strategy employs the simplest method to generate the centroid by only considering the center of *DPS*. However, when facing environmental changes, the objectives will change in a big or small way and, therefore, should be considered when looking for the centroid. In this way, the objective space can have a certain relationship with the last change, so as to predict more accurate solutions. Our algorithm is proposed based on this background, and the details of the algorithm are given in Algorithm 4.

At the beginning of the evolutionary process, the individuals of the population are initialized randomly, and the initial population is reevaluated. Then the change detection process described in Algorithm 3 is conducted. When an environmental change occurs, the population must be updated to respond to the change. As mentioned in Algorithm 4, the new population is composed of three kinds of individuals: the old solutions, the prediction solutions obtained by our proposed strategy and the random solutions around the prediction solutions. In most cases of real-world DMOPs, there are some similarities between the *DPS* of the consecutive DMOPs. Therefore, a certain percentage of old solutions can be reserved for the next population at the new time window. The old solutions from the last environment may perform better than reinitialized solutions. Besides, our prediction solutions and the old solutions are put into the new population uniformly, and the predicted centroid in the new environment can be described according to (7). Some random solutions around the predicted centroid are introduced as well, which adds diversity to the new population.

Algorithm 4 Second-order random strategy combined with NSGA-III

Input: the current population, P_T;
the time window, T;
the number of individuals in population, N;
the historic centroid points, C_{T-1};
the centroid of time window T, C_T;
Output: the next population P_{T+1};
Initialize population P_T and evaluate the inital population P_T;
Change detection (P_T);
if change is detected **then**
  while the maximum number of iterations is not reached **do**
    for $i = 1{:}N$ **do**
      if mod(i, 3) == 0 **then**
        $x_i^{T+1} = x_i^T + \overrightarrow{C_T' - C_{T-1}'}$;
        Random perturbation around x_i^{T+1};
      else
        $x_i^{T+1} = x_i^T$;
      end if
      Use NSGA-III to optimize x_i^{T+1} and get the next generation population P_{T+1};
    end for
  end while
end if
$T = T + 1$;
return P_{T+1}.

4. Experiments

All the experiments were conducted on MATLAB R2018a. Intel(R) Core (TM) i3-8100 CPU @ 3.60GHz was used as the hardware environment.

4.1. Benchmark Problems and Performance Metrics

In order to study the performance of our proposed strategies, two parts of the experiments were conducted for DMOPs. In this paper, six benchmarks with different change types were used for confirmation. The instances and definitions are listed in Table 1, and the functions have two objectives. The common formula of a bi-objective dynamic benchmark, which adds the time window parameter in forming the static ZDT [38] problems, can be described as follows:

$$\min f(x,t) = (f_1(x_I,t), g(x_{II},t) \cdot h(x_{III}, f_1(x_I,t), g(x_{II},t), t)) \quad (9)$$

where x_I, x_{II}, and x_{III} are three different subsets of design variables set x in decision space. For Fun1, as the environment changes, the *DPS* of Fun1 changes, whereas the moving *DPF* of Fun1 remains unchanged. In addition, the *DPF* of Fun1 is convex. Some parameters that do not require introduction include τ, which is the generation counter, and n_t, which represents the number of distinct steps in a fixed t. τ_T and n_t are two parameters that reflect the frequency and severity of an environmental change, respectively. In the benchmarks definitions, the time instance t can be computed by $t = (1/n_t) * (\tau/\tau_T)$. Different from Fun1, both *DPS* and *DPF* of Fun2 change over time, and *DPF* change from convex to nonconvex. The convergence speed and reactivity when the environment changes can be evaluated by Fun3 and Fun5 benchmarks. Specifically, the time-varying nonmonotonic dependencies between any two decision variables are introduced in Fun3, which is similar to the greenhouse system in the real world. As the time window increases, the dependency between two variables becomes more complicated, and the density of solutions also changes over time. Therefore, the relation between the decision variables and the diversity performance can also be assessed by Fun3 for DMOPs. Fun5 is a dynamic function with dynamic *DPF*s and *DPS*s, and the overall objective vectors change between several modes as the *DPS* changes. Fun5 is similar to the electric power supply system [39] in real-world applications. The *DPF* of Fun4 has a general change trend, which changes from convex to concave, and the values in both the *DPF* and *DPS* differ under different time windows. For Fun6, the characteristics of the *DPS* and *DPF* are totally different. The *DPS* is simpler than the *DPF* of Fun6, which specifically contains some locally linear, concave, or convex segments. What's more, the number of local segments is unfixed. The main characteristics of the six benchmarks mentioned above are summarized in Table 2.

Table 1. The instances and definitions of six benchmarks.

Instance	Definition
Fun1 [18]	$\begin{cases} f_1(x_I) = x_1 \\ g(x_{II}) = 1 + \sum\limits_{x_i \in x_{II}} (x_i - G(t))^2 \\ h(f_1, g) = 1 - \sqrt{\frac{f_1}{g}} \\ G(t) = \sin(0.5\pi t), t = \frac{1}{n_t}\left\lfloor \frac{\tau}{\tau_T} \right\rfloor \\ x_I = (x_1) \in [0,1], x_{II} = (x_2, \ldots, x_n) \in [-1,1]^{n-1} \end{cases}$
Fun2 [18]	$\begin{cases} f_1(x_I) = x_1 \\ g(x_{II}) = 1 + \sum\limits_{x_i \in x_{II}} x_i^2 \\ h(x_{III}, f_1, g) = 1 - \left(\frac{f_1}{g}\right)^{(H(t) + \sum\limits_{x_i \in x_{III}} (x_i - H(t))^2)} \\ H(t) = 0.75 + 0.7\sin(0.5\pi t), t = \frac{1}{n_t}\left\lfloor \frac{\tau}{\tau_T} \right\rfloor \\ x_I = (x_1) \in [0,1], x_{II}, x_{III} \in [-1,1]^{n-1} \end{cases}$
Fun3 [40]	$\begin{cases} \min F(x,t) = (f_1(x,t), f_2(x,t))^T \\ f_1(x,t) = (1 + g(x_{II}, t))(y_1 + A_t \sin(W_t \pi y_1)) \\ f_2(x,t) = (1 + g(x_{II}, t))(1 - y_1 + A_t \sin(W_t \pi y_1)) \\ g(x_{II}, t) = \sum_{x_i \in x_{II}} (y_i^2 - y_{i-1})^2, A(t) = 0.05 \\ W(t) = \lfloor 6\sin(0.5\pi(t-1)) \rfloor, \alpha = \lfloor 100\sin^2(0.5\pi t) \rfloor \\ y_1 = \lvert x_1 \sin((2\alpha + 0.5)\pi x_1) \rvert, y_i = x_i, i = 2, \ldots, n \\ x_I = (x_1) \in [0,1], x_{II} = (x_2, \ldots, x_n) \in [-1,1]^{n-1} \end{cases}$
Fun4 [41]	$\begin{cases} f_1(x_I) = x_1 \\ f_2(x_{II}) = g \cdot h \\ g = 1 + \sum_{i=2}^{m} (x_i - G(t))^2 \\ h = 1 - (\frac{f_1}{g})^{H(t)} \\ H(t) = 0.75 \cdot \sin(0.5\pi \cdot t) + 1.25 \\ G(t) = \sin(0.5\pi \cdot t) \\ x_I = (x_1) \in [0,1], x_{II} = (x_2, \ldots, x_n) \in [0,1] \end{cases}$
Fun5 [18]	$\begin{cases} \min F(x,t) = (f_1(x,t), f_2(x,t))^T \\ f_1(x,t) = (1 + g(x_{II}, t))(x_1 + A_t \sin(W_t \pi x_1)) \\ f_2(x,t) = (1 + g(x_{II}, t))(1 - x_1 + A_t \sin(W_t \pi x_1)) \\ g(x_{II}, t) = \sum_{x_i \in x_{II}} (x_i - G(t))^2, G(t) = \sin(0.05\pi t) \\ A(t) = 0.05, W(t) = \lfloor 6\sin(0.5\pi(t-1)) \rfloor \\ x_I = (x_1) \in [0,1], x_{II} = (x_2, \ldots, x_n) \in [-1,1]^{n-1} \end{cases}$

Table 1. *Cont.*

Instance	Definition
Fun6 [42]	$\begin{cases} \min F(x,t) = (f_1(x,t), f_2(x,t))^T \\ f_1(x,t) = g(x_{II},t)(x_1 + A_t \sin(W_t \pi x_1)) \\ f_2(x,t) = g(x_{II},t)(1 - x_1 + A_t \sin(W_t \pi x_1)) \\ g(x_{II},t) = 1 + \sum_{x_i \in x_{II}} (x_i^2 - G(t))^2, G(t) = \sin(0.05\pi t) \\ A(t) = 0.02, W_t = \lfloor 10G(t) \rfloor \\ x_I = (x_1) \in [0,1], x_{II} = (x_2, \ldots, x_n) \in [-1,1]^{n-1} \end{cases}$

Table 2. The dynamic characteristics of six benchmarks.

Problem	*DPS* Changes	*DPF* Changes
Fun1	Yes	No, the *DPF* is convex
Fun2	Yes	Yes, the *DPF* changes from convex to nonconvex
Fun3	Yes, the time-varying nonmonotonic dependencies are introduced	Yes
Fun4	Yes	Yes, the *DPF* changes from convex to concave
Fun5	Yes	Yes, the *DPF* changes its shape over time
Fun6	Yes, the *DPS* is rather simple	Yes, the *DPF* is sometimes linear, and sometimes concave or convex

Based on the benchmarks above, we tested our proposed algorithm under two different dynamic parameters. We set the change severity, n_t, to 5 and 10, and set the change frequency τ_T to 5, 10, and 20 respectively, which represents severe, moderate, and slight environmental changes. Therefore, we obtained six different sets of parameter values, namely (5, 5), (5, 10), (5, 20), (10, 5), (10, 10), and (10, 20). Each benchmark has six different cases with different parameters, and therefore, there are total of 36 cases in the six benchmarks. Then, six benchmark problems with different parameters were conducted, and three different dynamic metrics were employed to evaluate the performance of SDNSGA-III.

As we know, there are various performance metrics to assess the algorithm, such as inverted generational distance (*IGD*), generational distance (*GD*), and hypervolume (*HV*). For dynamic environment, a modified version of the *IGD*, *HV*, and *GD* is employed to evaluate the performance of DMOEAs. The *MIGD* metric proposed by Zhou et al. [43] is the dynamic version of *IGD*, which introduces the time window parameter, whereby the basic definition of *MIGD* metric is the average of *IGD* over a period of time. The definition of *IGD* can be described as follows. Let S be the solution set obtained by the algorithm, and P^* be a set of reference points that is uniformly sampled from *DPF*. Therefore, we can calculate *IGD* as:

$$IGD(S, P^*) = \frac{\sum_{x \in P^*} \min_{y \in S} dis(x, y)}{|P^*|} \tag{10}$$

where $dis(x, y)$ means the Euclidean distance from x in reference set P^* to y in solution set S. Both the convergence and diversity of an algorithm can be assessed by computing *IGD* at the same time. In addition, the smaller the *IGD* value is, the better the comprehensive performance of an algorithm. This is similar to a *MIGD* metric. The *MIGD* metric we used to evaluate DMOEAs can be formulated as:

$$MIGD = \frac{1}{|T|} \sum_{t \in T} IGD(S, P^*) \tag{11}$$

The HV is also one of the frequently used evaluation metrics, which represents the volume of the region that is rounded by the non-dominated solution set obtained by the algorithm and the reference points. HV can be described as:

$$HV(S^*, v_i) = \delta(\cup_{i=1}^{|S^*|} v_i) \tag{12}$$

where δ is a Lebesgue measure to caculate volume; $|S^*|$ is the number of non-dominated solutions sets; and v_i represents the super volume composed of the reference points and the i-th solution in the solution set. The MHV [44] metric is a modified dynamic version of the static HV metric, which is formulated as:

$$MHV = \frac{1}{|T|} \sum_{t \in T} HV(S^*, v_i) \tag{13}$$

The metric GD measures the diversity of an algorithm and describes the average of the minimum Euclidean distance from each point in the solution set S to the reference set P^*. Under dynamic environments, the MGD metric is proposed to measure the performance of an algorithm instead of GD. Similar to $MIGD$ and MHV, MGD is the modified version of GD, which can be defined as:

$$GD(S, P^*) = \frac{\sqrt{\sum_{y \in S} \min_{x \in P^*} dis(x, y)^2}}{|P|} \tag{14}$$

$$MGD = \frac{1}{|T|} \sum_{t \in T} GD(S, P^*) \tag{15}$$

In our experiments, we used $MIGD$, MHV, and MGD metrics to measure the performance of our proposed SDNSGA-III algorithm. T refers to a set of discrete time instances in a single run. Further, $|T|$ is the cardinality of T in the definitions of $MIGD$, MHV, and MGD. Our experiments were mainly conducted to measure the performance of NSGA-III combined with our proposed strategies. The experiments were divided into two parts: (1) the to compare the results of SDNSGA-III and the other four popular algorithms, and (2) to compare NSGA-III with different strategies to prove the effectiveness of the second-order difference strategy and random strategy. The general parameters in all algorithms are presented in Table 3.

Table 3. The general parameters of all algorithms.

Symbol	Meaning	Value
N	population size	100
M	the number of objectives	2
D	dimensions of decision vectors	10
FEs	fitness evaluation times	10,000
T	the time window	20
R	number of runs	30

4.2. Comparative Study for the Proposed Algorithm

Our proposed algorithm (SDNSGA-III) was compared with four state-of-the-art algorithms, including NSGA-III [17], DNSGA-II-A [21], MOEA/D-FD [20], and a multi-objective optimization framework (LSMOF) [45]. The comparison between SDNSGA-III and NSGA-III was made to prove the performance of our proposed strategies. Based on the differences in the performance metrics, we can determine the results of whether our second-order strategy and random strategy can obtain better convergence and diversity. DNSGA-II-A [21] and MOEA/D-FD [20] are DMOEAs that are primarily deal with dynamic problems. Particularly, DNSGA-II-A [21] introduces some new random individuals when a new population is generated. When merging the parent and child population into the next bigger population, all individuals are re-evaluated through the benchmarks. MOEA/D-FD [20] is a modified version of the original MOEA/D, which uses the first-order difference model. LSMOF reformulates the problems, tracks the Pareto optimal set directly, and also accelerates the computational efficiency of the multi-objective evolutionary algorithm.

The data in Table A1 (the first table in Appendix A) show the obtained *MGD* values and standard deviations over 30 runs. The last column in Table A1 means the percentage difference between SDNSGA-III and other four comparative algorithms. The positive (negative) value shows the performance of SDNSGA-III is better (worse) than the comparing algorithm. The performance evaluation was conducted at the 5% significance level. In Table A1, the results are recorded as "+", "−", and "=" for when SDNSGA-III performs significantly better than, worse than, and equivalent to the corresponding algorithm respectively. The bold font indicates that the algorithm has the best diversity on this benchmark. From Table A1, it is obvious that SDNSGA-III performed better than the other four algorithms in most cases, which means it has a better tracking ability of *DPS* and *DPF*. Moreover, the SDNSGA-III values in Table A1 are significantly better than those of NSGA-III, sufficiently proving that the second-order and random strategies improve the performance of NSGA-III for dealing with dynamic problems. Besides, from the perspective of different problems, SDNSGA-III performed best in Fun1, Fun4 and Fun6, while the performance of SDNSGA-III was roughly the same on Fun2 and Fun3. This means that our proposed strategy is more suitable for dynamic problems whose *DPS* changes. Although the performance on other problems was worse than on Fun1, Fun4 and Fun6, about 90% of the results are best among the five comparative algorithms. In general, our proposed strategy worked well for different DMOPs. What's more, when change severity was relatively smooth (n_t = 10), the five sixth values of SDNSGA-III were better than the other four algorithms. In addition, when the change frequency was fast (τ_T = 5), all results of the other comparative algorithms were worse than those of SDNSGA-III. Therefore, it can be suggested that our proposed strategy can obtain the best diversity performance of the population when the change is smooth and fast. The reason for this result is that when the change frequency is fast, our proposed strategies enhance the search efficiency and accuracy of tracking the moving *DPF*.

In order to investigate the convergence process of the algorithms, the *MIGD*, *MHV*, and *MGD* trends of Fun1–6 with a fixed n_t and τ_T were investigated, as shown in Figures 2–4. In Figure 2a–f, the overall trend was obviously downward. A total of 10 points were sampled randomly within a single run, and the evaluation numbers was set to 10,000. As the evaluations increased, the metrics of all five algorithms tend to become relatively stable. Among the compared algorithms, generally speaking, SDNSGA-III reached a better *MIGD* value in less time, as observed in Figure 2a–d,f. For Fun1 and Fun2, DNSGAII-A and MOEAD-FD obtained a lower *MIGD* at the beginning of the evaluations, and our proposed SDNSGA-III reached almost the same value in the later stage of evaluations. In Figure 3, contrary to *MIGD*, the general tendency of *MHV* exhibited an increase, while the *MHV* value of SDNSGA-III was higher than the other comparative algorithms observably on Fun1, Fun3, and Fun5. SDNSGA-III reached almost the same *MHV* as MOEAD-FD at the end of evaluations in Figure 3b,d,f, which means SDNSGA-III has a similar tracking ability with MOEAD-FD on Fun2, Fun4, and Fun6. The overall trend of *MGD* in Figure 4 is similar to *MIGD*, and SDNSGA-III performed more stable than the other comparative algorithms. Figures 2–4 further reveal that our proposed SDNSGA-III has a great improvement in tracking *DPF* and *DPS* compared with NSGA-III, and SDNSGA-III performed more stable than other algorithms with a steady tracking ability regardless of the environmental change.

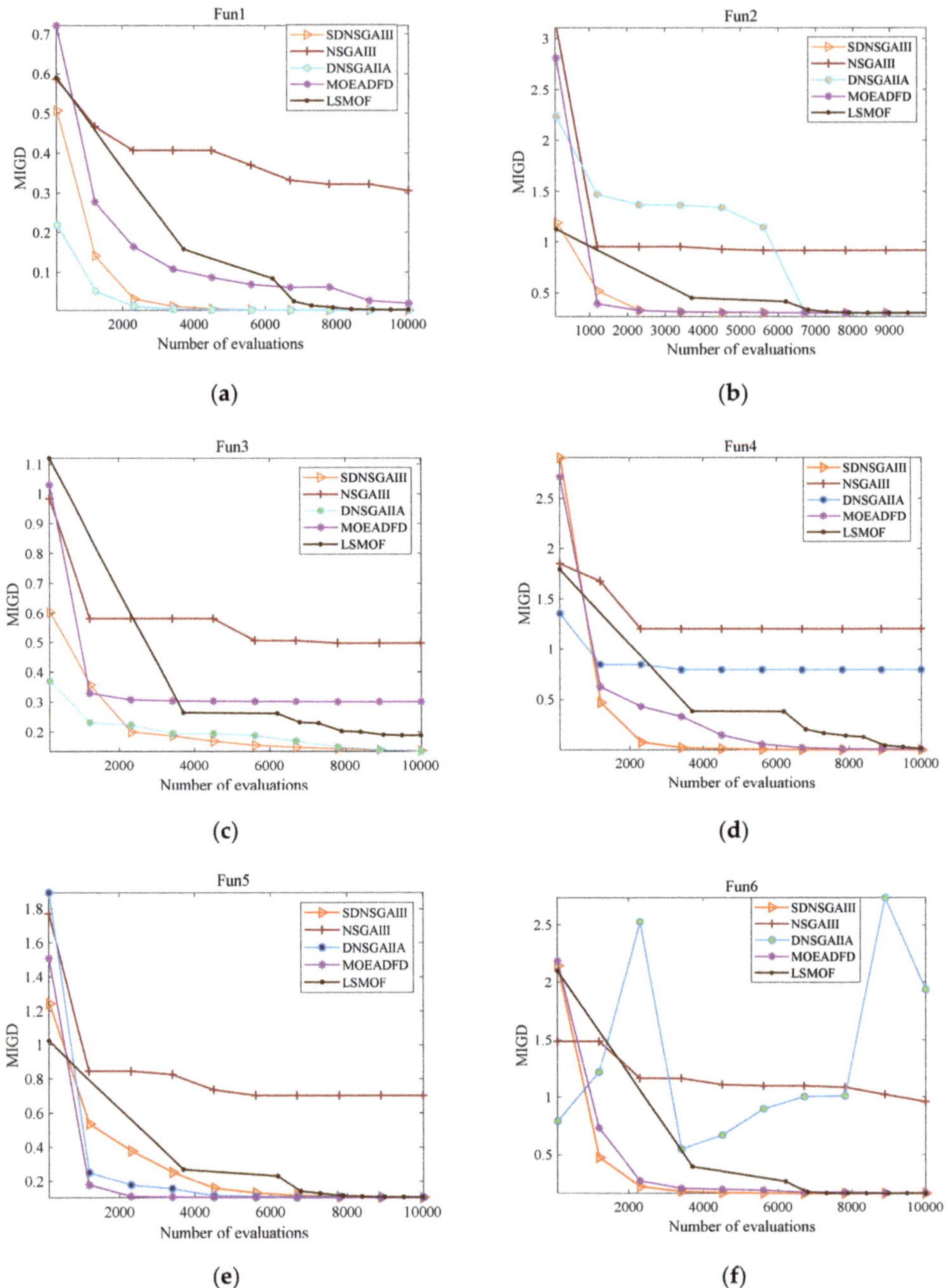

Figure 2. The *MIGD* trend of five algorithms with $n_t = 10$, $\tau_T = 20$. (**a**–**f**) are *MIGD* trend graphs of six functions, respectively.

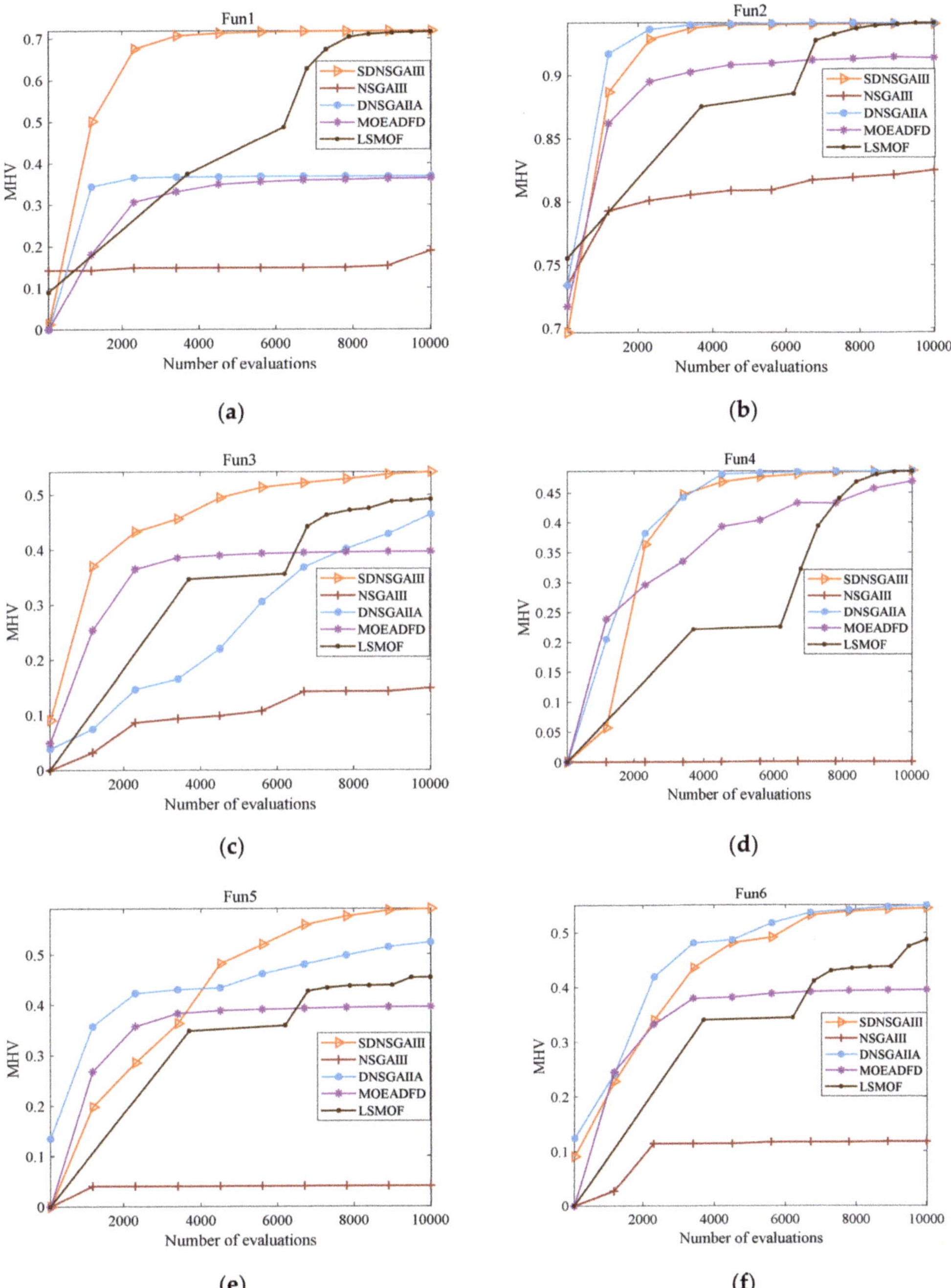

Figure 3. The *MHV* trend of five algorithms with $n_t = 10$, $\tau_T = 20$. (**a**–**f**) are *MHV* trend graphs of six functions, respectively.

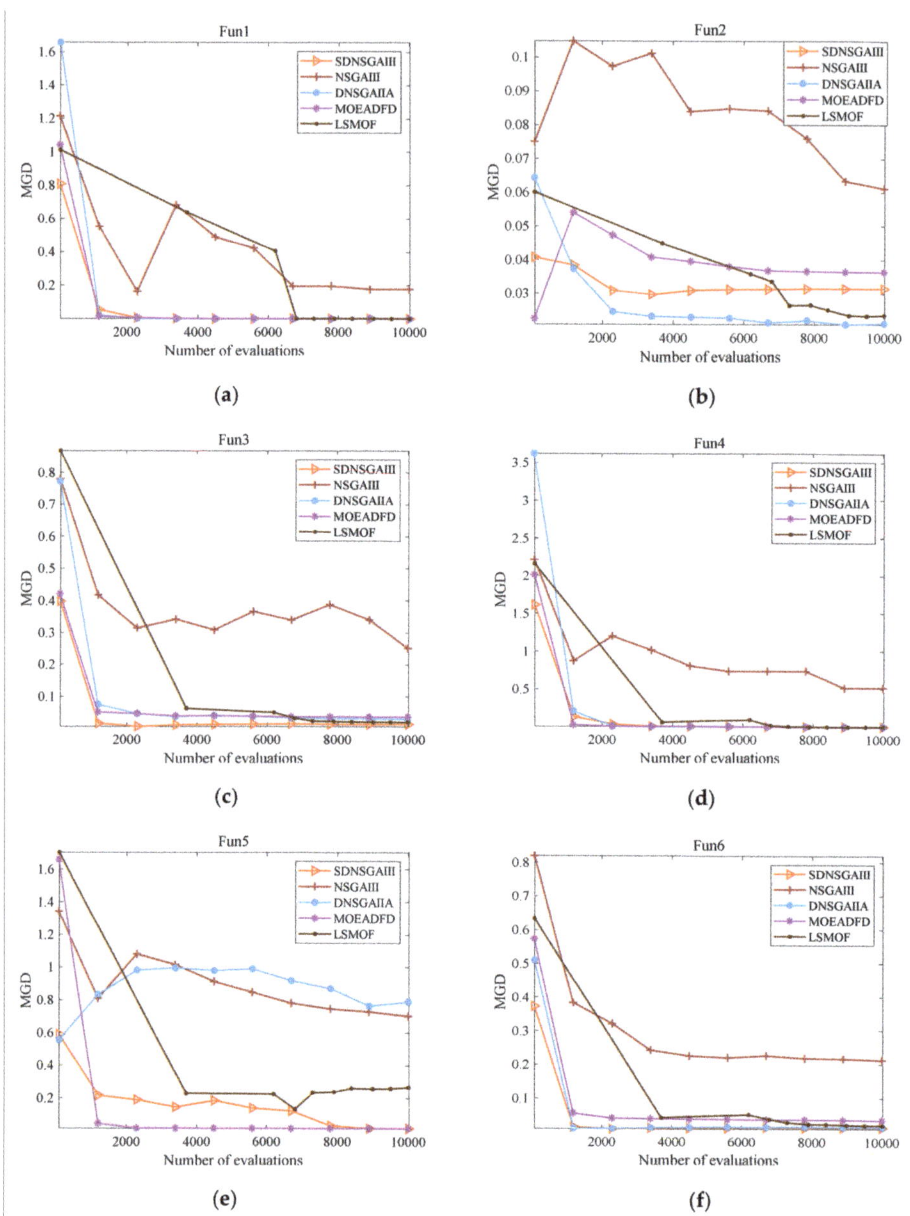

Figure 4. The *MGD* trend of five algorithms with n_t = 10, τ_T = 20. (**a–f**) are *MGD* trend graphs of six functions respectively.

The statistical results of six benchmarks and six parameters on *MIGD*, *MHV*, and *MGD* are recorded in Tables 4 and 5 respectively. Among the six different benchmarks, SDNSGA-III reached 18 best results on Fun1, which has dynamic *DPS* and static *DPF*. In terms of the six different parameters, our proposed SDNSGA-III achieved the best metric values on the severe change severity (n_t = 5) and the slow change frequency (τ_T = 20). The performance of SDNSGA-III was nearly the same on the rest of the five parameters. Generally, SDNSGA-III is more suitable for DMOPs, since it has a relatively smooth change or fixed *DPF*. This can be attributed to the prediction strategy that can achieve a centroid location more accurately with a slight change, while the prediction error increases with a severe and fast change. Even so, SDNSGA-III performed better than the other comparative algorithms on most test instances. To further evaluate the performances of the second-order strategy and random strategy, the comparisons between the two are discussed in the next section.

Table 4. The statistical results of six benchmarks on different metrics.

Metrics	Fun1	Fun2	Fun3	Fun4	Fun5	Fun6
MIGD	6	3	5	4	4	6
MHV	6	6	5	4	4	4
MGD	6	5	5	6	4	6
Total	18	14	15	14	12	16

Table 5. The statistical results of six parameters on different metrics.

Metrics	(5,5)	(5,10)	(5,20)	(10,5)	(10,10)	(10,20)
MIGD	3	5	6	5	5	4
MHV	5	5	6	5	4	5
MGD	5	5	5	5	6	6
Total	13	15	17	15	15	15

4.3. Comparative Study for Two Proposed Strategies

In the comparative study, the performances of the second-order difference strategy and random strategy were analyzed. NSGA-IIIs is the algorithm with the second-order difference strategy without random strategy, while NSGA-IIIr incorporates the random strategy without the second-order difference strategy. The main process of NSGA-IIIr is to first initialize the population, then use NSGA-III to optimize individuals at each generation. When a change is detected, random perturbation is conducted around the optimal solution of the previous generation. The comparative results show the efficiency of the second-order difference strategy and random strategy in SDNSGA-III.

Table A2 (the second table in Appendix A) depicts the *MIGD* values and standard deviations obtained by the four algorithms over 30 runs. In the table, about 67% of the results clearly reveal that the performance of SDNSGA-III is better than the other three comparative algorithms. In general, both NSGA-IIIs and NSGA-IIIr algorithms demonstrated little difference in performance. However, it is obvious that SDNSGA-III, which combines both strategies, performed best among all four algorithms. Besides, NSGA-IIIr performed mostly better than NSGA-IIIs in Fun3 and Fun6. In other words, the random strategy improves the diversity of SDNSGA-III in a dynamic environment. Furthermore, NSGA-IIIs exhibited better values than NSGA-IIIr with a smooth change severity and fast change frequency. This proves that our proposed second-order difference strategy could track the moving *DPS* and *DPF* directly, and the random strategy can improve the diversity ability of NSGA-III on DMOPs. Since random strategy can better adapt to dynamic characteristics for DMOPs, the SDNSGA-III algorithm exhibited the best comprehensive performance among all comparative algorithms.

Tables 6 and 7 include the statistical results from different perspectives, including the different benchmarks and parameters. In the tables, "a/b/c" represents the number of best *MIGD*, *MHV*, and *MGD* metric values, respectively. In Table 6, the data obviously show that SDNSGA-III combined with the two strategies performed best, and a single NSGA-III performed worst among the four comparative algorithms. NSGA-IIIs showed an obviously better performance than NSGAIIIr on Fun2 and Fun4. For the benchmarks of Fun3 and Fun6, NSGA-IIIr performed better than NSGA-IIIs, and both of algorithms performed the same on Fun1 and Fun5. These results demonstrate that SDNSGA-III effectively incorporates second-order difference strategy and random strategy to achieve the best metric values. It is pertinent to note that the former strategy is more applicable to DMOPs whose *DPS* and *DPF* changes from convex like Fun2 and Fun4. When dealing with more complex DMOPs, like Fun3 and Fun6, whose *DPS* and *DPF* change irregularly, the latter strategy behaves better. Table 7 represents the statistical results of the different change frequencies and change severities. When the change severity n_t was fixed to 10 and the change frequency τ_T was set to 20, SDNSGA-III reached the best metric values. In addition, the performance of SDNSGA-III with higher change severity and higher change frequency was better than the other comparative algorithms. When n_t and τ_T change, the random strategy can ensure the diversity in a population, while a single second-order strategy cannot track the moving *DPF* accurately. The statistical data in Tables 6 and 7 illustrate that combining the second-order difference strategy and random strategy is indeed effective to solve DMOPs better.

Table 6. The statistical results of six benchmarks on different metrics.

Algorithms	Fun1	Fun2	Fun3	Fun4	Fun5	Fun6
SDNSGA-III	5/5/4	3/3/4	4/5/3	3/4/3	3/2/3	4/3/3
NSGA-III	0/0/0	0/0/0	0/0/0	0/0/0	0/0/0	0/0/0
NSGA-IIIs	1/0/1	2/3/1	0/0/0	2/2/3	2/2/1	0/0/0
NSGA-IIIr	0/1/1	1/0/1	2/1/3	1/0/0	1/2/2	2/3/3

Table 7. The statistical results of six parameters on different metrics.

Algorithms	(5,5)	(5,10)	(5,20)	(10,5)	(10,10)	(10,20)
SDNSGA-III	2/2/2	2/5/4	4/3/2	4/5/4	5/3/2	5/4/6
NSGA-III	0/0/0	0/0/0	0/0/0	0/0/0	0/0/0	0/0/0
NSGA-IIIs	3/3/2	1/0/1	0/0/1	1/0/1	1/2/1	1/2/0
NSGA-IIIr	1/1/2	3/1/1	2/3/3	1/1/1	0/1/3	0/0/0

5. Conclusions

In this study, we propose a novel algorithm based on NSGA-III, which incorporates a second-order difference strategy and random strategy to solve DMOPs. These two strategies are specifically employed to predict the next centroid location based on its historical locations and random disturbances around the predicted centroid location when change is detected. Moreover, the performance of SDNSGA-III was validated using different benchmarks and different metrics via testing on different change frequencies and change severities. Compared with four other state-of-the-art evolutionary algorithms, our SDNSGA-III can obtain a better convergence speed and maintain diversity of a population when tracking the moving *DPS* and *DPF*. In addition, a comparison between the two proposed strategies was conducted to verify their effectiveness. It was found that the second-order difference strategy and random strategy have the ability to find the moving *DPF*, and SDNSGA-III can maintain the diversity of a population to respond to environmental change. In addition, the further innovations about prediction can be inspired through the proposed second difference strategy.

Despite our promising findings, some issues need to be further addressed. For example, more benchmarks should be employed to evaluate the performance of SDNSGA-III. Moreover, further studies are suggested for other state-of-the-art algorithms incorporated with second-order difference and random strategies to show their ability to enhance searching efficiency. The second-order difference strategy can be incorporated with other effective frameworks to increase the accuracy of prediction at the stage of change response. What's more, we tested our strategies only on two-objective benchmarks in this work. Therefore, we plan to focus on more than two-objective dynamic optimization problems in future studies.

Author Contributions: H.Z.: Conceptualization, methodology, software, visualization, investigation, writing—original draft preparation; G.-G.W.: supervision, validation, data curation, reviewing and editing; J.D.: visualization, reviewing and editing; A.H.G.: project administration, reviewing and editing. All authors have read and agreed to the published version of the manuscript.

Funding: This research was funded by National Natural Science Foundation of China, Grant Numbers U1706218, 41576011, 41706010, and 61503165.

Institutional Review Board Statement: Not applicable.

Informed Consent Statement: Not applicable.

Data Availability Statement: Not applicable.

Conflicts of Interest: The authors declare no conflict of interest.

Appendix A

Table A1. The *MGD* values and standard deviations obtained by five algorithms.

Problem	(n_t, τ_T)	SDNSGA-III	NSGA-III	DNSGA-II-A	MOEA/D-FD	LSMOF	Percentage Difference
Fun1	(5,5)	**9.7641e-5**	8.3088e−1	1.8830e−4	5.9543e−4	3.4402e−4	99.99%, 48.15%,
		(2.05e-5)	(2.10e−1) −	(2.55e−5) −	(1.26e−4) −	(8.51e−5) −	83.60%, 71.62%
	(5,10)	**1.0885e−4**	2.5769e−1	1.9373e−4	6.2599e−4	3.3416e−4	99.96%, 43.81%,
		(2.53e−5)	(1.35e−1) −	(3.34e−5) −	(1.63e−4) −	(5.25e−5) −	82.61%, 67.43%
	(5,20)	**7.2249e−5**	8.3223e−1	1.5780e−4	4.6507e−4	1.6833e−4	99.99%, 54.21%,
		(1.40e−5)	(2.49e−1) −	(2.44e−5) −	(9.94e−5) −	(3.98e−5) −	84.46%, 57.08%
	(10,5)	**9.2895e−5**	1.6212e−1	1.8772e−4	6.2374e−4	2.9641e−4	99.94%, 50.51%,
		(2.21e−5)	(3.82e−2) −	(3.37e−5) −	(1.69e−4) −	(7.04e−5) −	85.11%, 68.66%
	(10,10)	**9.4192e−5**	1.8184e−1	1.9987e−4	6.0922e−4	3.5548e−4	99.95%, 52.87%,
		(1.79e−5)	(5.75e−2) −	(3.70e−5) −	(1.70e−4) −	(9.81e−5) −	84.54%, 73.50%
	(10,20)	**9.6531e−5**	2.7691e−1	1.9788e−4	6.2510e−4	3.2505e−4	99.97%, 51.22%,
		(1.98e−5)	(9.54e−2) −	(3.50e−5) −	(1.48e−4) −	(7.25e−5) −	84.56%, 70.30%
Fun2	(5,5)	**1.5312e−2**	1.3209e+0	5.6395e−2	5.6148e−2	1.5518e−2	98.84%, 72.85%,
		(3.99e−5)	(3.64e−1) −	(7.95e−4) −	(2.25e−3) −	(2.13e−4) −	72.73%, 1.33%
	(5,10)	**2.9755e−2**	1.1544e+0	4.9951e−2	5.4658e−2	3.0794e−2	97.42%, 40.43%,
		(4.91e−4)	(2.13e−1) −	(3.24e−4) +	(4.02e−3) −	(4.69e−4) −	45.56%, 3.37%
	(5,20)	**1.2032e−4**	2.7577e−1	1.6799e−4	8.6767e−4	5.0816e−2	99.96%, 28.38%,
		(1.26e−4)	(7.57e−2) −	(7.21e−5) −	(7.26e−4) −	(1.39e−4) −	86.13%, 99.76%
	(10,5)	5.1443e−2	1.2241e+0	5.6449e−2	5.4444e−2	**7.9690e−3**	95.80%, 8.87%,
		(4.63e−6)	(3.38e−1) −	(5.87e−4) −	(2.26e−3) −	**(1.07e−4) +**	5.51%, −545.54%
	(10,10)	**1.5329e−2**	2.2435e−1	1.5834e−2	1.6239e−2	1.6044e−2	93.17%, 3.19%,
		(8.66e−5)	(4.74e−2) −	(1.26e−4) −	(3.13e−4) −	(2.18e−4) −	5.60%, 4.46%
	(10,20)	**2.9301e−2**	5.0671e−1	3.1420e−2	3.2043e−2	3.1609e−2	94.22%, 6.74%,
		(2.01e−5)	(1.11e−1) −	(3.74e−4) −	(5.79e−4) −	(3.73e−4) −	8.56%, 7.30%
Fun3	(5,5)	**1.4966e−2**	2.2557e−1	1.5307e−2	3.8589e−2	2.1146e−2	93.37%, 2.23%,
		(7.01e−3)	(3.44e−2) −	(7.32e−3) −	(6.69e−4) −	(4.48e−3) −	61.22%, 29.23%
	(5,10)	1.8975e−2	2.2439e−1	**1.8150e−2**	4.0220e−2	2.2850e−2	91.54%, −4.55%
		(3.09e−3)	(3.65e−2) −	**(5.22e−4) =**	(5.76e−4) −	(3.70e−3) −	52.82%, 16.96%
	(5,20)	**1.4379e−2**	2.2734e−1	4.0261e−2	3.4202e−2	2.0892e−2	93.68%, 64.29%,
		(3.98e−3)	(5.27e−2) −	(1.07e−3) −	(7.02e−3) −	(3.83e−3) −	57.96%, 31.17%
	(10,5)	**1.4928e−2**	2.0592e−1	3.7969e−2	3.2770e−2	2.0691e−2	92.75%, 60.68%,
		(5.67e−3)	(4.43e−2) −	(5.04e−4) −	(5.84e−3) −	(3.76e−3) −	54.45%, 27.85%
	(10,10)	**1.3473e−2**	2.1656e−1	1.6858e−2	3.2085e−2	2.8596e−2	93.78%, 20.08%,
		(6.74e−4)	(5.06e−2) −	(8.05e−3) −	(8.06e−3) −	(4.29e−3) −	58.01%, 52.89%
	(10,20)	**1.4165e−2**	2.2959e−1	1.6997e−2	3.5214e−2	2.1800e−2	93.83%, 16.66%,
		(5.29e−4)	(6.38e−2) −	(8.25e−3) =	(5.84e−3) −	(1.92e−3) −	59.77%, 35.02%

Table A1. *Cont.*

Problem	(n_t, τ_T)	SDNSGA-III	NSGA-III	DNSGA-II-A	MOEA/D-FD	LSMOF	Percentage Difference
Fun4	(5,5)	**5.7478e−5 (3.03e−5)**	3.0586e+0 (5.11e−1) −	1.2177e−4 (2.95e−5) −	5.9951e−4 (2.18e−4) −	1.2788e−4 (1.11e−4) −	100.00%, 52.80%, 90.41%, 55.05%
	(5,10)	**9.8560e−6 (6.66e−6)**	3.0143e+0 (5.17e−1) −	8.7865e−5 (3.11e−5) −	7.3340e−4 (2.66e−4) −	1.1863e−4 (3.68e−5) −	100.00%, 88.78%, 98.66%, 91.69%
	(5,20)	**1.7514e−4 (3.61e−5)**	1.1872e+0 (4.54e−1) −	2.0144e−4 (3.60e−5) −	1.3375e−3 (3.92e−4) −	3.3894e−4 (1.37e−4) −	99.99%, 13.06%, 86.91%, 48.33%
	(10,5)	**8.6286e−6 (4.66e−6)**	3.2288e+0 (7.88e−1) −	1.0346e−4 (3.86e−5) −	8.5105e−4 (4.65e−4) −	1.0690e−4 (3.37e−5) −	100.00%, 91.66%, 98.99%, 91.93%
	(10,10)	**1.7192e−4 (3.61e−5)**	8.0273e−1 (2.86e−1) −	1.9120e−4 (2.72e−5) −	1.2319e−3 (5.49e−4) −	3.0479e−4 (1.20e−4) −	99.98%, 10.08%, 86.04%, 43.59%
	(10,20)	**1.5884e−4 (4.62e−5)**	6.9277e−1 (2.27e−1) −	2.0395e−4 (2.67e−5) −	1.3210e−3 (3.97e−4) −	2.8649e−4 (8.88e−5) −	99.98%, 22.12%, 87.98%, 44.56%
Fun5	(5,5)	2.2038e−1 (2.09e−1)	6.9385e−1 (1.30e−1) −	8.8662e−1 (2.03e−1) −	**1.3465e−2 (5.94e−5) +**	9.6630e−1 (1.22e−1) −	68.24%, 75.14% −1536.69%, 77.19%
	(5,10)	**1.3050e−2 (5.39e−5)**	2.2280e−1 (5.30e−2) −	1.3435e−2 (2.36e−4) −	1.3153e−2 (2.04e−4) −	1.3589e−2 (1.58e−4) −	94.14%, 2.87%, 0.78%, 3.97%
	(5,20)	1.2587e−2 (2.49e−5)	9.3152e−1 (1.74e−1) −	**1.2475e−2 (2.03e−4) +**	1.2611e−2 (2.55e−5) −	1.2590e−2 (2.18e−4) =	98.65%, −0.90%, 0.19%, 0.02%
	(10,5)	**1.3049e−2 (4.49e−5)**	2.0971e−1 (3.26e−2) −	1.3518e−2 (1.89e−4) −	1.3240e−2 (2.93e−4) −	1.3537e−2 (2.81e−4) −	93.78%, 3.47%, 1.44%, 3.60%
	(10,10)	**8.0280e−3 (2.85e−6)**	1.4576e+0 (2.94e−1) −	1.0723e+0 (1.56e−1) −	8.0422e−3 (1.63e−5) −	4.3435e−1 (2.46e−1) −	99.45%, 99.25%, 0.18%, 98.15%
	(10,20)	**1.3176e−2 (1.01e−3)**	8.7853e−1 (1.33e−1) −	7.4932e−1 (8.50e−2) −	1.3472e−2 (9.12e−5) =	4.1628e−1 (2.99e−1) −	98.50%, 98.24%, 2.20%, 96.83%
Fun6	(5,5)	**1.4959e−2 (5.61e−3)**	2.2469e−1 (5.33e−2) −	1.7703e−2 (8.50e−3) −	3.7265e−2 (6.28e−4) −	2.1705e−2 (2.33e−3) −	93.34%, 15.50%, 59.86%, 31.08%
	(5,10)	**1.3725e−2 (5.89e−4)**	2.1227e−1 (3.26e−2) −	1.8282e−2 (9.14e−3) −	3.7403e−2 (6.66e−4) −	2.1408e−2 (1.84e−3) −	93.53%, 24.93%, 63.31%, 35.89%
	(5,20)	**1.3892e−2 (7.75e−4)**	2.2927e−1 (7.10e−2) −	1.8479e−2 (8.87e−3) −	3.6830e−2 (3.22e−3) −	2.1563e−2 (2.41e−3) −	93.94%, 24.82%, 62.28%, 35.57%
	(10,5)	**1.4122e−2 (1.49e−3)**	2.1570e−1 (3.32e−2) −	1.5653e−2 (5.02e−3) −	3.5765e−2 (5.70e−3) −	2.2094e−2 (2.74e−3) −	93.45%, 9.78%, 60.51%, 36.08%
	(10,10)	**1.4963e−2 (6.00e−3)**	2.1200e−1 (3.18e−2) −	1.6425e−2 (6.86e−3) −	3.6488e−2 (4.24e−3) −	2.1332e−2 (2.62e−3) −	92.94%, 8.90%, 58.99%, 29.86%
	(10,20)	**1.3778e−2 (6.56e−4)**	2.1564e−1 (3.12e−2) −	1.6983e−2 (7.13e−3) −	3.6915e−2 (2.32e−3) −	2.1979e−2 (2.30e−3) −	93.61%, 18.87%, 62.68%, 7.31%

Table A2. The *MIGD* values and standard deviations obtained by four algorithms.

Problem	(n_t, τ_T)	SDNSGA-III	NSGA-III	NSGA-IIIs	NSGA-IIIr	Percentage Difference
Fun1	(5,5)	**4.1173e−3 (7.59e−5)**	1.0619× 10+0 (3.04e−1) −	5.4190e−3 (7.05e−3) −	4.4643e−3 (2.05e−3) −	99.61%, 24.02%, 7.77%
	(5,10)	4.1621e−3 (1.09e−4)	3.1543e−1 (6.21e−2) −	**4.1246e−3 (6.60e−5) +**	4.7086e−3 (9.61e−4) −	98.68%, −0.91%, 11.61%
	(5,20)	**4.0346e−3 (6.50e−5)**	9.0279e−1 (2.77e−1) −	4.0621e−3 (1.80e−4) −	4.0917e−3 (2.52e−4) −	99.55%, 0.68%, 1.40%
	(10,5)	**4.1143e−3 (1.10e−4)**	3.1175e−1 (4.65e−2) −	4.1199e−3 (5.87e−5) −	4.1204e−3 (7.92e−5) −	98.68%, 0.14%, 0.15%
	(10,10)	**4.1051e−3 (7.77e−5)**	2.9676e−1 (4.29e−2) −	4.1788e−3 (2.68e−4) −	4.1059e−3 (6.46e−5) −	98.62%, 1.76%, 0.02%
	(10,20)	**4.1219e−3 (8.22e−5)**	3.2164e−1 (6.44e−2) −	4.1977e−3 (7.97e−5) −	4.1297e−3 (8.09e−5) −	98.72%, 1.81%, 0.19%
Fun2	(5,5)	5.2642e−1 (3.66e−6)	1.8119e+0 (4.16e−1) −	**5.2047e−1 (4.95e−3) +**	5.2795e−1 (5.98e−3) −	70.95%, −1.14%, 0.29%
	(5,10)	5.2642e−1 (7.13e−6)	1.5026e+0 (2.98e−1) −	3.0454e−1 (1.22e−3) +	**3.0438e−1 (1.73e−3) +**	64.97%, −72.86%,72.95%

Table A2. *Cont.*

Problem	(n_t, τ_T)	SDNSGA-III	NSGA-III	NSGA-IIIs	NSGA-IIIr	Percentage Difference
	(5,20)	**4.1727e−3 (1.60e−4)**	6.5787e−1 (1.70e−1) −	5.2028e−1 (4.53e−3) −	5.2708e−1 (6.02e−3) −	99.37%, 99.20%, 99.21%
	(10,5)	5.2642e−1 (3.09e−6)	1.8949e+0 (2.94e−1) −	**7.8633e−2 (1.47e−4)** +	7.8664e−2 (1.37e−4) +	72.22%, −569.46%, −569.20%
	(10,10)	**1.5659e−1 (1.74e−4)**	5.1829e−1 (6.53e−2) −	1.5659e−1 (1.79e−4) =	1.5660e−1 (1.74e−4) −	69.79%, 0.00%, 0.01%
	(10,20)	**3.0300e−1 (1.75e−4)**	9.4884e−1 (1.59e−1) −	3.0392e−1 (1.08e−4) −	3.0397e−1 (1.69e−4) −	68.07%, 0.30%, 0.32%
Fun3	(5,5)	1.3632e−1 (7.43e−3)	5.6611e−1 (4.71e−2) −	3.2547e−1 (5.79e−4) −	**1.3567e−1 (6.25e−3)** +	75.92%, 58.12%, −0.48%
	(5,10)	1.7548e−1 (4.89e−3)	6.6592e−1 (5.97e−2) −	2.8207e−1 (8.26e−4) −	**1.2592e−1 (6.82e−3)** +	73.65%, 37.79%, −39.36%
	(5,20)	**1.1587e−1 (3.98e−3)**	5.2969e−1 (6.02e−2) −	2.7057e−1 (5.31e−4) −	2.1101e−1 (4.50e−3) −	78.12%, 57.18%, 45.09%
	(10,5)	**1.2918e−1 (4.39e−3)**	5.6899e−1 (4.38e−2) −	3.2018e−1 (2.57e−4) −	1.6741e−1 (4.48e−2) −	77.30%, 59.65%, 22.84%
	(10,10)	**1.1544e−1 (3.43e−3)**	5.3525e−1 (4.65e−2) −	3.8872e−1 (3.76e−4) −	1.8442e−1 (2.78e−3) −	78.43%, 70.30%, 37.40%
	(10,20)	**1.3219e−1 (2.41e−3)**	5.2812e−1 (4.92e−2) −	3.0209e−1 (4.61e−4) −	1.3372e−1 (3.65e−3) −	74.97%, 56.24%, 1.14%
Fun4	(5,5)	5.5170e−3 (6.27e−3)	3.8735e+0 (1.09e+0) −	**5.0968e−3 (6.25e−3)** +	8.4618e−3 (1.82e−2)−	99.86%, −8.24%, 34.80%
	(5,10)	**3.8147e−3 (5.02e−6)**	3.9172e+0 (8.90e−1) −	7.7584e−3 (1.18e−2) −	3.8147e−3 (5.56e−6) =	99.90%, 50.83%, 0.00%
	(5,20)	4.4876e−3 (2.97e−4)	1.2906e+0 (4.14e−1) −	7.0802e−3 (1.09e−2) −	**4.4741e−3 (2.72e−4)** +	99.65%, 36.62%, −0.30%
	(10,5)	**3.8136e−3 (3.49e−6)**	3.8130e+0 (1.09e+0) −	7.0585e−3 (6.68e−3) −	3.8357e−3 (1.05e−4) −	99.90%, 45.97%, 0.58%
	(10,10)	6.7897e−3 (1.31e−2)	1.0657e+0 (2.65e−1) −	**4.7422e−3 (2.40e−3)** +	4.9961e−3 (2.52e−3) +	99.36%, −43.18%, −35.90%
	(10,20)	**4.2404e−3 (2.54e−4)**	9.7821e−1 (2.73e−1) −	4.3661e−3 (4.04e−4) −	4.7208e−3 (2.26e−3) −	99.57%, 2.88%, 10.18%
Fun5	(5,5)	1.3278e−1 (3.93e−2)	5.1958e−1 (9.60e−2) −	**1.0330e−1 (7.55e−4)** +	1.1789e−1 (1.56e−2) +	74.44%, −28.54%, −12.63%
	(5,10)	**1.0815e−1 (9.75e−5)**	4.6938e−1 (7.18e−2) −	1.0816e−1 (1.38e−4) −	1.0816e−1 (1.08e−4)−	76.96%, 0.01%, 0.01%
	(5,20)	**1.1799e−1 (8.65e−6)**	1.5532e+0 (2.29e−1) −	1.1814e−1 (7.71e−5) −	1.1899e−1 (3.72e−6) −	92.40%, 0.13%, 0.84%
	(10,5)	1.0815e−1 (1.04e−4)	4.6754e−1 (6.52e−2) −	1.0816e−1 (1.27e−4) −	**1.0813e−1 (1.07e−4)** +	76.87%, 0.01%, −0.02%
	(10,10)	**7.0561e−2 (3.46e−5)**	1.2441e+0 (3.16e−1) −	8.1913e−2 (3.05e−2) −	7.0571e−2 (4.43e−5) −	94.33%, 13.86%, 0.01%
	(10,20)	1.0723e−1 (1.59e−2)	6.1270e−1 (1.42e−1) −	**1.0480e−1 (7.60e−3)** +	1.0624e−1 (8.21e−3) +	82.50%, −2.32%, −0.93%
Fun6	(5,5)	**1.3247e−1 (2.20e−3)**	5.4546e−1 (5.54e−2) −	3.0250e−1 (1.09e−3) −	1.3344e−1 (4.59e−3) −	75.71%, 56.21%, 0.73%
	(5,10)	1.3335e−1 (2.65e−3)	5.5036e−1 (5.77e−2) −	3.0202e−1 (6.73e−4) −	**1.3311e−1 (3.58e−3)** +	75.77%, 55.85%, −0.18%
	(5,20)	1.3316e−1 (3.31e−3)	5.2619e−1 (3.90e−2) −	3.0203e−1 (1.57e−3) −	**1.3299e−1 (2.78e−3)** +	74.69%, 55.91%, −0.13%
	(10,5)	**1.3188e−1 (2.01e−3)**	5.4484e−1 (4.25e−2) −	3.0142e−1 (3.73e−3) −	1.3242e−1 (1.94e−3) −	75.79%, 56.25%, 0.41%
	(10,10)	**1.3393e−1 (3.40e−3)**	5.4706e−1 (4.81e−2) −	3.0204e−1 (5.06e−4) −	1.3588e−1 (1.54e−2) −	75.52%, 55.66%, 1.44%
	(10,20)	**1.3261e−1 (2.65e−3)**	5.3520e−1 (5.56e−2) −	3.0204e−1 (7.68e−4) −	1.3396e−1 (4.10e−3) −	75.22%, 56.10%, 1.01%

References

1. Zhang, Q.; Li, H. MOEA/D: A Multiobjective evolutionary algorithm based on decomposition. *IEEE Trans. Evol. Comput.* **2007**, *11*, 712–731. [CrossRef]
2. Deb, K.; Pratap, A.; Agarwal, S.; Meyarivan, T. A fast and elitist multiobjective genetic algorithm: NSGA-II. *IEEE Trans. Evol. Comput.* **2002**, *6*, 182–197. [CrossRef]
3. Wang, G.-G.; Tan, Y. Improving metaheuristic algorithms with information feedback models. *IEEE Trans. Cybern.* **2019**, *49*, 542–555. [CrossRef] [PubMed]
4. Wang, G.-G.; Guo, L.; Gandomi, A.H.; Hao, G.-S.; Wang, H. Chaotic krill herd algorithm. *Inf. Sci.* **2014**, *274*, 17–34. [CrossRef]
5. Gao, D.; Wang, G.-G.; Pedrycz, W. Solving fuzzy job-shop scheduling problem using DE algorithm improved by a selection mechanism. *IEEE Trans. Fuzzy Syst.* **2020**, *28*, 3265–3275. [CrossRef]
6. Wang, G.-G.; Deb, S.; Cui, Z. Monarch butterfly optimization. *Neural Comput. Appl.* **2019**, *31*, 1995–2014. [CrossRef]
7. Jing, S.; Zhuang, M.; Gong, D.; Zeng, X.; Li, J.; Wang, G.-G. Interval multi-objective optimization with memetic algorithms. *IEEE Trans. Cybern.* **2020**, *50*, 3444–3457.
8. Chen, S.; Chen, R.; Wang, G.-G.; Gao, J.; Sangaiah, A.K. An adaptive large neighborhood search heuristic for dynamic vehicle routing problems. *Comput. Electr. Eng.* **2018**, *67*, 596–607. [CrossRef]
9. Hu, Y.; Ou, J.; Zheng, J.; Zou, J.; Yang, S.; Ruan, G. Solving dynamic multi-objective problems with an evolutionary multi-directional search approach. *Knowl.-Based Syst.* **2020**, *194*, 105175. [CrossRef]
10. Luo, W.; Sun, J.; Bu, C.; Liang, H. Species-based Particle Swarm optimizer enhanced by memory for dynamic optimization. *Appl. Soft Comput.* **2016**, *47*, 130–140. [CrossRef]
11. Nakano, H.; Kojima, M.; Miyauchi, A. An artificial bee colony algorithm with a memory scheme for dynamic optimization problems. In Proceedings of the IEEE Congress on Evolutionary Computation (IEEE CEC), Sendai, Japan, 25–28 May 2015; pp. 2657–2663.
12. Rong, M.; Gong, D.; Zhang, Y.; Jin, Y.; Pedrycz, W. Multidirectional prediction approach for dynamic multiobjective optimization problems. *IEEE Trans. Cybern.* **2019**, *49*, 3362–3374. [CrossRef]
13. Rong, M.; Gong, D.; Pedrycz, W.; Wang, L. A multimodel prediction method for dynamic multiobjective evolutionary optimization. *IEEE Trans. Evol. Comput.* **2019**, *24*, 290–304. [CrossRef]
14. Wu, X.; Wang, S.; Pan, Y.; Shao, H. A knee point-driven multi-objective artificial flora optimization algorithm. *Wirel. Netw.* **2020**, 1–11. [CrossRef]
15. Peng, Z.; Zheng, J.; Zou, J. A population diversity maintaining strategy based on dynamic environment evolutionary model for dynamic multiobjective optimization. In Proceedings of the IEEE Congress on Evolutionary Computation (IEEE CEC), Beijing, China, 6–11 July 2014; pp. 274–281.
16. Liu, M.; Zheng, J.; Wang, J.; Liu, Y.; Lei, J. An adaptive diversity introduction method for dynamic evolutionary multiobjective optimization. In Proceedings of the 2014 IEEE Congress on Evolutionary Computation (IEEE CEC), Beijing, China, 6–11 July 2014; pp. 3160–3167.
17. Deb, K.; Jain, H. An evolutionary many-objective optimization algorithm using reference-point-based nondominated sorting approach, Part I: Solving problems with box constraints. *IEEE Trans. Evol. Comput.* **2013**, *18*, 577–601. [CrossRef]
18. Farina, M.; Deb, K.; Amato, P. Dynamic multiobjective optimization problems: Test cases, approximations, and applications. *IEEE Trans. Evol. Comput.* **2004**, *8*, 425–442. [CrossRef]
19. Emmerich, M.T.M.; Deutz, A.H. A tutorial on multiobjective optimization: Fundamentals and evolutionary methods. *Nat. Comput.* **2018**, *17*, 585–609. [CrossRef] [PubMed]
20. Cao, L.; Xu, L.; Goodman, E.D.; Li, H. A first-order difference model-based evolutionary dynamic multiobjective optimization. In Proceedings of the Asia-Pacific Conference on Simulated Evolution and Learning, Shenzhen, China, 10–13 November 2017; pp. 644–655.
21. Deb, K.; Karthik, S. Dynamic multi-objective optimization and decision-making using modified NSGA-II: A case study on hydro-thermal power scheduling. In Proceedings of the International Conference Evolutionary Multi-Criterion Optimization, Matsushima, Japan, 5–8 March 2007; pp. 803–817.
22. Yen, G.; Lu, H. Dynamic multiobjective evolutionary algorithm: Adaptive cell-based rank and density estimation. *IEEE Trans. Evol. Comput.* **2003**, *7*, 253–274. [CrossRef]
23. Wang, Y.; Yu, J.; Yang, S.; Jiang, S.; Zhao, S. Evolutionary dynamic constrained optimization: Test suite construction and algorithm comparisons. *Swarm Evol. Comput.* **2019**, *50*, 100559. [CrossRef]
24. Macias-Escobar, T.; Cruz-Reyes, L.; Fraire, H.; Dorronsoro, B. Plane separation: A method to solve dynamic multi-objective optimization problems with incorporated preferences. *Future Gener. Comp. Syst.* **2019**, *110*, 864–875. [CrossRef]
25. Zou, F.; Yen, G.G.; Tang, L. A knee-guided prediction approach for dynamic multi-objective optimization. *Inf. Sci.* **2020**, *509*, 193–209. [CrossRef]
26. Zhou, A.; Jin, Y.; Zhang, Q.; Sendhoff, B.; Tsang, E. Prediction-based population re-initialization for evolutionary dynamic multi-objective optimization. In Proceedings of the International Conference Evolutionary Multi-Criterion Optimization, Matsushima, Japan, 5–8 March 2007; pp. 832–846.
27. Hatzakis, I.; Wallace, D. Dynamic multi-objective optimization with evolutionary algorithms: A forward-looking approach. In Proceedings of the Genetic and Evolutionary Computation Conference, Seattle, WA, USA, 8–12 July 2006; pp. 1201–1208.

28. Liu, X.-F.; Zhou, Y.-R.; Yu, X.; Lin, Y. Dual-archive-based particle swarm optimization for dynamic optimization. *Appl. Soft Comput.* **2019**, *85*, 105876. [CrossRef]
29. Gong, D.; Xu, B.; Zhang, Y.; Guo, Y.; Yang, S. A similarity-based cooperative co-evolutionary algorithm for dynamic interval multiobjective optimization problems. *IEEE Trans. Evol. Comput.* **2020**, *24*, 142–156. [CrossRef]
30. Wang, C.; Yen, G.G.; Jiang, M. A grey prediction-based evolutionary algorithm for dynamic multiobjective optimization. *Swarm Evol. Comput.* **2020**, *56*, 100695. [CrossRef]
31. Chang, R.-I.; Hsu, H.-M.; Lin, S.-Y.; Chang, C.-C.; Ho, J.-M. Query-based learning for dynamic Particle Swarm optimization. *IEEE Access* **2017**, *5*, 7648–7658. [CrossRef]
32. Liang, Z.; Wu, T.; Ma, X.; Zhu, Z.; Yang, S. A dynamic multiobjective evolutionary algorithm based on decision variable classification. *IEEE Trans. Cybern.* **2020**, 1–14. [CrossRef]
33. Wang, Z.; Zhang, J.; Yang, S. An improved particle Swarm optimization algorithm for dynamic job shop scheduling problems with random job arrivals. *Swarm Evol. Comput.* **2019**, *51*, 100594. [CrossRef]
34. Luna, F.; Zapata-Cano, P.H.; González-Macías, J.C.; Valenzuela-Valdés, J.F. Approaching the cell switch-off problem in 5G ultra-dense networks with dynamic multi-objective optimization. *Future Gener. Comput. Syst.* **2020**, *110*, 876–891. [CrossRef]
35. Zhou, X.; Wang, X.; Huang, T.; Yang, C. Hybrid intelligence assisted sample average approximation method for chance constrained dynamic optimization. *IEEE Trans. Ind. Inform.* **2020**, 1. [CrossRef]
36. Chang, L.; Piao, S.; Leng, X.; Hu, Y.; Ke, W. Study on falling backward of humanoid robot based on dynamic multi objective optimization. *Peer Peer Netw. Appl.* **2020**, *13*, 1236–1247. [CrossRef]
37. Cabrera, A.; Acosta, A.; Almeida, F.; Blanco, V.; Perez, A.C. A dynamic multi-objective approach for dynamic load balancing in heterogeneous systems. *IEEE Trans. Parallel Distrib. Syst.* **2020**, *31*, 2421–2434. [CrossRef]
38. Zitzler, E.; Deb, K.; Thiele, L. Comparison of multiobjective evolutionary algorithms: Empirical results. *Evol. Comput.* **2000**, *8*, 173–195. [CrossRef] [PubMed]
39. Kong, W.; Chai, T.; Yang, S.; Ding, J. A hybrid evolutionary multiobjective optimization strategy for the dynamic power supply problem in magnesia grain manufacturing. *Appl. Soft Comput.* **2013**, *13*, 2960–2969. [CrossRef]
40. Jiang, S.; Yang, S. Evolutionary dynamic multiobjective optimization: Benchmarks and algorithm comparisons. *IEEE Trans. Cybern.* **2016**, *47*, 198–211. [CrossRef]
41. Goh, C.-K.; Tan, K.C. A competitive-cooperative coevolutionary paradigm for dynamic multiobjective optimization. *IEEE Trans. Evol. Comput.* **2008**, *13*, 103–127. [CrossRef]
42. Jiang, S.; Yang, S.; Yao, X.; Tan, K.C.; Kaiser, M.; Krasnogor, N. Benchmark problems for CEC2018 competition on dynamic multiobjective optimisation. In Proceedings of the IEEE Congress on Evolutionary Computation (IEEE CEC), Rio de Janeiro, Brazil, 8–13 July 2018; pp. 1–8.
43. Zhou, A.; Jin, Y.; Zhang, Q. A population prediction strategy for evolutionary dynamic multiobjective optimization. *IEEE Trans. Cybern.* **2014**, *44*, 40–53. [CrossRef]
44. Jiang, S.; Yang, S. A steady-state and generational evolutionary algorithm for dynamic multiobjective optimization. *IEEE Trans. Evol. Comput.* **2016**, *21*, 65–82. [CrossRef]
45. He, C.; Li, L.; Tian, Y.; Zhang, X.; Cheng, R.; Jin, Y.; Yao, X. Accelerating large-scale multiobjective optimization via problem reformulation. *IEEE Trans. Evol. Comput.* **2019**, *23*, 949–961. [CrossRef]

 processes

MDPI

Article

A Novel Evolutionary Arithmetic Optimization Algorithm for Multilevel Thresholding Segmentation of COVID-19 CT Images

Laith Abualigah [1,2], Ali Diabat [3,4], Putra Sumari [2] and Amir H. Gandomi [5,*]

1 Faculty of Computer Sciences and Informatics, Amman Arab University, Amman 11953, Jordan; Aligah.2020@gmail.com
2 School of Computer Sciences, Universiti Sains Malaysia, Pulau Pinang 11800, Malaysia; putras@usm.my
3 Division of Engineering, New York University Abu Dhabi, Saadiyat Island, Abu Dhabi 129188, United Arab Emirates; Diabat@nyu.edu
4 Department of Civil and Urban Engineering, Tandon School of Engineering, New York University, Brooklyn, NY 11201, USA
5 Faculty of Engineering and Information Technology, University of Technology Sydney, Ultimo, NSW 2007, Australia
* Correspondence: Gandomi@uts.edu.au

Citation: Abualigah, L.; Diabat, A.; Sumari, P.; Gandomi, A.H. A Novel Evolutionary Arithmetic Optimization Algorithm for Multilevel Thresholding Segmentation of COVID-19 CT Images. *Processes* **2021**, *9*, 1155. https://doi.org/10.3390/pr9071155

Academic Editor: Harvey Arellano-Garcia

Received: 5 May 2021
Accepted: 18 June 2021
Published: 2 July 2021

Abstract: One of the most crucial aspects of image segmentation is multilevel thresholding. However, multilevel thresholding becomes increasingly more computationally complex as the number of thresholds grows. In order to address this defect, this paper proposes a new multilevel thresholding approach based on the Evolutionary Arithmetic Optimization Algorithm (AOA). The arithmetic operators in science were the inspiration for AOA. DAOA is the proposed approach, which employs the Differential Evolution technique to enhance the AOA local research. The proposed algorithm is applied to the multilevel thresholding problem, using Kapur's measure between class variance functions. The suggested DAOA is used to evaluate images, using eight standard test images from two different groups: nature and CT COVID-19 images. Peak signal-to-noise ratio (PSNR) and structural similarity index test (SSIM) are standard evaluation measures used to determine the accuracy of segmented images. The proposed DAOA method's efficiency is evaluated and compared to other multilevel thresholding methods. The findings are presented with a number of different threshold values (i.e., 2, 3, 4, 5, and 6). According to the experimental results, the proposed DAOA process is better and produces higher-quality solutions than other comparative approaches. Moreover, it achieved better-segmented images, PSNR, and SSIM values. In addition, the proposed DAOA is ranked the first method in all test cases.

Keywords: Arithmetic Optimization Algorithm (AOA); meta-heuristics; Differential Evolution; Optimization Algorithms; engineering problems; optimization problems; real-world problems; multilevel thresholding; image segmentation

1. Introduction

One of the most often used image segmentation techniques is multilevel thresholding. It is divided into two types: bi-level and multilevel [1,2]. Multilevel thresholding is used to separate complex images, which can generate several thresholds, such as tri-level or quad-level thresholds, which break pixels into several identical parts depending on size. Bi-level thresholding divides the image into two levels, while multilevel thresholding divides the image into two classes [3,4]. When there are only two primary gray levels in an image, bi-level thresholding yields acceptable results; however, when it is expanded to multilevel thresholding, the main drawback is the time-consuming computation [5]. Bi-level thresholding cannot precisely find the optimum threshold, due to the slight variation between the target and the context of a complex image [6,7].

Medical imaging, machine vision, and satellite photography all use image segmentation [8–10]. The primary aim of image segmentation is to divide an image into relevant regions for a specific mission. The process of finding and isolating points of interest from the rest of the scene is known as the segmentation of pattern recognition systems [11,12]. Following image segmentation, certain features from objects are removed, and then objects are grouped into specific categories or classes, depending on the extracted features. Segmentation is used in medical applications to detect organs, such as the brain, heart, lungs, and liver, in CT or MR images [13,14]. It is also used to tell the difference between abnormal tissue, such as a tumor, and healthy tissue. Image segmentation techniques, such as image thresholding, edge detection, area expanding, stochastic models, Artificial Neural Network (ANN), and clustering techniques, have all been used, depending on the application [15,16].

Tsallis, Kapur, and Otsu procedures are the most widely used thresholding strategies [17,18]. The Otsu method maximizes the between-class variance function to find optimum thresholds, while the Kapur method maximizes the posterior entropy of the segmented groups. Due to exhaustive search, Tsallis and Otsu's computational complexity grows exponentially as the number of thresholds increases [19]. Many researchers have worked on image segmentation over the years. Image segmentation has been tackled using a variety of approaches and algorithms [20]. Examples of the used optimization algorithms are the Bat Algorithm (BA) [21], Firefly Algorithm (FA) [22], Genetic Algorithm (GA) [23], Gray Wolf Optimizer (GWO) [24,25], Dragonfly Algorithm (DA) [26], Moth–Flame Optimization Algorithm (MFO) [27], Marine Predators Algorithm (MPA) [28], Arithmetic Optimization Algorithm (AOA) [29], Aquila Optimizer (AO) [30], Krill Herd Optimizer (KHO) [31], Harris Hawks Optimizer (HHO) [32], Red Fox Optimization Algorithm (RFOA) [33], Artificial Bee Colony Algorithm (ABC) [34], and Artificial Ecosystem-based Optimization [35]. Many other optimizers can be found in [36,37].

The paper [38] used Kapur and Otsu's approaches to adjust the latest Elephant Herding Optimization Algorithm for multilevel thresholding. Its performance was compared to four other swarm intelligence algorithms, using regular benchmark images. The Elephant Herding Optimization Algorithm outperformed and proved more stable than other methods in the literature. Sahlol et al. in [39] introduced an improved hybrid method for COVID-19 images by merging the strengths of convolution neural networks (CNNs) to remove features and the MPA feature selection algorithm to choose the most important features. The proposed method exceeds several CNNs and other well-known methods on COVID-19 images.

The multi-verse optimizer (MVO), based on the multi-verse theorem, is a new algorithm for solving real-world multi-parameter optimization problems. A novel parallel multi-verse optimizer (PMVO) with a coordination approach is proposed in [40]. For each defined iteration, the parallel process is used to randomly split the original solutions into multiple groups and exchange the various groups' details. This significantly improves individual MVO algorithm cooperation and reduces the shortcomings of the original MVO algorithm, such as premature convergence, search stagnation, and easy trapping into the local optimal search. The PMVO algorithm was compared to methods under the CEC2013 test suite to validate the proposed scheme's efficiency. The experimental findings show that the PMVO outperforms the other algorithms under consideration. In addition, using minimum cross entropy thresholding, PMVO is used to solve complex multilevel image segmentation problems. In comparison with different related algorithms, the proposed PMVO algorithm seems to achieve better quality image segmentation.

For image segmentation, a modified artificial bee colony optimizer (MABC) is proposed [41], which balances the tradeoff between the search process by using a pool of optimal foraging strategies. MABC's main goal is to improve artificial bee foraging behaviors by integrating local search with detailed learning, using a multi-dimensional PSO-based equation. With detailed learning, the bees combine global best solution knowledge into the solution quest equation to increase exploration. Simultaneously, local search allows the bees to thoroughly exploit across the promising field, providing a good combination of exploration and exploitation. The proposed algorithm's feasibility was shown by the experimental findings comparing the MABC to several popular EA and SI algorithms on a series of benchmarks. The experimental findings verify the suggested algorithm's efficacy.

For solving the image segmentation problem, a novel multilevel thresholding algorithm based on a meta-heuristic Krill Herd Optimization (KHO) algorithm is proposed in [42]. The optimal threshold values are calculated, using the Krill Herd Optimization technique to maximize Kapur's or Otsu's objective function. The suggested method reduces the amount of time it takes to calculate the best multilevel thresholds. Various benchmark images are used to illustrate the applicability and numerical performance of the Krill Herd Optimization-based multilevel thresholding. To demonstrate the superior performance of the proposed method, a detailed comparison with other current bio-inspired techniques based on multilevel thresholding techniques, such as Bacterial Foraging (BF), Particle Swarm Optimization (PSO), Genetic Algorithm (GA), and Moth-Flame Optimization (MFO), was performed. The results confirmed that the proposed method achieved better results than other methods.

This paper presents a modified version of the Manta Ray Foraging Optimizer (MRFO) algorithm to deal with global optimization and multilevel image segmentation problems [43]. MRFO is a meta-heuristic technique that simulates the behaviors of manta rays to find food. The performance of the MRFO is improved by using fractional-order (FO) calculus during the exploitation phase. In this experiment, a variant of natural images is used to assess FO-MRFO. According to different performance measures, the FO-MRFO outperformed the compared algorithms in global optimization and image segmentation.

The concept "optimization" refers to the process of identifying the best solutions to a problem while keeping those constraints in mind [44,45]. The used optimization in solving the image segmentation problem is the method of finding the best threshold values for a given image. Swarm intelligence (SI) algorithms are used widely for multilevel thresholding problems to determine the optimal threshold values, using various objective functions to solve the problems of the computational inefficiency of traditional thresholding techniques. The primary motivation behind this paper is to find the optimal threshold values for image segmentation problems. At the same time, to address the weakness of the original AOA, it suffers from the local optimal problem and premature coverage in some cases. In this paper, an improved version of the Arithmetic Optimization Algorithm (AOA) by using Differential Evolution, called DAOA, is proposed. The proposed method uses Differential Evolution to tackle the conventional Arithmetic Optimization Algorithm's weaknesses, such as being trapped in local optima and fast convergence. Thus, Differential Evolution is used to enhance the performance of the Arithmetic Optimization Algorithm. The proposed DAOA assists by using eight standard test images from different groups: two-color images, two gray images, two normal CT COVID-19 images, and two confirmed COVID-19 CT images. Peak signal-to-noise ratio (PSNR), structural similarity index test (SSIM), and fitness function (Kapur's) are used to determine the accuracy of segmented images. The proposed DAOA method's efficiency is evaluated and compared to other multilevel thresholding methods. The findings are presented with a number of different threshold values (i.e., 2, 3, 4, 5, and 6). According to the experimental results, the proposed DAOA process is better and produces higher-quality solutions than other approaches. The encouraging findings suggest that using the DAOA-based thresholding strategy has potential and is helpful.

The rest of this paper is organized as follows. Section 2 presents the procedure of the proposed DAOA method. Section 3 presents the definitions and procedures of the image segmentation problem. The experiments and results are given in Section 4. Finally, in Section 5, the conclusions and potential future work directions are given.

2. The Proposed Method

In this section, we present the conventional Arithmetic Optimization Algorithm (AOA), Differential Evolution (DE), and the proposed Evolutionary Arithmetic Optimization Algorithm (DAOA).

2.1. Arithmetic Optimization Algorithm (AOA)

In this section, we describe the exploration and exploitation phases of the original AOA [29], which is motivated by the main operators in math science (i.e., multiplication (M), division (D), subtraction (S), and addition (A)). The main search methods of the AOA are presented in Figure 1, which are illustrated in the following subsections.

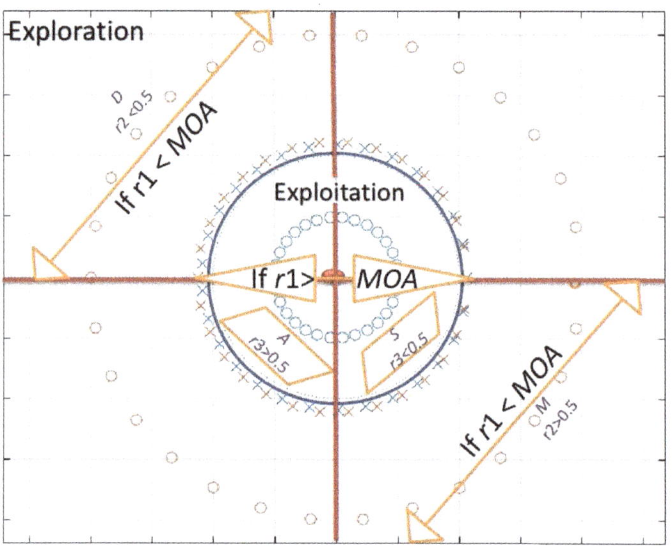

Figure 1. The search phases of the Arithmetic Optimization Algorithm.

The AOA should choose the search process before beginning its work (i.e., exploration or exploitation). So, in the following search steps, the Math Optimizer Accelerated (MOA) function is a coefficient determined by Equation (1).

$$MOA(C\_Iter) = Min + C\_Iter \times \left(\frac{Max - Min}{M\_Iter}\right) \tag{1}$$

where MOA(C_Iter) means the value at the tth iteration of MOA function, determined by Equation (1). C_Iter is the current iteration: $[1 \ldots\ldots\ldots M\_Iter]$. Min and Max are the accelerated function values (minimum and maximum), respectively.

2.1.1. Exploration Phase

The exploration operators of AOA are modeled in Equation (2). The exploration phase uses the D or M operators conditioned by $r1$ >MOA. The D operator is prepared by $r2 < 0.5$, or, otherwise, by the M operator. $r2$ is a random number. The position updating process is determined as follows.

$$x_{i,j}(C\_Iter+1) = \begin{cases} best(x_j) \div MOP \times ((UB_j - LB_j) \times \mu + LB_j), & r2 < 0.5 \\ best(x_j) \times MOP \times ((UB_j - LB_j) \times \mu + LB_j), & otherwise \end{cases} \tag{2}$$

where x_i(C_Iter+1) is the ith next solution, $x_{i,j}$(C_Iter) is the jth location of the ith solution, and $best(x_j)$ is the jth location in the best solution. μ is a control value (0.5) to tune the exploration search.

$$MOP(C\_Iter) = 1 - \frac{C\_Iter^{1/\alpha}}{M\_Iter^{1/\alpha}} \tag{3}$$

where MOP(C_Iter) denotes the coefficient value at the tth iteration. α is a control value (5) to tune the exploration search.

2.1.2. Exploitation Phase

The exploitation searching phase uses the S and A operators conditioned by the MOA function value. Subtraction (S) and addition (A) search strategies are represented in Equation (4).

$$x_{i,j}(C\_Iter+1) = \begin{cases} best(x_j) - MOP \times ((UB_j - LB_j) \times \mu + LB_j), & r3 < 0.5 \\ best(x_j) + MOP \times ((UB_j - LB_j) \times \mu + LB_j), & otherwise \end{cases} \tag{4}$$

The intuitive and detailed process of AOA is shown in Figure 2.

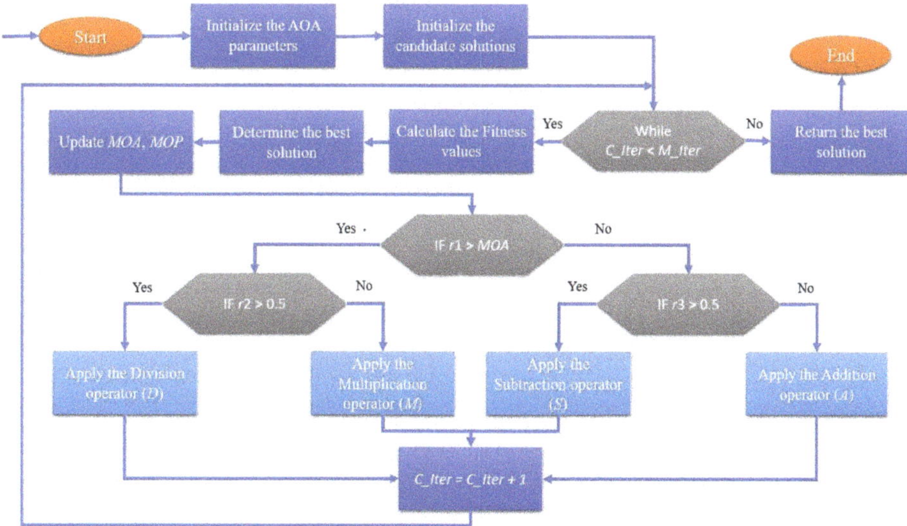

Figure 2. Flowchart of the conventional AOA.

2.2. *Differential Evolution (DE)*

In [46], Storn and Price introduced the DE as the first version to solve multiple optimization problems in 1997. DE stands out for its versatility, quick execution time, rapid acceleration pattern, and fast and accurate local operators [47,48]. In DE, the optimization process begins with a random selection of solutions for finding the majority of the points in the search space (initialization phase). The solutions can then be improved, using a series of operators called mutation and crossover. The new solution can be accepted if it has a higher objective value. For the current solution X_i, the mathematical model of the mutation operator Z_i^t can be applied as follows:

$$Z_{i,j} = XD_{rand_1} + F \times (XD_{r2} - XD_{r3}), \tag{5}$$

where r_1, r_2, and r_3 are random numbers, F is the mutation balancing factor, and F is greater than 0.

For the crossover operator, Equation (6) represents the new solution V_i, which is produced using the mutated operator through the crossover Z_i. The crossover is considered a mixture process among vectors Z_i and XD_i.

$$V_{i,j} = \begin{cases} Z_{i,j} & if\ rand \leq C_r \\ XD_{i,j} & otherwise \end{cases} \tag{6}$$

C_r is the crossover probability.

The DE algorithm improves its selected solutions according to the objective function values, where the generated V_i, C_Iter is replaced with the current one if it obtained a better fitness value, which is as follows.

$$XD_{i,j} = \begin{cases} V_{i,j} & if\ f(V_{i,j}) < f(XD_{i,j}) \\ XD_{i,j} & otherwise \end{cases} \tag{7}$$

2.3. The Proposed DAOA

In this section, the procedure of the proposed Evolutionary Arithmetic Optimization Algorithm (DAOA) is presented as follows.

2.3.1. Initialization Phase

When using the AOA, the optimization procedure begins with a number of random solutions (X) as designated in matrix (8). The best solution is taken in each iteration as the best-obtained solution.

$$X = \begin{bmatrix} x_{1,1} & \cdots & \cdots & x_{1,j} & x_{1,n-1} & x_{1,n} \\ x_{2,1} & \cdots & \cdots & x_{2,j} & \cdots & x_{2,n} \\ \cdots & \cdots & \cdots & \cdots & \cdots & \cdots \\ \vdots & \vdots & \vdots & \vdots & \vdots & \vdots \\ x_{N-1,1} & \cdots & \cdots & x_{N-1,j} & \cdots & x_{N-1,n} \\ x_{N,1} & \cdots & \cdots & x_{N,j} & x_{N,n-1} & x_{N,n} \end{bmatrix} \tag{8}$$

2.3.2. Phases of the Proposed DAOA

In this section, the main details and procedures of the proposed Evolutionary Arithmetic Optimization Algorithm (DAOA) are given.

The DAOA is introduced mainly to develop the original AOA's convergence ability, the quality of solutions, and the ability to avoid the local optima problem. Thus, the DE technique is introduced into the conventional AOA to form DAOA. This proposed DAOA method is introduced to perform the exploration search by the AOA and exploitation search by the DE. This also makes an excellent balance between the search strategies and guarantees that the proposed method averts the local optima.

Figure 3 depicts the proposed DAOA approach in this section. The DAOA procedure begins with (1) determining the values of the used algorithms' parameters, (2) generating candidate solutions, (3) calculating fitness functions, (4) selecting the best solution, (5) if a given condition is true, the AOA is executed to update the solutions; otherwise, the DE is executed to update the solutions, and (6) then another condition is given to stop or continue the optimization process. Figure 3 shows the flowchart for the proposed DAOA. The pseudo-code of the DAOA algorithm is given in Algorithm 1.

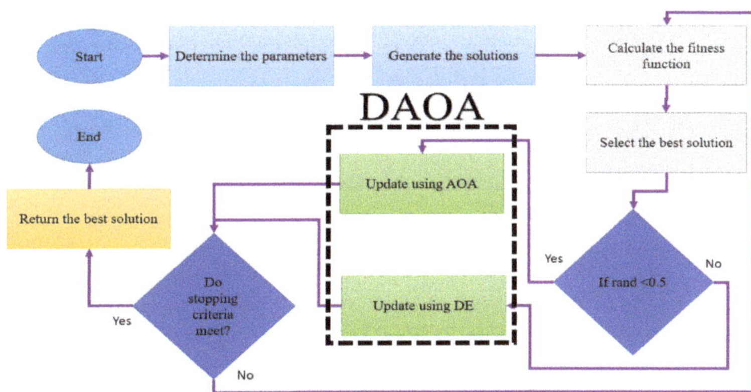

Figure 3. The flowchart of the proposed DAOA.

Algorithm 1 Pseudo-code of the DAOA algorithm

```
Initialize the Arithmetic Optimization Algorithm parameters α, μ.
Initialize the solutions' positions randomly. (Solutions: i = 1, ..., N.).
Calculate the Fitness values.
while (C_Iter < M_Iter) do
   Find the best solution (determined best so far).
   Update the MOA value using Equation (1).
   Update the MOP value using Equation (3).
   Calculate the Fitness Function (FF) for the given solutions.
   for (i = 1 to Solutions) do
      if rand < 0.5 then
         Generate a random values between [0, 1] (r1, r2, and r3)
         if r1 > MOA then
            if r2 > 0.5 then
               Update the ith solutions' positions using the first rule in Equation (2).
            else
               Update the ith solutions' positions using the second rule in Equation (2).
            end if
         else
            if r3 > 0.5 then
               Update the ith solutions' positions using the first rule in Equation (4).
            else
               Update the ith solutions' positions using the second rule in Equation (4).
            end if
         end if
      else
         if rand < 0.5 then
            Update the ith solutions' positions using Mutation operator as given in
            Equation (5).
         else
            Update the ith solutions' positions using Crossover operator as given in
            Equation (6).
         end if
      end if
   end for
   C_Iter = C_Iter + 1
end while
Return the best solution (x).
```

3. Definitions of the Multilevel Thresholding Image Segmentation Problems

In this section, we describe the main problem of multilevel thresholding. Let us suppose that I is a gray or color image that needs to be processed, where $K+1$ presents the classes that need to be produced. For segmenting the given image (I) into $K+1$ classes, the k thresholds' values are required to progress in the image segmentation procedure; $\{t_k, k = 1..........K\}$, and this can be expressed as follows [1,7,49].

$$\begin{aligned} C_0 &= \{I_{i,j} \mid 0 \leq I_{i,j} \leq t_1 - 1\}, \\ C_1 &= \{I_{i,j} \mid t_1 \leq I_{i,j} \leq t_2 - 1\}, \\ &\ldots \\ C_K &= \{I_{i,j} \mid t_K \leq I_{i,j} \leq L - 1\} \end{aligned} \tag{9}$$

where L indicates the highest gray levels and C_K indicates the kth class of the image I. The t_k is the k-th threshold, with $I_{i,j}$ being the gray level at the (i,j)th pixel. Furthermore, in Equation (10), multilevel thresholding is identified as a maximization optimization problem that needs to find the optimal threshold values.

K multilevel threshold values can be presented as follows.

$$t_1,*,t_2,*,\ldots,t_K,* = \arg\max_{t_1,\ldots,t_K} Fit(t_1,\ldots,t_K) \tag{10}$$

3.1. Fitness Function (Kapur's Entropy)

For the purpose of thresholding, consider a digital image I with N pixels and L gray levels. Via thresholds, these L number of gray levels are divided into classes: Class1, Class2, ..., Class$k$ [1].

In this proposed DAOA, Kapur's entropy is utilized for achieving optimum threshold values. Measurement of the bi-level thresholds needs the optimization process's objective function, as shown in Equation (11).

$$Fit(t_1,\ldots,t_K) = \sum_{k=1}, KH_i \tag{11}$$

$$H_k = -\sum_{i=0}^{L-1} \frac{p_i \times \mu_k(i)}{P_k} \times In(\frac{p_i \times \mu_k(i)}{P_k}), \tag{12}$$

$$P_k = \sum_{i=0}^{L-1} p_i \times \mu_k(i) \tag{13}$$

$$\mu_1(l) = \begin{cases} 1 & l \leq a_1 \\ \frac{l-c_1}{a_1-c_1} & a_1 \leq l \leq c_1 \\ 0 & l > c_1 \end{cases} \quad \mu_K(l) = \begin{cases} 1 & l \leq a_{K-1} \\ \frac{l-a_K}{c_K-a_K} & a_{K-1} < l \leq c_{K-1} \\ 0 & l > c_{K-1} \end{cases} \tag{14}$$

where p_i is the probability distribution, $h(i)$ is the numbers of pixels for the used gray level L, and N_p is the total numbers of pixels of the image I. p_i presents the probability value for the distribution, determined as $p_i = h(i)/N_p$ $(0 < i < L-1)$. $h(i)$ and N_k are the numbers of pixels for the used gray level L and total pixel of the image I. $a_1, c_1, \ldots, a_{k-1}, c_{k-1}$ are the used fuzzy parameters, and $0 \leq a_1 \leq c_1 \leq \ldots \leq a_{K-1} \leq c_{K-1}$.

Then, $t_1 = \frac{a_1+c_1}{2}, t_2 = \frac{a_2+c_2}{2}, ..., t_{K-1} = \frac{a_{K-1}+c_{K-1}}{2}$. The best fitness function obtained is the highest value.

3.2. Performance Measures

We assess the proposed DAOA method performance, using three performance measures: the fitness function value, the Structural Similarity Index (SSIM), and the Peak Signal-to-Noise Ratio (PSNR) [50,51]. The following equations compute SSIM and PSNR:

$$SSIM(I, I_S) = \frac{(2\mu_I\mu_{I_S} + c_1)(2\sigma_{I,I_S} + c_2)}{(\mu_I^2 + \mu_{I_S}^2 + c_1)(\sigma_I^2 + \sigma_{I_S}^2 + c_2)} \tag{15}$$

where μ_{I_S} (σ_{I_S}) and $\mu_I(\sigma_I)$ are the images' mean intensity of I_S and I, respectively, where σ_{I,I_S} is the governance of I and I_S, and c_1 and c_2 coefficient values are equal to 6.5025 and 58.52252, respectively [1].

$$PSNR = 20log_{10}(\frac{255}{RMSE}),\ RMSE = \sqrt{\frac{\sum_{i=1}^{N_r}\sum_{j=1}^{N_c}(I_{i,j} - I_S i, j)^2}{N_r \times N_c}} \tag{16}$$

where the $RMSE$ is the root-mean-squared error of each pixel, and $M \times N$ depicts the image's size. $I_{i,j}$ is the gray pixel value of the initial image, and Is_{ij} is the gray value of the pixel in the obtained segmented image.

4. Experiments and Results

4.1. Benchmark Images

In this section, the benchmark image data sets are presented in Figures 4 and 5. Two image types were used in this paper's experiments, taken from nature (as seen in Figure 4) and medical CT images (as seen in Figure 5). We chose eight images: two-color images from nature (i.e., Test 1 and Test 2), two gray images from nature (i.e., Test 3 and Test 4), two COVID-19 CT images (i.e., Test 5 and Test 6), and two normal COVID-19 CT images (i.e., Test 7 and Test 8). These benchmarks were taken from the Berkeley Segmentation Data Set: Images and BIMCV-COVID19 [52].

(a) Test 1 **(b)** Test 2 **(c)** Test 3 **(d)** Test 4

Figure 4. The nature benchmark images that have been used.

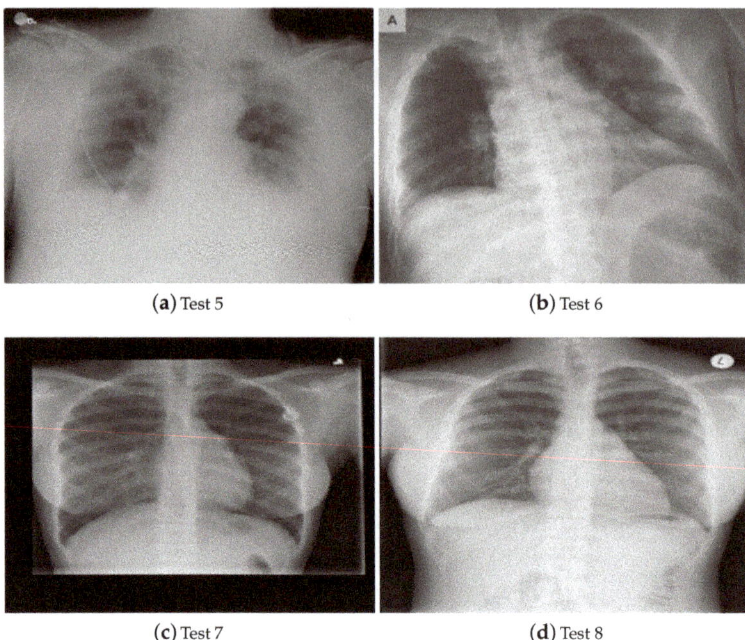

(**a**) Test 5 (**b**) Test 6

(**c**) Test 7 (**d**) Test 8

Figure 5. The CT benchmark images that were used.

4.2. Comparative Algorithms and Parameter Setting

The proposed DAOA is analyzed and compared with six recently well-known algorithms, including Aquila Optimizer (AO) [30], Whale Optimization Algorithm (WOA) [53], Salp Swarm Algorithm (SSA) [54], Arithmetic Optimization Algorithm (AOA) [29], Particle Swarm Optimization (PSO) [55], Marine Predators Algorithm (MPA) [56], and Differential Evolution (DE) [57].

These algorithms' parameters are set in the same way as they were in their original papers. The values of different parameter settings used in the tested algorithms are shown in Table 1. These sensitive parameters can be tuned for further investigation to show the effect of each parameter on the performance of the tested methods. The algorithms are executed by using the MATLAB 2015a software. These algorithms are run on an Intel Core i7 1.80 GHz 2.30 GHz processor with 16 GB RAM. The number of solutions used is twenty-five. For a systematic comparison, the maximum number of iterations is set to one hundred. Each competitor algorithm generates thirty independent runs.

Table 1. Parameter settings.

No.	Algorithm	Reference	Parameter	Value
1	AO	[30]	α	0.1
			$/\delta$	0.1
2	WOA	[53]	α	Decreased from 2 to 0
			b	2
3	SSA	[54]	v_0	0
4	AOA	[29]	α	5
			μ	0.5
5	PSO	[55]	Topology	Fully connected
			Cognitive and social constant	(C1, C2) 2, 2
			Inertia weight	Linear reduction values [0.9 0.1]
			Velocity limit	10% of dimension range
6	MPA	[56]	γ	$\gamma > 1$
			P	0.0
7	DE	[57]	Co	0.5
			Mu	0.5

4.3. Performance Evaluation

A comparison of the proposed DAOA for multilevel thresholding segmentation, using eight different images, is presented in this section. The following tables show the max, mean, min, and standard deviation of each test case's PSNR and SSIM. Moreover, the summation, mean rank, and final ranking are given, using the Friedman ranking test to prove the proposed method's significant improvement [58,59].

The PSNR and SSIM results of Test 1 are given in Tables 2 and 3. It is clear that the proposed DAOA obtained excellent results in almost all the test cases in terms of PSNR. For threshold 2, the proposed DAOA obtained the best results, and it ranked first when compared to all other comparative methods, followed by AOA, SSA, PSO, WOA, MPA, AO, and finally, DE. In addition, for threshold 6, the proposed method obtained promising results compared to other methods. DAOA obtained the first rank, followed by WOA, PSO, SSA, AOA, AO, SSA, MPA, and DE. Overall, we can see that the proposed method obtained the first ranking, followed by AOA, PSO, SSA, WOA, AO, MPA, and DE. The obtained results in this table prove the ability of the proposed DAOA to solve the given problems effectively.

For threshold 2 in Table 3, the proposed DAOA obtained the best results, and it ranked as the first method, compared to all other comparative methods, followed by PSO, SSA, DE, AOA, WOA, MPA, and finally, AO. In addition, for threshold 3, the proposed method obtained promising results, compared to other methods. DAOA obtained the first ranking, followed by PSO, SSA, DE, AOA, MPA, WOA, and AO. Overall, we can see that the proposed DAOA method obtained the first ranking, followed by PSO, AOA, SSA, DE, WOA, AO, and MPA. The obtained results in this table prove the ability of the proposed DAOA to solve the given problems effectively.

Table 2. The PSNR results of the test case 1.

Threshold	Metric	Comparative Methods							
		AO	WOA	SSA	AOA	PSO	MPA	DE	DAOA
2	Max	12.74479	13.20368	13.72521	13.27539	13.70085	12.02615	13.51551	14.55374
	Mean	11.43681	11.97746	12.58728	12.61579	12.51339	11.79407	11.25852	12.76203
	Min	10.60697	10.16222	10.98468	12.02184	11.85952	11.60000	10.33365	11.86358
	STD	1.14631	1.60401	1.42813	0.62935	1.03013	0.21560	0.65885	1.55167
	Ranking	7	5	3	2	4	6	8	1
3	Max	16.44011	14.27535	16.08140	16.89013	14.18813	15.13210	14.41440	15.73708
	Mean	15.19863	13.28605	14.43510	14.68456	13.49286	14.37895	12.22514	14.61015
	Min	14.28507	11.36665	12.18756	11.73528	12.53317	12.91041	11.02215	13.64585
	STD	1.11431	1.66251	2.01534	2.65668	0.85858	1.27194	0.56698	1.05506
	Ranking	1	7	4	2	6	5	8	3
4	Max	15.17656	17.95691	17.47235	17.16435	17.32833	16.30057	15.65854	17.94836
	Mean	14.00183	15.22041	16.87263	15.59121	16.52225	15.72838	14.25484	16.04748
	Min	13.30037	12.55354	15.90180	14.37544	16.09469	15.11861	13.95558	15.09213
	STD	1.02372	2.70236	0.84849	1.42839	0.69852	0.59188	0.47447	1.64622
	Ranking	8	6	1	5	2	4	7	3
5	Max	16.72622	16.42710	16.24110	17.90312	16.37420	16.30256	16.32254	18.67014
	Mean	15.54953	16.02096	15.54791	16.92259	15.84138	15.24571	15.22541	15.86760
	Min	14.49543	15.61807	14.88442	15.62385	15.34763	14.57955	14.02554	14.01993
	STD	1.12043	0.40452	0.67883	1.17248	0.51440	0.92557	0.65558	2.46778
	Ranking	5	2	6	1	4	7	8	3
6	Max	19.43582	20.61942	19.52344	20.43187	19.96838	18.86744	17.95101	20.03906
	Mean	18.38781	18.75391	17.85512	18.23439	18.71728	16.92716	16.25870	19.23425
	Min	16.37613	17.07040	14.78261	14.88956	17.44713	14.57855	15.33652	17.83410
	STD	1.74267	1.78149	2.66414	2.94391	1.26073	2.17340	1.25412	1.21708
	Ranking	4	2	6	5	3	7	8	1
Summation		25	22	20	15	19	29	39	11
Mean Rank		5	4.4	4	3	3.8	5.8	7.8	2.2
Final Ranking		6	5	4	2	3	7	8	1

Table 3. The SSIM results of the test case 1.

Threshold	Metric	Comparative Methods							
		AO	WOA	SSA	AOA	PSO	MPA	DE	DAOA
2	Max	0.269717	0.362146	0.380757	0.264672	0.454587	0.220516	0.374454	0.385465
	Mean	0.173575	0.223227	0.256072	0.231906	0.257714	0.1974	0.23555	0.277138
	Min	0.116657	0.019721	0.184588	0.204993	0.133367	0.1606	0.018985	0.122664
	STD	0.083731	0.180118	0.108367	0.030267	0.172455	0.032217	0.15415	0.137343
	Ranking	8	6	3	5	2	7	4	1
3	Max	0.417374	0.675072	0.673859	0.580374	0.631389	0.560641	0.64544	0.631938
	Mean	0.353685	0.44114	0.555818	0.511786	0.580212	0.504252	0.51445	0.588044
	Min	0.305838	0.217671	0.478485	0.417965	0.554552	0.438945	0.48554	0.53806
	STD	0.05743	0.22888	0.103854	0.084094	0.04432	0.061337	0.22252	0.047235
	Ranking	8	7	3	5	2	6	4	1
4	Max	0.417374	0.675072	0.631938	0.580374	0.631389	0.560641	0.58887	0.673859
	Mean	0.353685	0.44114	0.588044	0.511786	0.580212	0.504252	0.54414	0.555818
	Min	0.305838	0.217671	0.53806	0.417965	0.554552	0.438945	0.501141	0.478485
	STD	0.05743	0.22888	0.047235	0.084094	0.04432	0.061337	0.08885	0.103854
	Ranking	8	7	1	5	2	6	4	3
5	Max	0.577496	0.500592	0.535451	0.686032	0.685014	0.455519	0.55241	0.606479
	Mean	0.477899	0.461272	0.472833	0.604613	0.470727	0.406442	0.43525	0.483806
	Min	0.401935	0.390055	0.354047	0.545124	0.290476	0.317459	0.40125	0.387526
	STD	0.090135	0.061787	0.102922	0.072969	0.199459	0.077198	0.45452	0.111837
	Ranking	3	6	4	1	5	8	7	2
6	Max	0.716201	0.826943	0.751183	0.802334	0.768344	0.76727	0.59858	0.790498
	Mean	0.634075	0.674239	0.574338	0.641546	0.679048	0.575139	0.56555	0.736605
	Min	0.541354	0.522776	0.279256	0.394027	0.608891	0.388121	0.52555	0.669011
	STD	0.087904	0.152087	0.257223	0.217532	0.081431	0.189626	0.04414	0.061891
	Ranking	5	3	7	4	2	6	8	1
Summation		32	29	18	20	13	33	27	8
Mean Rank		6.4	5.8	3.6	4	2.6	6.6	5.4	1.6
Final Ranking		7	6	3	4	2	8	5	1

The PSNR and SSIM results of Test 2 are given in Tables 4 and 5. The proposed DAOA achieved excellent results in almost all the test cases in terms of PSNR. For threshold 4, the proposed DAOA obtained the best results, and it ranked as the first method, compared to all other comparative methods, followed by AO, MPA, AOA, SSA, DE, PSO, and finally, WOA. For threshold 5, the proposed method obtained promising results, compared to other methods. DAOA obtained the first rank, followed by AO, DE, PSO, WOA, AOA, SSA, and MPA. Overall, we can see that the proposed method obtained the first ranking, followed by AO, DE, SSA, PSO, MPA, WOA, and AOA. The achieved results in this table demonstrate the ability of the proposed DAOA to solve the given problems efficiently.

For threshold 4 in Table 5, the proposed DAOA obtained the best results, and it ranked as the first method, compared to all other comparative methods, followed by AO, AOA, MPA, SSA, DE, PSO, and finally, WOA. For threshold 3, the proposed method obtained promising results, compared to other methods. DAOA obtained the first ranking, followed by WOA, DE, AO, PSO, SSA, MPA, and AOA. Overall, we can see that the proposed DAOA method obtained the first ranking, followed by AO, MPA, SSA, DE, PSO, WOA, and AOA. The obtained results in this table confirm the performance of the proposed DAOA and its ability to solve the given problems efficiently.

Table 4. The PSNR results of the test case 2.

Threshold	Metric	Comparative Methods							
		AO	WOA	SSA	AOA	PSO	MPA	DE	DAOA
2	Max	13.77491	12.63881	14.5297	10.51233	12.64124	13.99214	13.25145	13.27729
	Mean	11.98281	11.5817	12.11305	10.25692	11.54775	13.59189	12.2221	12.59842
	Min	10.73612	10.40495	10.15998	9.823369	10.95149	13.24731	12.01211	12.09539
	STD	1.591121	1.121728	2.221443	0.377446	0.948281	0.375522	0.25212	0.610259
	Ranking	5	6	4	8	7	1	3	2
3	Max	15.86388	16.44113	15.82975	14.47644	14.87275	14.35493	15.32521	16.8866
	Mean	14.94664	15.4703	15.49832	12.8352	13.88411	13.22512	14.14191	14.508
	Min	13.15776	14.64112	14.9623	11.6457	13.24925	11.88844	13.95478	12.02299
	STD	1.549381	0.908328	0.468519	1.468445	0.867645	1.24619	2.25141	2.433551
	Ranking	3	2	1	8	6	7	5	4
4	Max	16.53801	13.20085	17.12718	16.28012	17.27728	16.84103	16.5474	16.96062
	Mean	15.69685	12.29933	15.28839	15.53055	14.89734	15.66251	15.25145	16.84254
	Min	14.14842	11.1013	13.34897	14.77009	12.92993	14.78322	14.25114	16.71792
	STD	1.342654	1.080725	1.891116	0.755074	2.202842	1.06104	0.25496	0.121484
	Ranking	2	8	5	4	7	3	6	1
5	Max	17.94799	16.77612	16.58759	17.72468	17.66078	15.7246	18.25641	20.50293
	Mean	17.3332	16.14486	15.69308	15.81531	16.97316	15.13329	17.14954	17.45356
	Min	16.85601	15.47308	14.74147	12.33775	16.23477	14.29046	16.25415	15.26742
	STD	0.55884	0.652461	0.924385	3.016486	0.714359	0.749426	2.33365	2.722404
	Ranking	2	5	7	6	4	8	3	1
6	Max	18.41641	16.73333	17.63323	17.86417	16.98896	17.69201	17.54845	20.23421
	Mean	18.35135	16.26801	15.56212	16.33655	16.40105	15.7609	16.36652	19.55858
	Min	18.23814	15.46543	13.57085	14.13927	15.2753	12.90929	14.95854	18.29391
	STD	0.098406	0.697999	2.032367	1.950656	0.975254	2.52073	1.36945	1.096094
	Ranking	2	6	8	5	3	7	4	1
Summation		14	27	25	31	27	26	21	9
Mean Rank		2.80	5.40	5.00	6.20	5.40	5.20	4.20	1.80
Final Ranking		2	6	4	8	6	5	3	1

Table 5. The SSIM results of the test case 2.

Threshold	Metric	Comparative Methods							
		AO	WOA	SSA	AOA	PSO	MPA	DE	DAOA
2	Max	0.406268	0.372786	0.488313	0.098776	0.381066	0.416608	0.35652	0.388003
	Mean	0.247171	0.228951	0.26402	0.062589	0.260623	0.406156	0.32541	0.299891
	Min	0.055504	0.119071	0.090368	0.028341	0.156191	0.397245	0.32336	0.153948
	STD	0.177636	0.130221	0.203748	0.035257	0.113289	0.009773	0.45485	0.127294
	Ranking	6	7	4	8	5	1	2	3
3	Max	0.542502	0.591926	0.62208	0.469152	0.556273	0.429674	0.55241	0.577128
	Mean	0.48568	0.515704	0.385801	0.280354	0.407758	0.322274	0.51254	0.543097
	Min	0.382238	0.463366	0.116914	0.133011	0.319945	0.116597	0.46524	0.479899
	STD	0.089729	0.067526	0.254157	0.171862	0.12933	0.17818	0.51425	0.054785
	Ranking	4	2	6	8	5	7	3	1
4	Max	0.643506	0.372327	0.649426	0.569008	0.616709	0.630261	0.53652	0.609209
	Mean	0.540682	0.251443	0.515642	0.536243	0.469759	0.533684	0.51414	0.591323
	Min	0.343984	0.050187	0.373123	0.484871	0.345458	0.483908	0.46585	0.565499
	STD	0.170405	0.175466	0.138359	0.045049	0.137036	0.083651	0.25854	0.022911
	Ranking	2	8	5	3	7	4	6	1
5	Max	0.568273	0.535196	0.563217	0.613346	0.611966	0.810183	0.58475	0.647781
	Mean	0.53032	0.527867	0.529725	0.488084	0.578773	0.65271	0.54541	0.618783
	Min	0.455118	0.523047	0.467038	0.239554	0.556387	0.507866	0.51245	0.604212
	STD	0.065128	0.006451	0.054331	0.215235	0.029323	0.151553	0.25414	0.025113
	Ranking	5	7	6	8	3	1	4	2
6	Max	0.728985	0.589624	0.791703	0.692676	0.55024	0.64645	0.42541	0.666054
	Mean	0.688055	0.543363	0.751029	0.58507	0.505091	0.45092	0.42545	0.490001
	Min	0.658136	0.498046	0.689368	0.4474	0.427197	0.183286	0.40121	0.306101
	STD	0.036686	0.045797	0.054299	0.125371	0.067742	0.239853	0.15424	0.180105
	Ranking	2	4	1	3	5	7	8	6
Summation		19	28	22	30	25	20	23	13
Mean Rank		3.8	5.6	4.4	6	5	4	4.6	2.6
Final Ranking		2	7	4	8	6	3	5	1

The PSNR and SSIM results of Test 3 are given in Tables 6 and 7. The proposed DAOA obtained new, promising results in almost all the test cases in terms of PSNR. For threshold 5, the proposed DAOA obtained the best results, and it ranked as the first method, compared to all other comparative methods, followed by SSA, AO, PSO, AOA, MPA, and finally, DE. For threshold 6, the proposed method obtained promising results compared to other methods. DAOA obtained the first rank, followed by AO, PSO, DE, WOA, AOA, SSA, and MPA. Overall we can see that the proposed method obtained the first ranking, followed by WOA, DE, AO, AOA, MPA, PSO, and SSA. The achieved results in this table demonstrate the ability of the proposed DAOA to solve the given problems efficiently. As well, it is clear the proposed DAOA has this ability at different threshold levels.

For threshold 2 in Table 7, the proposed DAOA obtained the best results, and it ranked as the first method, compared to all other comparative methods, followed by MPA, WOA, DE, AO, SSA, AOA, and finally, PSO. For threshold 5, the proposed method obtained promising results compared to other methods. DAOA obtained the first ranking, followed by PSO, AOA, MPA, WOA, AO, PSO, and DE. Overall, we can see that the proposed DAOA method obtained the first ranking, followed by MPA, AO, SSA, AOA, WOA, PSO, and DE. The obtained results in this table confirm the performance of the proposed DAOA and its ability to solve the given problems efficiently. The following results prove and support that the proposed algorithm's ability to solve such problems is strong and that it is capable of finding robust solutions in this field.

Table 6. The PSNR results of the test case 3.

Threshold	Metric	Comparative Methods							
		AO	WOA	SSA	AOA	PSO	MPA	DE	DAOA
2	Max	16.43851	15.52329	16.17178	13.34026	10.70417	13.80156	11.25454	17.02237
	Mean	11.77337	13.62698	11.4258	10.90943	8.940243	11.97697	12.54562	13.0293
	Min	6.919499	10.52621	8.829951	8.362172	7.32234	10.25363	11.65856	10.1937
	STD	4.762311	2.707536	4.116175	2.491085	1.695636	1.776132	2.66525	3.558437
	Ranking	5	1	6	7	8	4	3	2
3	Max	16.32218	14.37629	19.8809	17.08167	18.01695	16.32078	17.54548	18.7061
	Mean	15.52933	13.1609	18.08332	15.15354	14.40812	12.59758	15.36525	14.62128
	Min	14.29581	11.73748	17.06783	12.61578	11.10578	8.576877	13.52541	11.41963
	STD	1.082679	1.331649	1.561118	2.294509	3.465768	3.880514	3.25414	3.722657
	Ranking	2	7	1	4	6	8	3	5
4	Max	17.09702	17.55333	17.6847	19.73127	19.45281	19.19681	18.56958	20.754
	Mean	16.23917	14.16665	16.1249	18.28651	15.02841	15.17751	16.52565	18.02446
	Min	14.73755	10.43243	13.88539	16.01418	9.874085	12.62804	15.96841	13.89724
	STD	1.30484	3.573146	1.988767	1.991944	4.830901	3.522425	2.59716	3.635777
	Ranking	4	8	5	1	7	6	3	2
5	Max	20.47295	20.08702	20.21561	19.42765	17.8608	20.6646	18.49371	20.41022
	Mean	17.9702	17.76647	18.04407	16.94888	17.74178	16.39606	16.46743	18.58232
	Min	15.06634	14.62097	15.09986	14.78632	17.57167	13.9744	15.45547	16.23457
	STD	2.725534	2.824863	2.643952	2.336768	0.151186	3.707817	2.65478	2.135811
	Ranking	3	4	2	6	5	8	7	1
6	Max	21.16341	21.39185	21.70955	20.93593	23.09058	19.19694	20.12154	21.98094
	Mean	19.12374	19.85058	16.84504	17.56936	16.8876	16.95177	18.15414	20.41186
	Min	16.5207	18.82027	12.89516	13.91389	12.51858	15.33711	16.36987	19.1292
	STD	2.372071	1.359806	4.477811	3.519921	5.519455	2.005675	1.64856	1.447282
	Ranking	3	2	8	5	7	6	4	1
Summation		17	22	22	23	33	32	20	11
Mean Rank		3.40	4.40	4.40	4.60	6.60	6.40	4.00	2.20
Final Ranking		2	4	4	6	8	7	3	1

Table 7. The SSIM results of the test case 3.

Threshold	Metric	Comparative Methods							
		AO	WOA	SSA	AOA	PSO	MPA	DE	DAOA
2	Max	0.810797	0.784223	0.737701	0.807984	0.729805	0.784964	0.74548	0.810173
	Mean	0.701346	0.726586	0.679275	0.652562	0.652503	0.748363	0.71254	0.777743
	Min	0.51097	0.614391	0.642633	0.512459	0.544859	0.707409	0.69584	0.738525
	STD	0.165486	0.097176	0.051142	0.148357	0.096133	0.03896	0.02514	0.036303
	Ranking	5	3	6	7	8	2	4	1
3	Max	0.858715	0.84955	0.879072	0.869936	0.846474	0.810939	0.801454	0.862403
	Mean	0.829113	0.82117	0.846587	0.810768	0.803586	0.714867	0.74125	0.826664
	Min	0.794989	0.799088	0.811991	0.7567	0.72253	0.656383	0.70215	0.798218
	STD	0.032103	0.025814	0.033591	0.05679	0.070237	0.083855	0.02193	0.032708
	Ranking	2	4	1	5	6	8	7	3
4	Max	0.835799	0.786265	0.889158	0.89639	0.831634	0.877477	0.81256	0.842558
	Mean	0.816523	0.765247	0.835175	0.863757	0.802387	0.856942	0.76585	0.820005
	Min	0.780192	0.748503	0.782824	0.807497	0.751995	0.824078	0.71369	0.776304
	STD	0.031483	0.019241	0.053186	0.04893	0.043828	0.028755	0.21454	0.037853
	Ranking	5	8	3	1	6	2	7	4
5	Max	0.869899	0.862851	0.864441	0.895398	0.889066	0.86625	0.85645	0.901137
	Mean	0.846835	0.858303	0.821075	0.863123	0.872182	0.862527	0.81021	0.874055
	Min	0.830022	0.850494	0.788281	0.836486	0.854754	0.856039	0.75645	0.837004
	STD	0.02066	0.006793	0.039165	0.029858	0.017162	0.005639	0.021114	0.033208
	Ranking	6	5	7	3	2	4	8	1
6	Max	0.897095	0.882473	0.910729	0.920257	0.898452	0.884757	0.84145	0.890175
	Mean	0.893914	0.868849	0.869001	0.850437	0.825776	0.87171	0.79568	0.874336
	Min	0.892161	0.848929	0.845306	0.787637	0.758887	0.848342	0.76582	0.855048
	STD	0.002759	0.017636	0.036248	0.066588	0.069962	0.020283	0.029447	0.017816
	Ranking	1	5	4	6	7	3	8	2
Summation		19	25	21	22	29	19	34	11
Mean Rank		3.8	5	4.2	4.4	5.8	3.8	6.8	2.2
Final Ranking		2	6	4	5	7	2	8	1

The PSNR and SSIM results of Test 4 are given in Tables 8 and 9. The proposed DAOA obtained new promising results in almost all the test cases in terms of PSNR, as shown in Table 8. The proposed DAOA obtained the best results for two threshold values (i.e., 3 and 4 levels). For threshold 3, the proposed DAOA obtained the best results, and it ranked as the first method, compared to all other comparative methods, followed by PSO, SSA, WOA, AOA, MPA, DE, and finally, AO. For threshold 4, the proposed method obtained promising results, compared to other methods. DAOA obtained the first rank, followed by WOA, SSA, AOA, MPA, AO, DE, and AOA. Overall, we can see that the proposed method obtained the first ranking, followed by WOA, PSO, SSA, MPA, AO, AOA, and DE. The achieved results in this table demonstrate the ability of the proposed DAOA to solve the given problems efficiently. As well, it is clear the proposed DAOA has this ability at different threshold levels.

Table 9 shows that the proposed DAOA method obtained better results in almost all the test cases in terms of SSIM for Test 4. The proposed DAOA obtained the best results for two threshold values (i.e., 2 and 3 levels). For threshold 2, the proposed DAOA obtained the best results, and it ranked as the first method, compared to all other comparative methods, followed by DE, AOA, WOA, MPA, PSO, SSA, and finally, AOA. For threshold 3, the proposed method obtained promising results, compared to other methods. DAOA obtained the first ranking, followed by AOA, AO, PSO, WOA, MPA, SSA, and DE. Overall, we can see that the proposed DAOA method obtained the first ranking, followed by PSO, AO, WOA, AOA, MPA, SSA, and DE. The obtained results in this table confirm the performance of the proposed DAOA to solve the given problems efficiently. The following results prove and support that the proposed algorithm's ability to solve such problems is strong and that it is capable of finding robust solutions in this field.

Table 8. The PSNR results of the test case 4.

Threshold	Metric	Comparative Methods							
		AO	WOA	SSA	AOA	PSO	MPA	DE	DAOA
2	Max	0.810797	0.784223	0.737701	0.807984	0.729805	0.784964	0.74548	0.810173
	Mean	0.701346	0.726586	0.679275	0.652562	0.652503	0.748363	0.71254	0.777743
	Min	0.51097	0.614391	0.642633	0.512459	0.544859	0.707409	0.69584	0.738525
	STD	0.165486	0.097176	0.051142	0.148357	0.096133	0.03896	0.02514	0.036303
	Ranking	5	3	6	7	8	2	4	1
3	Max	0.858715	0.84955	0.879072	0.869936	0.846474	0.810939	0.801454	0.862403
	Mean	0.829113	0.82117	0.846587	0.810768	0.803586	0.714867	0.74125	0.826664
	Min	0.794989	0.799088	0.811991	0.7567	0.72253	0.656383	0.70215	0.798218
	STD	0.032103	0.025814	0.033591	0.05679	0.070237	0.083855	0.02193	0.032708
	Ranking	2	4	1	5	6	8	7	3
4	Max	0.835799	0.786265	0.889158	0.89639	0.831634	0.877477	0.81256	0.842558
	Mean	0.816523	0.765247	0.835175	0.863757	0.802387	0.856942	0.76585	0.820005
	Min	0.780192	0.748503	0.782824	0.807497	0.751995	0.824078	0.71369	0.776304
	STD	0.031483	0.019241	0.053186	0.04893	0.043828	0.028755	0.21454	0.037853
	Ranking	5	8	3	1	6	2	7	4
5	Max	0.869899	0.862851	0.864441	0.895398	0.889066	0.86625	0.85645	0.901137
	Mean	0.846835	0.858303	0.821075	0.863123	0.872182	0.862527	0.81021	0.874055
	Min	0.830022	0.850494	0.788281	0.836486	0.854754	0.856039	0.75645	0.837004
	STD	0.02066	0.006793	0.039165	0.029858	0.017162	0.005639	0.021114	0.033208
	Ranking	6	5	7	3	2	4	8	1
6	Max	0.897095	0.882473	0.910729	0.920257	0.898452	0.884757	0.84145	0.890175
	Mean	0.893914	0.868849	0.869001	0.850437	0.825776	0.87171	0.79568	0.874336
	Min	0.892161	0.848929	0.845306	0.787637	0.758887	0.848342	0.76582	0.855048
	STD	0.002759	0.017636	0.036248	0.066588	0.069962	0.020283	0.029447	0.017816
	Ranking	1	5	4	6	7	3	8	2
Summation		19	25	21	22	29	19	34	11
Mean Rank		3.8	5	4.2	4.4	5.8	3.8	6.8	2.2
Final Ranking		2	6	4	5	7	2	8	1

Table 9. The SSIM results of the test case 4.

Threshold	Metric	Comparative Methods							
		AO	WOA	SSA	AOA	PSO	MPA	DE	DAOA
2	Max	0.513176	0.446739	0.469347	0.303735	0.468033	0.472675	0.465855	0.493285
	Mean	0.420537	0.411645	0.379839	0.268241	0.395591	0.407972	0.432165	0.450537
	Min	0.297393	0.34853	0.312674	0.238732	0.353913	0.285825	0.415441	0.390067
	STD	0.111078	0.054773	0.080691	0.032912	0.062974	0.105845	0.065135	0.053842
	Ranking	3	4	7	8	6	5	2	1
3	Max	0.639969	0.518822	0.570567	0.642874	0.550295	0.516935	0.541685	0.638931
	Mean	0.539264	0.496794	0.476032	0.559045	0.522348	0.480253	0.451684	0.565094
	Min	0.483693	0.453509	0.406394	0.442666	0.507116	0.421783	0.401513	0.427189
	STD	0.087369	0.037488	0.084871	0.103997	0.024235	0.051181	0.165152	0.119529
	Ranking	3	5	7	2	4	6	8	1
4	Max	0.635584	0.520955	0.647437	0.629088	0.583309	0.567589	0.545438	0.59171
	Mean	0.558847	0.495081	0.624364	0.52697	0.542197	0.540923	0.484153	0.566602
	Min	0.496035	0.475477	0.612318	0.455541	0.49695	0.494102	0.351535	0.530519
	STD	0.070809	0.023378	0.019989	0.090753	0.043328	0.040678	0.91351	0.032038
	Ranking	3	7	1	6	4	5	8	2
5	Max	0.625459	0.722689	0.626088	0.678775	0.695846	0.682412	0.646849	0.752727
	Mean	0.574159	0.65505	0.608905	0.623735	0.681637	0.593452	0.568435	0.654905
	Min	0.544652	0.576303	0.589048	0.563386	0.661464	0.468945	0.515464	0.571449
	STD	0.044594	0.073823	0.018664	0.057878	0.01795	0.111084	0.51354	0.091489
	Ranking	7	2	5	4	1	6	8	3
6	Max	0.678699	0.761036	0.655062	0.693875	0.74922	0.721603	0.711543	0.765842
	Mean	0.655502	0.652531	0.558392	0.577277	0.735573	0.627711	0.658435	0.721667
	Min	0.613377	0.518519	0.460651	0.501179	0.727322	0.503922	0.615534	0.656329
	STD	0.036543	0.123255	0.09721	0.102534	0.011905	0.111878	0.153112	0.057742
	Ranking	4	5	8	7	1	6	3	2
Summation		20	23	28	27	16	28	29	9
Mean Rank		4	4.6	5.6	5.4	3.2	5.6	5.8	1.8
Final Ranking		3	4	6	5	2	6	8	1

In Tables 10 and 11, the PSNR and SSIM results of Test 5 are shown. As shown in Table 10, the proposed DAOA yielded new promising PSNR results in almost all test cases. For two threshold values, the proposed DAOA gave the best results (i.e., 2 and 5 levels). For threshold 2, the proposed DAOA produced the best results, placing it first among all other comparative methods, ahead of SSA, AOA, MPA, DE, AO, WOA, and PSO. In addition, when compared to other methods, the proposed method produced positive results for threshold 5. DAOA came first, followed by AOA, MPA, AO, WOA, PSO, and DE. Overall, we can see that DAOA came first, followed by SSA, DE, AO, AOA, PSO, MPA, and WOA. The obtained results in this table demonstrate the proposed DAOA's ability to solve the given problems efficiently. Furthermore, it is evident that the proposed DAOA has the potential to operate at various threshold levels.

In terms of SSIM for Test 5, Table 11 shows that the proposed DAOA system obtained better results in almost all test cases. For two threshold values, the proposed DAOA produced the best results (i.e., 2 and 5 levels). For threshold 2, the proposed DAOA received the best results, placing it first among SSA, AO, MPA, AOA, DE, PSO, and WOA. In addition, when compared to other methods, the proposed method produced positive results for threshold 5. The first-place winner was DAOA, followed by AO, WOA, SSA, AOA, DE, MPA, and PSO. Overall, we can see that the proposed DAOA method obtained the first ranking, followed by AO, SSA, WOA, PSO, MPA, DE, and AOA. The obtained results in this table confirm the performance of the proposed DAOA and its ability to solve the given problems efficiently. The following results prove and support that the proposed algorithm's ability to solve such problems is strong and that it is capable of finding robust solutions in this field.

Table 10. The PSNR results of the test case 5.

Threshold	Metric	Comparative Methods							
		AO	WOA	SSA	AOA	PSO	MPA	DE	DAOA
2	Max	15.93985	14.4854	15.99861	17.14705	15.0977	16.05831	16.55749	17.53916
	Mean	14.10613	13.00072	15.294	14.79149	12.72877	14.46087	14.35989	15.82038
	Min	12.39387	11.27298	14.72307	11.8401	11.51447	13.62663	11.9752	13.42614
	STD	1.776107	1.619943	0.648191	2.703178	2.051771	1.383874	2.296871	2.138085
	Ranking	6	7	2	3	8	4	5	1
3	Max	19.40981	18.86115	17.48266	16.55749	17.95279	17.69139	19.44663	19.93627
	Mean	18.72279	17.28213	16.52113	14.35989	16.27857	17.1941	19.26877	18.13099
	Min	18.29158	15.95137	15.92209	11.9752	14.07194	16.51551	19.01602	15.35808
	STD	0.60141	1.470694	0.841066	2.296871	1.994456	0.608547	0.224857	2.437657
	Ranking	2	4	6	8	7	5	1	3
4	Max	18.13386	19.1223	18.90861	17.45745	19.1043	18.23407	18.83735	18.38066
	Mean	16.94739	16.40329	17.94929	15.50893	18.1176	16.67171	17.28304	17.01377
	Min	15.98498	14.81291	16.68311	13.37735	16.4206	14.93103	15.51976	15.35199
	STD	1.091819	2.366025	1.14404	2.046204	1.476127	1.658724	1.494198	1.535719
	Ranking	5	7	2	8	1	6	3	4
5	Max	20.21398	20.02404	21.27214	20.18245	18.91496	20.6886	20.25546	21.56663
	Mean	18.99103	18.67004	20.53727	20.04683	18.26845	19.69356	18.05347	20.9167
	Min	17.36811	17.86499	19.6089	19.85296	17.79804	18.97903	17.51354	20.14592
	STD	1.464489	1.179571	0.848332	0.172297	0.578907	0.888634	0.15434	0.718025
	Ranking	5	6	2	3	7	4	8	1
6	Max	21.09476	21.57137	21.3434	22.18733	23.89824	19.44663	21.54999	22.99988
	Mean	20.41811	20.42529	20.47827	21.60813	22.10233	19.26877	20.25987	21.55413
	Min	19.49805	19.07294	19.90813	21.16347	21.02315	19.01602	19.64856	19.98576
	STD	0.825716	1.261926	0.761747	0.525019	1.565831	0.224857	0.16655	1.510799
	Ranking	6	5	4	2	1	8	7	3
Summation		24	29	16	24	24	27	24	12
Mean Rank		4.80	5.80	3.20	4.80	4.80	5.40	4.80	2.40
Final Ranking		3	8	2	3	3	7	3	1

Table 11. The SSIM results of the test case 5.

Threshold	Metric	Comparative Methods							
		AO	WOA	SSA	AOA	PSO	MPA	DE	DAOA
2	Max	0.670559	0.663185	0.670191	0.656707	0.638999	0.641328	0.625454	0.652777
	Mean	0.615808	0.577209	0.627177	0.602737	0.588212	0.607857	0.58944	0.638984
	Min	0.554493	0.497657	0.570064	0.537508	0.547586	0.575353	0.523565	0.622532
	STD	0.058311	0.082951	0.051531	0.060392	0.046546	0.032998	0.051351	0.015297
	Ranking	3	8	2	5	7	4	6	1
3	Max	0.728918	0.677149	0.714839	0.663648	0.667938	0.717054	0.646841	0.704409
	Mean	0.719125	0.660935	0.666536	0.59634	0.629253	0.664745	0.551844	0.671047
	Min	0.712656	0.629315	0.6219	0.510537	0.576046	0.622826	0.493545	0.639151
	STD	0.008626	0.027387	0.046578	0.078213	0.047636	0.047965	0.050315	0.032654
	Ranking	1	5	3	7	6	4	8	2
4	Max	0.748435	0.70822	0.687422	0.729796	0.706817	0.729298	0.715434	0.720751
	Mean	0.676713	0.668007	0.675521	0.661629	0.681985	0.674847	0.698434	0.678177
	Min	0.633531	0.627958	0.660419	0.590072	0.636336	0.601496	0.651354	0.614625
	STD	0.062543	0.040131	0.013783	0.069924	0.039584	0.065964	0.05134	0.056087
	Ranking	4	7	5	8	2	6	1	3
5	Max	0.760316	0.744595	0.760228	0.722269	0.701088	0.713294	0.715469	0.754593
	Mean	0.735256	0.721088	0.707533	0.706524	0.686357	0.694403	0.694685	0.740905
	Min	0.719429	0.693768	0.680723	0.680577	0.665994	0.665244	0.645135	0.720782
	STD	0.021952	0.025627	0.045637	0.022641	0.018212	0.025618	0.100351	0.0178
	Ranking	2	3	4	5	8	7	6	1
6	Max	0.778606	0.802338	0.787056	0.759268	0.775117	0.754872	0.714354	0.759996
	Mean	0.745632	0.767851	0.728188	0.743422	0.760561	0.715035	0.69456	0.757846
	Min	0.717157	0.728973	0.685219	0.730438	0.737718	0.682379	0.646758	0.755781
	STD	0.030971	0.036879	0.052747	0.014627	0.02003	0.036776	0.14353	0.002109
	Ranking	4	1	6	5	2	7	8	3
Summation		14	24	20	30	25	28	29	10
Mean Rank		2.8	4.8	4	6	5	5.6	5.8	2
Final Ranking		2	4	3	8	5	6	7	1

In Tables 12 and 13, the PSNR and SSIM results of Test 6 are shown. As shown in Table 12, the proposed DAOA yielded new promising PSNR results in nearly all test cases. For two threshold values, the proposed DAOA gave the best results (i.e., 5 and 6 levels). For threshold 5, the DE produced the best results, placing it first among all other comparative approaches, ahead of DAOA, SSA, PSO, AOA, WOA, MPA, and AO. In addition, when compared to other methods, the proposed method produced positive results for threshold 6. DAOAO came first, followed by SSA, WOA, AOA, MPA, DE, AO, and PSO. Overall, we can see that DAOA came first, followed by AOA, DE, PSO, SSA, WOA, AO, and MPA. The obtained results in this table demonstrate the proposed DAOA's ability to solve the given problems efficiently. Furthermore, it is evident that the proposed DAOA has the potential to operate at various threshold levels.

In terms of SSIM for Test 6, Table 13 shows that the proposed DAOA method obtained better results in almost all test cases. For one threshold value, the proposed DAOA produced the best results (i.e., three levels). For threshold 3, the proposed DAOA received the best results, placing it first, followed by AOA, AO, PSO, MPA, SSA, WOA, and DE. Overall, we can see that the proposed DAOA method obtained the first ranking, followed by AOA, SSA, MPA, AO, WOA, PSO, and DE. The obtained results in this table confirm the performance of the proposed DAOA to solve the given problems efficiently. The following results prove and support that the proposed algorithm's ability to solve such problems is strong and that it is capable of finding robust solutions in this field.

Table 12. The PSNR results of the test case 6.

Threshold	Metric	Comparative Methods							
		AO	WOA	SSA	AOA	PSO	MPA	DE	DAOA
2	Max	14.29003	14.99666	13.23076	14.2586	13.67276	13.88699	13.25669	13.62076
	Mean	12.13124	13.5679	12.00795	13.53452	12.69334	12.56354	12.76658	12.88169
	Min	11.02309	10.86625	10.71159	13.00777	11.90018	10.75974	11.81577	11.91395
	STD	1.86979	2.340995	1.261197	0.648342	0.900847	1.618015	0.727573	0.876084
	Ranking	7	1	8	2	5	6	4	3
3	Max	17.13609	13.26265	15.86297	16.70329	16.39937	16.77276	15.56435	16.70329
	Mean	14.80299	12.70421	14.27818	14.73781	15.54628	15.68245	14.56435	14.73781
	Min	11.88525	11.81721	13.19349	13.30319	14.95156	13.81152	13.54531	13.30319
	STD	2.673792	0.776716	1.40325	1.761111	0.757696	1.627652	0.35531	1.761111
	Ranking	3	8	7	4	2	1	6	4
4	Max	19.72094	16.65405	18.53668	18.36352	18.18219	16.8171	16.16153	17.29736
	Mean	17.42215	15.6372	16.4182	17.40203	16.17031	14.66404	15.61533	16.81609
	Min	14.47883	14.30732	14.08529	16.52227	14.97604	13.31745	14.65844	15.89593
	STD	2.679828	1.204274	2.233427	0.923342	1.752483	1.884058	0.513153	0.797164
	Ranking	1	6	4	2	5	8	7	3
5	Max	19.95971	18.41277	19.46272	20.77475	19.71927	18.09801	20.15615	21.7643
	Mean	16.21653	17.16753	18.12133	17.29533	17.30211	16.23449	19.56652	18.92589
	Min	12.15348	16.07533	17.44474	14.33996	15.60841	15.08771	18.91434	16.95589
	STD	3.912932	1.176211	1.161691	3.249246	2.148801	1.628114	0.44345	2.519088
	Ranking	8	6	3	5	4	7	1	2
6	Max	19.70338	18.99666	19.15752	19.97796	17.4372	20.54731	20.48618	21.16495
	Mean	17.55594	18.51687	18.95646	18.38941	16.73036	17.68373	17.67164	19.65434
	Min	14.92323	17.6852	18.62038	17.45989	16.33641	15.14494	14.56169	17.72153
	STD	2.426738	0.723084	0.292923	1.382353	0.613485	2.715789	3.51355	1.760107
	Ranking	7	3	2	4	8	5	6	1
Summation		26	24	24	17	24	27	24	13
Mean Rank		5.20	4.80	4.80	3.40	4.80	5.40	4.80	2.60
Final Ranking		7	3	3	2	3	8	3	1

Table 13. The SSIM results of the test case 6.

Threshold	Metric	Comparative Methods							
		AO	WOA	SSA	AOA	PSO	MPA	DE	DAOA
2	Max	0.585171	0.575848	0.507334	0.584204	0.554758	0.583274	0.545458	0.566448
	Mean	0.529789	0.511152	0.47466	0.558141	0.510173	0.517028	0.53251	0.551596
	Min	0.451429	0.421518	0.440468	0.528648	0.468501	0.418339	0.510512	0.540869
	STD	0.069769	0.08013	0.033459	0.027936	0.043202	0.087123	0.051651	0.013279
	Ranking	4	6	8	1	7	5	3	2
3	Max	0.633602	0.566445	0.593976	0.643107	0.624665	0.587316	0.584547	0.620738
	Mean	0.573218	0.53021	0.539426	0.586845	0.561226	0.554322	0.52548	0.6091
	Min	0.531603	0.476157	0.445784	0.557298	0.51424	0.500738	0.49522	0.590855
	STD	0.053527	0.047709	0.081464	0.048745	0.057022	0.046817	0.15479	0.015999
	Ranking	3	7	6	2	4	5	8	1
4	Max	0.625898	0.599759	0.688634	0.647084	0.588007	0.619283	0.60147	0.651858
	Mean	0.620455	0.584845	0.631581	0.584162	0.561247	0.577285	0.564549	0.631427
	Min	0.616739	0.561825	0.576099	0.541342	0.520971	0.552471	0.514625	0.594978
	STD	0.004818	0.020224	0.056284	0.055664	0.035503	0.036571	0.51556	0.031643
	Ranking	3	4	1	5	8	6	7	2
5	Max	0.639309	0.611599	0.670374	0.706368	0.662426	0.621919	0.61444	0.688438
	Mean	0.580397	0.593407	0.626018	0.659053	0.615559	0.601187	0.53255	0.657403
	Min	0.481486	0.57193	0.58953	0.596054	0.552583	0.585043	0.50144	0.634943
	STD	0.086179	0.020038	0.040992	0.056805	0.056666	0.018861	0.254516	0.027759
	Ranking	7	6	3	1	4	5	8	2
6	Max	0.732761	0.721116	0.702031	0.639447	0.669456	0.741237	0.62156	0.670238
	Mean	0.617062	0.675843	0.66037	0.614623	0.623098	0.698053	0.60156	0.621129
	Min	0.500442	0.628145	0.615404	0.582216	0.593419	0.632455	0.581685	0.555772
	STD	0.116162	0.046533	0.043408	0.029359	0.04067	0.057751	0.051617	0.058938
	Ranking	6	2	3	7	4	1	8	5
Summation		23	25	21	16	27	22	34	12
Mean Rank		4.6	5	4.2	3.2	5.4	4.4	6.8	2.4
Final Ranking		5	6	3	2	7	4	8	1

The PSNR and SSIM results of Test 7 are given in Tables 14 and 15. The proposed DAOA obtained new promising results in almost all the test cases in terms of PSNR, as shown in Table 14. The proposed DAOA obtained the best results for three threshold values (i.e., 3, 4, and 6 levels). For threshold 3, the proposed DAOA obtained the best results, and it ranked as the first method, compared to all other comparative methods, followed by PSO, WOA, MPA, PSO, SSA, AOA, and finally, DE. For threshold 6, the proposed method obtained promising results, compared to other methods. DAOA obtained the first rank, followed by MPA, AO, AOA, SSA, PSO, DE, and WOA. Overall, we can see that the proposed method obtained the first ranking, followed by MPA, AO, AOA, SSA, PSO, WOA, and DE. The achieved results in this table demonstrate the ability of the proposed DAOA to solve the given problems efficiently. Furthermore, it is obvious that the proposed DAOA has the potential to operate at various threshold levels.

Table 15 shows that the proposed DAOA method obtained better results in almost all the test cases in terms of SSIM for Test 7. The proposed DAOA obtained the best results for two threshold values (i.e., 2 and 4 levels). For threshold 2, the proposed DAOA obtained the best results, and it ranked as the first method, compared to all other comparative methods, followed by MPA, PSO, AOA, AO, WOA, SSA, and finally, DE. For threshold 4, the proposed method obtained promising results compared to other methods. DAOA obtained the first ranking, followed by PSO, SSA, AO, AOA, WOA, DE, and MPA. Overall, we can see that the proposed DAOA method obtained the first ranking, followed by PSO, AOA, AO, MPA, WOA, SSA, and DE. The obtained results in this table confirm the performance of the proposed DAOA to solve the given problems efficiently. The presented results demonstrate and declare the proposed algorithm's ability to solve such problems and find reliable solutions in this area.

Table 14. The PSNR results of the test case 7.

Threshold	Metric	Comparative Methods							
		AO	WOA	SSA	AOA	PSO	MPA	DE	DAOA
2	Max	13.20515	14.81541	13.46643	15.49531	15.62603	15.20099	10.56165	11.17017
	Mean	12.41868	11.78621	13.27459	12.55658	13.52688	13.84887	8.68468	9.014158
	Min	11.32961	9.315363	13.00626	9.442963	9.675792	12.68271	7.51654	6.574355
	STD	0.973697	2.792211	0.239434	3.029961	3.339657	1.269397	2.26558	2.311011
	Ranking	5	6	3	4	2	1	8	7
3	Max	16.56695	17.57886	16.2908	16.58067	16.53626	15.88719	16.98971	18.01761
	Mean	15.69994	15.46688	14.50615	14.19241	14.84154	15.17141	14.15556	16.2308
	Min	15.1533	14.25297	13.45161	10.44799	12.24822	14.03169	10.44162	13.04052
	STD	0.759328	1.835835	1.554045	3.283571	2.280885	0.997755	3.255033	2.769505
	Ranking	2	3	6	7	5	4	8	1
4	Max	16.90963	17.97752	17.78105	17.43689	16.94918	19.34317	17.45543	19.60255
	Mean	15.08816	14.69292	16.45214	14.2211	12.8884	16.99385	14.22669	17.64468
	Min	11.44578	11.37666	14.68077	11.04897	8.255611	13.29746	11.4265	16.42001
	STD	3.154397	3.300544	1.596797	3.194184	4.374921	3.240157	3.143737	1.71328
	Ranking	4	5	3	7	8	2	6	1
5	Max	19.95651	19.11383	17.60007	20.11243	19.91776	20.2333	19.11482	20.05669
	Mean	17.89666	15.96258	16.58689	18.8553	17.88667	18.66736	15.96018	18.17262
	Min	16.75255	10.41909	15.4047	16.91343	15.05058	17.34947	10.42944	16.15425
	STD	1.78753	4.815784	1.1074	1.705889	2.531472	1.457831	4.39728	1.954685
	Ranking	4	7	6	1	5	2	8	3
6	Max	21.4354	20.29606	21.05977	22.61835	20.03635	21.47684	20.24338	21.70695
	Mean	19.4399	17.25886	19.58102	19.89625	19.37838	18.84269	17.35445	20.29721
	Min	17.7851	14.28299	17.61442	17.48669	18.99359	15.24118	14.23249	19.17387
	STD	1.848842	3.007007	1.773725	2.580068	0.572543	3.22842	3.234234	1.290597
	Ranking	4	8	3	2	5	6	7	1
Summation		19	29	21	21	25	15	37	13
Mean Rank		3.80	5.80	4.20	4.20	5.00	3.00	7.40	2.60
Final Ranking		3	7	4	4	6	2	8	1

Table 15. The SSIM results of the test case 7.

Threshold	Metric	Comparative Methods							
		AO	WOA	SSA	AOA	PSO	MPA	DE	DAOA
2	Max	0.73278	0.712288	0.560122	0.72505	0.733305	0.723576	0.564865	0.730313
	Mean	0.643606	0.617796	0.485729	0.645712	0.676013	0.683473	0.448642	0.710998
	Min	0.567798	0.478926	0.34029	0.519034	0.567055	0.639678	0.348649	0.691961
	STD	0.0833	0.122848	0.125965	0.110867	0.094402	0.042071	0.184476	0.019178
	Ranking	5	6	7	4	3	2	8	1
3	Max	0.773342	0.738051	0.693211	0.729915	0.747102	0.706363	0.694864	0.760805
	Mean	0.770604	0.721144	0.67817	0.65064	0.677895	0.700246	0.676463	0.725786
	Min	0.768211	0.698601	0.665159	0.495505	0.593947	0.693075	0.666456	0.672091
	STD	0.002582	0.02032	0.014136	0.134362	0.077634	0.006707	0.017743	0.047213
	Ranking	1	3	5	8	6	4	7	2
4	Max	0.742182	0.74847	0.727827	0.732508	0.782284	0.746116	0.728807	0.790683
	Mean	0.687277	0.664643	0.701216	0.669205	0.739291	0.579421	0.652713	0.763686
	Min	0.610196	0.549827	0.659058	0.632616	0.701494	0.394956	0.62996	0.719773
	STD	0.068731	0.102883	0.036927	0.055044	0.040645	0.176253	0.050762	0.038362
	Ranking	4	6	3	5	2	8	7	1
5	Max	0.79007	0.805645	0.822652	0.819285	0.748588	0.740407	0.791501	0.798539
	Mean	0.76336	0.699754	0.772285	0.780482	0.714524	0.728473	0.774441	0.775382
	Min	0.733348	0.508825	0.736218	0.751391	0.65435	0.707817	0.732424	0.760992
	STD	0.028505	0.165678	0.044956	0.034973	0.052264	0.01796	0.036023	0.020251
	Ranking	5	8	4	1	7	6	3	2
6	Max	0.767482	0.807074	0.769107	0.784627	0.803515	0.797439	0.762609	0.800054
	Mean	0.748283	0.77661	0.748259	0.768866	0.791177	0.766987	0.745271	0.762369
	Min	0.713817	0.723776	0.710988	0.751543	0.768949	0.742903	0.718795	0.741915
	STD	0.029913	0.045934	0.032353	0.016597	0.019288	0.02782	0.030372	0.032675
	Ranking	6	2	7	3	1	4	8	5
Summation		21	25	26	21	19	24	33	11
Mean Rank		4.2	5	5.2	4.2	3.8	4.8	6.6	2.2
Final Ranking		3	6	7	3	2	5	8	1

The PSNR and SSIM results of Test 8 are given in Tables 16 and 17. The proposed DAOA obtained got new promising results in almost all the test cases in terms of PSNR, as shown in Table 16. The proposed DAOA obtained the best results for four threshold values (i.e., 2, 3, 4, and 5 levels). For threshold 2, the proposed DAOA obtained the best results, and it ranked as the first method, compared to all other comparative methods, followed by DE, AOA, PSO, SSA, AO, MPA, and finally, WOA. For threshold 5, the proposed method obtained promising results, compared to other methods. DAOA obtained the first rank, followed by AO, DE, PSO, WOA, AOA, MPA, and SSA. Overall, we can see that the proposed method obtained the first ranking, followed by AO, AOA, PSO, DE, WOA, MPA, and SSA. The achieved results in this table demonstrate the ability of the proposed DAOA to solve the given problems efficiently. Furthermore, it is obvious that the proposed DAOA has the potential to operate at various threshold levels.

Table 17 shows that the proposed DAOA method obtained better results in almost all the test cases in terms of SSIM for Test 8. The proposed DAOA obtained the best results for two threshold values (i.e., 3 and 5 levels). For threshold 3, the proposed DAOA obtained the best results, and it ranked as the first method, compared to all other comparative methods, followed by AOA, SSA, DE, WOA, AO, PSO, and finally, MPA. For threshold 5, the proposed method obtained promising results compared to other methods. DAOA obtained the first ranking, followed by AO, DE, SSA, PSO, AOA, WOA, and MPA. Overall, we can see that the proposed DAOA method obtained the first ranking, followed by AOA, AO, WOA, SSA, PSO, DE, and MPA. The obtained results in this table confirm the performance of the proposed DAOA and its ability to solve the given problems efficiently. The presented results demonstrate and declare the proposed algorithm's ability to solve such problems and find reliable solutions in this area. We added a Wilcoxon sign test to show the significant improvement for test case number 8 as shown in Tables 16 and 17. It is clear that the proposed method is more effective and better than the other methods.

Table 16. The PSNR results of the test case 8.

Threshold	Metric	Comparative Methods							
		AO	WOA	SSA	AOA	PSO	MPA	DE	DAOA
2	Max	13.33503	12.70418	13.72589	13.87541	15.0179	13.22514	13.93463	14.2634
	Mean	11.87001	11.35382	11.90455	12.30024	12.0788	11.78679	11.15836	13.26591
	Min	9.627695	9.504035	10.85728	11.2966	9.738495	9.930564	10.85364	11.58005
	STD	1.972111	1.657497	1.583232	1.381142	2.690154	1.686571	1.547691	1.468158
	Ranking	5	7	4	2	3	6	8	1
3	Max	17.28625	16.57081	13.49278	14.92705	17.54352	15.13415	16.09049	16.35347
	Mean	14.74539	15.01544	12.50533	14.48463	14.21932	14.43299	15.04035	15.21107
	Min	9.822409	12.05905	11.30417	14.18863	12.3542	13.62575	12.05713	13.80324
	STD	4.264167	2.561471	1.109849	0.390375	2.885987	0.759773	2.554852	1.295669
	Ranking	4	3	8	5	7	6	2	1
4	Max	17.00622	17.0363	16.18631	17.26998	15.66957	17.89161	15.60819	18.921
	Mean	14.98358	12.94104	14.49141	15.41649	13.31057	14.87131	13.39887	15.69837
	Min	13.05113	8.268197	13.05194	13.97806	8.738069	11.04877	8.879347	13.14273
	STD	1.979087	4.412494	1.582717	1.684754	3.96057	3.491248	3.703273	2.946311
	Ranking	3	8	5	2	7	4	6	1
5	Max	19.33652	17.74905	19.68408	19.66732	20.9309	17.50957	17.83407	19.59432
	Mean	17.51972	16.46752	14.99745	15.68612	17.03271	15.38835	17.15497	17.61533
	Min	16.56115	14.43826	9.169088	8.179696	14.13458	12.63953	15.98558	15.74747
	STD	1.574189	1.777538	5.349667	6.504749	3.506793	2.494939	1.29808	1.925829
	Ranking	2	5	8	6	4	7	3	1
6	Max	22.35209	17.96208	20.45875	19.73508	19.56292	19.22911	19.91049	21.32792
	Mean	19.18047	17.28236	15.36816	17.47522	17.49288	16.85966	15.75105	18.93027
	Min	14.9198	15.96895	12.60476	14.82897	13.72243	12.6153	8.963192	16.39383
	STD	3.833964	1.137678	4.413974	2.475771	3.270534	3.684026	6.585767	2.46997
	Ranking	1	5	8	4	3	6	7	2
Summation		15	28	33	19	24	29	26	6
Mean Rank		3.00	5.60	6.60	3.80	4.80	5.80	5.20	1.20
Final Ranking		2	6	8	3	4	7	5	1
***p*-value**		2.254×10^{-3}	3.455×10^{-2}	4.254×10^{-2}	2.368×10^{-2}	6.589×10^{-2}	4.554×10^{-2}	3.887×10^{-2}	NaN
Wilcoxon sign		1	1	1	1	1	1	1	NaN

Table 17. The SSIM results of the test case 8.

Threshold	Metric	Comparative Methods							
		AO	WOA	SSA	AOA	PSO	MPA	DE	DAOA
2	Max	0.576023	0.454896	0.60648	0.629978	0.625702	0.567363	0.699381	0.633192
	Mean	0.448062	0.416027	0.47257	0.542842	0.482223	0.501881	0.444351	0.525409
	Min	0.313677	0.343605	0.371043	0.424905	0.33731	0.400359	0.354733	0.383819
	STD	0.131291	0.062776	0.121013	0.10595	0.144201	0.089145	0.142674	0.128078
	Ranking	6	8	5	1	4	3	7	2
3	Max	0.62594	0.649127	0.606039	0.621	0.673756	0.482675	0.653015	0.670905
	Mean	0.521445	0.5597	0.577973	0.59104	0.509775	0.443829	0.570487	0.644174
	Min	0.328192	0.41791	0.533724	0.562574	0.411628	0.394558	0.52212	0.615842
	STD	0.167546	0.124184	0.038779	0.029241	0.142927	0.044974	0.020342	0.027566
	Ranking	6	5	3	2	7	8	4	1
4	Max	0.659553	0.689073	0.651031	0.688422	0.619171	0.654426	0.609938	0.702364
	Mean	0.585789	0.577215	0.572652	0.650246	0.538835	0.569699	0.444351	0.581725
	Min	0.457314	0.421803	0.524473	0.602193	0.438635	0.46478	0.373331	0.415275
	STD	0.111669	0.138856	0.068471	0.043955	0.091893	0.096422	0.144427	0.148926
	Ranking	2	4	5	1	7	6	8	3
5	Max	0.71179	0.653103	0.679673	0.728578	0.739026	0.654866	0.710614	0.723238
	Mean	0.667834	0.608033	0.628176	0.612995	0.625972	0.569295	0.65234	0.680701
	Min	0.645318	0.518308	0.529437	0.405271	0.535025	0.49256	0.543264	0.618906
	STD	0.038071	0.077705	0.085537	0.180276	0.103782	0.081513	0.020135	0.054767
	Ranking	2	7	4	6	5	8	3	1
6	Max	0.749897	0.758664	0.700491	0.739996	0.733031	0.674032	0.621232	0.726155
	Mean	0.700148	0.702408	0.625182	0.675597	0.681783	0.590369	0.60327	0.683228
	Min	0.614634	0.666246	0.567525	0.585931	0.60607	0.43936	0.513341	0.620501
	STD	0.074387	0.049377	0.068218	0.08008	0.066923	0.131032	0.780799	0.05554
	Ranking	2	1	6	5	4	8	7	3
Summation		18	25	23	15	27	33	29	10
Mean Rank		3.6	5	4.6	3	5.4	6.6	5.8	2
Final Ranking		3	5	4	2	6	8	7	1
***p*-value**		3.856×10^{-2}	2.669×10^{-2}	2.665×10^{-2}	2.814×10^{-2}	6.665×10^{-2}	3.854×10^{-2}	3.225×10^{-2}	NaN
Wilcoxon sign		1	1	1	1	1	1	1	NaN

The segmentation results (segmented images) of the proposed DAOA and the other comparative methods for Test 8 are shown in Figures 6–10. Figures 6–10 show the segmented images for all the tested methods, when the threshold values are 2, 3, 4, 5, and 6, respectively. According to these figures, we can recognize that the proposed DAOA showed good segmented images for various images (CT COVID-19 medical images) under different thresholds. Additionally, these figures prove that the segmented images are better in terms of quality when the threshold value is higher.

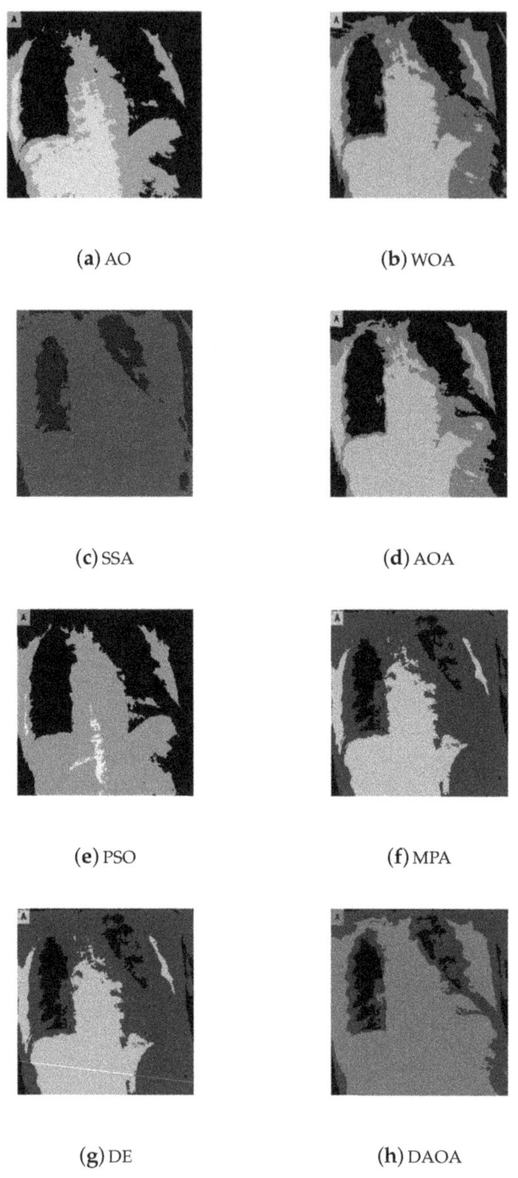

(a) AO (b) WOA (c) SSA (d) AOA (e) PSO (f) MPA (g) DE (h) DAOA

Figure 6. The segmented image (Test 8) by the comparative methods when the threshold value is 2.

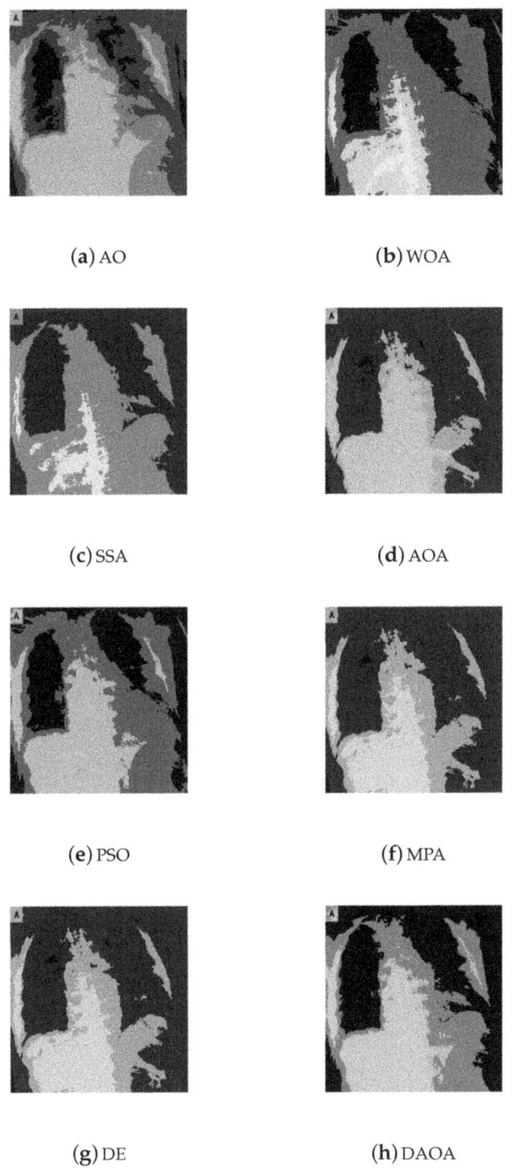

(a) AO **(b)** WOA **(c)** SSA **(d)** AOA **(e)** PSO **(f)** MPA **(g)** DE **(h)** DAOA

Figure 7. The segmented image (Test 8) by the comparative methods when the threshold value is 3.

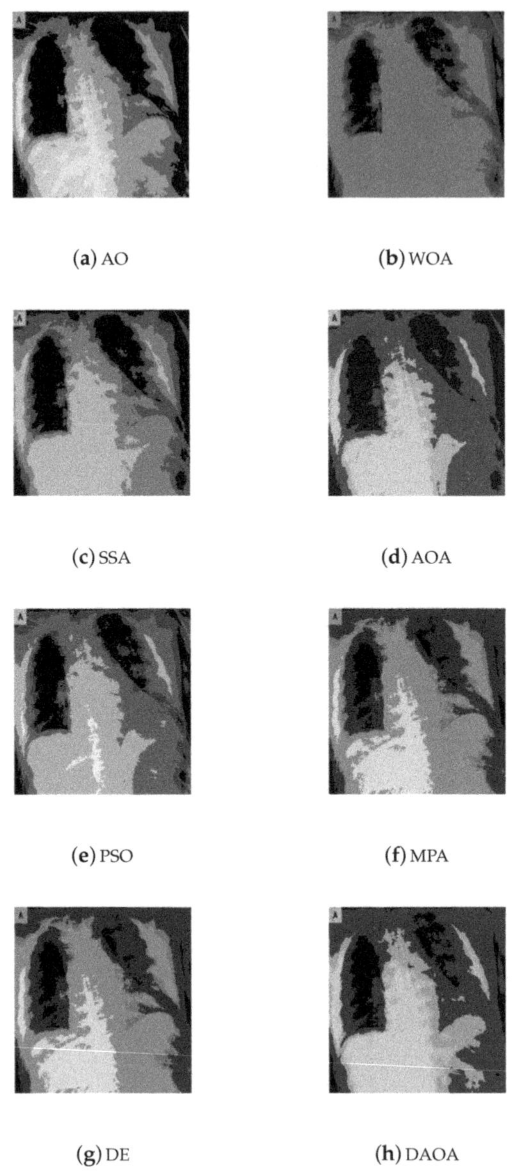

(**a**) AO (**b**) WOA (**c**) SSA (**d**) AOA (**e**) PSO (**f**) MPA (**g**) DE (**h**) DAOA

Figure 8. The segmented image (Test 8) by the comparative methods when the threshold value is 4.

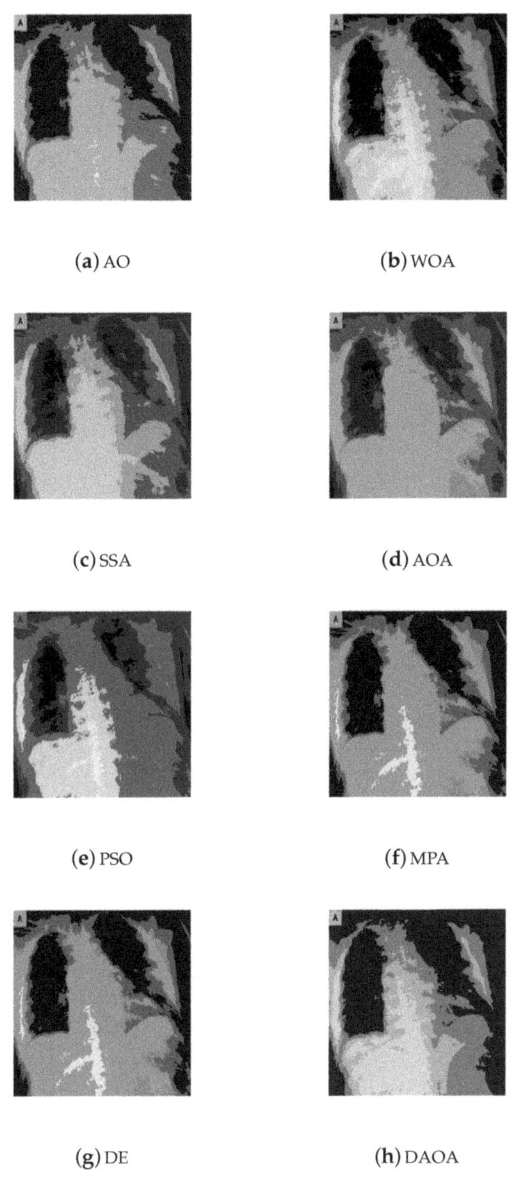

(a) AO **(b)** WOA **(c)** SSA **(d)** AOA **(e)** PSO **(f)** MPA **(g)** DE **(h)** DAOA

Figure 9. The segmented image (Test 8) by the comparative methods when the threshold value is 5.

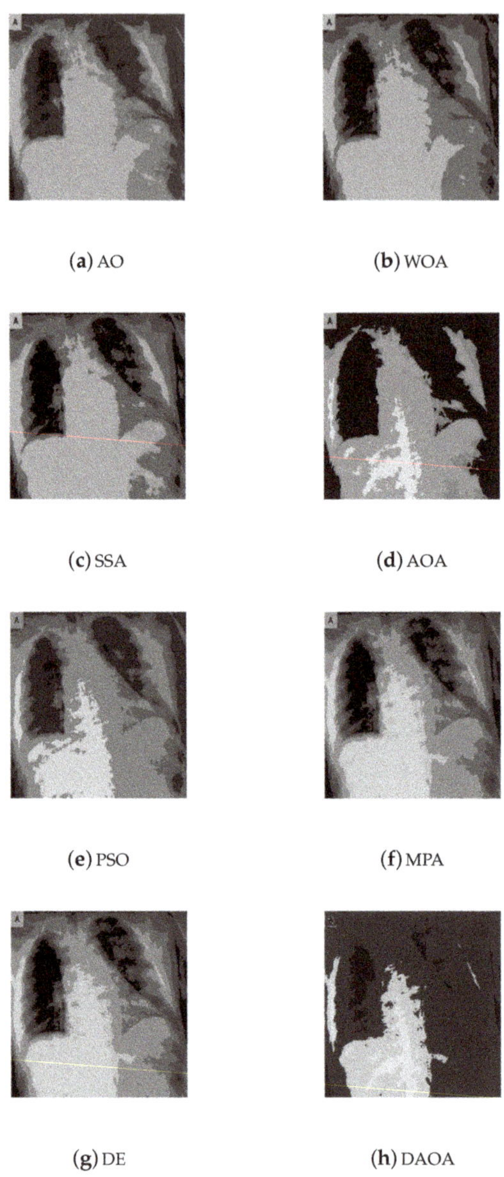

(a) AO (b) WOA (c) SSA (d) AOA (e) PSO (f) MPA (g) DE (h) DAOA

Figure 10. The segmented image (Test 8) by the comparative methods when the threshold value is 6.

The thresholds are shown in Figures 11–15, applied over the selected images. In Figures 11–15, the histogram images are given with the best threshold values obtained by the comparative methods for Test 8, where the threshold values are taken (i.e., 2, 3, 4, 5, and 6). The X and Y axes present the threshold values and Kapur measure values, respectively. It is feasible to recognize that the histogram classes are uniformly created, even in complex situations from such images. This means that the proposed method has an excellent ability to find always the same threshold values. The complexity is different from case to case because of the various peaks displayed in the pixels' distribution, which could create multiple classes or even carefully obtain the selection of the optimal thresholds.

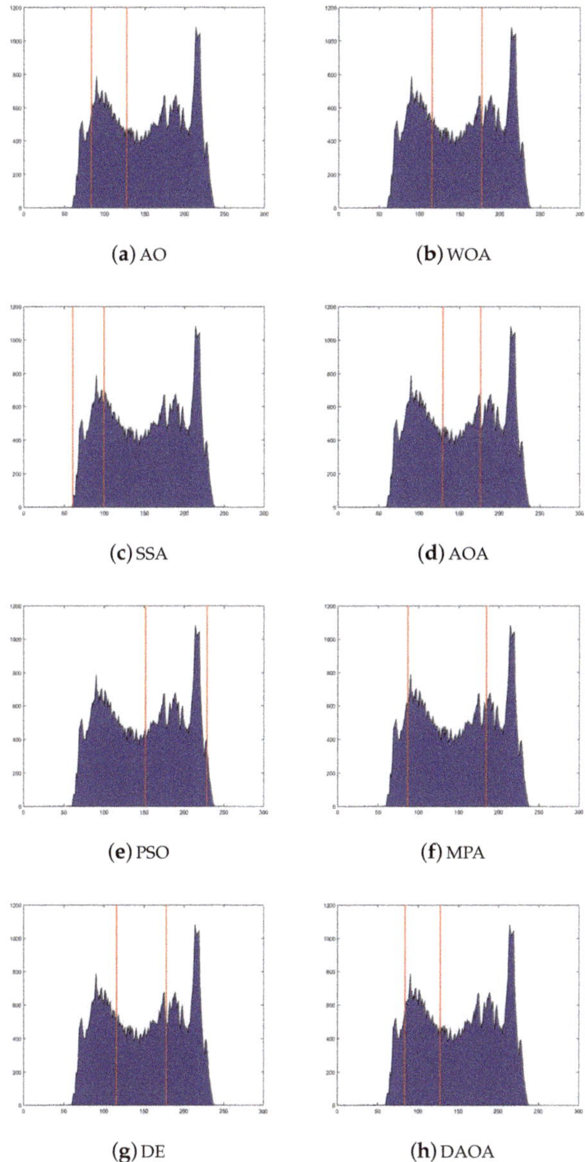

Figure 11. The histogram image (Test 8) by the comparative methods when the threshold value is 2.

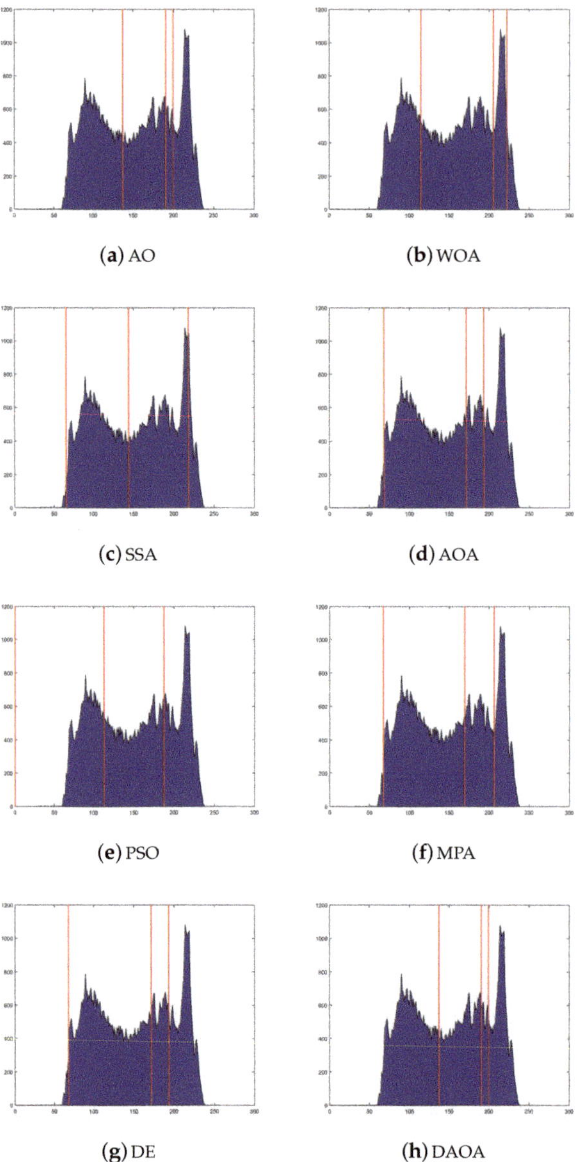

Figure 12. The histogram image (Test 8) by the comparative methods when the threshold value is 3.

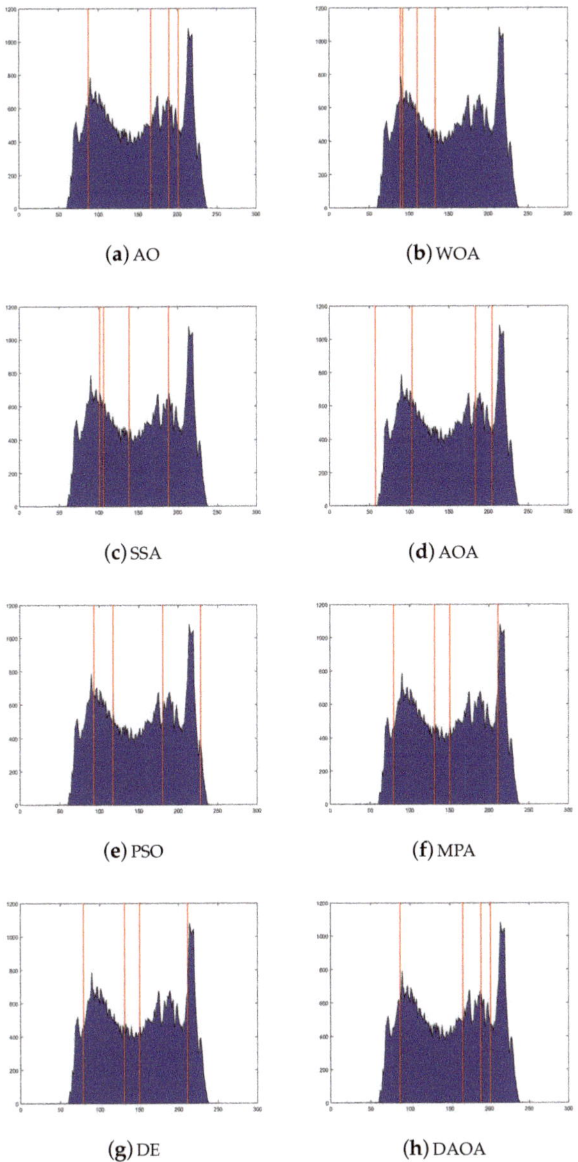

Figure 13. The histogram image (Test 8) by the comparative methods when the threshold value is 4.

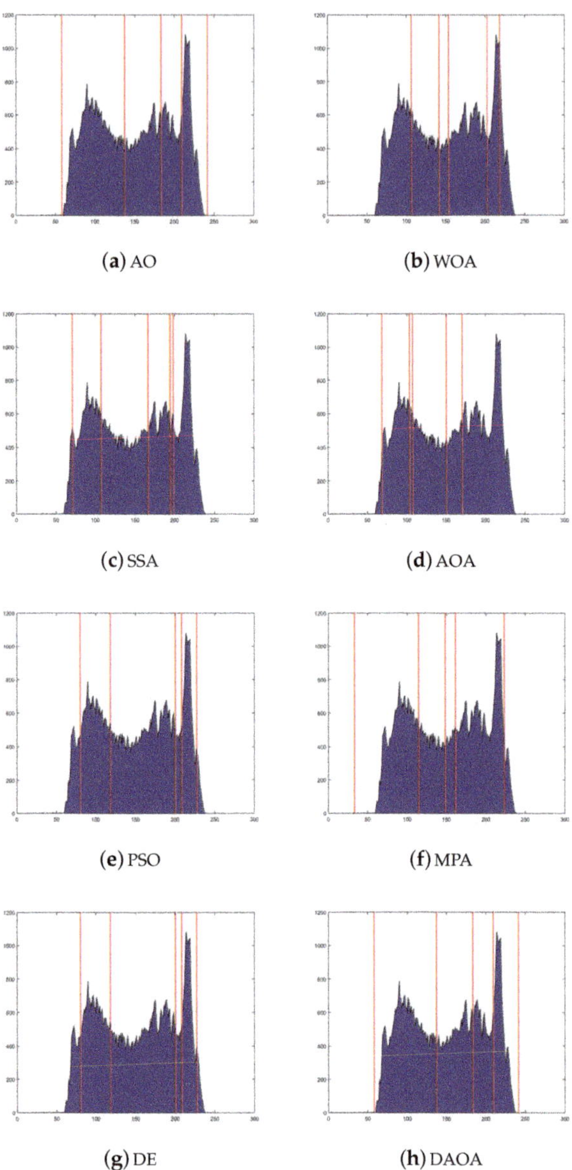

Figure 14. The histogram image (Test 8) by the comparative methods when the threshold value is 5.

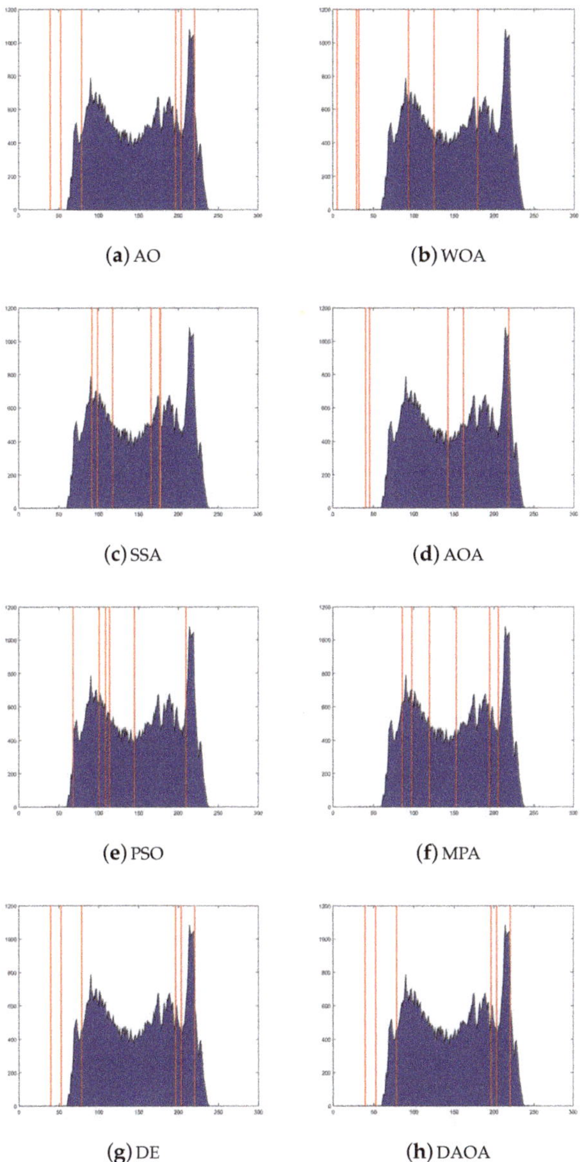

Figure 15. The histogram image (Test 8) by the comparative methods when the threshold value is 6.

Figure 16 shows the convergence curves of the proposed DAOA and its comparative optimization algorithms on eight tested images (i.e., Test 1 to Test 8); it can be seen that the proposed DAOA performs better than all involved other optimization methods in Test 8 when the threshold value is 6. For almost all the test images, the excellent optimized performance with accelerated convergence and more reliable accuracy achieved by the proposed DAOA can be seen as being remarkably smoothing behavior in the convergence curve. Moreover, we recognize that the curves of the proposed method always converge smoothly, reflecting the proposed DAOA's ability to avoid the common problem (local optima). In the end, the proposed DAOA reached the best solutions almost in all the tested cases, compared to the other comparative methods, as clearly shown in Figure 16.

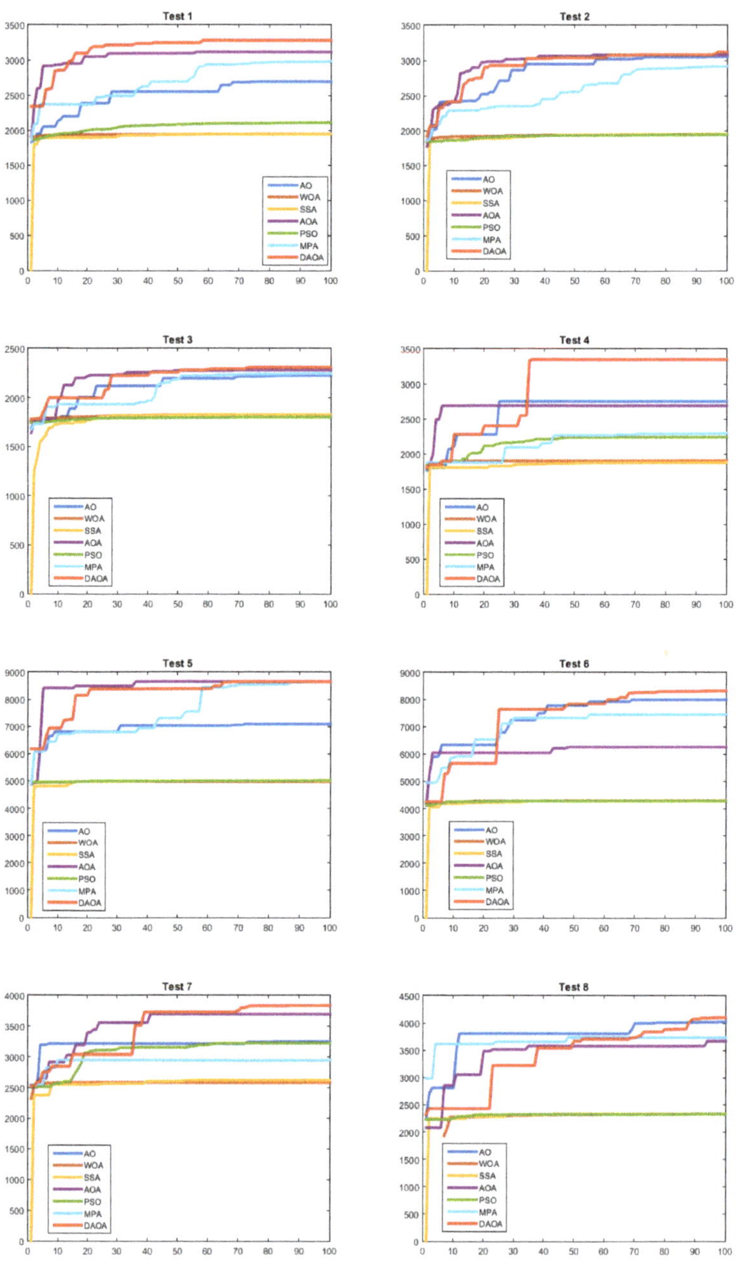

Figure 16. The convergence behavior of the comparative methods in solving Test 8 when the threshold value is 6

5. Conclusions and Future Works

The most crucial aspect of image segmentation is multilevel thresholding. However, multilevel thresholding displays require increasingly more computational complexity as the number of thresholds grows. In order to address this weakness, this paper proposes a new multilevel thresholding approach based on using an improved optimization-based evolutionary method.

The Arithmetic Optimization Algorithm (AOA) is a recently proposed optimization technique to solve different complex optimization problems. An enhanced version of the Arithmetic Optimization Algorithm is proposed in this paper to solve multilevel thresholding image segmentation problems. The proposed method combines the conventional Arithmetic Optimization Algorithm with the Differential Evolution technique, called DAOA. The main aim of the proposed DAOA is to improve the local search of the Arithmetic Optimization Algorithm and to establish an equilibrium among the search methods (exploration and exploitation).

The proposed DAOA method was applied to the multilevel thresholding problem, using Kapur's measure between class variance functions as a fitness function. The proposed DAOA evaluated eight standard test images from two different groups: nature images and CT medical images (i.e., COVID-19). The Peak Signal-to-Noise Ratio (PSNR) and Structural Similarity Index Test (SSIM) were used to determine the segmented images' accuracy. The proposed DAOA method's efficiency was evaluated and compared to other multilevel thresholding methods, including the Aquila Optimizer (AO), Whale Optimization Algorithm (WOA), Salp Swarm Algorithm (SSA), Arithmetic Optimization Algorithm (AOA), Particle Swarm Optimization (PSO), and Marine Predators Algorithm (MPA). The findings were presented, using a number of different threshold values (i.e., 2, 3, 4, 5, and 6). According to the experimental results, the proposed DAOA produced higher quality solutions than the other approaches. It achieved better results in almost all the tested cases, compared to other methods.

For future work, other fitness functions, evaluation measures, and benchmark images can be used. The conventional Arithmetic Optimization Algorithm can be improved, using other different optimization operations to enhance its performance further. As well, the proposed DAOA method can be used to solve other problems, such as text clustering, feature selection, photovoltaic parameter estimations, task scheduling in fog and cloud computing, appliances management in smart homes, advanced benchmark functions, text classification, text summarization, data clustering, engineering design problems, industrial problems, image construction, short-term wind speed forecasting, fuel cell modeling, damage identification, the prediction of the software vulnerability, knapsack problems, and others.

Author Contributions: L.A.: data curation, formal analysis, investigation, methodology, resources, software, supervision, validation, visualization, writing—original draft, and writing—review and editing. A.D.: formal analysis, investigation, supervision, writing—original draft, and writing—review and editing. P.S.: investigation, supervision, and writing—review and editing. A.H.G.: investigation, supervision, and writing—review and editing. All authors have read and agreed to the published version of the manuscript.

Funding: This research received no external funding.

Conflicts of Interest: The authors declare no conflict of interest.

References

1. Abd El Aziz, M.; Ewees, A.A.; Hassanien, A.E. Whale optimization algorithm and moth-flame optimization for multilevel thresholding image segmentation. *Expert Syst. Appl.* **2017**, *83*, 242–256. [CrossRef]
2. Shubham, S.; Bhandari, A.K. A generalized Masi entropy based efficient multilevel thresholding method for color image segmentation. *Multimed. Tools Appl.* **2019**, *78*, 17197–17238. [CrossRef]
3. Abd Elaziz, M.; Lu, S. Many-objectives multilevel thresholding image segmentation using knee evolutionary algorithm. *Expert Syst. Appl.* **2019**, *125*, 305–316. [CrossRef]
4. Pare, S.; Bhandari, A.K.; Kumar, A.; Singh, G.K. An optimal color image multilevel thresholding technique using grey-level co-occurrence matrix. *Expert Syst. Appl.* **2017**, *87*, 335–362. [CrossRef]
5. Bao, X.; Jia, H.; Lang, C. A novel hybrid harris hawks optimization for color image multilevel thresholding segmentation. *IEEE Access* **2019**, *7*, 76529–76546. [CrossRef]
6. Abd Elaziz, M.; Ewees, A.A.; Oliva, D. Hyper-heuristic method for multilevel thresholding image segmentation. *Expert Syst. Appl.* **2020**, *146*, 113201. [CrossRef]
7. Houssein, E.H.; Helmy, B.E.d.; Oliva, D.; Elngar, A.A.; Shaban, H. A novel Black Widow Optimization algorithm for multilevel thresholding image segmentation. *Expert Syst. Appl.* **2021**, *167*, 114159. [CrossRef]
8. Gill, H.S.; Khehra, B.S.; Singh, A.; Kaur, L. Teaching-learning-based optimization algorithm to minimize cross entropy for Selecting multilevel threshold values. *Egypt. Inform. J.* **2019**, *20*, 11–25. [CrossRef]
9. Tan, Z.; Zhang, D. A fuzzy adaptive gravitational search algorithm for two-dimensional multilevel thresholding image segmentation. *J. Ambient. Intell. Humaniz. Comput.* **2020**, *11*, 4983–4994. [CrossRef]
10. Yousri, D.; Abd Elaziz, M.; Abualigah, L.; Oliva, D.; Al-Qaness, M.A.; Ewees, A.A. COVID-19 X-ray images classification based on enhanced fractional-order cuckoo search optimizer using heavy-tailed distributions. *Appl. Soft Comput.* **2021**, *101*, 107052. [CrossRef] [PubMed]
11. Srikanth, R.; Bikshalu, K. Multilevel thresholding image segmentation based on energy curve with harmony Search Algorithm. *Ain Shams Eng. J.* **2021**, *12*, 1–20. [CrossRef]
12. Duan, L.; Yang, S.; Zhang, D. Multilevel thresholding using an improved cuckoo search algorithm for image segmentation. *J. Supercomput.* **2021**, *77*, 6734–6753. [CrossRef]
13. Jia, H.; Lang, C.; Oliva, D.; Song, W.; Peng, X. Hybrid grasshopper optimization algorithm and differential evolution for multilevel satellite image segmentation. *Remote Sens.* **2019**, *11*, 1134. [CrossRef]
14. Khairuzzaman, A.K.M.; Chaudhury, S. Multilevel thresholding using grey wolf optimizer for image segmentation. *Expert Syst. Appl.* **2017**, *86*, 64–76. [CrossRef]
15. He, L.; Huang, S. Modified firefly algorithm based multilevel thresholding for color image segmentation. *Neurocomputing* **2017**, *240*, 152–174. [CrossRef]
16. Li, Y.; Bai, X.; Jiao, L.; Xue, Y. Partitioned-cooperative quantum-behaved particle swarm optimization based on multilevel thresholding applied to medical image segmentation. *Appl. Soft Comput.* **2017**, *56*, 345–356. [CrossRef]
17. Manic, K.S.; Priya, R.K.; Rajinikanth, V. Image multithresholding based on Kapur/Tsallis entropy and firefly algorithm. *Indian J. Sci. Technol.* **2016**, *9*, 89949. [CrossRef]
18. Bhandari, A.K.; Kumar, A.; Singh, G.K. Modified artificial bee colony based computationally efficient multilevel thresholding for satellite image segmentation using Kapur's, Otsu and Tsallis functions. *Expert Syst. Appl.* **2015**, *42*, 1573–1601. [CrossRef]
19. Liang, H.; Jia, H.; Xing, Z.; Ma, J.; Peng, X. Modified grasshopper algorithm-based multilevel thresholding for color image segmentation. *IEEE Access* **2019**, *7*, 11258–11295. [CrossRef]
20. Abualigah, L. Group search optimizer: A nature-inspired meta-heuristic optimization algorithm with its results, variants, and applications. *Neural Comput. Appl.* **2020**, *33*, 2949–2972. [CrossRef]
21. Alsalibi, B.; Abualigah, L.; Khader, A.T. A novel bat algorithm with dynamic membrane structure for optimization problems. *Appl. Intell.* **2021**, *51*, 1992–2017. [CrossRef]
22. Ewees, A.A.; Abualigah, L.; Yousri, D.; Algamal, Z.Y.; Al-qaness, M.A.; Ibrahim, R.A.; Abd Elaziz, M. Improved Slime Mould Algorithm based on Firefly Algorithm for feature selection: A case study on QSAR model. *Eng. Comput.* **2021**, 1–15. [CrossRef]
23. Şahin, C.B.; Dinler, Ö.B.; Abualigah, L. Prediction of software vulnerability based deep symbiotic genetic algorithms: Phenotyping of dominant-features. *Appl. Intell.* **2021**, 1–17. [CrossRef]
24. Mirjalili, S.; Mirjalili, S.M.; Lewis, A. Grey wolf optimizer. *Adv. Eng. Softw.* **2014**, *69*, 46–61. [CrossRef]
25. Safaldin, M.; Otair, M.; Abualigah, L. Improved binary gray wolf optimizer and SVM for intrusion detection system in wireless sensor networks. *J. Ambient. Intell. Humaniz. Comput.* **2021**, *12*, 1559–1576. [CrossRef]
26. Alshinwan, M.; Abualigah, L.; Shehab, M.; Abd Elaziz, M.; Khasawneh, A.M.; Alabool, H.; Al Hamad, H. Dragonfly algorithm: A comprehensive survey of its results, variants, and applications. *Multimed. Tools Appl.* **2021**, *80*, 14979–15016. [CrossRef]
27. Shehab, M.; Abualigah, L.; Al Hamad, H.; Alabool, H.; Alshinwan, M.; Khasawneh, A.M. Moth–flame optimization algorithm: Variants and applications. *Neural Comput. Appl.* **2020**, *32*, 9859–9884. [CrossRef]
28. Al-Qaness, M.A.; Ewees, A.A.; Fan, H.; Abualigah, L.; Abd Elaziz, M. Marine predators algorithm for forecasting confirmed cases of COVID-19 in Italy, USA, Iran and Korea. *Int. J. Environ. Res. Public Health* **2020**, *17*, 3520. [CrossRef]
29. Abualigah, L.; Diabat, A.; Mirjalili, S.; Abd Elaziz, M.; Gandomi, A.H. The arithmetic optimization algorithm. *Comput. Methods Appl. Mech. Eng.* **2021**, *376*, 113609. [CrossRef]

30. Abualigah, L.; Yousri, D.; Abd Elaziz, M.; Ewees, A.A.; Al-qaness, M.A.; Gandomi, A.H. Aquila Optimizer: A novel meta-heuristic optimization algorithm. *Comput. Ind. Eng.* **2021**, *157*, 107250. [CrossRef]
31. Gandomi, A.H.; Alavi, A.H. Krill herd: A new bio-inspired optimization algorithm. *Commun. Nonlinear Sci. Numer. Simul.* **2012**, *17*, 4831–4845. [CrossRef]
32. Jouhari, H.; Lei, D.; Al-qaness, M.A.; Elaziz, M.A.; Damaševičius, R.; Korytkowski, M.; Ewees, A.A. Modified Harris Hawks Optimizer for Solving Machine Scheduling Problems. *Symmetry* **2020**, *12*, 1460. [CrossRef]
33. Połap, D.; Woźniak, M. Red fox optimization algorithm. *Expert Syst. Appl.* **2021**, *166*, 114107. [CrossRef]
34. Mernik, M.; Liu, S.H.; Karaboga, D.; Črepinšek, M. On clarifying misconceptions when comparing variants of the artificial bee colony algorithm by offering a new implementation. *Inf. Sci.* **2015**, *291*, 115–127. [CrossRef]
35. Sahlol, A.T.; Abd Elaziz, M.; Tariq Jamal, A.; Damaševičius, R.; Farouk Hassan, O. A novel method for detection of tuberculosis in chest radiographs using artificial ecosystem-based optimisation of deep neural network features. *Symmetry* **2020**, *12*, 1146. [CrossRef]
36. Abualigah, L.; Diabat, A. Advances in sine cosine algorithm: A comprehensive survey. *Artif. Intell. Rev.* **2021**, *54*, 2567–2608. [CrossRef]
37. Abualigah, L.; Diabat, A. A comprehensive survey of the Grasshopper optimization algorithm: Results, variants, and applications. *Neural Comput. Appl.* **2020**, *32*, 15533–15556. [CrossRef]
38. Tuba, E.; Alihodzic, A.; Tuba, M. Multilevel image thresholding using elephant herding optimization algorithm. In Proceedings of the 2017 14th International Conference on Engineering of Modern Electric Systems (EMES), Oradea, Romania, 1–2 June 2017; pp. 240–243.
39. Sahlol, A.T.; Yousri, D.; Ewees, A.A.; Al-Qaness, M.A.; Damasevicius, R.; Abd Elaziz, M. COVID-19 image classification using deep features and fractional-order marine predators algorithm. *Sci. Rep.* **2020**, *10*, 1–15. [CrossRef] [PubMed]
40. Wang, X.; Pan, J.S.; Chu, S.C. A parallel multi-verse optimizer for application in multilevel image segmentation. *IEEE Access* **2020**, *8*, 32018–32030. [CrossRef]
41. Gao, Y.; Li, X.; Dong, M.; Li, H.P. An enhanced artificial bee colony optimizer and its application to multi-level threshold image segmentation. *J. Cent. South Univ.* **2018**, *25*, 107–120. [CrossRef]
42. Resma, K.B.; Nair, M.S. Multilevel thresholding for image segmentation using Krill Herd Optimization algorithm. *J. King Saud. Univ.-Comput. Inf. Sci.* **2018**, *33*, 528–541.
43. Abd Elaziz, M.; Yousri, D.; Al-qaness, M.A.; AbdelAty, A.M.; Radwan, A.G.; Ewees, A.A. A Grunwald–Letnikov based Manta ray foraging optimizer for global optimization and image segmentation. *Eng. Appl. Artif. Intell.* **2021**, *98*, 104105. [CrossRef]
44. Sörensen, K. Metaheuristics—The metaphor exposed. *Int. Trans. Oper. Res.* **2015**, *22*, 3–18. [CrossRef]
45. García-Martínez, C.; Gutiérrez, P.D.; Molina, D.; Lozano, M.; Herrera, F. Since CEC 2005 competition on real-parameter optimisation: A decade of research, progress and comparative analysis's weakness. *Soft Comput.* **2017**, *21*, 5573–5583. [CrossRef]
46. Storn, R.; Price, K. Differential evolution—A simple and efficient heuristic for global optimization over continuous spaces. *J. Glob. Optim.* **1997**, *11*, 341–359. [CrossRef]
47. Tanabe, R.; Fukunaga, A. Success-history based parameter adaptation for differential evolution. In Proceedings of the 2013 IEEE Congress on Evolutionary Computation, Cancun, Mexico, 20–23 June 2013; pp. 71–78.
48. Črepinšek, M.; Liu, S.H.; Mernik, M. Replication and comparison of computational experiments in applied evolutionary computing: Common pitfalls and guidelines to avoid them. *Appl. Soft Comput.* **2014**, *19*, 161–170. [CrossRef]
49. Maitra, M.; Chatterjee, A. A hybrid cooperative–comprehensive learning based PSO algorithm for image segmentation using multilevel thresholding. *Expert Syst. Appl.* **2008**, *34*, 1341–1350. [CrossRef]
50. Yin, P.Y. Multilevel minimum cross entropy threshold selection based on particle swarm optimization. *Appl. Math. Comput.* **2007**, *184*, 503–513. [CrossRef]
51. Zhou, W. Image quality assessment: From error measurement to structural similarity. *IEEE Trans. Image Process.* **2004**, *13*, 600–613.
52. Sumari, P.; Syed, S.J.; Abualigah, L. A Novel Deep Learning Pipeline Architecture based on CNN to Detect Covid-19 in Chest X-ray Images. *Turk. J. Comput. Math. Educ. (TURCOMAT)* **2021**, *12*, 2001–2011.
53. Mirjalili, S.; Lewis, A. The whale optimization algorithm. *Adv. Eng. Softw.* **2016**, *95*, 51–67. [CrossRef]
54. Mirjalili, S.; Gandomi, A.H.; Mirjalili, S.Z.; Saremi, S.; Faris, H.; Mirjalili, S.M. Salp Swarm Algorithm: A bio-inspired optimizer for engineering design problems. *Adv. Eng. Softw.* **2017**, *114*, 163–191. [CrossRef]
55. Eberhart, R.; Kennedy, J. A new optimizer using particle swarm theory. In Proceedings of the MHS'95. Proceedings of the Sixth International Symposium on Micro Machine and Human Science, Nagoya, Japan, 4–6 October 1995; pp. 39–43.
56. Faramarzi, A.; Heidarinejad, M.; Mirjalili, S.; Gandomi, A.H. Marine Predators Algorithm: A nature-inspired metaheuristic. *Expert Syst. Appl.* **2020**, *152*, 113377. [CrossRef]
57. Price, K.; Storn, R.M.; Lampinen, J.A. *Differential Evolution: A Practical Approach to Global Optimization*; Springer Science & Business Media: Berlin/Heidelberg, Germany, 2006.
58. Derrac, J.; García, S.; Molina, D.; Herrera, F. A practical tutorial on the use of nonparametric statistical tests as a methodology for comparing evolutionary and swarm intelligence algorithms. *Swarm Evol. Comput.* **2011**, *1*, 3–18. [CrossRef]
59. Abualigah, L.M.Q. *Feature Selection and Enhanced Krill Herd Algorithm for Text Document Clustering*; Springer: Berlin/Heidelberg, Germany, 2019.

MDPI

Article

An Improved Hybrid Aquila Optimizer and Harris Hawks Algorithm for Solving Industrial Engineering Optimization Problems

Shuang Wang [1], Heming Jia [1,*], Laith Abualigah [2,3,4], Qingxin Liu [5] and Rong Zheng [1]

1 School of Information Engineering, Sanming University, Sanming 365004, China; shuang.wang@fjsmu.edu.cn (S.W.); zhengr@fjsmu.edu.cn (R.Z.)
2 Research and Innovation Department, Skyline University College, Sharjah 1797, United Arab Emirates; Aligah.2020@gmail.com
3 Faculty of Computer Sciences and Informatics, Amman Arab University, Amman 11953, Jordan
4 School of Computer Science, Universiti Sains Malaysia, Gelugor 11800, Pulau Pinang, Malaysia
5 School of Computer Science and Technology, Hainan University, Haikou 570228, China; lqxass@fjsmu.edu.cn
* Correspondence: jiaheming@fjsmu.edu.cn

Abstract: Aquila Optimizer (AO) and Harris Hawks Optimizer (HHO) are recently proposed meta-heuristic optimization algorithms. AO possesses strong global exploration capability but insufficient local exploitation ability. However, the exploitation phase of HHO is pretty good, while the exploration capability is far from satisfactory. Considering the characteristics of these two algorithms, an improved hybrid AO and HHO combined with a nonlinear escaping energy parameter and random opposition-based learning strategy is proposed, namely IHAOHHO, to improve the searching performance in this paper. Firstly, combining the salient features of AO and HHO retains valuable exploration and exploitation capabilities. In the second place, random opposition-based learning (ROBL) is added in the exploitation phase to improve local optima avoidance. Finally, the nonlinear escaping energy parameter is utilized better to balance the exploration and exploitation phases of IHAOHHO. These two strategies effectively enhance the exploration and exploitation of the proposed algorithm. To verify the optimization performance, IHAOHHO is comprehensively analyzed on 23 standard benchmark functions. Moreover, the practicability of IHAOHHO is also highlighted by four industrial engineering design problems. Compared with the original AO and HHO and five state-of-the-art algorithms, the results show that IHAOHHO has strong superior performance and promising prospects.

Keywords: Aquila Optimizer; Harris Hawks Optimizer; hybrid algorithm; nonlinear escaping energy parameter; random opposition-based learning

Citation: Wang, S.; Jia, H.; Abualigah, L.; Liu, Q.; Zheng, R. An Improved Hybrid Aquila Optimizer and Harris Hawks Algorithm for Solving Industrial Engineering Optimization Problems. *Processes* **2021**, *9*, 1551. https://doi.org/10.3390/pr9091551

Academic Editor: Jean-Pierre Corriou

Received: 20 July 2021
Accepted: 27 August 2021
Published: 30 August 2021

1. Introduction

Meta-heuristic optimization algorithms inspired by nature are becoming more and more popular in real-world applications [1]. Meta-heuristics usually mimic biological or physical phenomena and only consider inputs and outputs, making them flexible and straightforward. Furthermore, meta-heuristics is a kind of stochastic optimization technique. This property assists them to effectively avoid local optima, which usually occurs in real problems. Because of the advantages of simplicity, flexibility, and ability to avoid local optima, meta-heuristic optimization algorithms outperform heuristic optimization algorithms to solve various complex and tricky optimization problems in the real world [2].

Three dominant categories are divided from meta-heuristic optimization algorithms: evolutionary, physics-based, and swarm intelligence techniques. Evolutionary algorithms are inspired by the laws of evolution in nature. The randomly generated population evolves over subsequent generations as the number of iterations increases. Each generation of individuals is always formed by the combination of best individuals so that the population can be optimized over several generations of evolution. The most popular evolutionary technique is Genetic Algorithms (GA) [3], which simulates Darwin's theory of evolution. There are several other popular evolutionary algorithms, such as Differential Evolution Algorithm (DE) [4], Genetic Programming (GP) [5], Evolution Strategy (ES) [6], Biogeography-Based Optimizer (BBO) [7], Evolutionary Deduction Algorithm (ED) [8], and Probability-Based Incremental Learning (PBIL) [9]. Physics-based methods are inspired by the physical rules of the universe. The most popular algorithms in this category are Simulated Annealing (SA) [10], Big-Bang Big-Crunch (BBBC) [11], Gravity Search Algorithm (GSA) [12], Gravitational Local Search (GLSA) [13], Heat Transfer Relation-based Optimization Algorithm (HTOA) [14], Charged System Search (CSS) [15], Artificial Chemical Reaction Optimization Algorithm (ACROA) [16], Central Force Optimization (CFO) [17], Ray Optimization (RO) [18] algorithm, Black Hole (BH) [19] algorithm, Small-World Optimization Algorithm (SWOA) [20], Galaxy-based Search Algorithm (GbSA) [21], Curved Space Optimization (CSO) [22], Multi-Verse Optimizer (MVO) [23], Sine Cosine Algorithm (SCA) [24], and Arithmetic Optimization Algorithm (AOA) [25].

The third category is swarm intelligence algorithms, which simulate the behaviour of swarms of creatures in nature. The most well-known swarm intelligence technique is Particle Swarm Optimization (PSO), first proposed by Kennedy and Eberhart [26]. PSO mimics the behaviour of bird flocks in navigating and foraging, and the birds achieve the optimal position through collective cooperation. Particles update positions not only considering their own best positions but also according to the best position of the swarm obtained so far. Other representative algorithms include Ant Colony Optimization Algorithm (ACO) [27], Monkey Search [28], Firefly Algorithm [29], Bat Algorithm (BA) [30], Krill Herd (KH) [31], Grey Wolf Optimizer (GWO) [32], Cuckoo Search (CS) Algorithm [33], Fruit Fly Optimization (FFO) [34], Dolphin Partner Optimization (DPO) [35], Ant Lion Optimizer (ALO) [36], Remora Optimization Algorithm (ROA) [37], Whale Optimization Algorithm (WOA) [38], Salp Swarm Algorithm (SSA) [39], Bald Eagle Search (BES) algorithm [40], and Slime Mould Algorithm (SMA) [41].

As one of the swarm intelligence algorithms, the Harris Hawks Optimizer (HHO) [42] was proposed in 2019. HHO simulates several hunting strategies of Harris's hawk and attracted several researchers to apply it to solve practical problems [43–47]. The exploitation phase of HHO includes four strategies, but the exploration phase is insufficient, and the balance between the exploration and exploitation phases is not good enough. Therefore, many improved and hybrid researches have been proposed to enhance the performance of HHO. Yousri et al. [48] proposed an enhanced algorithm based on the fractional calculus (FOC) memory concept to improve the performance of exploration phase, which is known as FMHHO. The hawk moves with a fractional-order velocity, and the escaping energy of the prey is adaptively adjusted based on FOC parameters to avoid local optima stagnation. Gupta et al. [49] introduced a nonlinear energy parameter, different settings for rapid dives, opposition-based learning strategy, and a greedy selection mechanism into HHO to enhance the search efficiency and avoid premature convergence. Hussien and Amin [50] proposed an improved HHO called IHHO to enhance the performance of HHO. The proposed IHHO applied opposition-based learning (OBL) in the initialization phase to diverse the initial population as well as Chaotic Local Search (CLS) strategy and a self-adaptive technique to improve its performance and speed up the convergence of the algorithm. Sihwail et al. [51] proposed a new search mechanism and then applied it and elite opposite-based learning (EOBL) technique to HHO. The improved HHO raised the search capabilities by mutation, mutation neighborhood search (MNS), and rollback strategy. It can avoid local optimum entrapment and improve population diversity, convergence accuracy, and rate. Bao et al. [52] proposed HHO-DE by hybridizing HHO and Differential Evolution (DE) algorithms. HHO and DE were used to update the positions of two equal subpopulations respectively. The proposed HHO-DE has high accuracy, ability to avoid local optima, and remarkable stability. Houssein et al. [53] combined HHO with cuckoo search (CS) and chaotic maps to propose a hybrid algorithm called CHHO-CS. CS was used to control the main position vectors of HHO to achieve a better balance between exploration and exploitation phases, and chaotic maps were adopted to update the control energy parameters to avoid premature convergence. Kaveh et al. [54] proposed an effective algorithm called ICHHO by hybridizing HHO with Imperialist Competitive Algorithm (ICA). Combination of the exploration strategy of ICA and exploitation technique of HHO helps to achieve a better search strategy. These improved and hybrid algorithms have proven that HHO is a valuable algorithm. Aquila Optimizer (AO) [55] is the latest swarm intelligence algorithm, proposed in 2021. This algorithm simulates different hunting methods of Aquila for different kinds of prey. The hunting methods for fast-moving prey reflect the global exploration ability of the algorithm, and the hunting methods for slow-moving prey reflect the local exploitation ability of the algorithm. AO algorithm possesses strong global exploration ability, high search efficiency, and fast convergence speed, but its local exploitation ability is insufficient, so it is easy to fall into local optima. Due to the short time that has elapsed since the algorithm has been proposed, there is no research on the improvement of AO yet.

Therefore, we tried a kind of hybridization to improve the performance of HHO and AO. As far as we know, this kind of hybridization of HHO with AO has not been used before. We propose a new, improved hybrid Aquila Optimizer and Harris Hawks Optimization (IHAOHHO) by combining the salient features of AO and HHO. In this paper, we integrate the exploitation strategy of HHO into the AO algorithm, which is added random opposition-based learning (ROBL) in the exploitation phase to avoid local optima stagnation. At the same time, the nonlinear escaping energy parameter balances the exploration and exploitation phases of the algorithm. The 23 standard benchmark functions and four engineering design problems were applied to test the exploration and exploitation capabilities of IHAOHHO. The proposed algorithm is compared with original AO, HHO, and several well-known algorithms, including SMA, SSA, WOA, GWO, and PSO. The experimental results show that the proposed IHAOHHO algorithm performs better than the state-of-the-art meta-heuristic algorithms.

The rest of this paper is organized as follows (Figure 1): Section 2 provides a brief overview of the related work: original Harris Hawks Optimization algorithm and Aquila Optimizer, as well as two improvement strategies. Section 3 describes in detail the proposed hybrid algorithm. Section 4 conducts simulation experiments and results analysis. Finally, Section 5 concludes the paper.

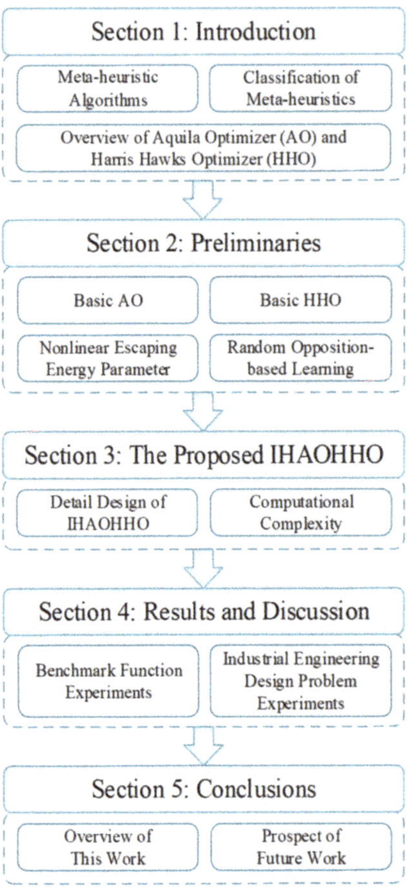

Figure 1. The overview sketch of this paper.

2. Preliminaries

2.1. Aquila Optimizer (AO)

AO is a latest novel swarm intelligence algorithm proposed by Abualigah et al. in 2021. There are four hunting behaviors of Aquila for different kinds of prey. Aquila can switch hunting strategies flexibly for different prey and then uses its fast speed united with sturdy feet and claws to attack prey. The brief description of mathematical model can be described as follows.

Step 1: Expanded exploration (X_1): high soar with a vertical stoop

In this method, the Aquila flies high over the ground and explores the search space widely, and then a vertical dive is taken once the Aquila determines the area of the prey. The mathematical representation of this behaviour is written as:

$$X_1(t+1) = X_{best}(t) \times (1 - \frac{t}{T}) + (X_M(t) - X_{best}(t) \times r_1) \quad (1)$$

$$X_M(t) = \frac{1}{N}\sum_{i=1}^{N} X_i(t) \quad (2)$$

where $X_{best}(t)$ represents the best position obtained so far, and $X_M(t)$ denotes the average position of all Aquilas in current iteration. t and T is the current iteration and the maximum number of iterations, respectively. N is the population size, and r_1 is a random number between 0 and 1.

Step 2: Narrowed exploration (X_2): contour flight with short glide attack

This is the most commonly used hunting method for Aquila. It uses short gliding to attack the prey after descending within the selected area and flying around the prey. The position update formula is represented as:

$$X_2(t+1) = X_{best}(t) \times LF(D) + X_R(t) + (y-x) \times r_2 \tag{3}$$

where $X_R(t)$ represents a random position of the Aquila, D is the dimension size, and r_2 is a random number within (0, 1). $LF(D)$ represents Levy flight function, which is presented as follows:

$$LF(D) = s \times \frac{u \times \sigma}{|v|^{\frac{1}{\beta}}} \tag{4}$$

$$\sigma = \left(\frac{\Gamma(1+\beta) \times \sin(\frac{\pi\beta}{2})}{\Gamma(\frac{1+\beta}{2}) \times \beta \times 2^{(\frac{\beta-1}{2})}} \right) \tag{5}$$

where s and β are constant values equal to 0.01 and 1.5, respectively, and u and v are random numbers between 0 and 1. y and x are used to present the spiral shape in the search, which are calculated as follows:

$$\begin{cases} x = r \times \sin(\theta) \\ y = r \times \cos(\theta) \\ r = r_3 + 0.00565 \times D_1 \\ \theta = -\omega \times D_1 + \frac{3 \times \pi}{2} \end{cases} \tag{6}$$

where r_3 means the number of search cycles between 1 and 20, D_1 is composed of integer numbers from 1 to the dimension size (D), and ω is equal to 0.005.

Step 3: Expanded exploitation (X_3): low flight with a slow descent attack

In the third method, when the area of prey is roughly determined, the Aquila descends vertically to do a preliminary attack. AO exploits the selected area to get close and attack the prey. This behaviour is presented as follows:

$$X_3(t+1) = (X_{best}(t) - X_M(t)) \times \alpha - r_4 + ((UB - LB) \times r_5 + LB) \times \delta \tag{7}$$

where $X_{best}(t)$ denotes to the best position obtained so far, and $X_M(t)$ means the average value of the current positions. α and δ are the exploitation adjustment parameters fixed to 0.1, UB and LB are the upper and lower bound of the problem, and r_4 and r_5 are random numbers within (0, 1).

Step 4: Narrowed exploitation (X_4): walking and grabbing prey

In this method, the Aquila chases the prey in the light of its escape trajectory and then attacks the prey on the ground. The mathematical representation of this behaviour is as follows:

$$\begin{cases} X_4(t+1) & = QF \times X_{best}(t) - (G_1 \times X(t) \times r_6) \\ & -G_2 \times LF(D) + r_7 \times G_1 \\ QF(t) = & t^{\frac{2 \times rand() - 1}{(1-T)^2}} \\ G_1 = 2 \times & r_8 - 1 \\ G_2 = 2 \times & (1 - \frac{t}{T}) \end{cases} \tag{8}$$

where $X(t)$ is the current position, and $QF(t)$ represents the quality function value, which used to balance the search strategy. G_1 denotes the movement parameter of Aquila during tracking prey, which is a random number between [−1,1]. G_2 denotes the flight slope when chasing prey, which decreases linearly from 2 to 0. r_6, r_7, and r_8 are random numbers between 0 and 1.

2.2. Harris's Hawks Optimizer (HHO)

HHO is a new meta-heuristic optimization algorithm proposed by Heidari et al. in 2019. It is inspired by the unique cooperative foraging activities of Harris's hawk. Harris's hawk can show a variety of chasing patterns according to the dynamic nature of the environment and the escaping patterns of the prey. These switching activities are conducive to confuse the running prey, and these cooperative strategies can help Harris's hawk chase the detected prey to exhaustion, which increases its vulnerability. The brief description of mathematical model is as follows.

2.2.1. Exploration Phase

The Harris's hawks usually perch on some random locations, wait, and monitor the desert to detect the prey. There are two perching strategies based on the positions of other family members and the prey or random tall trees, which is selected in accordance with the random q value.

$$X(t+1) = \begin{cases} X_r(t) - r_1|X_r(t) - 2r_2X(t)| & q \geq 0.5 \\ (X_{prey}(t) - X_m(t)) - r_3(LB + r_4(UB - LB)) & q < 0.5 \end{cases} \tag{9}$$

$$X_m(t) = \frac{1}{N}\sum_{i=1}^{N} X_i(t) \tag{10}$$

where $X_r(t)$ is the position of a random hawk, $X_{prey}(t)$ represents the position of the prey, that is the best position obtained so far, and $X_m(t)$ denotes the average position of the current population. N is total number of hawks, UB and LB are the upper and lower bound of the problem, and q, r_1, r_2, r_3, and r_4 are random numbers between 0 and 1.

2.2.2. Transition from Exploration to Exploitation Phase

The HHO algorithm has a transition mechanism from exploration to exploitation phase based on the escaping energy of the prey and then changes the different exploitative behaviors. The energy of the prey is modelled as follows, which decreases during the escaping behaviour.

$$E = 2E_0(1 - \frac{t}{T}) \tag{11}$$

where E represents the escaping energy of the prey, E_0 is the initial state of the energy, and t and T are the current and maximum number of iterations, respectively. When $|E| \geq 1$, the algorithm performs the exploration stage, and when $|E| < 1$, the algorithm performs the exploitation phase.

2.2.3. Exploitation Phase

In this phase, four different chasing and attacking strategies are proposed on the basis of the escaping energy of the prey and chasing styles of the Harris's hawks. Except for the escaping energy, parameter r is also utilized to choose the chasing strategy, which indicates the chance of the prey in successfully escaping ($r < 0.5$) or not ($r \geq 0.5$) before attack.

Soft besiege

When $r \geq 0.5$ and $|E| \geq 0.5$, the prey still has enough energy and tries to escape, so the Harris's hawks encircle it softly to make the prey more exhausted and then attack it. This behaviour is modeled as follows:

$$X(t+1) = \Delta X(t) - E|JX_{prey}(t) - X(t)| \tag{12}$$

$$\Delta X(t) = X_{prey}(t) - X(t) \tag{13}$$

$$J = 2(1 - r_5) \tag{14}$$

where $\Delta X(t)$ indicates the difference between the position of the prey and the current position, J represents the random jump strength of the prey, $X_{prey}(t)$ represents the position of the prey, $X(t)$ is the current position, and r_5 is a random number within (0, 1).

Hard besiege

When $r \geq 0.5$ and $|E| < 0.5$, the prey has a low escaping energy, and the Harris's hawks encircle the prey readily and finally attack it. In this situation, the positions are updated as follows:

$$X(t+1) = X_{prey}(t) - E|\Delta X(t)| \tag{15}$$

Soft besiege with progressive rapid dives

When $|E| \geq 0.5$ and $r < 0.5$, the prey has enough energy to successfully escape, so the Harris's hawks perform soft besiege with several rapid dives around the prey and try to progressively correct its position and direction. This behaviour is modeled as follows:

$$Y = X_{prey}(t) - E|JX_{prey}(t) - X(t)| \tag{16}$$

$$Z = Y + S \times LF(D) \tag{17}$$

$$LF(x) = 0.01 \times \frac{u \times \sigma}{|v|^{\frac{1}{\beta}}} \tag{18}$$

$$\sigma = \left(\frac{\Gamma(1+\beta) \times sin(\frac{\pi\beta}{2})}{\Gamma(\frac{1+\beta}{2}) \times \beta \times 2^{(\frac{\beta-1}{2})}} \right)^{\frac{1}{\beta}} \tag{19}$$

$$X(t+1) = \begin{cases} Y & if\ F(Y) < F(X(t)) \\ Z & if\ F(Z) < F(X(t)) \end{cases} \tag{20}$$

where D is the dimension size of the problem, and S is a random vector. LF is Levy flight function, which is utilized to mimic the deceptive motions of the prey. u and v are random values between 0 and 1, and β is a constant number equal to 1.5. Note that only the better position between Y and Z is selected as the next position.

Hard besiege with progressive rapid dives

When $|E| < 0.5$ and $r < 0.5$, the prey does not have enough energy to escape, so the hawks perform a hard besiege to decrease the distance between their average position and the prey and finally attack and kill the prey. The mathematical representation of this behaviour is as follows:

$$Y = X_{prey}(t) - E|JX_{prey}(t) - X_m(t)| \tag{21}$$

$$Z = Y + S \times LF(D) \tag{22}$$

$$X(t+1) = \begin{cases} Y & if\ F(Y) < F(X(t)) \\ Z & if\ F(Z) < F(X(t)) \end{cases} \tag{23}$$

Note that only the better position between Y and Z will be the next position for the new iteration.

2.3. Nonlinear Escaping Energy Parameter

In the original HHO algorithm, the escaping energy E is used to control the transition from exploration to exploitation phase. The parameter E is linearly reduced from 2 to 0, that is, only local search is performed in the second half of the iterations, which is easy to fall into local optima. In order to overcome this shortcoming of the algorithm, another way to update the escaping energy E is utilized [56]:

$$E = E_1(2 \times \text{rand}-1) \tag{24}$$

$$E_1 = 2 \times (1 - \left(\frac{t}{T}\right)^{1/3}{}^{1/3}) \tag{25}$$

where t and T are the current and maximum number of iterations, respectively. It can be seen from Figure 2a that E_1 decreases rapidly in the early stage of the iterations, which controls the global search ability of the algorithm and changes slowly in the middle of the iterations. E_1 also balances the global and local search capabilities and decreases rapidly in the later stage of the iterations to speed up the local search. E can perform global search and local search in the whole iterative process. It mainly performs global search in the early stage and retains the possibility of global search while mainly performing local search in the later stage, as shown in Figure 2b.

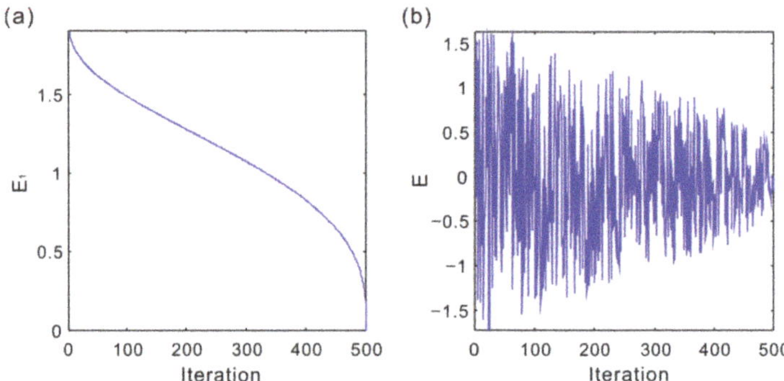

Figure 2. (**a**) E_1 curve and (**b**) E curve.

2.4. Random Opposition-Based Learning (ROBL)

Opposition-based learning (OBL) is a powerful optimization tool proposed by Tizhoosh [57]. The main idea of OBL is simultaneously considering the fitness of an estimate and its corresponding opposite estimate to obtain a better candidate solution. The OBL concept has successfully been used in varieties of meta-heuristics algorithms [58–62] to improve the convergence speed. Different from the original OBL, this paper utilizes an improved OBL strategy, called random opposition-based learning (ROBL) [63], which is defined by:

$$\hat{x}_j = l_j + u_j - rand \times x_j, \quad j = 1,2,\ldots,n \tag{26}$$

where $\hat{x}_j$ represents the opposite solution, l_j and u_j are the lower and upper bound of the problem in jth dimension, and *rand* is a random number within (0, 1). The opposite solution described by Equation (26) is more random than the original OBL and can effectively help the population jump out of the local optima.

3. The Proposed IHAOHHO Algorithm

3.1. The Detail Design of IHAOHHO

The exploration phase of AO mimics the hunting behaviour for fast-moving prey with a wide flying area, making AO have a strong global search ability and fast convergence speed. However, the selected search space is not exhaustively searched during the exploitation phase. The effect of Levy flight is relatively weak, leading to premature convergence. In a word, the AO algorithm possesses strong randomness and fast convergence speed in the global exploration phase. However, it is easy to fall into local optima in the local exploitation stage. For the HHO algorithm, the transition from global to local search is realized based on the energy attenuation of the prey. In the early iterations, which reflect the exploration phase, the diversity of the population is insufficient, and the convergence speed is slow. As the number of iterations increases, the energy of prey decreases, and the algorithm enters the stage of local exploitation. Four different hunting strategies are adopted in the light of the energy and escape probability of the prey. The Levy flight term is added in the exploitation phase. Whether to use Levy flight to update positions is decided by fitness values so that the algorithm can jump out of the local optima to a certain extent.

Therefore, we combine the global exploration phase of AO and the local exploitation phase of HHO to give full play to the advantages of these two algorithms. The global search capability, faster convergence speed, and the ability to jump out of the local optima of the algorithm are all retained. Meanwhile, a nonlinear escaping energy mechanism is utilized to control the transition from exploration to exploitation phase, which retains the possibility of global search in the later iterations. ROBL strategy is added to the exploitation phase to enhance further the ability to jump out of the local optima. All these strategies improve the convergence speed and accuracy of the hybrid algorithm and effectively enhance the overall optimization performance of the algorithm. This improved hybrid Aquila Optimizer and Harris Hawks Optimization algorithm is named IHAOHHO. Different phases of IHAOHHO are shown in Figure 3. The pseudo-code of IHAOHHO is given in Algorithm 1, and the summarized flowchart is illustrated in Figure 4.

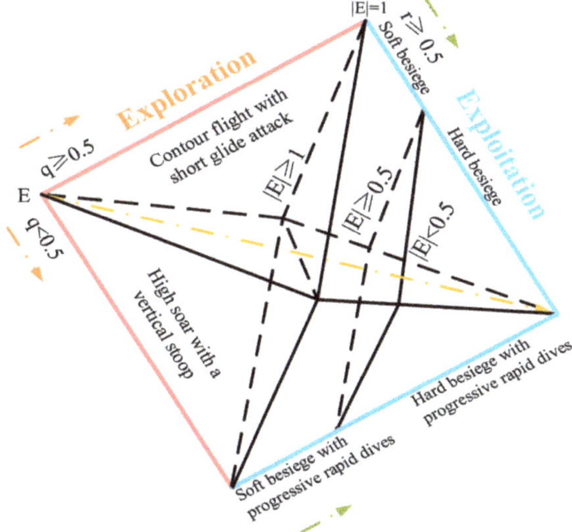

Figure 3. Different phases of IHAOHHO.

3.2. Computational Complexity of IHAOHHO

Computation complexity is a key metric for an algorithm to evaluate its time consumption during operation. The computational complexity of the IHAOHHO algorithm depends on three processes: initialization, evaluation of fitness, and updating of hawks. In the initialization stage, the computational complexity of positions generated of N hawks is O(N × D), where D is dimension size of the problem. Then, the computational complexity of fitness evaluation for the best solution is O(N) during the iteration process. Considering the worst condition, the computational complexities of position updating of hawks and fitness comparison are O(3 × N × D) and O(3 × N), respectively. In a word, the total computational complexity of the proposed IHAOHHO algorithm is O(N × D + (3 × D + 4) × N × T).

Algorithm 1 Pseudo-code of IHAOHHO.

```
Set initial values of the population size N and the maximum number of iterations T
Initialize positions of the population X
While t < T
  For i = 1 to N
    Check if the position goes out of the search space boundary, and bring it back.
    Calculate the fitness of X_i
    Update X_best
  End for
  Update x, y, QF, G_1, G_2, E_1
  For i = 1 to N
    Update E using Equation (24)      % Nonlinear escaping energy parameter
    If |E| >= 1                       % Exploration part of AO
      If rand < 0.5
        Update the position of Xnew_i using Equation (1)
          If f(Xnew_i) < f(X_i)
            X_i = Xnew_i
          End if
        Else
        Update the position of Xnew_i using Equation (3)
          If f(Xnew_i) < f(X_i)
            X_i = Xnew_i
          End if
      End if
    Else                              % Exploitation part of HHO
      If r >= 0.5 and |E| >= 0.5
        Update the position of X_i using Equation (12)
      End if
      If r >= 0.5 and |E| < 0.5
        Update the position of X_i using Equation (15)
      End if
      If r < 0.5 and |E| >= 0.5
        Update the position of Xnew_i using Equation (16)
        If f(Xnew_i) < f(X_i)
          X_i = Xnew_i
        Else
          Update the position of Xnew_i using Equation (17)
          If f(Xnew_i) < f(X_i)
            X_i = Xnew_i
          End if
        End if
      End if
      If r < 0.5 and |E| < 0.5
        Update the position of Xnew_i using Equation (21)
        If f(Xnew_i) < f(X_i)
          X_i = Xnew_i
        Else
          Update the position of Xnew_i using Equation (22)
          If f(Xnew_i) < f(X_i)
            X_i = Xnew_i
          End if
        End if
      End if
      Update the position of Xnew_i using Equation (26)   % ROBL
      If f(Xnew_i) < f(X_i)
        X_i = Xnew_i
      End if
    End if
    t = t + 1
  End for
End while
Return X_best
```

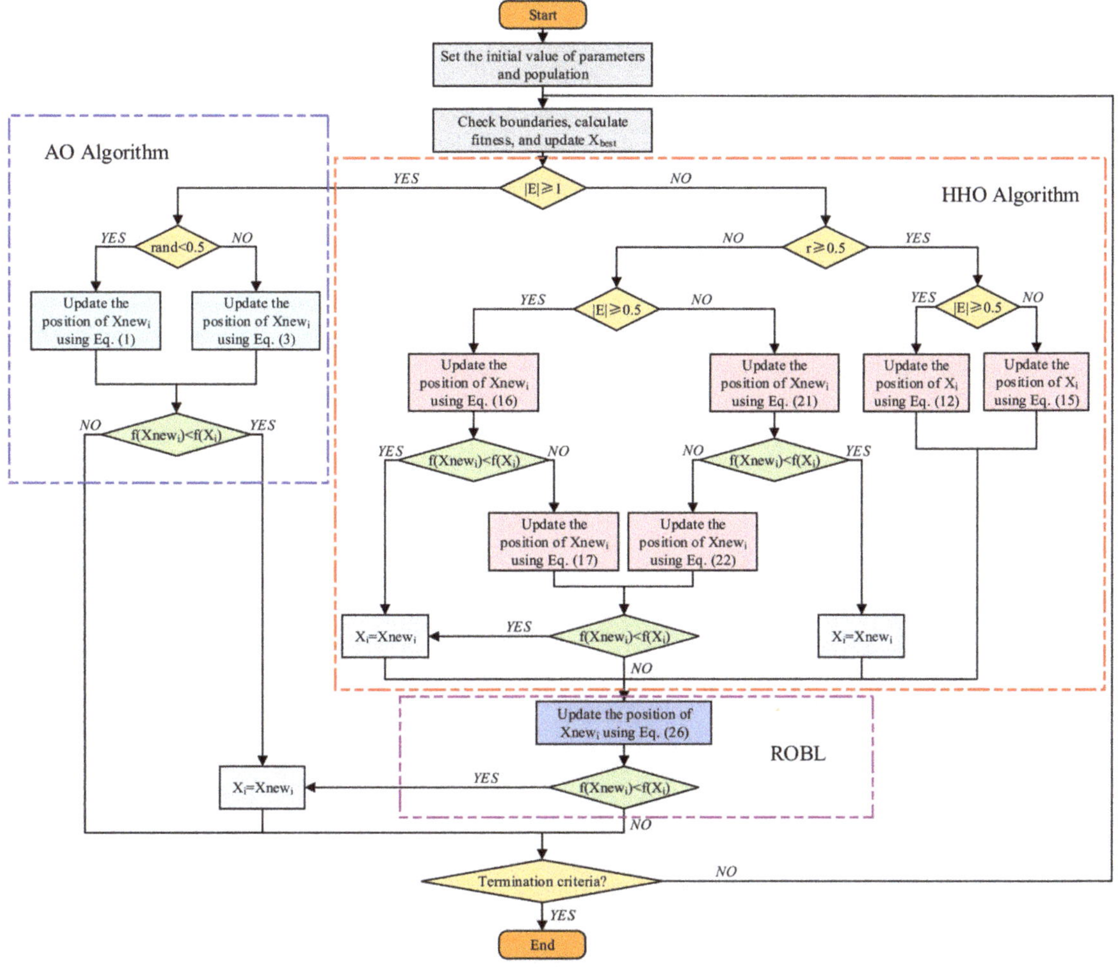

Figure 4. IHAOHHO algorithm flowchart.

4. Results and Discussion

In this section, two main experiments were carried out to evaluate the performance of the IHAOHHO algorithm. The first kind of experiments is benchmark function experiments, which aimed to evaluate the performance of IHAOHHO in solving 23 numerical optimization problems. The second experiment is industrial engineering design problems, which aimed to evaluate the performance of IHAOHHO in solving real-world problems. All experiments are implemented in MATLAB R2016a on a PC with Intel (R) core (TM) i5-9500 CPU @ 3.00 GHz and RAM 2 GB memory on OS windows 10.

4.1. Benchmark Function Experiments

To investigate the performance of the IHAOHHO algorithm, 23 standard benchmark functions of three different types were utilized for testing [64]. The main characteristic of the first type, namely unimodal benchmark functions, is that there is only one global optimum with no local optima. These test functions can be used to evaluate the exploitation capability and convergence rate of an algorithm. The second type, namely multimodal benchmark functions, has a global optimum and multiple local optima, which includes general and fixed-dimension multimodal test functions. This type of functions was utilized to evaluate the exploitation and local optima avoidance capability of the algorithm. The benchmark function details, including dimensions, ranges, and optima, are listed in Tables 1–3.

Table 1. Unimodal benchmark functions.

Function	Dim	Range	Fmin
$F_1(x) = \sum_{i=1}^{n} x_i^2$	30	(−100, 100)	0
$F_2(x) = \sum_{i=1}^{n} \lvert x_i \rvert + \prod_{i=1}^{n} \lvert x_i \rvert$	30	(−10, 10)	0
$F_3(x) = \sum_{i=1}^{n} \left(\sum_{j-1}^{i} x_j\right)^2$	30	(−100, 100)	0
$F_4(x) = \max_i\{\lvert x_i \rvert, 1 \leq i \leq n\}$	30	(−100, 100)	0
$F_5(x) = \sum_{i=1}^{n-1} \left[100(x_{i+1} - x_i^2)^2 + (x_i - 1)^2\right]$	30	(−30,30)	0
$F_6(x) = \sum_{i=1}^{n} (x_i + 5)^2$	30	(−100, 100)	0
$F_7(x) = \sum_{i=1}^{n} i x_i^4 + random[0, 1)$	30	(−1.28, 1.28)	0

Table 2. Multimodal benchmark functions.

Function	Dim	Range	Fmin
$F_8(x) = \sum_{i=1}^{n} -x_i \sin(\sqrt{\lvert x_i \rvert})$	30	(−500, 500)	−418.9829 × 30
$F_9(x) = \sum_{i=1}^{n} \left[x_i^2 - 10\cos(2\pi x_i) + 10\right]$	30	(−5.12, 5.12)	0
$F_{10}(x) = -20\exp(-0.2\sqrt{\frac{1}{n}\sum_{i=1}^{n} x_i^2}) - \exp(\frac{1}{n}\sum_{i=1}^{n}\cos(2\pi x_i)) + 20 + e$	30	(−32, 32)	0
$F_{11}(x) = \frac{1}{4000}\sum_{i=1}^{n} x_i^2 - \prod_{i=1}^{n}\cos(\frac{x_i}{\sqrt{i}}) + 1$	30	(−600, 600)	0
$F_{12}(x) = \frac{\pi}{n}\{10\sin(\pi y_1) + \sum_{i=1}^{n-1}(y_i - 1)^2[1 + 10\sin^2(\pi y_{i+1})] + (y_n - 1)^2\}$ $+\sum_{i=1}^{n} u(x_i, 10, 100, 4)$, where $y_i = 1 + \frac{x_i + 1}{4}$, $u(x_i, a, k, m) = \begin{cases} k(x_i - a)^m & x_i > a \\ 0 & -a < x_i < a \\ k(-x_i - a)^m & x_i < -a \end{cases}$	30	(−50, 50)	0
$F_{13}(x) = 0.1(\sin^2(3\pi x_1) + \sum_{i=1}^{n}(x_i - 1)^2[1 + \sin^2(3\pi x_i + 1)]$ $+(x_n - 1)^2[1 + \sin^2(2\pi x_n)]) + \sum_{i=1}^{n} u(x_i, 5, 100, 4)$	30	(−50, 50)	0

Table 3. Fixed-dimension multimodal benchmark functions.

Function	Dim	Range	Fmin
$F_{14}(x) = \left(\frac{1}{500} + \sum_{j=1}^{25}\frac{1}{j + \sum_{i=1}^{2}(x_i - a_{ij})^6}\right)^{-1}$	2	(−65, 65)	1
$F_{15}(x) = \sum_{i=1}^{11}\left[a_i - \frac{x_1(b_i^2 + b_i x_2)}{b_i^2 + b_i x_3 + x_4}\right]^2$	4	(−5, 5)	0.00030
$F_{16}(x) = 4x_1^2 - 2.1x_1^4 + \frac{1}{3}x_1^6 + x_1 x_2 - 4x_2^2 + x_2^4$	2	(−5, 5)	−1.0316
$F_{17}(x) = \left(x_2 - \frac{5.1}{4\pi^2}x_1^2 + \frac{5}{\pi}x_1 - 6\right)^2 + 10\left(1 - \frac{1}{8\pi}\right)\cos x_1 + 10$	2	(−5, 5)	0.398
$F_{18}(x) = [1 + (x_1 + x_2 + 1)^2(19 - 14x_1 + 3x_1^2 - 14x_2 + 6x_1x_2 + 3x_2^2)]$ $\times[30 + (2x_1 - 3x_2)^2 \times (18 - 32x_2 + 12x_1^2 + 48x_2 - 36x_1x_2 + 27x_2^2)]$	2	(−2, 2)	3
$F_{19}(x) = -\sum_{i=1}^{4} c_i \exp(-\sum_{j=1}^{3} a_{ij}(x_j - p_{ij})^2)$	3	(−1, 2)	−3.86
$F_{20}(x) = -\sum_{i=1}^{4} c_i \exp(-\sum_{j=1}^{6} a_{ij}(x_j - p_{ij})^2)$	6	(0, 1)	−3.32
$F_{21}(x) = -\sum_{i=1}^{5}\left[(X - a_i)(X - a_i)^T + c_i\right]^{-1}$	4	(0, 10)	−10.1532
$F_{22}(x) = -\sum_{i=1}^{7}\left[(X - a_i)(X - a_i)^T + c_i\right]^{-1}$	4	(0, 10)	−10.4028
$F_{23}(x) = -\sum_{i=1}^{10}\left[(X - a_i)(X - a_i)^T + c_i\right]^{-1}$	4	(0, 10)	−10.5363

For verification of the results, the IHAOHHO algorithm was compared with the original AO and HHO; SMA as one of the recent algorithms; SSA, WOA, and GWO as several classical meta-heuristic algorithms; and PSO as the most well-known swarm intelligence algorithm. For all these algorithms, we set the population size N = 30, dimension size D = 30, maximum number of iterations T = 500, and ran them 30 times independently. The parameter settings of each algorithm are shown in Table 4. After all, the average and standard deviation results of these test functions are exhibited in Tables 4–6. Figure 5 shows the convergence curves of 23 test functions. The partial search history, trajectory and average fitness maps are represented in Figure 6. Wilcoxon signed-rank test results are also listed in Table 6. The detailed data analysis is given in the following subsections.

Table 4. Parameter settings for the comparative algorithms.

Algorithm	Parameters
AO	$U = 0.00565$; $r1 = 10$; $\omega = 0.005$; $\alpha = 0.1$; $\delta = 0.1$; $G1 \in [-1, 1]$; $G2 = [2, 0]$
HHO	$q \in [0, 1]$; $r \in [0, 1]$; $E0 \in [-1, 1]$; $E1 = [2, 0]$; $E \in [-2, 2]$;
SMA	$z = 0.03$
SSA	$c1 = [1, 0]$; $c2 \in [0, 1]$; $c3 \in [0, 1]$
WOA	$a1 = [2, 0]$; $a2 = [-1, -2]$; $b = 1$
GWO	$a = [2, 0]$
PSO	$c1 = 2$; $c2 = 2$; $vmax = 6$

4.1.1. Evaluation of Exploitation Capability (Functions F1–F7)

Functions F1–F7 are used to investigate the exploitation capability of the algorithm since they have only one global optimum and no local optima. It can be seen from Table 5 that IHAOHHO can achieve much better results than other meta-heuristic algorithms excluding F6. For F1 and F3, IHAOHHO can find the theoretical optimum. For all unimodal functions excluding F6, IHAOHHO gets the smallest average values and standard deviations compared to other algorithms, which indicate the best accuracy and stability. Hence, the exploitation capability of the proposed IHAOHHO algorithm is excellent.

4.1.2. Evaluation of Exploration Capability (Functions F8–F23)

Multimodal functions F8–F23 contain plentiful local optima whose number increases exponentially with the dimension size of the problem. This kind of functions is very useful to evaluate the exploration ability of the algorithm. From the results shown in Table 5, IHAOHHO outperforms other algorithms in most of the multimodal and fixed-dimension multimodal functions. For multimodal functions F8–F13, IHAOHHO almost obtains all the best average values and standard deviations. Among ten fixed-dimensions multimodal functions F14–F23, IHAOHHO achieves the best accuracy of eight functions and best stability of four functions. These results indicate that IHAOHHO also provides robust exploitation capability.

Table 5. Results of algorithms on 23 benchmark functions.

F		IHAOHHO	AO	HHO	SMA	SSA	WOA	GWO	PSO
F1	Avg	0.0000×10^{0}	2.5120×10^{-128}	1.7359×10^{-98}	6.7559×10^{-287}	2.0918×10^{-7}	7.0172×10^{-75}	2.7553×10^{-27}	1.7920×10^{-4}
	Std	0.0000×10^{0}	1.3759×10^{-127}	3.8748×10^{-98}	0.0000×10^{0}	2.5521×10^{-7}	2.0985×10^{-74}	7.4745×10^{-27}	2.1473×10^{-4}
F2	Avg	3.1773×10^{-283}	3.0714×10^{-51}	3.6162×10^{-49}	1.7722×10^{-136}	2.1400×10^{0}	2.1103×10^{-49}	7.2224×10^{-17}	2.2676×10^{-1}
	Std	0.0000×10^{0}	1.6823×10^{-50}	1.9747×10^{-48}	9.7069×10^{-136}	1.5737×10^{0}	1.1221×10^{-48}	4.3158×10^{-17}	2.0215×10^{-2}
F3	Avg	0.0000×10^{0}	2.3884×10^{-101}	7.9368×10^{-70}	2.7958×10^{-305}	1.5707×10^{3}	4.8346×10^{4}	1.9688×10^{-5}	8.7992×10^{1}
	Std	0.0000×10^{0}	9.262×10^{-101}	4.3417×10^{-69}	0.0000×10^{0}	1.0057×10^{3}	1.5295×10^{4}	8.5080×10^{-5}	3.7192×10^{1}
F4	Avg	1.1105×10^{-281}	1.0656×10^{-53}	1.2768×10^{-49}	1.0217×10^{-160}	1.1623×10^{1}	5.4222×10^{1}	9.2533×10^{-7}	1.0783×10^{0}
	Std	0.0000×10^{0}	5.8309×10^{-53}	4.4293×10^{-49}	5.5961×10^{-160}	3.3373×10^{0}	2.9852×10^{1}	9.1688×10^{-7}	2.1854×10^{-1}
F5	Avg	2.8203×10^{-3}	6.4303×10^{-3}	1.1390×10^{-2}	9.4019×10^{0}	3.1709×10^{2}	2.7969×10^{1}	2.7412×10^{1}	1.0424×10^{2}
	Std	4.4716×10^{-3}	9.1289×10^{-3}	1.2058×10^{-2}	1.2466×10^{1}	8.0601×10^{2}	4.5551×10^{-1}	8.8086×10^{-1}	9.9130×10^{1}
F6	Avg	4.2411×10^{-6}	1.1861×10^{-4}	1.1430×10^{-4}	5.2584×10^{-3}	3.5188×10^{-7}	3.6078×10^{-1}	8.0826×10^{-1}	1.1828×10^{-4}
	Std	6.2092×10^{-6}	2.1625×10^{-4}	1.4084×10^{-4}	3.1160×10^{-3}	7.3563×10^{-7}	1.8848×10^{-1}	3.3042×10^{-1}	1.3013×10^{-4}
F7	Avg	7.1381×10^{-5}	9.2969×10^{-5}	1.4408×10^{-4}	2.2317×10^{-4}	1.7310×10^{-1}	2.6756×10^{-3}	2.2547×10^{-3}	1.8040×10^{-1}
	Std	7.6852×10^{-5}	1.1466×10^{-4}	1.5482×10^{-4}	1.6750×10^{-4}	7.8997×10^{-2}	2.3949×10^{-3}	1.1317×10^{-3}	7.5627×10^{-2}
F8	Avg	−12,447.8654	−7073.9882	−12,568.7811	−12,568.9426	−7591.3246	−10,430.3986	−6049.3246	−5317.3115
	Std	4.5359×10^{2}	3.5511×10^{3}	1.3999×10^{0}	4.0261×10^{-1}	6.9106×10^{2}	1.9097×10^{3}	8.0214×10^{2}	1.5005×10^{3}
F9	Avg	0.0000×10^{0}	0.0000×10^{0}	0.0000×10^{0}	0.0000×10^{0}	5.5253×10^{1}	1.8948×10^{-15}	4.8419×10^{0}	5.6659×10^{1}
	Std	0.0000×10^{0}	0.0000×10^{0}	0.0000×10^{0}	0.0000×10^{0}	1.9037×10^{1}	1.0378×10^{-14}	6.2042×10^{0}	1.5111×10^{1}
F10	Avg	8.8818×10^{-16}	8.8818×10^{-16}	8.8818×10^{-16}	8.8818×10^{-16}	2.7561×10^{0}	3.9672×10^{-15}	1.0356×10^{-13}	2.0903×10^{-1}
	Std	0.0000×10^{0}	0.0000×10^{0}	0.0000×10^{0}	0.0000×10^{0}	1.9773×10^{0}	2.4210×10^{-15}	2.1323×10^{-14}	4.4871×10^{-1}
F11	Avg	0.0000×10^{0}	0.0000×10^{0}	0.0000×10^{0}	0.0000×10^{0}	1.7030×10^{-2}	5.9385×10^{-3}	2.5384×10^{-3}	4.9459×10^{-3}
	Std	0.0000×10^{0}	0.0000×10^{0}	0.0000×10^{0}	0.0000×10^{0}	1.9430×10^{-2}	3.2527×10^{-2}	8.7348×10^{-3}	9.4682×10^{-3}
F12	Avg	5.3164×10^{-7}	4.6513×10^{-6}	9.2636×10^{-6}	5.0331×10^{-3}	7.0564×10^{0}	2.5778×10^{-2}	3.7730×10^{-2}	6.9126×10^{-3}
	Std	9.6698×10^{-7}	8.9371×10^{-6}	1.2911×10^{-5}	6.3463×10^{-3}	3.0595×10^{0}	2.0942×10^{-2}	1.8369×10^{-2}	2.6301×10^{-2}
F13	Avg	1.1694×10^{-5}	3.3938×10^{-5}	1.2604×10^{-4}	7.3800×10^{-3}	1.7887×10^{1}	5.8549×10^{-1}	6.1135×10^{-1}	4.4120×10^{-3}
	Std	1.7961×10^{-5}	3.2363×10^{-5}	1.5375×10^{-4}	8.9329×10^{-3}	1.5307×10^{1}	2.9719×10^{-1}	1.7136×10^{-1}	6.6275×10^{-3}
F14	Avg	1.7919×10^{0}	1.5940×10^{0}	1.1635×10^{0}	9.9800×10^{-1}	1.1637×10^{0}	5.0748×10^{0}	5.2681×10^{0}	3.5906×10^{0}
	Std	9.1746×10^{-1}	2.1763×10^{0}	4.5784×10^{-1}	1.1156×10^{-12}	3.7678×10^{-1}	4.4603×10^{0}	4.6022×10^{0}	2.904×10^{0}
F15	Avg	3.5291×10^{-4}	5.5590×10^{-4}	4.0350×10^{-4}	5.1576×10^{-4}	2.8218×10^{-3}	6.6118×10^{-4}	6.3719×10^{-3}	9.3864×10^{-4}
	Std	4.8766×10^{-5}	1.1640×10^{-4}	2.3353×10^{-4}	3.0066×10^{-4}	5.9580×10^{-3}	7.1226×10^{-4}	1.2424×10^{-2}	2.6081×10^{-4}
F16	Avg	-1.0316×10^{0}	-1.0311×10^{0}	-1.0316×10^{0}	-1.0316×10^{0}	-1.0316×10^{0}	-1.0316×10^{0}	-1.0316×10^{0}	-1.0316×10^{0}
	Std	1.0379×10^{-10}	3.7614×10^{-4}	2.5745×10^{-9}	4.3934×10^{-10}	2.0489×10^{-14}	6.1164×10^{-10}	1.4772×10^{-8}	6.4539×10^{-16}
F17	Avg	3.9789×10^{-1}	3.9812×10^{-1}	3.9790×10^{-1}	3.9789×10^{-1}	3.9789×10^{-1}	3.9789×10^{-1}	3.9789×10^{-1}	3.9789×10^{-1}
	Std	5.4022×10^{-7}	2.2378×10^{-4}	2.4237×10^{-5}	2.4814×10^{-8}	1.4663×10^{-14}	8.5493×10^{-6}	8.9987×10^{-7}	0.0000×10^{0}
F18	Avg	3.0000×10^{0}	3.0439×10^{0}	3.0000×10^{0}	3.0000×10^{0}	3.0000×10^{0}	3.0000×10^{0}	3.0000×10^{0}	3.0000×10^{0}
	Std	2.714×10^{-7}	6.4693×10^{-2}	1.6198×10^{-7}	4.7705×10^{-10}	9.5042×10^{-14}	2.6269×10^{-4}	4.7607×10^{-5}	1.639×10^{-15}

Table 5. *Cont.*

F		IHAOHHO	AO	HHO	SMA	SSA	WOA	GWO	PSO
F19	Avg	-3.8628×10^{0}	-3.8539×10^{0}	-3.8616×10^{0}	-3.8628×10^{0}	-3.8628×10^{0}	-3.8597×10^{0}	-3.8593×10^{0}	-3.8628×10^{0}
	Std	1.8351×10^{-4}	6.0669×10^{-3}	1.7013×10^{-3}	3.0254×10^{-7}	8.1972×10^{-13}	3.1652×10^{-3}	4.2427×10^{-3}	2.6823×10^{-15}
F20	Avg	-3.1298×10^{0}	-3.1572×10^{0}	-3.0533×10^{0}	-3.2425×10^{0}	-3.2215×10^{0}	-3.2391×10^{0}	-3.2442×10^{0}	-3.2665×10^{0}
	Std	1.1264×10^{-1}	1.0448×10^{-1}	1.1671×10^{-1}	5.7177×10^{-2}	5.1720×10^{-2}	1.3596×10^{-1}	9.0427×10^{-2}	6.0328×10^{-2}
F21	Avg	-1.0152×10^{1}	-1.0142×10^{1}	-5.5370×10^{0}	-1.0152×10^{1}	-7.3774×10^{0}	-9.0891×10^{0}	-9.1419×10^{0}	-6.7868×10^{0}
	Std	5.6352×10^{-4}	1.8288×10^{-2}	1.484×10^{0}	2.2592×10^{-3}	2.9079×10^{0}	2.0545×10^{0}	2.3491×10^{0}	3.2622×10^{0}
F22	Avg	-1.0402×10^{1}	-1.0388×10^{1}	-5.2528×10^{0}	-1.0402×10^{1}	-8.1232×10^{0}	-7.5395×10^{0}	-1.0401×10^{1}	-8.1542×10^{0}
	Std	6.3272×10^{-4}	2.4782×10^{-2}	9.3628×10^{-1}	7.5981×10^{-4}	3.3371×10^{0}	3.1570×10^{0}	8.9128×10^{-4}	3.2898×10^{0}
F23	Avg	-1.0535×10^{1}	-1.0525×10^{1}	-5.2858×10^{0}	-1.0535×10^{1}	-7.6861×10^{0}	-6.6213×10^{0}	-1.0535×10^{1}	-1.0087×10^{1}
	Std	9.8617×10^{-4}	6.9516×10^{-3}	8.8012×10^{-1}	1.3006×10^{-3}	3.6004×10^{0}	3.0127×10^{0}	9.0143×10^{-4}	1.7472×10^{0}

Table 6. *p*-Values from the Wilcoxon signed-rank test for the results in Table 5.

F	IHAOHHO vs. AO	IHAOHHO vs. HHO	IHAOHHO vs. SMA	IHAOHHO vs. SSA	IHAOHHO vs. WOA	IHAOHHO vs. GWO	IHAOHHO vs. PSO
F1	6.1035×10^{-5}	6.1035×10^{-5}	N/A	6.1035×10^{-5}	6.1035×10^{-5}	6.1035×10^{-5}	6.1035×10^{-5}
F2	6.1035×10^{-5}	6.1035×10^{-5}	6.1035×10^{-5}	6.1035×10^{-5}	6.1035×10^{-5}	6.1035×10^{-5}	6.1035×10^{-5}
F3	6.1035×10^{-5}	6.1035×10^{-5}	N/A	6.1035×10^{-5}	6.1035×10^{-5}	6.1035×10^{-5}	6.1035×10^{-5}
F4	6.1035×10^{-5}	6.1035×10^{-5}	1.2207×10^{-4}	6.1035×10^{-5}	6.1035×10^{-5}	6.1035×10^{-5}	6.1035×10^{-5}
F5	6.7877×10^{-1}	6.3867×10^{-1}	6.1035×10^{-5}	6.1035×10^{-5}	6.1035×10^{-5}	6.1035×10^{-5}	6.1035×10^{-5}
F6	1.5076×10^{-2}	8.5449×10^{-4}	6.1035×10^{-5}	8.5449×10^{-4}	6.1035×10^{-5}	6.1035×10^{-5}	6.1035×10^{-5}
F7	8.0396×10^{-1}	4.2725×10^{-3}	3.0518×10^{-4}	6.1035×10^{-5}	1.2207×10^{-4}	6.1035×10^{-5}	6.1035×10^{-5}
F8	1.0699×10^{-3}	5.5359×10^{-3}	6.7139×10^{-3}	8.5449×10^{-4}	7.2998×10^{-2}	6.1035×10^{-5}	6.1035×10^{-5}
F9	N/A	N/A	N/A	6.1035×10^{-5}	N/A	6.1035×10^{-5}	6.1035×10^{-5}
F10	N/A	N/A	N/A	6.1035×10^{-5}	4.8828×10^{-4}	6.1035×10^{-5}	6.1035×10^{-5}
F11	N/A	N/A	N/A	6.1035×10^{-5}	N/A	6.2500×10^{-2}	6.1035×10^{-5}
F12	9.7797×10^{-1}	2.7686×10^{-1}	6.1035×10^{-5}	6.1035×10^{-5}	6.1035×10^{-5}	6.1035×10^{-5}	5.2448×10^{-1}
F13	8.9038×10^{-1}	3.5339×10^{-2}	6.1035×10^{-5}	6.1035×10^{-5}	6.1035×10^{-5}	6.1035×10^{-5}	2.1545×10^{-2}
F14	3.5339×10^{-2}	3.8940×10^{-2}	1.2207×10^{-4}	4.7913×10^{-2}	6.1035×10^{-4}	1.0699×10^{-1}	2.1545×10^{-2}
F15	3.3569×10^{-3}	7.1973×10^{-1}	4.7913×10^{-2}	6.1035×10^{-5}	1.1597×10^{-3}	2.1545×10^{-2}	6.1035×10^{-5}
F16	6.1035×10^{-5}	3.0151×10^{-2}	8.5449×10^{-4}	4.0283×10^{-3}	4.2725×10^{-3}	6.1035×10^{-5}	1.2207×10^{-4}
F17	6.1035×10^{-5}	3.0280×10^{-1}	1.0254×10^{-2}	6.1035×10^{-5}	6.7139×10^{-3}	2.5574×10^{-2}	6.1035×10^{-5}
F18	6.1035×10^{-5}	8.3618×10^{-3}	8.3618×10^{-3}	3.0518×10^{-4}	1.2207×10^{-4}	6.1035×10^{-5}	6.1035×10^{-5}
F19	N/A	N/A	N/A	N/A	N/A	N/A	6.1035×10^{-5}
F20	7.2998×10^{-2}	1.8762×10^{-1}	7.2998×10^{-2}	1.0699×10^{-2}	2.7686×10^{-1}	1.0254×10^{-2}	3.3569×10^{-3}
F21	1.8762×10^{-1}	6.1035×10^{-5}	4.8871×10^{-1}	4.2120×10^{-1}	8.5449×10^{-4}	5.9949×10^{-3}	2.5574×10^{-2}
F22	4.7913×10^{-2}	6.1035×10^{-5}	1.8066×10^{-2}	8.0396×10^{-1}	1.2207×10^{-4}	2.0776×10^{-1}	8.3618×10^{-3}
F23	6.1035×10^{-5}	6.1035×10^{-5}	5.5359×10^{-2}	8.3252×10^{-2}	6.1035×10^{-5}	8.3252×10^{-2}	8.3252×10^{-2}

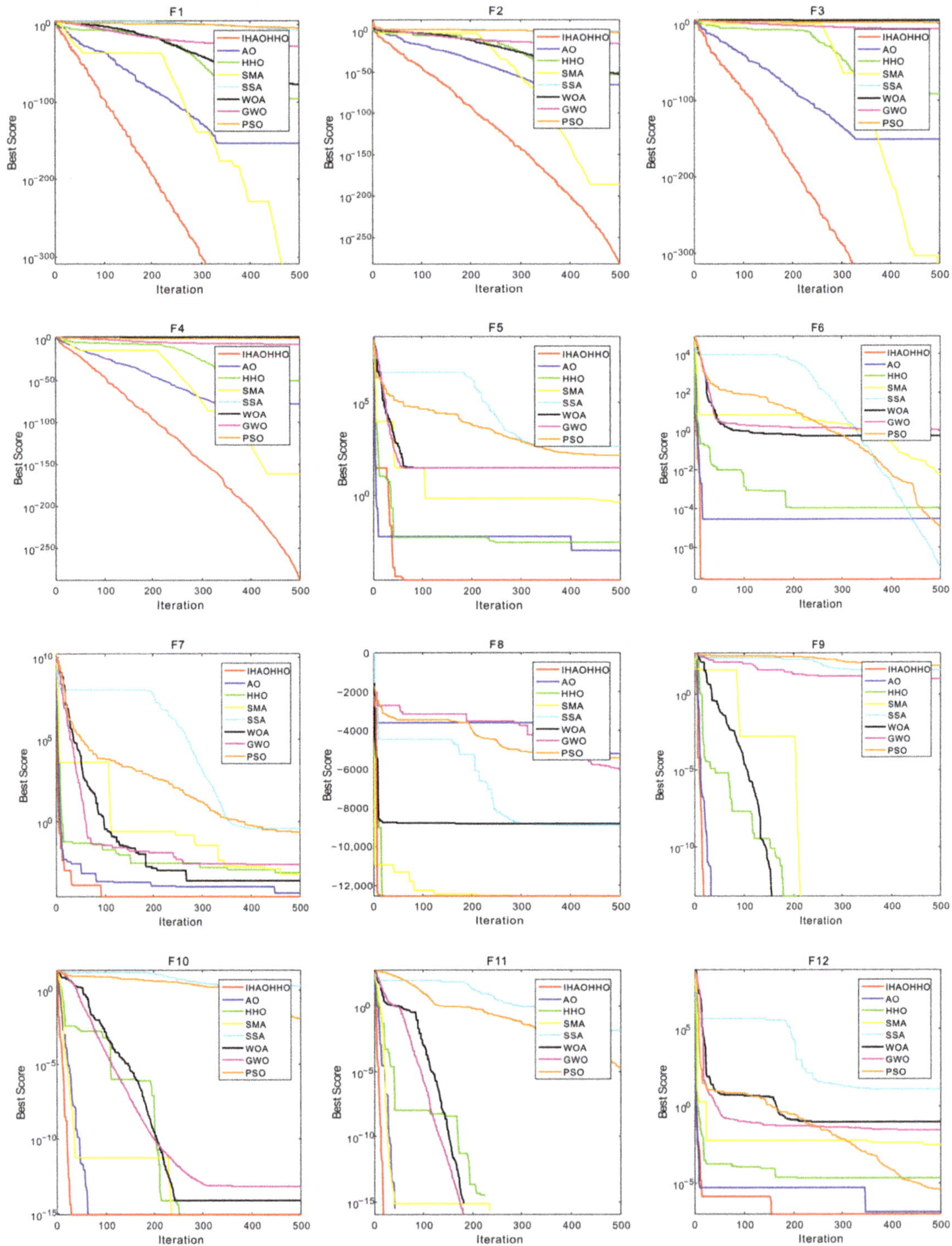

Figure 5. *Cont.*

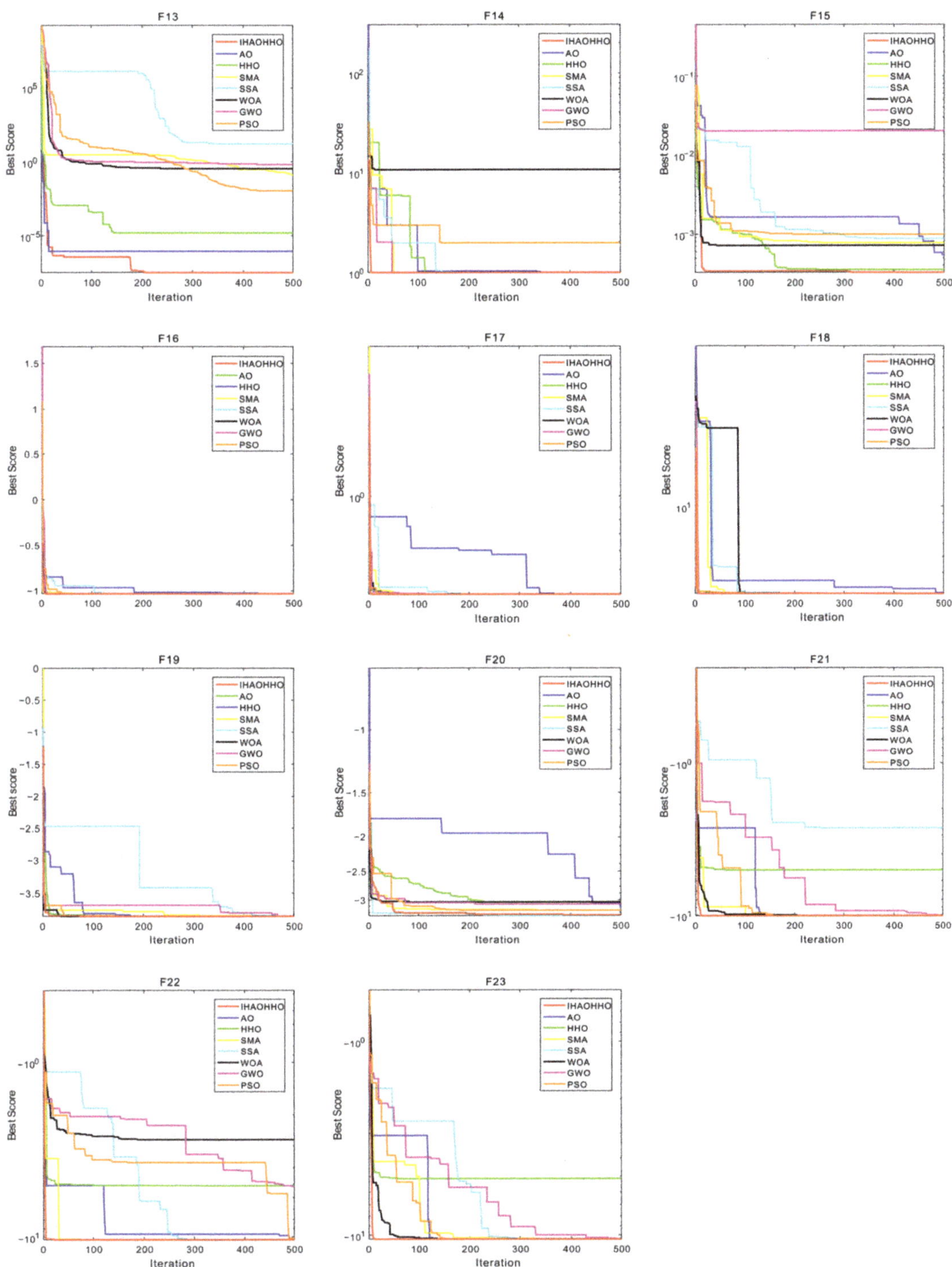

Figure 5. Convergence curves of 23 benchmark functions.

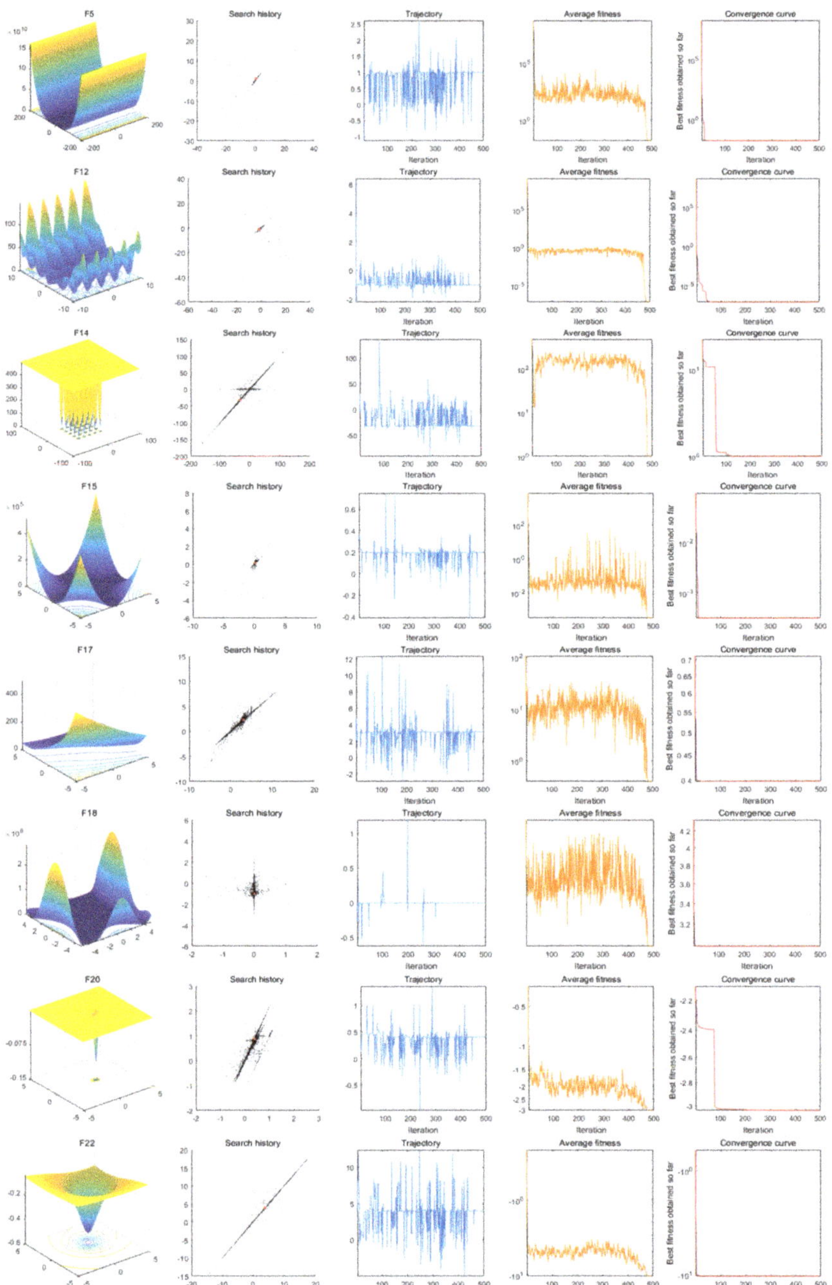

Figure 6. Parameter space, search history, trajectory, average fitness, and convergence curve of IHAOHHO.

4.1.3. Analysis of Convergence Behavior

In the light of the mathematical formula of the IHAOHHO algorithm, search agents tend to investigate promising regions of the search space widely and then exploit it in detail. Search agents change drastically in early iterations and then converge gradually as the number of iterations increases. Convergence curves of the proposed IHAOHHO and AO, HHO, SMA, SSA, WOA, GWO, and PSO for 23 benchmark functions are provided in Figure 5, which shows the convergence rate of algorithms. It can be seen that IHAOHHO shows great superiority compared to other state-of-the-art algorithms. The IHAOHHO algorithm presents three different convergence behaviors during optimization processes. Firstly, for F1-F4, IHAOHHO gradually converges to the optimal values at a faster speed than other algorithms, and the optimal value is better than the others in three of the functions. The second behaviour is extremely fast convergence speed, as observed in F6, F8–F11, F14–F19, and F21–F23. For these functions, IHAOHHO can find the optimum at an extremely fast speed within 20 iterations, and the accurate approximation of the global optimum is almost the best. The last behaviour is observed in F5, F7, F12, F13, and F20 and shows the local optimum avoidance capability of IHAOHHO. The proposed algorithm jumps out of the local optimum after several times of stagnation. This is probably due to the effect of nonlinear escaping energy parameter. Overall, IHAOHHO can efficiently achieve great solutions for all these 23 standard benchmark functions.

In addition, the search history, trajectory, and average fitness figures of several functions are given in Figure 6. Search history figures show us how the algorithm explores and exploits the search space while solving optimization problems. Trajectory figures reveal the order in which an algorithm explores and exploits the search space. Meanwhile, average fitness presents if exploration and exploitation are conducive to improve the first random population, and an accurate approximation of the global optimum can be found in the end. Inspecting Figure 6, the IHAOHHO algorithm samples the most promising areas observed from search histories. Because of the fast convergence, the vast majority of search agents are concentrated near the global optimum. From trajectory and average fitness maps, it can be noticed that exploration almost spread throughout the iterative process until the last 50 iterations focused on exploitation, and average fitness decreased abruptly and leveled off accordingly. The average fitness figures also show the great improvement of the first random population and the acquisition of the final global optimal accurate approximation.

4.1.4. The Wilcoxon Test

Furthermore, the Wilcoxon signed-rank test results are listed in Table 6, which is used to evaluate the statistical performance differences between the proposed IHAOHHO algorithm with other algorithms. It is worth noting that a p-value less than 0.05 means that there is a significant difference between the two compared algorithms. In the light of this criterion, IHAOHHO outperforms all other algorithms in varying degrees. This superiority is statistically significant on unimodal functions F1–F7, which indicates that IHAOHHO benefits from high exploitation. IHAOHHO also shows better results on multimodal function F8–F23, from which we may conclude that IHAOHHO has a high capability of exploration to investigate the most promising regions in the search space. To sum up, the IHAOHHO algorithm can provide better results on almost all benchmark functions than other comparative algorithms.

4.1.5. Computation Time

The computation time is useful to assess the efficiency for an algorithm in solving optimization problems. From the computation time results of all algorithms shown in Table 7, it is obvious that IHAOHHO spent more time in solving these benchmark functions than other comparative algorithms, especially the earlier classic methods of SSA, WOA, GWO, and PSO. The computation time of IHAOHHO is also slightly longer than the basic AO and HHO, which may be ascribed to the ROBL strategy. ROBL produces one more candidate solution in each iteration, increasing the computation time. However, the IHAOHHO took less time than SMA on most test functions. In view of the best search performance of IHAOHHO and the rapid development of the computational machines, it is acceptable for the proposed algorithm to improve the optimization performance.

Table 7. Computation time results of algorithms on 23 benchmark functions.

F	IHAOHHO	AO	HHO	SMA	SSA	WOA	GWO	PSO
F1	2.8539×10^{-1}	2.3253×10^{-1}	1.3713×10^{-1}	8.8997×10^{-1}	8.5420×10^{-2}	7.5875×10^{-2}	1.1491×10^{-1}	6.5132×10^{-2}
F2	2.8946×10^{-1}	2.5214×10^{-1}	1.4672×10^{-1}	9.1203×10^{-1}	1.0346×10^{-1}	1.1982×10^{-1}	1.2761×10^{-1}	7.3814×10^{-2}
F3	1.6030×10^{0}	9.2890×10^{-1}	9.3324×10^{-1}	1.2673×10^{0}	4.6382×10^{-1}	3.9400×10^{-1}	4.2700×10^{-1}	3.9204×10^{-1}
F4	2.8070×10^{-1}	1.9787×10^{-1}	1.5712×10^{-1}	9.5399×10^{-1}	8.2341×10^{-2}	7.3767×10^{-2}	1.1442×10^{-1}	6.4915×10^{-2}
F5	3.3725×10^{-1}	2.2214×10^{-1}	2.2123×10^{-1}	1.0204×10^{0}	9.8470×10^{-2}	8.7778×10^{-2}	1.2667×10^{-1}	7.8503×10^{-2}
F6	2.7707×10^{-1}	2.0399×10^{-1}	1.7800×10^{-1}	9.0977×10^{-1}	8.2725×10^{-2}	7.4251×10^{-2}	1.1248×10^{-1}	6.5708×10^{-2}
F7	5.0109×10^{-1}	3.0078×10^{-1}	2.8662×10^{-1}	9.5443×10^{-1}	1.3880×10^{-1}	1.2862×10^{-1}	1.6701×10^{-1}	1.1976×10^{-1}
F8	3.9395×10^{-1}	2.3581×10^{-1}	2.3276×10^{-1}	9.7695×10^{-1}	1.0531×10^{-1}	9.7443×10^{-2}	1.3674×10^{-1}	9.1720×10^{-2}
F9	3.2379×10^{-1}	1.9907×10^{-1}	1.9594×10^{-1}	9.5132×10^{-1}	9.5204×10^{-2}	7.9254×10^{-2}	1.1801×10^{-1}	7.4441×10^{-2}
F10	3.5602×10^{-1}	2.3037×10^{-1}	2.3125×10^{-1}	9.4870×10^{-1}	1.0399×10^{-1}	9.0064×10^{-2}	1.2725×10^{-1}	8.3986×10^{-2}
F11	4.0659×10^{-1}	2.4303×10^{-1}	2.4198×10^{-1}	9.3026×10^{-1}	1.1382×10^{-1}	1.0089×10^{-1}	1.3566×10^{-1}	9.2499×10^{-2}
F12	1.0131×10^{0}	6.0006×10^{-1}	6.9400×10^{-1}	1.1939×10^{0}	2.6401×10^{-1}	2.5229×10^{-1}	3.4237×10^{-1}	2.4517×10^{-1}
F13	1.0300×10^{0}	5.6112×10^{-1}	6.1205×10^{-1}	1.1549×10^{0}	2.7393×10^{-1}	2.7208×10^{-1}	3.3915×10^{-1}	2.4746×10^{-1}
F14	2.3159×10^{0}	1.2173×10^{0}	1.5168×10^{0}	8.9676×10^{-1}	5.9818×10^{-1}	6.0722×10^{-1}	5.9328×10^{-1}	5.5450×10^{-1}
F15	2.6135×10^{-1}	1.7086×10^{-1}	1.7031×10^{-1}	3.4136×10^{-1}	9.9034×10^{-2}	7.5482×10^{-2}	6.4104×10^{-2}	4.2546×10^{-2}
F16	2.0719×10^{-1}	1.4146×10^{-1}	1.3859×10^{-1}	2.7170×10^{-1}	5.9081×10^{-2}	4.9666×10^{-2}	6.0033×10^{-2}	4.1193×10^{-2}
F17	1.8138×10^{-1}	1.3529×10^{-1}	1.5833×10^{-1}	2.7311×10^{-1}	5.2979×10^{-2}	4.1321×10^{-2}	4.1556×10^{-2}	2.3066×10^{-2}
F18	1.8108×10^{-1}	1.3183×10^{-1}	1.2693×10^{-1}	2.7041×10^{-1}	5.4471×10^{-2}	4.0487×10^{-2}	4.1752×10^{-2}	2.2830×10^{-2}
F19	3.5119×10^{-1}	2.4635×10^{-1}	2.4016×10^{-1}	3.4125×10^{-1}	9.8544×10^{-2}	8.5903×10^{-2}	9.2049×10^{-2}	7.0253×10^{-2}
F20	3.7106×10^{-1}	2.2549×10^{-1}	2.4656×10^{-1}	4.0519×10^{-1}	1.0270×10^{-1}	9.0294×10^{-2}	9.8020×10^{-2}	7.2095×10^{-2}
F21	5.8451×10^{-1}	3.2169×10^{-1}	3.6385×10^{-1}	4.1713×10^{-1}	1.4920×10^{-1}	1.3664×10^{-1}	1.3885×10^{-1}	1.2170×10^{-1}
F22	7.2861×10^{-1}	3.9414×10^{-1}	4.3943×10^{-1}	4.9071×10^{-1}	1.8789×10^{-1}	1.7154×10^{-1}	1.7638×10^{-1}	1.5215×10^{-1}
F23	9.4549×10^{-1}	4.9412×10^{-1}	5.7464×10^{-1}	4.9527×10^{-1}	2.3551×10^{-1}	2.7717×10^{-1}	2.2549×10^{-1}	2.0513×10^{-1}

4.2. Experiments on Industrial Engineering Design Problems

Most optimization problems have constraints in the real world, so considering equality and inequality constraints during optimization is a necessary process. In this subsection, four well-known constrained industrial engineering design problems, which include pressure vessel design problem, speed reducer design problem, tension/compression spring design problem, and three-bar truss design problem, were solved to further verify the performance of the proposed IHAOHHO algorithm. The results of IHAOHHO were compared to various classical optimizers proposed in previous studies. The parameter settings were as same as the previous experiments.

4.2.1. Pressure Vessel Design Problem

The objective of this problem was to minimize the fabrication cost of the cylindrical pressure vessel to meet the pressure requirements. As shown in Figure 7, four structural parameters in this problem needed to be minimized, including the thickness of the shell (T_s), the thickness of the head (T_h), inner radius (R), and the length of the cylindrical section without the head (L). The formulation of four optimization constraints can be described as follows:

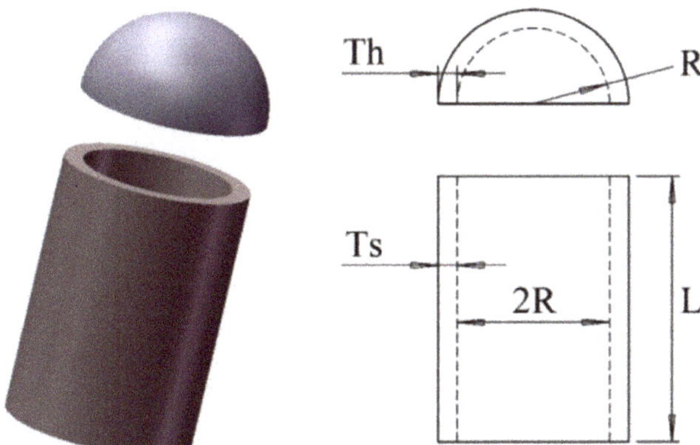

Figure 7. Pressure vessel design problem.

Consider

$$\vec{x} = [x_1\ x_2\ x_3\ x_4] = [T_s\ T_h\ \mathrm{R}\ L], \tag{27}$$

Minimize

$$f(\vec{x}) = 0.6224x_1x_3x_4 + 1.7781x_2x_3^2 + 3.1661x_1^2x_4 + 19.84x_1^2x_3, \tag{28}$$

Subject to

$$\begin{aligned} g_1(\vec{x}) &= -x_1 + 0.0193x_3 \leq 0, \\ g_2(\vec{x}) &= -x_3 + 0.00954x_3 \leq 0, \\ g_3(\vec{x}) &= -\pi x_3^2 x_4 - \tfrac{4}{3}\pi x_3^3 + 1296000 \leq 0, \\ g_4(\vec{x}) &= x_4 - 240 \leq 0, \end{aligned} \tag{29}$$

Variable range

$$\begin{aligned} 0 &\leq x_1 \leq 99, \\ 0 &\leq x_2 \leq 99, \\ 10 &\leq x_3 \leq 200, \\ 10 &\leq x_4 \leq 200, \end{aligned} \tag{30}$$

From the results in Table 8, it is obvious that IHAOHHO can obtain superior optimal values compared to AO, HHO, SMA, WOA, GWO, MVO, GA, ES, and CPSO [65].

Table 8. Comparison of IHAOHHO results with other competitors for the pressure vessel design problem.

Algorithm	Optimum Variables				Optimum Cost
	T_s	T_h	R	L	
IHAOHHO	0.8363559	0.4127868	45.08462	142.9202	5932.3392
AO [55]	1.0540	0.182806	59.6219	38.8050	5949.2258
HHO [42]	0.81758383	0.4072927	42.09174576	176.7196352	6000.46259
SMA [41]	0.7931	0.3932	40.6711	196.2178	5994.1857
WOA [38]	0.8125	0.4375	42.0982699	176.638998	6059.7410
GWO [32]	0.8125	0.4345	42.0892	176.7587	6051.5639
MVO [23]	0.8125	0.4375	42.090738	176.73869	6060.8066
GA [3]	0.8125	0.4375	42.097398	176.65405	6059.94634
ES [6]	0.8125	0.4375	42.098087	176.640518	6059.74560
CPSO [65]	0.8125	0.4375	42.091266	176.7465	6061.0777

4.2.2. Speed Reducer Design Problem

This problem aims to optimize seven variables to minimize the speed reducer's total weights, which include the face width (x_1), module of teeth (x_2), a discrete design variable on behalf of the teeth in the pinion (x_3), length of the first shaft between bearings (x_4), length of the second shaft between bearings (x_5), diameters of the first shaft (x_6), and diameters of the second shaft (x_7). Four constraints—covering stress, bending stress of the gear teeth, stresses in the shafts, and transverse deflections of the shafts, as shown in Figure 8—should be satisfied. The mathematical formulation is represented as follows:

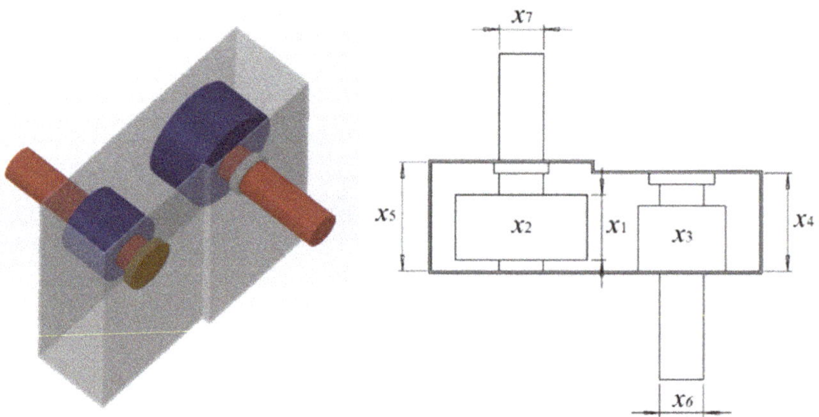

Figure 8. Speed reducer design problem.

Minimize

$$f(\vec{x}) = 0.7854x_1x_2^2(3.3333x_3^2 + 14.9334x_3 - 43.0934) - 1.508x_1(x_6^2 + x_7^2) + 7.4777(x_6^3 + x_7^3), \tag{31}$$

Subject to

$$
\begin{aligned}
g_1(\vec{x}) &= \frac{27}{x_1 x_2^2 x_3} - 1 \le 0, \\
g_2(\vec{x}) &= \frac{397.5}{x_1 x_2^2 x_3^2} - 1 \le 0, \\
g_3(\vec{x}) &= \frac{1.93 x_4^3}{x_2 x_3 x_6^4} - 1 \le 0, \\
g_4(\vec{x}) &= \frac{1.93 x_5^3}{x_2 x_3 x_7^4} - 1 \le 0, \\
g_5(\vec{x}) &= \frac{\sqrt{\left(\frac{745 x_4}{x_2 x_3}\right)^2 + 16.9 \times 10^6}}{110.0 x_6^3} - 1 \le 0, \\
g_6(\vec{x}) &= \frac{\sqrt{\left(\frac{745 x_4}{x_2 x_3}\right)^2 + 157.5 \times 10^6}}{85.0 x_6^3} - 1 \le 0, \\
g_7(\vec{x}) &= \frac{x_2 x_3}{40} - 1 \le 0, \\
g_8(\vec{x}) &= \frac{5 x_2}{x_1} - 1 \le 0, \\
g_9(\vec{x}) &= \frac{x_1}{12 x_2} - 1 \le 0, \\
g_{10}(\vec{x}) &= \frac{1.5 x_6 + 1.9}{x_4} - 1 \le 0, \\
g_{11}(\vec{x}) &= \frac{1.1 x_7 + 1.9}{x_5} - 1 \le 0,
\end{aligned}
\tag{32}
$$

Variable range

$$
\begin{aligned}
&2.6 \le x_1 \le 3.6, \\
&0.7 \le x_2 \le 0.8, \\
&17 \le x_3 \le 28, \\
&7.3 \le x_4 \le 8.3, \\
&7.8 \le x_5 \le 8.3, \\
&2.9 \le x_6 \le 3.9, \\
&5.0 \le x_7 \le 5.5,
\end{aligned}
\tag{33}
$$

Compared to AO, PSO, AOA, MFO [66], GA, SCA, HS [67], FA [68], and MDA [69], IHAOHHO can obviously achieve better results in the speed reducer design problem, as shown in Table 9.

Table 9. Comparison of IHAOHHO results with other competitors for the speed reducer design problem.

Algorithm	Optimum Variables							Optimum Weight
	x_1	x_2	x_3	x_4	x_5	x_6	x_7	
IHAOHHO	3.49924	0.7	17	7.3	7.8191	3.35006	5.28531	2996.0935
AO [55]	3.5021	0.7	17	7.3099	7.7476	3.3641	5.2994	3007.7328
PSO [26]	3.5001	0.7	17.0002	7.5177	7.7832	3.3508	5.2867	3145.922
AOA [25]	3.50384	0.7	17	7.3	7.72933	3.35649	5.2867	2997.9157
MFO [66]	3.49745	0.7	17	7.82775	7.71245	3.35178	5.28635	2998.9408
GA [3]	3.51025	0.7	17	8.35	7.8	3.36220	5.28772	3067.561
SCA [24]	3.50875	0.7	17	7.3	7.8	3.46102	5.28921	3030.563
HS [67]	3.52012	0.7	17	8.37	7.8	3.36697	5.28871	3029.002
FA [68]	3.50749	0.7001	17	7.71967	8.08085	3.35151	5.28705	3010.13749
MDA [69]	3.5	0.7	17	7.3	7.67039	3.54242	5.24581	3019.58336

4.2.3. Tension/Compression Spring Design Problem

In this case, the intention is to minimize the weight of the tension/compression spring shown in Figure 9. Constraints on surge frequency, shear stress, and deflection must be satisfied during optimum design. There are three parameters that needed to be minimized, including the wire diameter (d), mean coil diameter (D), and the number of active coils (N). The mathematical form of this problem can be written as follows:

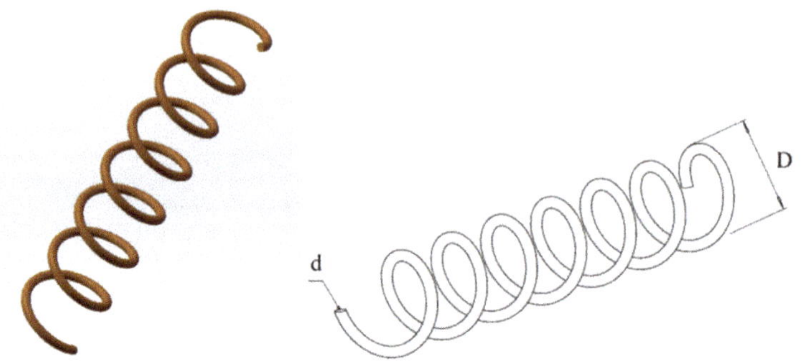

Figure 9. Tension/compression spring design problem.

Consider

$$\vec{x} = [x_1\ x_2\ x_3\ x_4] = [\mathrm{d\ D\ N}], \tag{34}$$

Minimize

$$f(\vec{x}) = (x_3 + 2)x_2 x_1^2, \tag{35}$$

Subject to

$$\begin{aligned} g_1(\vec{x}) &= 1 - \tfrac{x_2^3 x_3}{71{,}785 x_1^4} \le 0, \\ g_2(\vec{x}) &= \tfrac{4x_2^2 - x_1 x_2}{12{,}566(x_2 x_1^3 - x_1^4)} + \tfrac{1}{5108 x_1^2} \le 0, \\ g_3(\vec{x}) &= 1 - \tfrac{140.45 x_1}{x_2^2 x_3} \le 0, \\ g_4(\vec{x}) &= \tfrac{x_1 + x_2}{1.5} - 1 \le 0, \end{aligned} \tag{36}$$

Variable range

$$\begin{aligned} 0.05 &\le x_1 \le 2.00, \\ 0.25 &\le x_2 \le 1.30, \\ 2.00 &\le x_3 \le 15.00, \end{aligned} \tag{37}$$

The proposed IHAOHHO is compared with AO, HHO, SSA, WOA, GWO, PSO, MVO, GA, and HS algorithms. Results are listed in Table 10 and show that the IHAOHHO can attain the best weight values compared to all other algorithms. Additionally, it is clear that the proposed method found a more accurate design with new parameter values.

Table 10. Comparison of IHAOHHO results with other competitors for the tension/compression spring design problem.

Algorithm	**Optimum Variables**			**Optimum Weight**
	d	**D**	**N**	
IHAOHHO	0.055883	0.52784	4.7603	0.011144
AO [55]	0.0502439	0.35262	10.5425	0.011165
HHO [42]	0.051796393	0.359305355	11.138859	0.012665443
SSA [39]	0.051207	0.345215	12.004032	0.0126763
WOA [38]	0.051207	0.345215	12.004032	0.0126763
GWO [32]	0.05169	0.356737	11.28885	0.012666
PSO [26]	0.051728	0.357644	11.244543	0.0126747
MVO [23]	0.05251	0.37602	10.33513	0.012790
GA [3]	0.051480	0.351661	11.632201	0.01270478
HS [67]	0.051154	0.349871	12.076432	0.0126706

4.2.4. Three-Bar Truss Design Problem

The three-bar truss design problem is a classical optimization application in civil engineering field. The main intention of this case is to minimize the weight of a truss with three bars by considering two structural parameters as illustrated in Figure 10. Deflection, stress, and buckling are the three main constrains. The mathematical formulation of this problem is given:

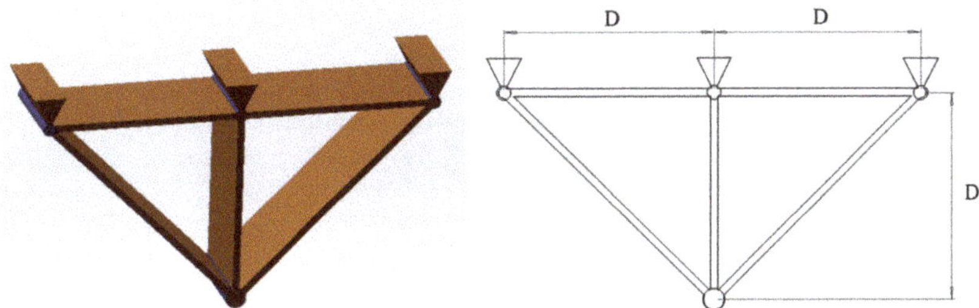

Figure 10. Three-bar truss design problem.

Consider

$$\vec{x} = [x_1\ x_2] = [A_1\ A_2], \tag{38}$$

Minimize

$$f(\vec{x}) = (2\sqrt{2}x_1 + x_2) * l, \tag{39}$$

Subject to

$$\begin{aligned} g_1(\vec{x}) &= \frac{\sqrt{2}x_1 + x_2}{\sqrt{2}x_1^2 + 2x_1x_2} P - \sigma \leq 0, \\ g_2(\vec{x}) &= \frac{x_2}{\sqrt{2}x_1^2 + 2x_1x_2} P - \sigma \leq 0, \\ g_3(\vec{x}) &= \frac{1}{\sqrt{2}x_2 + x_1} P - \sigma \leq 0, \end{aligned} \tag{40}$$

Variable range

$$0 \leq x_1, x_2 \leq 1, \tag{41}$$

where $l = 100\text{cm}, P = 2\text{KN}/\text{cm}^2, \sigma = 2\text{KN}/\text{cm}^2$. Results of IHAOHHO for solving three-bar truss design problem are listed in Table 11, which are compared with AO, HHO, SSA, AOA, MVO, MFO, and GOA [70]. It can be observed that IHAOHHO outperforms other optimization algorithms published in the literature.

Table 11. Comparison of IHAOHHO results with other competitors for the three-bar truss design problem.

Algorithm	Optimum Variables		Optimum Weight
	x_1	x_2	
IHAOHHO	0.79002	0.40324	263.8622
AO [55]	0.7926	0.3966	263.8684
HHO [42]	0.788662816	0.408283133832900	263.8958434
SSA [39]	0.78866541	0.408275784	263.89584
AOA [25]	0.79369	0.39426	263.9154
MVO [23]	0.78860276	0.408453070000000	263.8958499
MFO [66]	0.788244771	0.409466905784741	263.8959797
GOA [70]	0.788897555578973	0.407619570115153	263.895881496069

As a summary, this section demonstrates the superiority of the proposed IHAOHHO algorithms in different characteristics and real case studies. IHAOHHO is able to outperform the original AO and HHO and other well-known algorithms with very competitive results, which were derived from the robust exploration and exploitation capabilities of IHAOHHO. Excellent performance in solving industrial engineering design problems indicates that IHAOHHO can be widely used in real-world optimization problems.

5. Conclusions

This study proposed an improved hybrid Aquila Optimizer and Harris Hawks Optimization by combining the exploration part of AO with the exploitation part of HHO and a nonlinear escaping energy parameter and random opposition-based learning (ROBL) strategy. The proposed method integrated the mentioned search methods to tackle the weakness of the traditional search methods. The proposed IHAOHHO algorithm was tested using 23 mathematical benchmark functions to analyze its exploration, exploitation, local optima avoidance capabilities, and convergence behaviors. Results show competitive results compared to other state-of-the-art meta-heuristic algorithms. To further verify the superiority of IHAOHHO, four industrial engineering design problems were solved. The results are also competitive with other meta-heuristic algorithms.

As future perspectives, binary and multi-objective versions of IHAOHHO will be considered. More applications of this algorithm in different fields are valuable works, including text clustering, scheduling problems, appliances management, parameters estimation, multi-objective engineering problems, feature selection, test classification, image segmentation problems, network applications, sentiment analysis, etc.

Author Contributions: Conceptualization, H.J. and L.A.; methodology, S.W.; software, R.Z.; validation, S.W., Q.L. and R.Z.; formal analysis, S.W.; writing—original draft preparation, S.W.; writing—review and editing, S.W.; visualization, Q.L.; supervision, L.A.; project administration, H.J.; funding acquisition, S.W. and H.J. All authors have read and agreed to the published version of the manuscript.

Funding: This research was funded by the Sanming University Introduces High-level Talents to Start Scientific Research Funding Support Project, grant number 20YG01, 20YG14; the Guiding Science and Technology Projects in Sanming City, grant number 2020-S-39, 2020-G-61, 2021-S-8; the Educational Research Projects of Young and Middle-aged Teachers in Fujian Province, grant number JAT200638, JAT200618; the Scientific Research and Development Fund of Sanming University, grant number B202029, B202009; Collaborative education project of industry university cooperation of the Ministry of Education, grant number 202002064014; School level education and teaching reform project of Sanming University, grant number J2010306, J2010305; and Higher education research project of Sanming University, grant number SHE2102, SHE2013.

Institutional Review Board Statement: Not applicable.

Informed Consent Statement: Not applicable.

Data Availability Statement: Not applicable.

Conflicts of Interest: The authors declare no conflict of interest.

References

1. Abualigah, L.; Diabat, A. Advances in sine cosine algorithm: A comprehensive survey. *Artif. Intell. Rev.* **2021**, *54*, 2567–2608. [CrossRef]
2. Abualigah, L.; Diabat, A. A comprehensive survey of the Grasshopper optimization algorithm: Results, variants, and applications. *Neural Comput. Appl.* **2020**, *32*, 15533–15556. [CrossRef]
3. Holland, J.H. Genetic algorithms. *Sci. Am.* **1992**, *267*, 66–72. [CrossRef]
4. Storn, R.; Price, K. Differential evolution-a simple and efficient heuristic for global optimization over continuous spaces. *J. Glob. Optim.* **1997**, *11*, 341–359. [CrossRef]
5. Koza, J.R. *Genetic Programming: On the Programming of Computers by Means of Natural Selection*; MIT Press: Cambridge, MA, USA, 1992.
6. Rechenberg, I. Evolutionsstrategien. In *Simulationsmethoden in der Medizin und Biologie*; Springer: Berlin/Heidelberg, Germany, 1978; Volume 8, pp. 83–114.
7. Simon, D. Biogeography-based optimization. *IEEE Trans. Evol. Comput.* **2008**, *12*, 702–713. [CrossRef]
8. Yao, X.; Liu, Y.; Lin, G. Evolutionary programming made faster. *IEEE Trans. Evol. Comput.* **1999**, *3*, 82–102. [CrossRef]
9. Dasgupta, D.; Michalewicz, Z. *Evolutionary Algorithms in Engineering Applications*; DBLP: Trier, Germany, 1997.

10. Kirkpatrick, S.; Gelatt, C.D.; Vecchi, M.P. Optimization by simmulated annealing. *Science* **1983**, *220*, 671–680. [CrossRef] [PubMed]
11. Erol, O.K.; Eksin, I. A new optimization method: Big bang-big crunch. *Adv. Eng. Softw.* **2006**, *37*, 106–111. [CrossRef]
12. Rashedi, E.; Nezamabadi-Pour, H.; Saryazdi, S. GSA: A Gravitational Search Algorithm. *Inf. Sci.* **2009**, *179*, 2232–2248. [CrossRef]
13. Webster, B.; Bernhard, P.J. A local search optimization algorithm based on natural principles of gravitation. In *Information & Knowledge Engineering, Proceedings of the 2003 International Conference on Information and Knowledge Engineering (IKE'03), Las Vegas, NV, USA, 23–26 June 2003*; DBLP: Trier, Germany, 2003.
14. Asef, F.; Majidnezhad, V.; Feizi-Derakhshi, M.R.; Parsa, S. Heat transfer relation-based optimization algorithm (HTOA). *Soft Comput.* **2021**, 1–30. [CrossRef]
15. Kaveh, A.; Talatahari, S. A novel heuristic optimization method: Charged system search. *Acta Mech.* **2010**, *213*, 267–289. [CrossRef]
16. Alatas, B. ACROA: Artificial Chemical Reaction Optimization Algorithm for global optimization. *Expert Syst. Appl.* **2011**, *38*, 13170–13180. [CrossRef]
17. Formato, R.A. Central force optimization: A new metaheuristic with applications in applied electromagnetics. *Prog. Electromag. Res.* **2007**, *77*, 425–491. [CrossRef]
18. Kaveh, A.; Khayatazad, M. A new meta-heuristic method: Ray optimization. *Comput. Struct.* **2012**, *112*, 283–294. [CrossRef]
19. Hatamlou, A. Black hole: A new heuristic optimization approach for data clustering. *Inf. Sci.* **2013**, *222*, 175–184. [CrossRef]
20. Du, H.; Wu, X.; Zhuang, J. Small-world optimization algorithm for function optimization. In *Advances in Natural Computation, Advances in Natural Computation, Second International Conference*; ICNC: Xi'an, China, 2006.
21. Shah-Hosseini, H. Principal components analysis by the galaxy-based search algorithm: A novel metaheuristic for continuous optimisation. *Int. J. Comput. Sci. Eng.* **2011**, *6*, 132–140. [CrossRef]
22. Moghaddam, F.F.; Moghaddam, R.F.; Cheriet, M. Curved space optimization: A random search based on general relativity theory. *arXiv* **2012**, arXiv:1208.2214.
23. Mirjalili, S.; Mirjalili, S.M.; Hatamlou, A. Multi-Verse Optimizer: A nature-inspired algorithm for global optimization. *Neural Comput. Appl.* **2015**, *27*, 495–513. [CrossRef]
24. Mirjalili, S. SCA: A Sine Cosine Algorithm for Solving Optimization Problems. *Knowl.-Based Syst.* **2016**, *96*. [CrossRef]
25. Abualigah, L.; Diabat, A.; Mirjalili, S.; Elaziz, M.A.; Gandomi, A.H. The Arithmetic Optimization Algorithm. *Comput. Methods Appl. Mech. Eng.* **2021**, *376*, 113609. [CrossRef]
26. Kennedy, J.; Eberhart, R. Particle swarm optimization. In Proceedings of the 1995 IEEE International Conference on Neural Networks (ICNN '93), Perth, WA, Australia, 27 November–1 December 1995; IEEE: Piscataway, NJ, USA, 1995.
27. Dorigo, M.; Birattari, M.; Stutzle, T. Ant colony optimization. *IEEE Comput. Intell.* **2006**, *1*, 28–39. [CrossRef]
28. Mucherino, A.; Seref, O.; Seref, O.; Kundakcioglu, O.E.; Pardalos, P. Monkey search: A novel metaheuristic search for global optimization. *Am. Inst. Phys.* **2007**, *953*, 162–173. [CrossRef]
29. Yang, X.S. Firefly algorithm, stochastic test functions and design optimization. *Int. J. Bio-Inspired Comput.* **2010**, *2*, 78–84. [CrossRef]
30. Yang, X.S. A new metaheuristic bat-inspired algorithm. In *Nature Inspired Cooperative Strategies for Optimization (NICSO)*; Springer: Berlin/Heidelberg, Germany, 2010.
31. Gandomi, A.H.; Alavi, A.H. Krill Herd: A new bio-inspired optimization algorithm. *Commun. Nonlinear Sci. Numer. Simul.* **2012**, *17*, 4831–4845. [CrossRef]
32. Mirjalili, S.; Mirjalili, S.M.; Lewis, A. Grey wolf optimizer. *Adv. Eng. Softw.* **2014**, *69*, 46–61. [CrossRef]
33. Gandomi, A.H.; Yang, X.S.; Alavi, A.H. Cuckoo search algorithm: A metaheuristic approach to solve structural optimization problems. *Eng. Comput.* **2013**, *29*, 17–35. [CrossRef]
34. Pan, W.T. A new fruit fly optimization algorithm: Taking the financial distress model as an example. *Knowl.-Based Syst.* **2012**, *26*, 69–74. [CrossRef]
35. Yang, S.; Jiang, J.; Yan, G. A dolphin partner optimization. In Proceedings of the 2009 WRI Global Congress on Intelligent Systems (GCIS 2009), Xiamen, China, 19–21 May 2009; IEEE: Piscataway, NJ, USA, 2009.
36. Mirjalili, S. The Ant Lion optimizer. *Adv. Eng. Softw.* **2015**, *83*, 80–98. [CrossRef]
37. Jia, H.; Peng, X.; Lang, C. Remora optimization algorithm. *Expert Syst. Appl.* **2021**, *185*, 115665. [CrossRef]
38. Mirjalili, S.; Lewis, A. The whale optimization algorithm. *Adv. Eng. Softw.* **2016**, *95*, 51–67. [CrossRef]
39. Mirjalili, S.; Gandomi, A.H.; Mirjalili, S.Z.; Saremi, S.; Faris, H.; Mirjalili, S.M. Salp swarm algorithm: A bio-inspired optimizer for engineering design problems. *Adv. Eng. Softw.* **2017**, *114*, 163–191. [CrossRef]
40. Alsattar, H.A.; Zaidan, A.A.; Zaidan, B.B. Novel meta-heuristic bald eagle search optimisation algorithm. *Artif. Intell. Rev.* **2020**, *53*, 2237–2264. [CrossRef]
41. Li, S.M.; Chen, H.L.; Wang, M.J.; Heidari, A.A.; Mirjalili, S. Slime mould algorithm: A new method for stochastic optimization. *Future Gener. Comput. Syst.* **2020**, *111*, 300–323. [CrossRef]
42. Heidari, A.A.; Mirjalili, S.; Faris, H.; Aljarah, I.; Mafarja, M.; Chen, H.L. Harris Hawks optimization: Algorithm and applications. *Future Gener. Comput. Syst.* **2019**, *97*, 849–872. [CrossRef]
43. Yousri, D.; Fathy, A.; Thanikanti, S.B. Recent methodology based Harris Hawks optimizer for designing load frequency control incorporated in multi-interconnected renewable energy plants. *Sustain. Energy Grids Netw.* **2020**, *22*, 100352. [CrossRef]
44. Bui, D.T.; Moayedi, H.; Kalantar, B.; Osouli, A.; Rashid, A. A Novel Swarm Intelligence Technique Harris Hawks Optimization for Spatial Assessment of Landslide Susceptibility. *Sensors* **2019**, *19*, 3590. [CrossRef]

45. Golilarz, N.A.; Gao, H.; Demirel, H. Satellite image de-noising with Harris Hawks meta heuristic optimization algorithm and improved adaptive generalized gaussian distribution threshold function. *IEEE Access* **2019**, *7*, 57459–57468. [CrossRef]
46. Jia, H.; Peng, X.; Kang, L.; Li, Y.; Sun, K. Pulse coupled neural network based on Harris Hawks optimization algorithm for image segmentation. *Multimed Tools Appl.* **2020**, *79*, 28369–28392. [CrossRef]
47. Jia, H.; Lang, C.; Oliva, D.; Song, W.; Peng, X. Dynamic Harris Hawks Optimization with Mutation Mechanism for Satellite Image Segmentation. *Remote Sens.* **2019**, *11*, 1421. [CrossRef]
48. Yousri, D.; Mirjalili, S.; Machado, J.A.T.; Thanikantie, S.B.; Elbaksawi, O.; Fathy, A. Efficient fractional-order modified Harris Hawks optimizer for proton exchange membrane fuel cell modeling. *Eng. Appl. Artif. Intell.* **2021**, *100*, 104193. [CrossRef]
49. Gupta, S.; Deep, K.; Heidari, A.A.; Moayedi, H.; Wang, M. Opposition-based Learning Harris Hawks Optimization with Advanced Transition Rules: Principles and Analysis. *Expert Syst. Appl.* **2020**, *158*, 113510. [CrossRef]
50. Hussien, A.G.; Amin, M. A self-adaptive Harris Hawks optimization algorithm with opposition-based learning and chaotic local search strategy for global optimization and feature selection. *Int. J. Mach. Learn. Cyber.* **2021**, 1–28. [CrossRef]
51. Sihwail, R.; Omar, K.; Ariffin, K.; Tubishat, M. Improved Harris Hawks Optimization Using Elite Opposition-Based Learning and Novel Search Mechanism for Feature Selection. *IEEE Access* **2020**, *8*, 121127–121145. [CrossRef]
52. Bao, X.; Jia, H.; Lang, C. A Novel Hybrid Harris Hawks Optimization for Color Image Multilevel Thresholding Segmentation. *IEEE Access* **2019**, *7*, 76529–76546. [CrossRef]
53. Houssein, E.H.; Hosney, M.E.; Elhoseny, M.; Oliva, D.; Hassaballah, M. Hybrid Harris Hawks Optimization with Cuckoo Search for Drug Design and Discovery in Chemoinformatics. *Sci. Rep.* **2020**, *10*, 14439. [CrossRef] [PubMed]
54. Kaveh, A.; Rahmani, P.; Eslamlou, A.D. An efficient hybrid approach based on Harris Hawks optimization and imperialist competitive algorithm for structural optimization. *Eng. Comput.* **2021**, 4598. [CrossRef]
55. Abualigah, L.; Yousri, D.; Elaziz, M.A.; Ewees, A.A.; Al-qaness, M.A.A.; Gandomi, A.H. Aquila Optimizer: A novel meta-heuristic optimization algorithm. *Comput. Ind. Eng.* **2021**, *157*, 107250. [CrossRef]
56. Tang, A.D.; Han, T.; Xu, D.W.; Xie, L. Chaotic Elite Harris Hawk Optimization Algorithm. *J. Comput. Appl.* **2021**, 1–10. Available online: https://kns.cnki.net/kcms/detail/detail.aspx?dbcode=CAPJ&dbname=CAPJLAST&filename=JSJY2021011300H&v=5lc3RO%25mmd2BEUUC%25mmd2FhVq8jnE%25mmd2BxfkAnjCOOEL7xcSF5jPQfItuqOALm2aHD2u1aGLhSpw1 (accessed on 15 January 2021).
57. Tizhoosh, H. Opposition-based learning: A new scheme for machine intelligence. In *Control and Automation, Proceedings of the International Conference on Computational Intelligence for Modeling, Vienna, Austria, 28–30 November 2005*; IEEE: Piscataway, NJ, USA, 2005.
58. Rahnamayan, S.; Tizhoosh, H.R.; Salama, M.M.A. Opposition-based differential evolution. *IEEE Trans. Evol. Comput.* **2014**, *12*, 64–79. [CrossRef]
59. Jia, Z.; Li, L.; Hui, S. Artificial Bee Colony Using Opposition-Based Learning. *Adv. Intell. Syst. Comput.* **2015**, *329*, 3–10.
60. Elaziz, M.A.; Oliva, D.; Xiong, S. An improved Opposition-Based Sine Cosine Algorithm for global optimization. *Expert Syst. Appl.* **2017**, *90*, 484–500. [CrossRef]
61. Ewees, A.A.; Elaziz, M.A.; Houssein, E.H. Improved Grasshopper Optimization Algorithm using Opposition-based Learning. *Expert Syst. Appl.* **2018**, *112*, 156–172. [CrossRef]
62. Fan, C.; Zheng, N.; Zheng, J.; Xiao, L.; Liu, Y. Kinetic-molecular theory optimization algorithm using opposition-based learning and varying accelerated motion. *Soft Comput.* **2020**, *24*, 12709–12730. [CrossRef]
63. Long, W.; Jiao, J.; Liang, X.; Cai, S.; Xu, M. A Random Opposition-Based Learning Grey Wolf Optimizer. *IEEE Access* **2019**, *7*, 113810–113825. [CrossRef]
64. Molga, M.; Smutnicki, C. Test Functions for Optimization Needs. 2005. Available online: http://www.robertmarks.org/Classes/ENGR5358/Papers/functions.pdf (accessed on 1 January 2005).
65. He, Q.; Wang, L. An effective co-evolutionary particle swarm optimization for constrained engineering design problems. *Eng. Appl. Artif. Intell.* **2007**, *20*, 89–99. [CrossRef]
66. Mirjalili, S. Moth-flame optimization algorithm: A novel nature-inspired heuristic paradigm. *Knowl.-Based Syst.* **2015**, *89*, 228–249. [CrossRef]
67. Geem, Z.W.; Kim, J.H.; Loganathan, G.V. A new heuristic optimization algorithm: Harmony search. *Simulation* **2001**, *76*, 60–68. [CrossRef]
68. Baykasoğlu, A.; Ozsoydan, F.B. Adaptive firefly algorithm with chaos for mechanical design optimization problems. *Appl. Soft Comput.* **2015**, *36*, 152–164. [CrossRef]
69. Lu, S.; Kim, H.M. A regularized inexact penalty decomposition algorithm for multidisciplinary design optimization problems with complementarity constraints. *J. Mech. Des.* **2010**, *132*, 041005. [CrossRef]
70. Saremi, S.; Mirjalili, S.; Lewis, A. Grasshopper optimisation algorithm: Theory and application. *Adv. Eng. Softw.* **2017**, *105*, 30–47. [CrossRef]

 processes

MDPI

Article

Deep Ensemble of Slime Mold Algorithm and Arithmetic Optimization Algorithm for Global Optimization

Rong Zheng [1,*], Heming Jia [1,*], Laith Abualigah [2,3,4], Qingxin Liu [5] and Shuang Wang [1]

1 School of Information Engineering, Sanming University, Sanming 365004, China; wang_shuang@fjsmu.edu.cn
2 Research and Innovation Department, Skyline University College, Sharjah 1797, United Arab Emirates; Aligah.2020@gmail.com
3 Faculty of Computer Sciences and Informatics, Amman Arab University, Amman 11953, Jordan
4 School of Computer Science, Universiti Sains Malaysia, Gelugor 11800, Malaysia
5 School of Computer Science and Technology, Hainan University, Haikou 570228, China; qxliu@hainanu.edu.cn
* Correspondence: zhengr@fjsmu.edu.cn (R.Z.); jiaheming@fjsmu.edu.cn (H.J.)

Abstract: In this paper, a new hybrid algorithm based on two meta-heuristic algorithms is presented to improve the optimization capability of original algorithms. This hybrid algorithm is realized by the deep ensemble of two new proposed meta-heuristic methods, i.e., slime mold algorithm (SMA) and arithmetic optimization algorithm (AOA), called DESMAOA. To be specific, a preliminary hybrid method was applied to obtain the improved SMA, called SMAOA. Then, two strategies that were extracted from the SMA and AOA, respectively, were embedded into SMAOA to boost the optimizing speed and accuracy of the solution. The optimization performance of the proposed DESMAOA was analyzed by using 23 classical benchmark functions. Firstly, the impacts of different components are discussed. Then, the exploitation and exploration capabilities, convergence behaviors, and performances are evaluated in detail. Cases at different dimensions also were investigated. Compared with the SMA, AOA, and another five well-known optimization algorithms, the results showed that the proposed method can outperform other optimization algorithms with high superiority. Finally, three classical engineering design problems were employed to illustrate the capability of the proposed algorithm for solving the practical problems. The results also indicate that the DESMAOA has very promising performance when solving these problems.

Keywords: slime mold algorithm; arithmetic optimization algorithm; meta-heuristics algorithm; global optimization; engineering design problem

Citation: Zheng, R.; Jia, H.; Abualigah, L.; Liu, Q.; Wang, S. Deep Ensemble of Slime Mold Algorithm and Arithmetic Optimization Algorithm for Global Optimization. *Processes* **2021**, *9*, 1774. https://doi.org/10.3390/pr9101774

Academic Editor: Jae-Yoon Jung

Received: 18 August 2021
Accepted: 2 October 2021
Published: 4 October 2021

1. Introduction

Nowadays, optimization problems exist in various scenarios, for instance, the engineering design problems. The objective of these optimization problems is to find the extreme values with determined constraint conditions. Then, commonly, the cost is reduced as much as possible. To tackle these problems, researchers have proposed many optimization algorithms [1–4]. Generally speaking, traditional optimization algorithms, such as gradient-based methods, are susceptible to the initial positions and have difficulties to deal with the non-convex problems that may contain a mass of local optimums. In practice, when we are faced with complex constraint conditions in the real world, it is more essential to obtain the optimal solutions within limited time and cost. At this point, the important thing is not to find the theoretical optimal result but to obtain as good an approximate solution as possible under restricted conditions. For this purpose, many stochastic optimizers have been developed and employed to solve complex optimization problems. As its name implies, the random operator is the main feature for a stochastic optimizer, which allows the algorithms to avoid the stagnation and search the whole search region for global optimization result.

The meta-heuristic algorithms (MAs) have shown very powerful capability in the fields of computational sciences. In general, MAs have four types according to the sources of inspiration, namely, physics-inspired (PI), evolution-inspired (EI), swarm-inspired (SI), and human-inspired (HI). Some representative algorithms are shown below:

- Physics-inspired: multi-verse optimizer (MVO) [5], gravitational search algorithm (GSA) [6], thermal exchange optimization (TEO) [7], heat transfer relation-based optimization algorithm (HTOA) [8].
- Evolution-inspired: genetic algorithm (GA) [9], differential evolution (DE) [10], evolutionary programming (EP) [11].
- Swarm-inspired: particle swarm optimization (PSO) [12], emperor penguin optimizer (EPO) [13], Aquila optimizer (AO) [14], remora optimization algorithm (ROA) [4], marine predators algorithm (MPA) [15].
- Human-inspired: teaching–learning-based optimization (TLBO) [16], social group optimization (SGO) [17], β-hill climbing (βHC) [18], coronavirus optimization algorithm (COA) [19].

For a meta-heuristic algorithm, one important thing is to balance of the global search and local search [20]. It is known that the search agents are first randomly generated within the search spaces. Then, positions of these search agents are updated according to the formulas in the algorithm. In the early stage, drastic exploration in the search space should be performed as much as possible in the early stage. Then, in the later phase, more local exploitation should be conducted to improve the accuracy of obtained optimal solution. Although hundreds of MAs have been proposed for the optimization problems, there is still need for new algorithms to solve these optimization problems. According to the No-Free-Lunch (NFL) theory [21], on one optimization algorithm can solve all the optimization problems. Generally speaking, it is common that MAs suffer from local optimum stagnation and poor convergence speed as a result of poor optimization ability. Thus, it is very important to develop new optimization algorithms or improve existing MAs by taking some effective measures. Up until now, there have been three primary methods for the improvements of the existing algorithms, which are listed in Table 1.

The slime mold algorithm (SMA) [33] and arithmetic optimization algorithm (AOA) [34] are two newly proposed MAs, and both have the merits of simplicity, efficiency, and flexibility. The SMA has good population diversity and stable performance when solving optimization problems. However, it gets stuck in local optima sometimes for the limited global search capability. On the contrary, the AOA has powerful exploration capability by using the arithmetic operators. However, the performance of AOA is not stable because of the poor population diversity. Therefore, the SMA and AOA are considered to be hybridized together in this paper for solving the global optimization problems. To evaluate the performance of proposed algorithm, we employed 23 classical benchmark functions and 4 constrained engineering design problems. The main contributions of this works are as follows:

1. Hybridizing the slime mold algorithm (SMA) [33] and arithmetic optimization algorithm (AOA) [34] named SMAOA to improve the exploration capability of original SMA.
2. Applying the random contraction strategy (RCS), which is inspired from SMA to help the SMAOA jump out from local optimum.
3. Applying the subtraction and addition strategy (SAS), which is extracted from AOA to enhance the exploitation ability of SMAOA.
4. When the RCS and SAS were applied on SMAOA, the DESMAOA was finally obtained. By comparing seven well-known optimization algorithms, we identified the proposed DESMAOA to be powerful according to the experimental results.

Table 1. A summary of methods for improving the optimization algorithms developed in the literature.

Name of Method	Representative Algorithm	Description
Hybridize two or more algorithms	Hybrid sperm swarm optimization and gravitational search algorithm (HSSOGSA) [22]	The capability of exploitation in SSO and the capability of exploration in GSA are combined for better performance.
	Imperialist competitive Harris hawks optimization (ICHHO) [23]	The exploration of ICA is utilized to improve the HHO for global optimization.
	Hybrid particle swarm and spotted hyena optimizer (HPSSHO) [24]	Particle swarm algorithm is used to improve the hunting strategy of spotted hyena optimizer.
Add one or more strategies onto an algorithm	Sine-cosine and spotted hyena-based chimp optimization algorithm (SSC) [25]	Sine-cosine functions and attacking strategy of SHO are embedded in ChoA for better exploration and exploitation.
	Representative-based grey wolf optimizer (R-GWO) [26]	A search strategy named representative-based hunting (RH) is utilized to improve the exploration and diversity of the population
	Reinforced salp swarm algorithm (CMSRSSSA) [27]	An ensemble/composite mutation strategy (CMS) is applied to boost the exploitation and exploration speed of SSA, while restart strategy (RS) is used to get away from local optimum.
	Boosting quantum rotation gate embedded slime mold algorithm (WQSMA) [28]	The quantum rotation gate mechanism and the operation from water cycle are applied to balance the exploration and exploitation inclinations.
	Enhanced salp swarm algorithm (ESSA) [29]	Orthogonal learning, quadratic interpolation, and generalized oppositional learning are embedded into SSA to boost the global exploration and local exploitation.
Hybridize two or more algorithms that are further improved by one or more strategies	Whale optimization with seagull algorithm (WSOA) [30]	WOA's contraction surrounding mechanism and SOA's spiral attack behavior work together, and then levy flight strategy is employed on the search process of SOA.
	Chaotic sine-cosine firefly (CSCF) algorithm [31]	Chaotic form of SCA and FA are integrated together to improve the convergence speed and efficiency.
	Hybrid grasshopper optimization algorithm with bat algorithm (BGOA) [32]	In BGOA, Levy fight, local search part of BA, and random strategy are introduced into basic GOA.

The rest of this paper is organized as follows: The basics of SMA and AOA are described in Section 2. Then, the hybrid method is presented in Section 3, including two strategies that are obtained from these two algorithms. In Section 4, a series of experimental tests are conducted to evaluate the performance of proposed DESMAOA. In Section 5, three engineering design problems are employed to assess the applicability of proposed algorithm in practice. Finally, Section 6 concludes this paper and provides some directions for meaningful future research.

2. Preliminaries

2.1. Slime Mold Algorithm (SMA)

The slime mold algorithm (SMA) is a recent meta-heuristic algorithm proposed by Li et al. in 2020 [33]. The basic idea of SMA is based on the foraging behavior of slime mold, which have different feedback characteristics according to the food quality. Three special behaviors of the slime mold are mathematical formulated in the SMA, i.e., approaching food, wrapping food, and finally grabbling food. First, the process of approaching food can be expressed as

$$Xi(t+1) = \begin{cases} Xb(t) + vb \cdot (W \cdot XA(t) - XB(t)), & r1 < p \\ vc \cdot Xi(t), & r1 \geq p \end{cases} \tag{1}$$

where t is the number of current iteration, $X_i(t+1)$ is the newly generated position, $X_b(t)$ denotes the best position found by slime mold in iteration t, $X_A(t)$ and $X_B(t)$ are two random positions selected from the population of slime mold, and r_1 is a random value in [0, 1].

vb and vc are the coefficients that simulate the oscillation and contraction mode of slime mold, respectively, and vc is designed to linearly decrease from one to zero during the iterations. The range of vb is from $-a$ to a, and the computational formula of a is

$$a = \operatorname{arctanh}(1 - \frac{t}{T}) \tag{2}$$

where T is the maximum number of iterations.

According to Equations (1) and (2), it can be seen that as the number of iterations increases, the slime mold will wrap the food.

W is a very important factor that indicates the weight of slime mold, and it is calculated as follows:

$$W(SmellIndex(i)) = \begin{cases} 1 + rand \cdot \log(\frac{bF-S(i)}{bF-wF} + 1),\ i \leq N/2 \\ 1 - rand \cdot \log(\frac{bF-S(i)}{bF-wF} + 1),\ i > N/2 \end{cases} \tag{3}$$

$$SmellIndex(i) = sort(S(i)) \tag{4}$$

where *rand* means a random value between 0 and 1; bF and wF are the best and worst fitness values, respectively, obtained by far; $S(i)$ is the fitness value of ith slime mold; N is the popsize of the population; and *SmellIndex* is a ranking of fitness values for individuals in the population.

In Equation (1), it is also worth noting that p is the probability of determining the update location for slime mold, which is related to the fitness values of slime mold and food and can be calculated as follows:

$$p = \tanh|S(i) - DF| \tag{5}$$

where DF denotes the best fitness obtained by population.

Finally, when the slime mold has found the food (i.e., grabble food), it still has a certain chance (z) to search other new food, which is formulated as

$$X(t+1) = rand \cdot (UB - LB) + LB,\ r_2 < z \tag{6}$$

where UB and LB are the upper boundary and lower boundary, respectively, and r_2 implies a random value in the region [0, 1].

In general, z should be very small; thus, it is set to 0.03 in SMA. Finally, the pseudo-code of SMA is given in Algorithm 1.

Algorithm 1. Pseudo-code of SMA

Initialize the parameters popsize (N) and maximum iterations (T)
Initialize the positions of all slime mold X_i (i = 1, 2, ..., N)
While ($t \leq T$)
Calculate the fitness of all slime mold
Update *bestFitness*, X_b
Calculate the weight W by Equation (3) and (4)
For each search agent
If $r_2 < z$
Update position by Equation (6)
Else
Update p, vb, and vc
Update position by Equation (1)
End if
End for
$t = t + 1$
End While
Return *bestFitness*, X_b

2.2. Arithmetic Optimization Algorithm (AOA)

Arithmetic optimization algorithm (AOA) is a very new meta-heuristic method proposed by Abualigah and others in 2021 [34]. The main inspiration of this algorithm is to combine the four traditional arithmetic operators in mathematics, i.e., multiplication (M), division (D), subtraction (S), and addition (A). Similar to sine-cosine algorithm (SCA) [35], AOA also has a very simple structure and low computation complexity. Considering the

M and D operators can produce large steps in the iterations, M and D are hence mainly conducted in the exploration phase. The expression is as follows:

$$Xi(t+1) = \begin{cases} Xb(t)/(MOP+eps)\cdot((UB-LB)\mu+LB), & rand < 0.5 \\ Xb(t)\cdot MOP\cdot((UB-LB)\mu+LB), & rand \geq 0.5 \end{cases} \tag{7}$$

where *eps* is a very small positive number, and μ is a constant coefficient (0.499) that is carefully designed for this algorithm.

MOP is non-linearly decreased from 1 to 0 during the iterations, and the expression is as follows:

$$MOP = 1-(\frac{t}{T})^{1/\alpha} \tag{8}$$

where α is a constant value, which is set to 5 according to the AOA.

From Equation (7), it can be seen that both M and D operators can generate very stochastic positions for the search agent on the basis of the best position. By contrast, S and A operators are applied to emphasize the local exploitation that will generate smaller steps in the search space. The mathematical expression is defined as

$$Xi(t+1) = \begin{cases} Xb(t)-MOP\cdot((UB-LB)\mu+LB), & rand < 0.5 \\ Xb(t)+MOP\cdot((UB-LB)\mu+LB), & rand \geq 0.5 \end{cases} \tag{9}$$

There is no doubt that the importance of balance between exploration and exploitation for an optimization algorithm. In AOA, the parameter *MOA* is utilized to switch the exploration and exploitation over the course of iterations, which is expressed as

$$MOA(t) = Min + t(\frac{Max-Min}{T}) \tag{10}$$

where *Min* and *Max* are constant values.

According to Equation (10), *MOA* increases from *Min* to *Max*. Thus, in the early phase, search agent has more chance to perform exploration in the search space, while in the later stage, search agent will be more likely to conduct search near the best position. The pseudo-code of AOA is shown in Algorithm 2.

Algorithm 2. Pseudo-code of AOA

Initialize the parameters popsize (N) and maximum iterations (T)
Initialize the positions of all search agents X_i ($i = 1, 2, \ldots, N$)
Set the parameters α, μ, *Min*, and *Max*
While ($t \leq T$)
Calculate the fitness of all search agents
Update *bestFitness*, X_b
Calculate the *MOP* by Equation (8)
Calculate the *MOA* by Equation (10)
For each search agent
If *rand* > *MOA*
Update position by Equation (7)
Else
Update position by Equation (9)
End if
End for
$t = t + 1$
End While
Return *bestFitness*, X_b

3. The Proposed Hybridized Algorithm (DESMAOA)

It is well known that MAs have the merits of concision, flexibility, and especially utility. Hence, many scholars are working on developing new meta-heuristic-based approaches for optimization problems. However, several optimization algorithms such as slime mold algorithm and arithmetic optimization algorithm still have some drawbacks. For instance, when dealing with complex optimization problems, SMA tends to drop into local best, and also converges slowly. Similarly, AOA only utilizes the information of best position in the population, which may suffer the problem of low precision. Therefore, this paper aimed to develop a new hybridization algorithm composed of SMA and AOA for better optimization performance.

In this paper, the SMA and AOA are firstly integrated to form a hybridized style named SMAOA. Then, the preliminary hybrid algorithm is further enhanced by adding two strategies. One is the random contraction strategy (RCS), which is an improved version of contraction formula in SMA. The other is the subtraction and addition strategy (SAS), which is extracted from the local search in AOA. Finally, the deep ensemble of SMA and AOA is accomplished, and the hybridized algorithm (i.e., DESMAOA) is obtained. The detailed implement of proposed algorithm is delineated in the following.

3.1. The Hybridization of SMA and AOA

In SMA, the contraction formula (see Equations (1) and (2)) is utilized to help slime mold jump out of local minima, which will tend to zero in the later iterations. Thus, it will not play the role of global exploration. On the other hand, the multiplication and division methods in AOA display a powerful capability in global exploration. Thus, the formulas of multiplication and division (see in Equation (7)) are considered to replace the contraction equation. Therefore, the hybrid algorithm SMAOA will perform good global search in the whole stage. To be specific, for the search agent that is close to best position, the multiplication and division operators will make it more likely to search other spaces.

3.2. Random Contraction Strategy (RCS)

In this work, we present the RCS on the basis of the mathematical formula of contraction mode in SMA, which is applied to expand exploration space and avoid local optimum. The coefficient vc is replaced by a random value lying between -1 and 1. The position update formula is calculated as follows:

$$Vi2(t+1) = (2rand - 1)Xi(t) \tag{11}$$

From Equation (11), we should note that the generated position of RCS is within the range $[-|X_i(t)|, |X_i(t)|]$ with uniform distribution, which adds more flexibility for the search agents in the proposed algorithm. Note that the generated position of RCS is taken as a candidate solution.

3.3. Subtraction and Addition Strategy (SAS)

The other strategy proposed here is the SAS, which is also the exploitation method of AOA. According to the AOA, SAS can be performed locally and increase the accuracy of solutions effectively. It is worth mentioning here that the SAS is conducted behind the RCS.

In the same way, the position generated by SAS is treated as a candidate solution, and if a better position is found, then it will be adopted.

3.4. The Deep Ensemble of SMA and AOA

As mentioned above, the SMA and AOA are hybridized together firstly to achieve the SMAOA. Then, two strategies are introduced in the SMAOA, namely, random contraction strategy and subtraction and addition strategy. In order to perform a better balance effect between exploration and exploitation, we utilize a parameter, b, that is related with iterations to represent the probability of conducting the strategies. Its computational formula is given below:

$$b = 1 - \frac{t}{T} \tag{12}$$

The pseudo-code of DESMAOA is shown in Algorithm 3. Moreover, the flowchart of proposed method is shown in Figure 1.

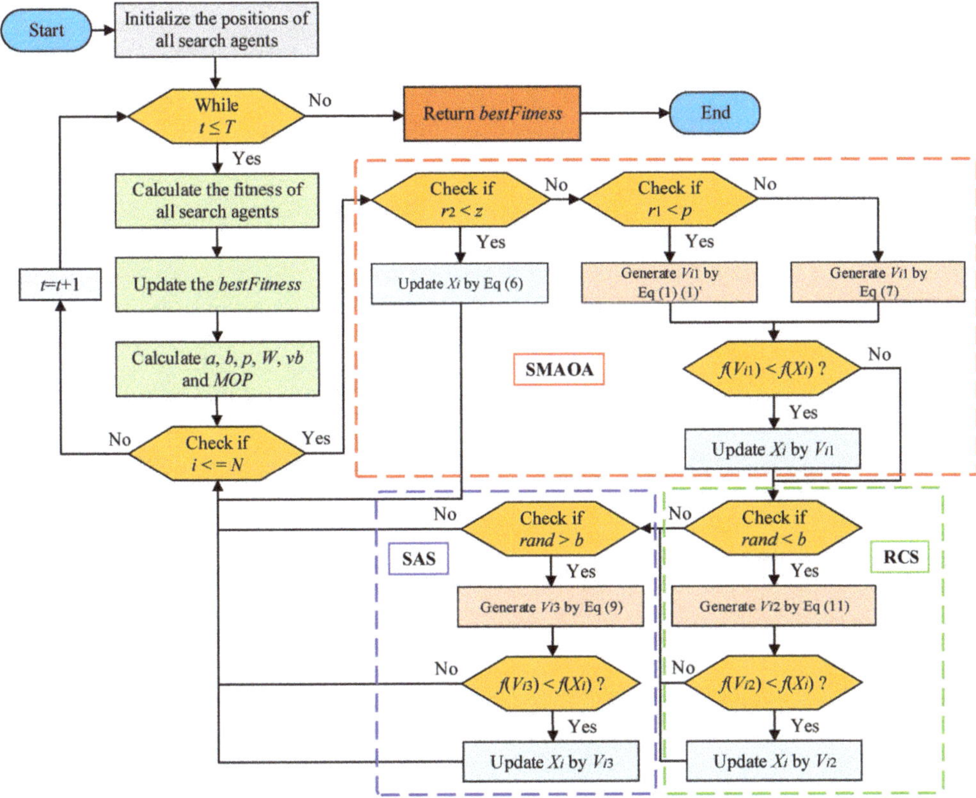

Figure 1. Flowchart of the proposed DESMAOA.

3.5. The Computational Complexity of DESMAOA

The computational complexity of DESMAOA depends on the population size (N), dimension size (D), and maximum iterations (T). First, the computational complexity of initialization is $O(N \times D)$. Then, in the iterations, the computational complexity of calculating the fitness values of all search agents is $O(N)$. The computational complexity of sorting is $O(N \times \log N)$. Moreover, the computational complexity of updating the positions of search agents in SMAOA is $O(N \times D)$. Considering the worst cases, the computational complexity of RCS and SAS is $O(2N \times D)$. In summary, the final computational complexity of the DESMAOA is $O(N \times D + T \times N(1 + \log N + 3D))$.

Algorithm 3. Pseudo-code of DESMAOA

Initialize the parameters popsize (N) and maximum iterations (T)
Initialize the positions of all search agents X_i ($i = 1, 2, \ldots, N$)
Set the parameters α, μ, Min, and Max
While ($t \leq T$)
Calculate the fitness of all search agents
Update $bestFitness$, X_b
Calculate a, b, p, and W by Equation (2)–(5)
Calculate the MOP by Equation (8)
Update vb
For each search agent
If $r_2 < z$
Update position by Equation (6)
Else
If $r_1 < p$
Update position V_{i1} by Equation (1) (1)′
Else
Update position V_{i1} by Equation (7)
End if
If $f(V_{i1}) < f(X_i)$
$X_i = V_{i1}$
End if
If $rand < b$
Apply RCS and generate candidate position V_{i2} by Equation (11)
If $f(V_{i2}) < f(X_i)$
$X_i = V_{i2}$
End if
End if
If $rand > b$
Apply SAS and generate candidate position V_{i3} by Equation (9)
If $f(V_{i3}) < f(X_i)$
$X_i = V_{i3}$
End if
End if
End if
End for
$t = t + 1$
End While
Return $bestFitness$, X_b

4. Experimental Results and Discussions

In this section, we provide the results of a series of comparative experiments that were conducted by using 23 classical benchmark functions and 10 IEEE CEC2021 single objective optimization functions to evaluate the performance of proposed DESMAOA [36,37]. Table 2 lists the detailed parameter values of these test functions. It can be seen that these classical test functions included unimodal functions (F1–F7), multimodal functions (F8–F13), and also fixed-dimension multimodal functions (F14–F23). Moreover, the CEC2021 test functions contained four types of functions: unimodal function, basic functions, hybrid functions, and composition functions. The unimodal functions are suitable for testing the exploitation capability of algorithms, while the other types of test functions that contain a large number of local minimas can reveal the exploration capability and stability of algorithms.

In the experiments of test functions, the impacts of two applied strategies were firstly analyzed by using the classical test functions. Then, the test results of DESMAOA in classical test functions were compared with seven well-known algorithms. Multiple aspects of the analysis including exploitation capability, exploration capability, qualitative analysis, and convergence behavior are described. Moreover, the results of CEC2021 test functions were also analyzed to investigate the performance of proposed algorithm.

Table 2. Benchmark function properties (D indicates dimension).

Function Type	Function	Dimension	Range	Theoretical Optimization Value
Unimodal test functions	F1	30, 50, 200, 1000	[−100, 100]	0
	F2	30, 50, 200, 1000	[−10, 10]	0
	F3	30, 50, 200, 1000	[−100, 100]	0
	F4	30, 50, 200, 1000	[−100, 100]	0
	F5	30, 50, 200, 1000	[−30, 30]	0
	F6	30, 50, 200, 1000	[−100, 100]	0
	F7	30, 50, 200, 1000	[−1.28, 1.28]	0
Multimodal test functions	F8	30, 50, 200, 1000	[−500, 500]	$-418.9829 \times D$
	F9	30, 50, 200, 1000	[−5.12, 5.12]	0
	F10	30, 50, 200, 1000	[−32, 32]	0
	F11	30, 50, 200, 1000	[−600, 600]	0
	F12	30, 50, 200, 1000	[−50, 50]	0
	F13	30, 50, 200, 1000	[−50, 50]	0
Fixed-dimension multimodal test functions	F14	2	[−65, 65]	0.998004
	F15	4	[−5, 5]	0.0003075
	F16	2	[−5, 5]	−1.03163
	F17	2	[−5, 5]	0.398
	F18	2	[−2, 2]	3
	F19	3	[−1, 2]	−3.8628
	F20	6	[0, 1]	−3.3220
	F21	4	[0, 10]	−10.1532
	F22	4	[0, 10]	−10.4028
	F23	4	[0, 10]	−10.5363
CEC2021 unimodal test functions	CEC_01	10	[−100, 100]	100
CEC2021 basic test functions	CEC_02	10	[−100, 100]	1100
	CEC_03	10	[−100, 100]	700
	CEC_04	10	[−100, 100]	1900
CEC2021 hybrid test functions	CEC_05	10	[−100, 100]	1700
	CEC_06	10	[−100, 100]	1600
	CEC_07	10	[−100, 100]	2100
CEC2021 composition test functions	CEC_08	10	[−100, 100]	2200
	CEC_09	10	[−100, 100]	2400
	CEC_10	10	[−100, 100]	2500

4.1. Impacts of Components

The impacts of different versions are investigated in this section. SMA showed very outstanding performance in optimization problems. However, it still had the problems of premature convergence and local optima. According to the works of Abualigah [34], AOA shows powerful global exploration and local exploitation capability. Hence, we first hybridized the SMA and AOA to obtain the SMAOA. Then, in order to help the search agent jump out of local optima, we integrated the RCS into SMAOA. Moreover, SAS was introduced into SMAOA to improve the local search capability. Different combinations between SMAOA and two strategies are listed below:

- SMAOA;
- SMAOA combined with RCS (SMAOA1);
- SMAOA combined with SAS (SMAOA2);
- SMAOA combined with RCS and SAS (DESMAOA).

For impartial comparison, the number of iterations and population size for all tests were set as 500 and 30, respectively. Moreover, we conducted independent tests 30 times for each algorithm. The averages and standard deviations were utilized for analysis and comparison between these algorithms. The results are listed in Table 3. Note that the dimension of F1–F13 was set to 30.

Table 3. Comparison of the SMAOA, SMAOA1, SMAOA2, and DESMAOA.

Function	SMAOA		SMAOA1		SMAOA2		DESMAOA	
	Mean	Std	Mean	Std	Mean	Std	Mean	Std
F1	0.00×10^{0}	0.00×10^{0}	0.00×10^{0}	0.00×10^{0}	0.00×10^{0}	0.00×10^{0}	0.00×10^{0}	0.00×10^{0}
F2	0.00×10^{0}	0.00×10^{0}	0.00×10^{0}	0.00×10^{0}	0.00×10^{0}	0.00×10^{0}	0.00×10^{0}	0.00×10^{0}
F3	0.00×10^{0}	0.00×10^{0}	0.00×10^{0}	0.00×10^{0}	0.00×10^{0}	0.00×10^{0}	0.00×10^{0}	0.00×10^{0}
F4	0.00×10^{0}	0.00×10^{0}	0.00×10^{0}	0.00×10^{0}	0.00×10^{0}	0.00×10^{0}	0.00×10^{0}	0.00×10^{0}
F5	2.82×10^{0}	7.00×10^{0}	5.85×10^{-1}	1.07×10^{0}	2.46×10^{-1}	1.33×10^{0}	1.17×10^{-3}	1.55×10^{-3}
F6	2.44×10^{-2}	2.34×10^{-2}	7.77×10^{-3}	1.21×10^{-2}	5.80×10^{-6}	1.92×10^{-6}	4.95×10^{-6}	2.01×10^{-6}
F7	1.16×10^{-4}	9.72×10^{-5}	5.67×10^{-5}	5.61×10^{-5}	6.72×10^{-5}	6.77×10^{-5}	4.27×10^{-5}	4.76×10^{-5}
F8	−12,569.2361	1.89×10^{-1}	−12,569.3229	1.38×10^{-1}	−12,569.4866	4.96×10^{-6}	−12,569.4866	4.11×10^{-6}
F9	0.00×10^{0}	0.00×10^{0}	0.00×10^{0}	0.00×10^{0}	0.00×10^{0}	0.00×10^{0}	0.00×10^{0}	0.00×10^{0}
F10	8.8818×10^{-16}	0.00×10^{0}	8.8818×10^{-16}	0.00×10^{0}	8.8818×10^{-16}	0.00×10^{0}	8.8818×10^{-16}	0.00×10^{0}
F11	2.13×10^{-1}	2.92×10^{-1}	0.00×10^{0}	0.00×10^{0}	8.85×10^{-3}	2.30×10^{-2}	0.00×10^{0}	0.00×10^{0}
F12	2.63×10^{-3}	4.65×10^{-3}	5.08×10^{-5}	8.33×10^{-5}	4.84×10^{-8}	7.54×10^{-8}	1.09×10^{-7}	1.59×10^{-7}
F13	1.70×10^{-2}	3.67×10^{-2}	8.82×10^{-4}	1.15×10^{-3}	2.50×10^{-3}	6.04×10^{-3}	5.91×10^{-7}	1.06×10^{-6}
F14	9.98×10^{-1}	2.60×10^{-11}	9.98×10^{-1}	9.58×10^{-12}	9.98×10^{-1}	9.91×10^{-16}	9.98×10^{-1}	8.05×10^{-16}
F15	4.16×10^{-4}	1.56×10^{-4}	3.63×10^{-4}	9.61×10^{-5}	4.07×10^{-4}	1.90×10^{-4}	3.34×10^{-4}	8.63×10^{-5}
F16	-1.0316×10^{0}	3.97×10^{-8}	-1.0316×10^{0}	9.46×10^{-8}	-1.0316×10^{0}	1.67×10^{-11}	-1.0316×10^{0}	1.87×10^{-11}
F17	3.9789×10^{-1}	6.82×10^{-7}	3.9789×10^{-1}	3.97×10^{-7}	3.9789×10^{-1}	5.63×10^{-12}	3.9789×10^{-1}	5.94×10^{-12}
F18	3.00×10^{0}	2.79×10^{-9}	3.00×10^{0}	6.69×10^{-10}	3.00×10^{0}	8.56×10^{-11}	3.00×10^{0}	9.42×10^{-11}
F19	-3.8627×10^{0}	4.32×10^{-5}	-3.8628×10^{0}	4.68×10^{-5}	-3.8628×10^{0}	5.61×10^{-5}	-3.8627×10^{0}	7.91×10^{-5}
F20	-3.25×10^{0}	5.98×10^{-2}	-3.2859×10^{0}	5.59×10^{-2}	-3.2583×10^{0}	6.06×10^{-2}	-3.286×10^{0}	5.59×10^{-2}
F21	-1.01528×10^{1}	4.26×10^{-4}	-1.01529×10^{1}	3.67×10^{-4}	-1.01531×10^{1}	9.07×10^{-5}	-1.01531×10^{1}	1.42×10^{-4}
F22	-1.04023×10^{1}	4.61×10^{-4}	-1.04025×10^{1}	4.47×10^{-4}	-1.04028×10^{1}	8.31×10^{-5}	-1.04028×10^{1}	7.38×10^{-5}
F23	-1.0536×10^{1}	3.47×10^{-4}	-1.05362×10^{1}	2.47×10^{-4}	-1.05363×10^{1}	9.44×10^{-5}	-1.05363×10^{1}	8.26×10^{-5}

From Table 3, it can be seen that these four improved algorithms could obtain the same optimal fitness in F1–F4, F9, F10, F14, and F16–F18. In particular, the theoretical optimization values were obtained in F1–F4, F9, and F18. Compared to SMAOA, SMAOA1, and SMAOA2, DESMAOA won in F5–F8, F11, F13, F15, and F20–F24. This demonstrates that the significant effect with the combination of RCS and SAS. In addition, it is worth mentioning here that the results of DESMAOA in F12 and F19 were very close to the best ones. From the values of standard deviations, it was also shown that DESMAOA had good stability and strong robustness in solving these test functions.

4.2. The Classical Benchmark Functions

This section outlines the 23 classical test functions that were employed for experiments. The performance of DESMAOA was compared with two newly proposed algorithms (SMA and AOA) and another five very famous optimization algorithms (GWO [38], WOA [39], SSA [40], MVO [5], and PSO [12]). Table 4 lists the main parameter values used in each algorithm. Note that the parameter values used in DESMAOA were the same as those used in two original algorithms. Therefore, the stable performance could be guaranteed to some extent for the proposed algorithm. In addition, the test conditions were the same as previously for equal comparison.

Table 4. Parameter values for the optimization algorithms.

Algorithm	Parameter Settings
DESMAOA	$z = 0.03$; $\alpha = 5$; $\mu = 0.499$
SMA [33]	$z = 0.03$
AOA [34]	$\alpha = 5$; $\mu = 0.499$; $Min = 0.2$; $Max = 1$
GWO [38]	$a = [2, 0]$
WOA [39]	$a_1 = [2, 0]$; $a_2 = [-2, -1]$; $b = 1$
SSA [40]	$c_1 \in [0, 1]$; $c_2 \in [0, 1]$
MVO [5]	$WEP \in [0.2, 1]$; $TDR \in [0, 1]$; $r_1, r_2, r_3 \in [0, 1]$
PSO [12]	$c_1 = 2$; $c_2 = 2$; $W \in [0.2, 0.9]$; $vMax = 6$

4.2.1. Exploration and Exploitation Capability Analysis

Table 5 lists the experimental results of these algorithms. It was shown that the performance of DESMAOA is not only better than the original SMA and AOA but also superior to other comparative algorithms on 20 out of 23 benchmark functions. In F1–F5, F7–F15, F22, and F23, DESMAOA had the lowest average values and stand deviations. This reveals that DESMAOA possesses very good stability and also can find the optimal solution. It is worth noting that the proposed algorithm can obtain the theoretical optimization values in test functions F1–F4, F9, F11, and F18. In F6, F19, and F20, the results of DESMAOA were very close to the best ones. Therefore, these results demonstrated the remarkable effect of the proposed hybrid method. With the help of RCS, the proposed algorithm can jump out of the local minima and obtain the global optimal solution. In the meantime, high precision results could be obtained by using SAS.

In addition, the Wilcoxon signed-rank test was utilized to confirm the statistical superiority of DESMAOA [41], which revealing the statistical differences between two algorithms. The results are given in Table 6. On the basis of these results and the results in Table 5, DESMAOA outperformed SMA for 15 benchmark functions (except F1, F3, F7, F9, F10, F11, F15, and F20) and AOA for 20 benchmark functions (except F7, F15, and F17). Moreover, DESMAOA was found to be better than other comparative algorithms in most of the functions. Furthermore, the results of test functions were also evaluated using the Friedman ranking test [42], which can reveal the overall performance ranking of the comparative algorithms to the test functions. As can be seen from Figure 2, the proposed DESMAOA achieved the first rank among these algorithms. In summary, DESMAOA had excellent optimization performance that was significantly better than SMA and AOA.

Table 5. The result statistics of benchmark functions for the DESMAOA and competitor algorithms.

Function	Metric	DESMAOA	SMA	AOA	GWO	WOA	SSA	MVO	PSO
F1	Mean	0.00×10^{0}	9.93×10^{-302}	5.37×10^{-6}	7.21×10^{-28}	2.42×10^{-73}	3.96×10^{-7}	1.34×10^{0}	1.76×10^{-4}
	Std	0.00×10^{0}	0.00×10^{0}	2.14×10^{-6}	1.17×10^{-27}	8.81×10^{-73}	9.50×10^{-7}	5.38×10^{-1}	1.82×10^{-4}
F2	Mean	0.00×10^{0}	5.05×10^{-138}	1.74×10^{-3}	8.26×10^{-17}	8.81×10^{-52}	2.07×10^{0}	2.20×10^{0}	7.05×10^{0}
	Std	0.00×10^{0}	2.77×10^{-137}	2.08×10^{-3}	6.54×10^{-17}	2.46×10^{-51}	1.33×10^{0}	7.31×10^{0}	7.01×10^{0}
F3	Mean	0.00×10^{0}	5.43×10^{-323}	1.24×10^{-3}	1.55×10^{-5}	4.41×10^{4}	1.66×10^{3}	2.04×10^{2}	7.93×10^{1}
	Std	0.00×10^{0}	0.00×10^{0}	8.14×10^{-4}	3.50×10^{-5}	1.08×10^{4}	9.24×10^{2}	6.63×10^{1}	2.57×10^{1}
F4	Mean	0.00×10^{0}	7.56×10^{-154}	1.53×10^{-2}	8.03×10^{-7}	4.68×10^{1}	1.15×10^{1}	2.16×10^{0}	1.12×10^{0}
	Std	0.00×10^{0}	4.14×10^{-153}	1.06×10^{-2}	6.71×10^{-7}	2.77×10^{1}	4.04×10^{0}	8.66×10^{-1}	2.40×10^{-1}
F5	Mean	1.17×10^{-3}	8.56×10^{0}	2.79×10^{1}	2.71×10^{1}	2.82×10^{1}	2.90×10^{2}	7.89×10^{2}	8.16×10^{1}
	Std	1.55×10^{-3}	1.21×10^{1}	3.01×10^{-1}	8.49×10^{-1}	4.97×10^{-1}	4.77×10^{2}	8.74×10^{2}	7.03×10^{1}
F6	Mean	4.95×10^{-6}	5.74×10^{-3}	3.06×10^{0}	7.58×10^{-1}	3.72×10^{-1}	1.78×10^{-7}	1.34×10^{0}	1.37×10^{-4}
	Std	2.01×10^{-6}	3.38×10^{-3}	2.69×10^{-1}	4.94×10^{-1}	2.18×10^{-1}	1.51×10^{-7}	3.43×10^{-1}	1.65×10^{-4}
F7	Mean	4.27×10^{-5}	1.24×10^{-4}	6.74×10^{-5}	1.69×10^{-3}	3.15×10^{-3}	1.73×10^{-1}	3.21×10^{-2}	2.55×10^{0}
	Std	4.76×10^{-5}	1.07×10^{-4}	7.11×10^{-5}	8.95×10^{-4}	3.61×10^{-3}	5.61×10^{-2}	1.32×10^{-2}	4.54×10^{0}
F8	Mean	−12,569.4866	−12,569.1799	-5.48×10^{3}	-6.01×10^{3}	-1.06×10^{4}	-7.47×10^{3}	-7.55×10^{3}	-4.69×10^{3}
	Std	4.11×10^{-6}	2.66×10^{-1}	3.69×10^{2}	6.42×10^{2}	1.69×10^{3}	8.76×10^{2}	6.27×10^{2}	1.21×10^{3}
F9	Mean	0.00×10^{0}	0.00×10^{0}	1.66×10^{-6}	2.26×10^{0}	3.79×10^{-15}	5.53×10^{1}	1.20×10^{2}	1.02×10^{2}
	Std	0.00×10^{0}	0.00×10^{0}	1.27×10^{-6}	3.27×10^{0}	2.08×10^{-14}	1.83×10^{1}	3.29×10^{1}	3.19×10^{1}
F10	Mean	8.8818×10^{-16}	8.8818×10^{-16}	4.36×10^{-4}	1.01×10^{-13}	3.85×10^{-15}	2.56×10^{0}	2.03×10^{0}	1.69×10^{-2}
	Std	0.00×10^{0}	0.00×10^{0}	1.62×10^{-4}	1.81×10^{-14}	2.10×10^{-15}	6.94×10^{-1}	5.47×10^{-1}	1.30×10^{-2}
F11	Mean	0.00×10^{0}	0.00×10^{0}	8.42×10^{-4}	6.19×10^{-3}	1.68×10^{-2}	1.88×10^{-2}	8.60×10^{-1}	4.39×10^{-3}
	Std	0.00×10^{0}	0.00×10^{0}	3.12×10^{-3}	8.92×10^{-3}	6.38×10^{-2}	1.46×10^{-2}	8.21×10^{-2}	6.85×10^{-3}
F12	Mean	1.09×10^{-7}	5.81×10^{-3}	7.44×10^{-1}	4.42×10^{-2}	2.81×10^{-2}	7.54×10^{0}	2.43×10^{0}	2.07×10^{-2}
	Std	1.59×10^{-7}	6.50×10^{-3}	3.03×10^{-2}	1.86×10^{-2}	2.18×10^{-2}	3.43×10^{0}	1.39×10^{0}	4.22×10^{-2}
F13	Mean	5.91×10^{-7}	6.35×10^{-3}	2.96×10^{0}	6.94×10^{-1}	6.36×10^{-1}	1.38×10^{1}	1.96×10^{-1}	5.55×10^{-3}
	Std	1.06×10^{-6}	7.05×10^{-3}	1.03×10^{-2}	2.45×10^{-1}	3.53×10^{-1}	1.10×10^{1}	1.26×10^{-1}	9.01×10^{-3}
F14	Mean	9.98×10^{-1}	9.98×10^{-1}	9.87×10^{0}	4.16×10^{0}	2.54×10^{0}	1.36×10^{0}	9.98×10^{-1}	2.97×10^{0}
	Std	8.05×10^{-16}	3.93×10^{-13}	3.89×10^{0}	4.28×10^{0}	2.91×10^{0}	8.82×10^{-1}	4.31×10^{-11}	2.55×10^{0}
F15	Mean	3.34×10^{-4}	4.84×10^{-4}	8.39×10^{-3}	3.15×10^{-3}	8.25×10^{-4}	2.91×10^{-3}	5.24×10^{-3}	7.22×10^{-3}
	Std	8.63×10^{-5}	2.15×10^{-4}	1.29×10^{-2}	6.88×10^{-3}	5.40×10^{-4}	5.93×10^{-3}	1.26×10^{-2}	9.03×10^{-3}
F16	Mean	-1.0316×10^{0}	-1.0316×10^{0}	-1.0316×10^{0}	-1.0316×10^{0}	-1.0316×10^{0}	-1.0316×10^{0}	-1.0316×10^{0}	-1.0316×10^{0}
	Std	1.87×10^{-11}	8.36×10^{-10}	2.28×10^{-11}	3.13×10^{-8}	1.68×10^{-9}	3.89×10^{-14}	4.19×10^{-7}	6.25×10^{-16}
F17	Mean	3.9789×10^{-1}	3.9789×10^{-1}	4.0217×10^{-1}	3.9789×10^{-1}	3.9789×10^{-1}	3.9789×10^{-1}	3.9789×10^{-1}	3.9789×10^{-1}
	Std	5.94×10^{-12}	5.40×10^{-9}	1.52×10^{-2}	8.10×10^{-7}	6.71×10^{-6}	7.99×10^{-15}	1.27×10^{-7}	0.00×10^{0}
F18	Mean	3.0000×10^{0}	3.0000×10^{0}	4.8000×10^{0}	5.7000×10^{0}	3.0001×10^{0}	3.0000×10^{0}	3.0000×10^{0}	3.0000×10^{0}
	Std	9.42×10^{-11}	1.17×10^{-9}	6.85×10^{0}	1.48×10^{1}	8.14×10^{-5}	2.13×10^{-13}	3.49×10^{-6}	1.79×10^{-15}
F19	Mean	-3.8627×10^{0}	-3.8628×10^{0}	-3.8627×10^{0}	-3.8605×10^{0}	-3.8572×10^{0}	-3.8628×10^{0}	-3.8628×10^{0}	-3.8628×10^{0}
	Std	7.91×10^{-5}	1.58×10^{-7}	2.62×10^{-4}	4.08×10^{-3}	1.02×10^{-2}	1.17×10^{-12}	7.73×10^{-6}	2.58×10^{-15}
F20	Mean	-3.286×10^{0}	-3.2503×10^{0}	-3.2942×10^{0}	-3.2339×10^{0}	-3.2225×10^{0}	-3.2255×10^{0}	-3.2454×10^{0}	-3.2402×10^{0}
	Std	5.59×10^{-2}	5.95×10^{-2}	5.12×10^{-2}	7.30×10^{-2}	1.10×10^{-1}	5.45×10^{-2}	5.93×10^{-2}	8.13×10^{-2}
F21	Mean	-1.01531×10^{1}	-1.01531×10^{1}	-7.8781×10^{0}	-8.8066×10^{0}	-8.4438×10^{0}	-6.9755×10^{0}	-7.048×10^{0}	-6.3883×10^{0}
	Std	1.42×10^{-4}	1.05×10^{-4}	2.68×10^{0}	2.54×10^{0}	2.44×10^{0}	3.35×10^{0}	3.28×10^{0}	3.27×10^{0}
F22	Mean	-1.04028×10^{1}	-1.04028×10^{1}	-7.2814×10^{0}	-1.02239×10^{1}	-7.0271×10^{0}	-8.938×10^{0}	-9.0327×10^{0}	-8.71250×10^{0}
	Std	7.38×10^{-5}	1.82×10^{-4}	3.52×10^{0}	9.70×10^{-1}	3.08×10^{0}	2.99×10^{0}	2.83×10^{0}	2.91×10^{0}
F23	Mean	-1.05363×10^{1}	-1.05363×10^{1}	-6.6743×10^{0}	-1.05349×10^{1}	-7.7815×10^{0}	-8.1138×10^{0}	-8.5201×10^{0}	-9.1233×10^{0}
	Std	8.26×10^{-5}	9.71×10^{-5}	3.31×10^{0}	8.48×10^{-4}	3.28×10^{0}	3.51×10^{0}	3.20×10^{0}	2.93×10^{0}

Table 6. p-values of the Wilcoxon signed-rank test between DESMAOA and other competitor algorithms.

Function	DESMAOA vs. SMA	DESMAOA vs. AOA	DESMAOA vs. GWO	DESMAOA vs. WOA	DESMAOA vs. SSA	DESMAOA vs. MVO	DESMAOA vs. PSO
F1	1.00×10^{0}	6.10×10^{-5}	6.10×10^{-5}	6.10×10^{-5}	6.10×10^{-5}	6.10×10^{-5}	6.10×10^{-5}
F2	6.10×10^{-5}	6.10×10^{-5}	6.10×10^{-5}	6.10×10^{-5}	6.10×10^{-5}	6.10×10^{-5}	6.10×10^{-5}
F3	1.00×10^{0}	6.10×10^{-5}	6.10×10^{-5}	6.10×10^{-5}	6.10×10^{-5}	6.10×10^{-5}	6.10×10^{-5}
F4	6.10×10^{-5}	6.10×10^{-5}	6.10×10^{-5}	6.10×10^{-5}	6.10×10^{-5}	6.10×10^{-5}	6.10×10^{-5}
F5	6.10×10^{-5}	6.10×10^{-5}	6.10×10^{-5}	6.10×10^{-5}	6.10×10^{-5}	6.10×10^{-5}	6.10×10^{-5}
F6	1.22×10^{-4}	6.10×10^{-5}	6.10×10^{-5}	6.10×10^{-5}	6.10×10^{-5}	6.10×10^{-5}	8.54×10^{-4}
F7	2.52×10^{-1}	1.88×10^{-1}	6.10×10^{-5}	6.10×10^{-5}	6.10×10^{-5}	6.10×10^{-5}	6.10×10^{-5}
F8	6.10×10^{-5}	6.10×10^{-5}	6.10×10^{-5}	6.10×10^{-5}	6.10×10^{-5}	6.10×10^{-5}	6.10×10^{-5}
F9	1.00×10^{0}	1.22×10^{-4}	6.10×10^{-5}	1.00×10^{0}	6.10×10^{-5}	6.10×10^{-5}	6.10×10^{-5}
F10	1.00×10^{0}	6.10×10^{-5}	6.10×10^{-5}	9.77×10^{-4}	6.10×10^{-5}	6.10×10^{-5}	6.10×10^{-5}
F11	1.00×10^{0}	6.10×10^{-5}	2.50×10^{-1}	1.00×10^{0}	6.10×10^{-5}	6.10×10^{-5}	6.10×10^{-5}
F12	6.10×10^{-5}	6.10×10^{-5}	6.10×10^{-5}	6.10×10^{-5}	6.10×10^{-5}	6.10×10^{-5}	8.36×10^{-3}
F13	6.10×10^{-5}	6.10×10^{-5}	6.10×10^{-5}	6.10×10^{-5}	6.10×10^{-5}	6.10×10^{-5}	1.22×10^{-4}
F14	6.10×10^{-5}	6.10×10^{-5}	6.10×10^{-5}	6.10×10^{-5}	8.14×10^{-2}	6.10×10^{-5}	7.93×10^{-3}
F15	3.30×10^{-1}	5.54×10^{-2}	4.54×10^{-1}	5.54×10^{-2}	6.10×10^{-5}	1.16×10^{-3}	1.22×10^{-4}
F16	2.56×10^{-2}	3.36×10^{-3}	6.10×10^{-5}	2.52×10^{-1}	6.10×10^{-5}	6.10×10^{-5}	6.10×10^{-5}
F17	6.10×10^{-4}	6.39×10^{-1}	6.10×10^{-5}	6.10×10^{-5}	6.10×10^{-5}	6.10×10^{-5}	6.10×10^{-5}
F18	1.81×10^{-2}	7.62×10^{-1}	8.36×10^{-3}	8.36×10^{-3}	6.10×10^{-5}	8.36×10^{-3}	6.10×10^{-5}
F19	6.10×10^{-5}	1.03×10^{-2}	2.56×10^{-2}	6.10×10^{-5}	6.10×10^{-5}	4.27×10^{-3}	6.10×10^{-5}
F20	6.39×10^{-1}	1.81×10^{-2}	2.52×10^{-1}	2.52×10^{-1}	1.51×10^{-2}	5.99×10^{-1}	2.08×10^{-1}
F21	3.05×10^{-4}	3.02×10^{-2}	6.10×10^{-5}	1.22×10^{-4}	1.88×10^{-1}	1.53×10^{-3}	8.04×10^{-1}
F22	6.10×10^{-5}	4.27×10^{-3}	6.10×10^{-4}	1.22×10^{-4}	8.04×10^{-1}	3.03×10^{-1}	7.62×10^{-1}
F23	6.10×10^{-4}	1.21×10^{-1}	1.22×10^{-4}	6.10×10^{-5}	3.30×10^{-1}	6.79×10^{-1}	6.79×10^{-1}

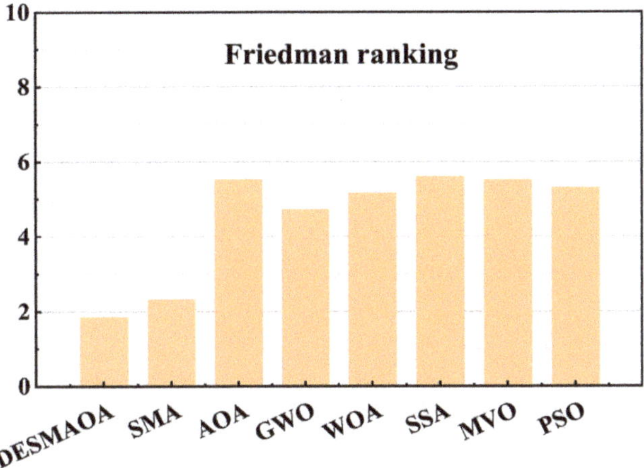

Figure 2. Average Friedman ranking values of DESMAOA and other comparative algorithms on 30 dimensions.

4.2.2. Qualitative Analysis

Figure 3 shows the qualitative results of proposed algorithm in F4, F5, F6, F8, F12, F13, F15, and F21. From the scatter plot of the search history, we were able to see that search agents were distributed in the whole search space in the early stage. During the iteration progresses, they concentrated in a quick time. The density of distribution for different functions indicates that DESMAOA had balanced performance between exploration and exploitation. Moreover, some sudden changes in the amplitude were observed clearly in the trajectory of the first search agent, which revealed that DESMAOA had strong exploration capability over the course of iterations when handing these test functions. The drastic fluctuation of average fitness also showed that DESMAOA can jump out of local optima and explore more spaces when dealing with different types of optimization issues. Hence, the local optimal solution can be avoided effectively. Finally, the DESMAOA was able to find better solutions in most of the functions compared with SMA and AOA, which demonstrates the effectiveness of the proposed method.

4.2.3. Analysis of Convergence Behavior

It is important to study the convergence behavior of optimization algorithms when they are searching for the optimal solution. In general, fast convergence speed is required in the early exploration, which implies the algorithm has powerful exploration capability. On the other hand, local optima also should be avoided, which can be seen from the convergence curve. Figure 4 shows the convergence curves of DESMAOA and other compared algorithms on 30 dimensions. Some benchmark functions are used for analysis including F4, F5, F6, F8, F12, F13, F15, and F21. From these functions, it can be seen that the initial convergence speed of DESMAOA is the fastest in most cases. In Figure 5, step-like or cliff-like declines in the convergence curves of DESMAOA can be observed. This suggests that the DESMAOA has a prominent exploration capability. From F5, F6, F12, and F13, the precision of solutions for DESMAOA is further improved with the help of SAS during the iteration. In sum, DESMAOA achieved the best solutions in these functions.

4.2.4. Scalability Test

The performance fluctuations of optimization algorithms can be revealed according to the scalability test. In this work, the performance of DESMAOA in different dimensions (D = 50, 200, 1000) were also tested. It is easy to understand that the higher dimension will make it harder for the algorithm to find the global optimal solution. Note that only F1–F13 in the 23 benchmark functions were selected for this test. As mentioned previously, F1–F7 are single-mode functions that only have one locally optimal solution. In contrast, F8–F13 are multimode functions that have many locally optimal solutions. Moreover, the experimental parameters were kept the same as previous experiments. Tables 7 and 8 show the results of DESMAOA and other algorithms in different dimensions.

Both results of unimodal and multimode functions indicated that DESMAOA had excellent performance in the conditions of high dimensions. Compared with SMA, AOA, and other well-known algorithms, DESMAOA was the first in all functions except F7. In F7, AOA became better and had more stable results in different dimensions. It is also noted that these comparative algorithms (GWO, WOA, SSA, MVO, and PSO) presented poor optimization capability in some cases, especially in higher dimensions. Furthermore, the Wilcoxon signed-rank test and Friedman ranking test were utilized to analyze the differences between DESMAOA and other algorithms, as listed in Tables 9–12. From Tables 9–11, it can be seen that DESMAOA had significant differences compared with these comparative algorithms. Moreover, in Table 12, the proposed DESMAOA ranked first compared to other algorithms in different dimensions. It is noted that the distance between the first and second was evident. In summary, the proposed DESMAOA had better optimization behavior and stability in dealing with high-dimensional problems.

4.3. The IEEE CEC2021 Standard Test Functions

This section describes the IEEE CEC2021 test functions that were employed to further analyze the performance of proposed DESMAOA on solving global optimization problems. The comparative algorithms included the SMA, AOA, GWO, WOA, SSA, MVO, and PSO. To achieve the statistical results, 30 repeated independent tests were conducted for each function. The experimental results are given in Table 13.

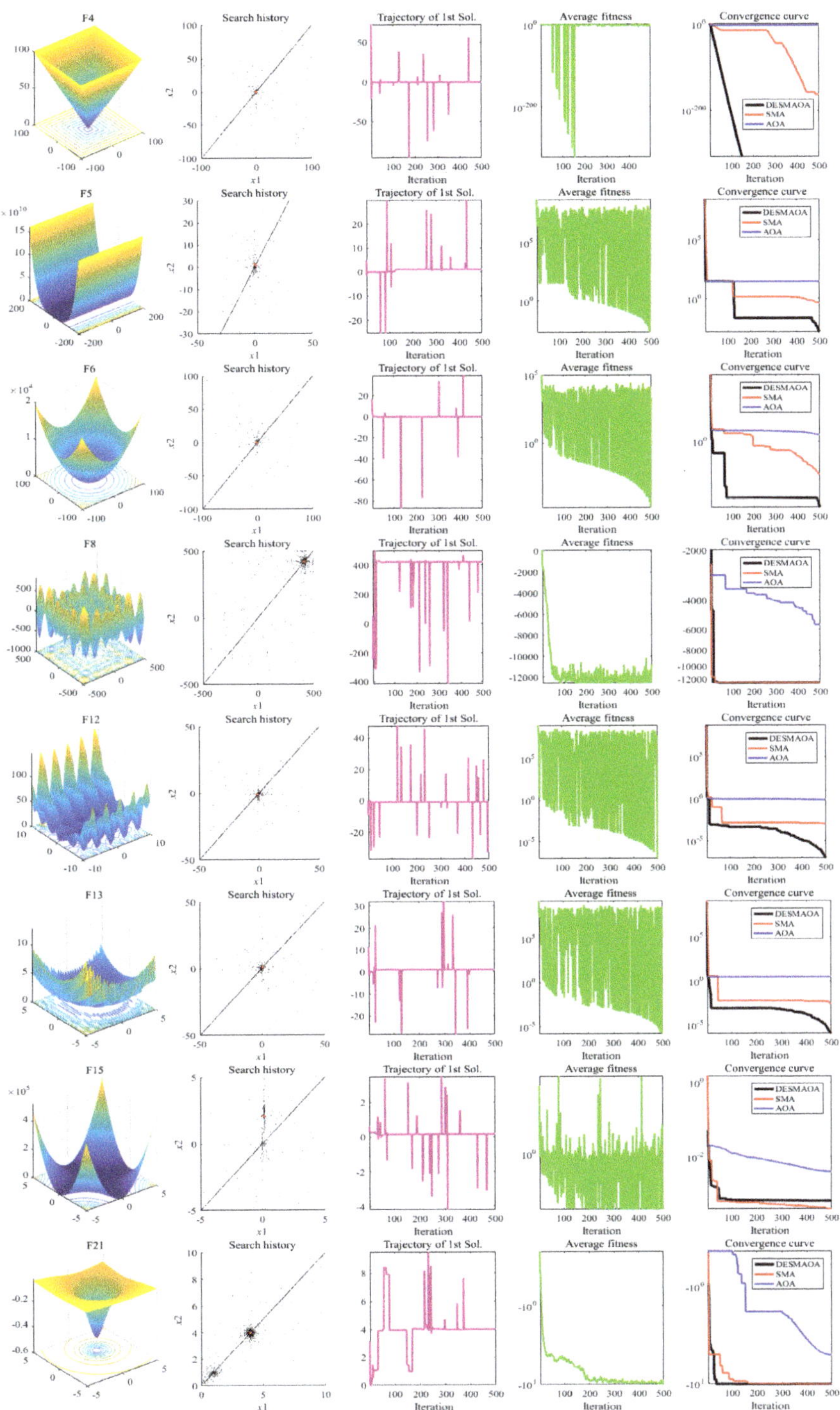

Figure 3. Qualitative results for the benchmark functions F4, F5, F6, F8, F12, F13, F15, and F21.

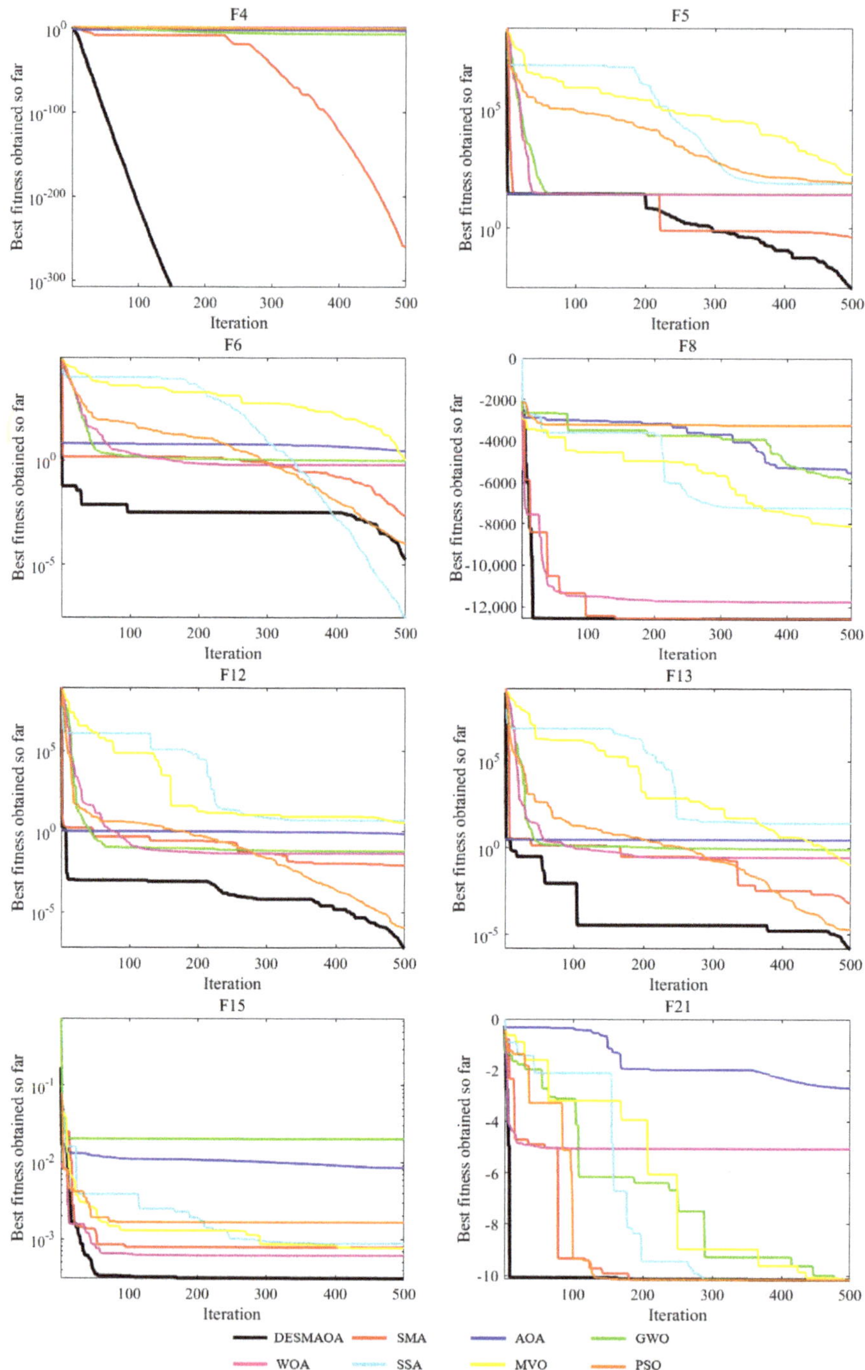

Figure 4. The convergence curves of F4, F5, F6, F8, F12, F13, F15, and F21.

Table 7. Unimodal benchmark function result statistics of the DESMAOA and competitor algorithms in different dimensions.

Function	D	Metric	DESMAOA	SMA	AOA	GWO	WOA	SSA	MVO	PSO
F1	50	Mean	0.00×10^{0}	3.94×10^{-310}	4.05×10^{-5}	5.96×10^{-20}	3.97×10^{-71}	6.27×10^{-1}	9.07×10^{0}	2.05×10^{-1}
		Std	0.00×10^{0}	0.00×10^{0}	1.33×10^{-5}	5.86×10^{-20}	2.17×10^{-70}	5.32×10^{-1}	2.48×10^{0}	1.82×10^{-1}
	200	Mean	0.00×10^{0}	5.15×10^{-244}	4.63×10^{-2}	1.09×10^{-7}	2.33×10^{-71}	1.76×10^{4}	2.84×10^{3}	3.31×10^{2}
		Std	0.00×10^{0}	0.00×10^{0}	1.24×10^{-2}	7.08×10^{-8}	9.24×10^{-71}	1.60×10^{3}	3.15×10^{2}	4.11×10^{1}
	1000	Mean	0.00×10^{0}	2.20×10^{-246}	1.50×10^{0}	2.53×10^{-1}	3.57×10^{-68}	2.29×10^{5}	7.94×10^{5}	4.11×10^{4}
		Std	0.00×10^{0}	0.00×10^{0}	4.79×10^{-2}	5.65×10^{-2}	1.95×10^{-67}	1.19×10^{4}	2.70×10^{4}	2.32×10^{3}
F2	50	Mean	0.00×10^{0}	1.50×10^{-145}	6.70×10^{-3}	2.60×10^{-12}	1.56×10^{-49}	9.29×10^{0}	3.50×10^{3}	2.58×10^{1}
		Std	0.00×10^{0}	8.24×10^{-145}	3.05×10^{-3}	1.52×10^{-12}	7.19×10^{-49}	3.59×10^{0}	1.73×10^{4}	1.89×10^{1}
	200	Mean	0.00×10^{0}	7.90×10^{-138}	7.23×10^{-2}	3.25×10^{-5}	2.93×10^{-48}	1.55×10^{2}	5.08×10^{77}	4.66×10^{2}
		Std	0.00×10^{0}	3.91×10^{-137}	1.18×10^{-2}	7.69×10^{-6}	1.17×10^{-47}	1.44×10^{1}	2.73×10^{78}	6.40×10^{1}
	1000	Mean	0.00×10^{0}	5.92×10^{-1}	1.58×10^{0}	6.78×10^{-1}	1.44×10^{-47}	1.19×10^{3}	3.59×10^{278}	1.41×10^{3}
		Std	0.00×10^{0}	2.92×10^{0}	1.08×10^{-1}	5.77×10^{-1}	7.74×10^{-47}	2.48×10^{1}	Inf	6.51×10^{1}
F3	50	Mean	0.00×10^{0}	1.03×10^{-293}	1.81×10^{-2}	3.84×10^{-1}	2.01×10^{5}	9.10×10^{3}	6.50×10^{3}	1.48×10^{3}
		Std	0.00×10^{0}	0.00×10^{0}	8.60×10^{-3}	1.01×10^{0}	4.50×10^{4}	4.69×10^{3}	1.95×10^{3}	4.64×10^{2}
	200	Mean	0.00×10^{0}	3.94×10^{-219}	7.32×10^{-1}	1.98×10^{4}	4.55×10^{6}	2.05×10^{5}	3.16×10^{5}	8.34×10^{4}
		Std	0.00×10^{0}	0.00×10^{0}	1.86×10^{-1}	9.23×10^{3}	1.51×10^{6}	7.25×10^{4}	3.10×10^{4}	2.24×10^{4}
	1000	Mean	0.00×10^{0}	4.81×10^{-125}	3.35×10^{1}	1.53×10^{6}	1.36×10^{8}	5.19×10^{6}	7.98×10^{6}	2.27×10^{6}
		Std	0.00×10^{0}	2.64×10^{-124}	6.49×10^{0}	3.16×10^{5}	6.32×10^{7}	2.46×10^{6}	8.76×10^{5}	4.56×10^{5}
F4	50	Mean	0.00×10^{0}	3.80×10^{-158}	3.56×10^{-2}	7.25×10^{-4}	7.35×10^{1}	1.94×10^{1}	1.67×10^{1}	3.74×10^{0}
		Std	0.00×10^{0}	2.06×10^{-157}	7.14×10^{-3}	1.42×10^{-3}	1.99×10^{1}	3.14×10^{0}	4.24×10^{0}	7.18×10^{-1}
	200	Mean	0.00×10^{0}	3.11×10^{-114}	9.10×10^{-2}	2.39×10^{1}	8.41×10^{1}	3.52×10^{1}	8.32×10^{1}	1.93×10^{1}
		Std	0.00×10^{0}	1.19×10^{-113}	1.18×10^{-2}	5.51×10^{0}	1.89×10^{1}	3.49×10^{0}	3.80×10^{0}	1.45×10^{0}
	1000	Mean	0.00×10^{0}	3.86×10^{-101}	1.54×10^{-1}	7.88×10^{1}	7.94×10^{1}	4.43×10^{1}	9.81×10^{1}	3.31×10^{1}
		Std	0.00×10^{0}	1.90×10^{-100}	7.55×10^{-3}	3.25×10^{0}	2.09×10^{1}	3.19×10^{0}	6.42×10^{-1}	1.52×10^{0}
F5	50	Mean	4.84×10^{0}	1.89×10^{1}	4.83×10^{1}	4.72×10^{1}	4.83×10^{1}	1.64×10^{3}	7.66×10^{2}	4.21×10^{2}
		Std	1.47×10^{1}	1.93×10^{1}	1.40×10^{-1}	7.35×10^{-1}	4.05×10^{-1}	3.66×10^{3}	7.47×10^{2}	2.18×10^{2}
	200	Mean	1.29×10^{1}	6.23×10^{1}	1.98×10^{2}	1.98×10^{2}	1.98×10^{2}	3.79×10^{6}	3.91×10^{5}	5.98×10^{5}
		Std	3.68×10^{1}	7.20×10^{1}	7.40×10^{-2}	4.19×10^{-1}	1.65×10^{-1}	9.40×10^{5}	1.21×10^{5}	1.04×10^{5}
	1000	Mean	1.11×10^{2}	4.01×10^{2}	1.00×10^{3}	1.05×10^{3}	9.94×10^{2}	1.21×10^{8}	2.33×10^{9}	2.95×10^{8}
		Std	2.50×10^{2}	4.15×10^{2}	2.71×10^{-1}	2.55×10^{1}	1.03×10^{0}	1.12×10^{7}	1.85×10^{8}	4.28×10^{7}
F6	50	Mean	1.66×10^{-4}	8.78×10^{-2}	7.29×10^{0}	2.57×10^{0}	1.20×10^{0}	8.30×10^{-1}	9.61×10^{0}	1.97×10^{-1}
		Std	5.60×10^{-5}	6.54×10^{-2}	4.04×10^{-1}	4.02×10^{-1}	5.39×10^{-1}	7.25×10^{-1}	1.88×10^{0}	1.74×10^{-1}
	200	Mean	3.73×10^{-2}	8.26×10^{0}	3.59×10^{1}	2.91×10^{1}	1.11×10^{1}	1.76×10^{4}	2.99×10^{3}	3.21×10^{2}
		Std	3.21×10^{-2}	8.04×10^{0}	1.19×10^{0}	1.28×10^{0}	2.90×10^{0}	2.45×10^{3}	4.10×10^{2}	4.66×10^{1}
	1000	Mean	3.34×10^{0}	6.80×10^{1}	2.42×10^{2}	2.02×10^{2}	6.68×10^{1}	2.37×10^{5}	8.04×10^{5}	4.02×10^{4}
		Std	4.80×10^{0}	8.81×10^{1}	1.23×10^{0}	2.57×10^{0}	1.52×10^{1}	1.11×10^{4}	2.79×10^{4}	2.17×10^{3}
F7	50	Mean	1.36×10^{-4}	1.96×10^{-4}	6.63×10^{-5}	3.54×10^{-3}	3.97×10^{-3}	4.86×10^{-1}	1.07×10^{-1}	3.97×10^{1}
		Std	1.18×10^{-4}	1.69×10^{-4}	5.90×10^{-5}	1.90×10^{-3}	4.79×10^{-3}	1.53×10^{-1}	2.32×10^{-2}	2.76×10^{1}
	200	Mean	1.29×10^{-4}	4.32×10^{-4}	5.35×10^{-5}	1.63×10^{-2}	4.14×10^{-3}	1.72×10^{1}	5.40×10^{0}	2.95×10^{3}
		Std	1.52×10^{-4}	3.04×10^{-4}	5.09×10^{-5}	5.34×10^{-3}	4.22×10^{-3}	4.14×10^{0}	7.17×10^{-1}	4.68×10^{2}
	1000	Mean	1.17×10^{-4}	6.93×10^{-4}	8.25×10^{-5}	1.55×10^{-1}	3.29×10^{-3}	1.74×10^{3}	2.88×10^{4}	2.39×10^{5}
		Std	1.28×10^{-4}	4.85×10^{-4}	6.96×10^{-5}	3.32×10^{-2}	3.73×10^{-3}	1.75×10^{2}	2.72×10^{3}	7.66×10^{3}

Table 8. Unimodal benchmark function result statistics of the DESMAOA and competitor algorithms in different dimensions.

Function	D	Metric	DESMAOA	SMA	AOA	GWO	WOA	SSA	MVO	PSO
F8	50	Mean	-2.0949×10^{4}	-2.0947×10^{4}	-8.3989×10^{3}	-9.1468×10^{3}	-1.8030×10^{4}	-1.2107×10^{4}	-1.2350×10^{4}	-7.6172×10^{3}
		Std	1.54×10^{-4}	2.27×10^{0}	5.06×10^{2}	1.59×10^{3}	2.78×10^{3}	1.00×10^{3}	1.11×10^{3}	2.18×10^{3}
	200	Mean	-8.3796×10^{4}	-8.3757×10^{4}	-2.1657×10^{4}	-2.7533×10^{4}	-7.1281×10^{4}	-3.4381×10^{4}	-4.0399×10^{4}	-1.5591×10^{4}
		Std	6.62×10^{-1}	6.42×10^{1}	1.27×10^{3}	5.65×10^{3}	1.28×10^{4}	2.25×10^{3}	2.30×10^{3}	6.41×10^{3}
	1000	Mean	-4.1892×10^{5}	-4.1862×10^{5}	-5.4566×10^{4}	-8.4602×10^{4}	-3.5941×10^{5}	-8.8084×10^{4}	-1.1041×10^{5}	-3.3236×10^{4}
		Std	1.25×10^{2}	5.75×10^{2}	2.29×10^{3}	2.28×10^{4}	5.92×10^{4}	7.25×10^{3}	3.94×10^{3}	1.51×10^{4}
F9	50	Mean	0.00×10^{0}	0.00×10^{0}	1.61×10^{-5}	5.24×10^{0}	1.89×10^{-15}	8.82×10^{1}	2.54×10^{2}	2.84×10^{2}
		Std	0.00×10^{0}	0.00×10^{0}	4.42×10^{-6}	7.71×10^{0}	1.04×10^{-14}	2.32×10^{1}	5.65×10^{1}	5.01×10^{1}
	200	Mean	0.00×10^{0}	0.00×10^{0}	1.34×10^{-3}	2.41×10^{1}	7.58×10^{-15}	8.27×10^{2}	1.90×10^{3}	2.02×10^{3}
		Std	0.00×10^{0}	0.00×10^{0}	1.82×10^{-4}	9.14×10^{0}	4.15×10^{-14}	8.74×10^{1}	1.30×10^{2}	1.25×10^{2}
	1000	Mean	0.00×10^{0}	0.00×10^{0}	3.79×10^{-2}	2.06×10^{2}	0.00×10^{0}	7.63×10^{3}	1.46×10^{4}	1.41×10^{4}
		Std	0.00×10^{0}	0.00×10^{0}	1.94×10^{-3}	5.67×10^{1}	0.00×10^{0}	2.12×10^{2}	2.44×10^{2}	2.98×10^{2}
F10	50	Mean	8.8818×10^{-16}	8.8818×10^{-16}	1.14×10^{-3}	4.3720×10^{-11}	4.3225×10^{-15}	4.83×10^{0}	3.56×10^{0}	1.69×10^{0}
		Std	0.00×10^{0}	0.00×10^{0}	1.93×10^{-4}	2.44×10^{-11}	2.38×10^{-15}	1.23×10^{0}	3.13×10^{0}	5.70×10^{-1}
	200	Mean	8.8818×10^{-16}	8.8818×10^{-16}	1.06×10^{-2}	2.18×10^{-5}	4.09×10^{-15}	1.30×10^{1}	2.04×10^{1}	6.61×10^{0}
		Std	0.00×10^{0}	0.00×10^{0}	1.01×10^{-3}	6.01×10^{-6}	2.70×10^{-15}	4.61×10^{-1}	2.15×10^{-1}	3.38×10^{-1}
	1000	Mean	8.8818×10^{-16}	8.8818×10^{-16}	3.32×10^{-2}	1.89×10^{-2}	4.91×10^{-15}	1.45×10^{1}	2.10×10^{1}	1.60×10^{1}
		Std	0.00×10^{0}	0.00×10^{0}	7.69×10^{-4}	3.23×10^{-3}	2.42×10^{-15}	1.94×10^{-1}	3.27×10^{-2}	2.33×10^{-1}
F11	50	Mean	0.00×10^{0}	0.00×10^{0}	7.70×10^{-3}	2.94×10^{-3}	1.34×10^{-2}	5.55×10^{-1}	1.09×10^{0}	1.62×10^{-2}
		Std	0.00×10^{0}	0.00×10^{0}	1.91×10^{-2}	6.06×10^{-3}	5.11×10^{-2}	2.74×10^{-1}	2.30×10^{-2}	1.19×10^{-2}
	200	Mean	0.00×10^{0}	0.00×10^{0}	7.85×10^{0}	6.27×10^{-3}	0.00×10^{0}	1.46×10^{2}	2.73×10^{1}	2.28×10^{0}
		Std	0.00×10^{0}	0.00×10^{0}	1.19×10^{1}	1.48×10^{-2}	0.00×10^{0}	1.88×10^{1}	3.08×10^{0}	2.70×10^{0}
	1000	Mean	0.00×10^{0}	0.00×10^{0}	1.33×10^{4}	2.37×10^{-2}	0.00×10^{0}	2.12×10^{3}	7.25×10^{3}	2.74×10^{2}
		Std	0.00×10^{0}	0.00×10^{0}	2.64×10^{3}	3.67×10^{-2}	0.00×10^{0}	8.35×10^{1}	3.01×10^{2}	1.86×10^{1}
F12	50	Mean	6.67×10^{-7}	6.02×10^{-3}	9.06×10^{-1}	1.22×10^{-1}	3.46×10^{-2}	1.26×10^{1}	5.51×10^{0}	8.03×10^{-2}
		Std	6.28×10^{-7}	1.22×10^{-2}	2.39×10^{-2}	7.35×10^{-2}	1.86×10^{-2}	4.57×10^{0}	1.37×10^{0}	1.53×10^{-1}
	200	Mean	2.01×10^{-5}	5.76×10^{-3}	8.41×10^{-1}	5.42×10^{-1}	7.03×10^{-2}	7.55×10^{3}	2.30×10^{3}	4.84×10^{1}
		Std	1.57×10^{-5}	8.11×10^{-3}	5.60×10^{-2}	6.68×10^{-2}	3.52×10^{-2}	1.01×10^{4}	2.88×10^{3}	3.71×10^{1}
	1000	Mean	1.53×10^{-4}	9.67×10^{-3}	1.04×10^{0}	1.26×10^{0}	1.05×10^{-1}	1.16×10^{7}	4.19×10^{9}	9.21×10^{6}
		Std	2.46×10^{-4}	1.70×10^{-2}	1.12×10^{-2}	2.99×10^{-1}	5.30×10^{-2}	4.67×10^{6}	4.67×10^{8}	2.30×10^{6}
F13	50	Mean	1.08×10^{-3}	2.52×10^{-2}	4.94×10^{0}	2.03×10^{0}	1.14×10^{0}	8.07×10^{1}	7.29×10^{0}	1.84×10^{-1}
		Std	4.27×10^{-3}	3.02×10^{-2}	6.56×10^{-4}	2.80×10^{-1}	4.84×10^{-1}	1.61×10^{1}	1.15×10^{1}	1.16×10^{-1}
	200	Mean	1.85×10^{-3}	4.73×10^{-1}	1.97×10^{1}	1.67×10^{1}	6.18×10^{0}	1.61×10^{6}	1.11×10^{5}	5.27×10^{3}
		Std	1.29×10^{-3}	7.17×10^{-1}	9.97×10^{-2}	4.32×10^{-1}	1.75×10^{0}	7.60×10^{5}	1.03×10^{5}	2.58×10^{3}
	1000	Mean	6.06×10^{-2}	3.82×10^{0}	1.00×10^{2}	1.21×10^{2}	4.02×10^{1}	1.47×10^{8}	9.13×10^{9}	8.26×10^{7}
		Std	8.56×10^{-2}	3.59×10^{0}	3.49×10^{-1}	7.98×10^{0}	1.12×10^{1}	2.91×10^{7}	8.64×10^{8}	1.28×10^{7}

Table 9. p-values of the Wilcoxon signed-rank test between DESMAOA and other competitor algorithms on 50 dimensions.

Function	DESMAOA vs. SMA	DESMAOA vs. AOA	DESMAOA vs. GWO	DESMAOA vs. WOA	DESMAOA vs. SSA	DESMAOA vs. MVO	DESMAOA vs. PSO
F1	5.00×10^{-1}	6.10×10^{-5}	6.10×10^{-5}	6.10×10^{-5}	6.10×10^{-5}	6.10×10^{-5}	6.10×10^{-5}
F2	6.10×10^{-5}	6.10×10^{-5}	6.10×10^{-5}	6.10×10^{-5}	6.10×10^{-5}	6.10×10^{-5}	6.10×10^{-5}
F3	5.00×10^{-1}	6.10×10^{-5}	6.10×10^{-5}	6.10×10^{-5}	6.10×10^{-5}	6.10×10^{-5}	6.10×10^{-5}
F4	6.10×10^{-5}	6.10×10^{-5}	6.10×10^{-5}	6.10×10^{-5}	6.10×10^{-5}	6.10×10^{-5}	6.10×10^{-5}
F5	5.37×10^{-3}	6.10×10^{-5}	1.22×10^{-4}	6.10×10^{-5}	6.10×10^{-5}	6.10×10^{-5}	6.10×10^{-5}
F6	6.10×10^{-5}	6.10×10^{-5}	6.10×10^{-5}	6.10×10^{-5}	6.10×10^{-5}	6.10×10^{-5}	6.10×10^{-5}
F7	8.33×10^{-2}	7.30×10^{-2}	6.10×10^{-5}	6.10×10^{-5}	6.10×10^{-5}	6.10×10^{-5}	6.10×10^{-5}
F8	6.10×10^{-5}	6.10×10^{-5}	6.10×10^{-5}	6.10×10^{-5}	6.10×10^{-5}	6.10×10^{-5}	6.10×10^{-5}
F9	1.00×10^{0}	6.10×10^{-5}	6.10×10^{-5}	1.00×10^{0}	6.10×10^{-5}	6.10×10^{-5}	6.10×10^{-5}
F10	1.00×10^{0}	6.10×10^{-5}	6.10×10^{-5}	2.44×10^{-4}	6.10×10^{-5}	6.10×10^{-5}	6.10×10^{-5}
F11	1.00×10^{0}	6.10×10^{-5}	2.50×10^{-1}	1.00×10^{0}	6.10×10^{-5}	6.10×10^{-5}	6.10×10^{-5}
F12	6.10×10^{-5}	6.10×10^{-5}	6.10×10^{-5}	6.10×10^{-5}	6.10×10^{-5}	6.10×10^{-5}	6.10×10^{-5}
F13	2.01×10^{-3}	6.10×10^{-5}	6.10×10^{-5}	6.10×10^{-5}	6.10×10^{-5}	6.10×10^{-5}	6.10×10^{-5}

Table 10. p-Values of the Wilcoxon signed-rank test between DESMAOA and other competitor algorithms on 200 dimensions.

Function	DESMAOA vs. SMA	DESMAOA vs. AOA	DESMAOA vs. GWO	DESMAOA vs. WOA	DESMAOA vs. SSA	DESMAOA vs. MVO	DESMAOA vs. PSO
F1	2.50×10^{-1}	6.10×10^{-5}	6.10×10^{-5}	6.10×10^{-5}	6.10×10^{-5}	6.10×10^{-5}	6.10×10^{-5}
F2	6.10×10^{-5}	6.10×10^{-5}	6.10×10^{-5}	6.10×10^{-5}	6.10×10^{-5}	6.10×10^{-5}	6.10×10^{-5}
F3	6.10×10^{-5}	6.10×10^{-5}	9.77×10^{-4}	6.10×10^{-5}	6.10×10^{-5}	6.10×10^{-5}	6.10×10^{-5}
F4	6.10×10^{-5}	6.10×10^{-5}	6.10×10^{-5}	6.10×10^{-5}	6.10×10^{-5}	6.10×10^{-5}	6.10×10^{-5}
F5	6.10×10^{-5}	6.10×10^{-5}	6.10×10^{-5}	6.10×10^{-5}	8.54×10^{-4}	6.10×10^{-5}	6.10×10^{-5}
F6	6.10×10^{-5}	6.10×10^{-5}	6.10×10^{-5}	6.10×10^{-5}	6.10×10^{-5}	1.22×10^{-4}	6.10×10^{-5}
F7	6.10×10^{-5}	6.10×10^{-5}	6.10×10^{-5}	6.10×10^{-5}	6.10×10^{-5}	6.10×10^{-5}	6.10×10^{-5}
F8	2.56×10^{-2}	6.10×10^{-5}	1.22×10^{-4}	6.10×10^{-5}	6.10×10^{-5}	6.10×10^{-5}	6.10×10^{-5}
F9	1.00×10^{0}	6.10×10^{-5}	6.10×10^{-5}	6.10×10^{-5}	6.10×10^{-5}	6.10×10^{-5}	6.10×10^{-5}
F10	1.00×10^{0}	6.10×10^{-5}	6.10×10^{-5}	6.10×10^{-5}	6.10×10^{-5}	6.10×10^{-5}	6.10×10^{-5}
F11	1.00×10^{0}	6.10×10^{-5}	6.10×10^{-5}	1.00×10^{0}	6.10×10^{-5}	4.88×10^{-4}	6.10×10^{-5}
F12	6.10×10^{-5}	6.10×10^{-5}	6.10×10^{-5}	6.10×10^{-5}	6.10×10^{-5}	6.10×10^{-5}	1.22×10^{-4}
F13	6.10×10^{-5}	6.10×10^{-5}	6.10×10^{-5}	6.10×10^{-5}	6.10×10^{-5}	6.10×10^{-5}	6.10×10^{-5}

Table 11. p-Values of the Wilcoxon signed-rank test between DESMAOA and other competitor algorithms on 1000 dimensions.

Function	DESMAOA vs. SMA	DESMAOA vs. AOA	DESMAOA vs. GWO	DESMAOA vs. WOA	DESMAOA vs. SSA	DESMAOA vs. MVO	DESMAOA vs. PSO
F1	3.91×10^{-3}	6.10×10^{-5}	6.10×10^{-5}	6.10×10^{-5}	6.10×10^{-5}	6.10×10^{-5}	6.10×10^{-5}
F2	6.10×10^{-5}	6.10×10^{-5}	6.10×10^{-5}	6.10×10^{-5}	6.10×10^{-5}	6.10×10^{-5}	6.10×10^{-5}
F3	6.10×10^{-5}	6.10×10^{-5}	1.22×10^{-4}	6.10×10^{-5}	6.10×10^{-5}	6.10×10^{-5}	6.10×10^{-5}
F4	6.10×10^{-5}	6.10×10^{-5}	6.10×10^{-5}	6.10×10^{-5}	6.10×10^{-5}	6.10×10^{-5}	6.10×10^{-5}
F5	6.10×10^{-5}	6.10×10^{-5}	6.10×10^{-5}	6.10×10^{-5}	6.10×10^{-4}	6.10×10^{-5}	6.10×10^{-5}
F6	6.10×10^{-5}	6.10×10^{-5}	6.10×10^{-5}	6.10×10^{-5}	6.10×10^{-5}	1.22×10^{-4}	6.10×10^{-5}
F7	6.10×10^{-5}	6.10×10^{-5}	6.10×10^{-5}	6.10×10^{-5}	6.10×10^{-5}	6.10×10^{-5}	8.36×10^{-3}
F8	4.21×10^{-1}	6.10×10^{-5}	1.22×10^{-4}	6.10×10^{-5}	6.10×10^{-5}	6.10×10^{-5}	6.10×10^{-5}
F9	1.00×10^{0}	6.10×10^{-5}	6.10×10^{-5}	1.00×10^{0}	6.10×10^{-5}	6.10×10^{-5}	6.10×10^{-5}
F10	1.00×10^{0}	6.10×10^{-5}	6.10×10^{-5}	6.10×10^{-5}	6.10×10^{-5}	6.10×10^{-5}	6.10×10^{-5}
F11	1.00×10^{0}	6.10×10^{-5}	6.10×10^{-5}	1.00×10^{0}	6.10×10^{-5}	9.77×10^{-4}	6.10×10^{-5}
F12	6.10×10^{-5}	6.10×10^{-5}	6.10×10^{-5}	6.10×10^{-5}	6.10×10^{-5}	6.10×10^{-5}	1.16×10^{-3}
F13	6.10×10^{-5}	6.10×10^{-5}	6.10×10^{-5}	6.10×10^{-5}	6.10×10^{-5}	6.10×10^{-5}	6.10×10^{-5}

Table 12. Experimental results of Friedman test on 50, 200, and 1000 dimensions.

Algorithm	$D = 50$		$D = 200$		$D = 1000$	
	Mean	Rank	Mean		Mean	Rank
DESMAOA	1.1923	1	1.2308	1	1.2692	1
SMA	1.9615	2	2.0000	2	2.1923	2
AOA	4.7692	5	4.5385	5	4.4615	4
GWO	4.3077	3	4.3846	4	4.6923	5
WOA	4.3846	4	3.7692	3	3.4615	3
SSA	6.7692	7	7.0000	8	6.1538	7
MVO	6.8462	8	6.7692	7	7.4615	8
PSO	5.7692	6	6.3077	6	6.3077	6

Table 13. The result statistics of CEC2021 test functions for the DESMAOA and competitor algorithms.

Function	Metric	DESMAOA	SMA	AOA	GWO	WOA	SSA	MVO	PSO
CEC_01	Mean	3.4126×10^3	8.4790×10^3	1.4400×10^{10}	8.5800×10^7	8.8200×10^7	3.0497×10^3	2.0700×10^4	2.6509×10^3
	Std	3.2484×10^3	4.3942×10^3	5.4800×10^9	2.0600×10^8	1.0800×10^8	2.6586×10^3	1.2300×10^4	2.8147×10^3
CEC_02	Mean	1.6460×10^3	1.7230×10^3	2.3794×10^3	1.7650×10^3	2.3408×10^3	1.9202×10^3	1.7883×10^3	2.0353×10^3
	Std	1.8064×10^2	2.0356×10^2	2.5500×10^2	4.0002×10^2	2.8907×10^2	3.0299×10^2	2.8914×10^2	3.3248×10^2
CEC_03	Mean	7.4197×10^2	7.3268×10^2	8.0255×10^2	7.3186×10^2	7.9430×10^2	7.4133×10^2	7.3257×10^2	7.3096×10^2
	Std	1.3713×10^1	9.9242×10^0	8.5999×10^0	1.0358×10^1	2.9899×10^1	1.5608×10^1	9.4278×10^0	1.1527×10^1
CEC_04	Mean	1.9024×10^3	1.9015×10^3	3.9200×10^5	1.9028×10^3	1.9116×10^3	1.9015×10^3	1.9014×10^3	1.9011×10^3
	Std	1.5132×10^0	4.7478×10^{-1}	2.0400×10^5	1.1137×10^0	8.0854×10^0	4.6367×10^{-1}	6.4272×10^{-1}	6.8410×10^{-1}
CEC_05	Mean	4.7672×10^3	2.0900×10^4	4.6200×10^5	1.1600×10^5	5.8000×10^5	3.2100×10^4	6.7593×10^3	5.0581×10^3
	Std	4.2784×10^3	4.6800×10^4	1.1100×10^5	1.9300×10^5	8.5200×10^5	7.2900×10^4	4.2953×10^3	3.3481×10^3
CEC_06	Mean	1.7312×10^3	1.7684×10^3	2.2222×10^3	1.7776×10^3	1.8431×10^3	1.7573×10^3	1.7587×10^3	1.8632×10^3
	Std	1.2285×10^2	9.1820×10^1	2.0670×10^2	1.1171×10^2	1.0223×10^2	8.6052×10^1	1.0387×10^2	1.0503×10^2
CEC_07	Mean	7.0311×10^3	6.5603×10^3	2.9700×10^6	1.8000×10^4	3.4500×10^5	7.3733×10^3	7.5164×10^3	6.0549×10^3
	Std	7.7063×10^3	6.4319×10^3	3.8500×10^6	3.7100×10^4	5.3700×10^5	4.9714×10^3	6.1260×10^3	2.6469×10^3
CEC_08	Mean	2.2974×10^3	2.4032×10^3	3.5700×10^3	2.3384×10^3	2.4289×10^3	2.3012×10^3	2.3861×10^3	2.4193×10^3
	Std	2.2368×10^1	3.1043×10^2	3.7790×10^2	9.1745×10^1	3.3946×10^2	1.4399×10^1	2.6287×10^2	3.7983×10^2
CEC_09	Mean	2.7018×10^3	2.7599×10^3	2.9038×10^3	2.7449×10^3	2.7752×10^3	2.7330×10^3	2.7514×10^3	2.7922×10^3
	Std	1.0302×10^2	1.0211×10^1	9.9242×10^1	4.4570×10^1	6.2479×10^1	6.4081×10^1	9.5351×10^0	1.0509×10^2
CEC_10	Mean	2.9231×10^3	2.9323×10^3	3.6576×10^3	2.9366×10^3	2.9545×10^3	2.9289×10^3	2.9290×10^3	2.9234×10^3
	Std	2.2799×10^1	3.1577×10^1	4.0387×10^2	2.4483×10^1	6.9125×10^1	2.4243×10^1	2.9085×10^1	2.3865×10^1
Average rank		2.3	3.9	7.9	4.6	6.9	3.3	3.7	3.4
Rank		1	5	8	6	7	2	4	3

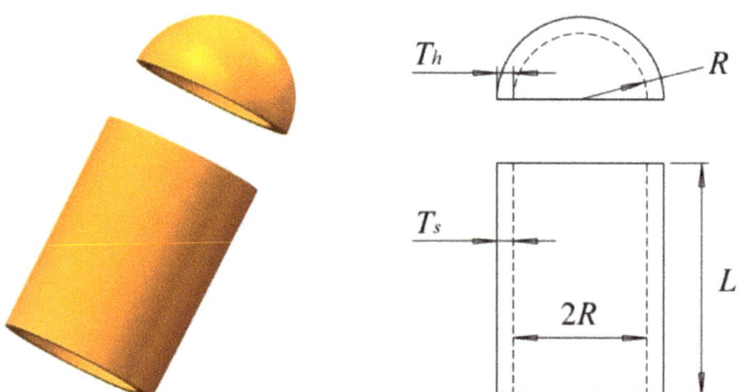

Figure 5. Pressure vessel design problem: model diagram (**left**) and structure parameters (**right**).

From Table 13, it can be observed that the DESMAOA was able to obtain the best results in six functions: CEC_02, CEC_05, CEC_06, CEC_08, CEC_09, and CEC_10. Thus, we can find that the DESMAOA has good performance in hybrid and composition test functions. By comparing it with other optimization algorithms, we found that DESMAOA showed very competitive performance for these CEC2021 test functions. Moreover, the Friedman's ranking test was also used to evaluate the performance of DESMAOA. The average rank and rank were also given in Table 13. It can be seen that DESMAOA obtained the best statistical ranking result among these algorithms.

Therefore, the results of CEC2021 test functions also showed the high performance for solving optimization problems.

5. Applicability for Solving Engineering Design Problems

This section reports the three classical engineering design problems we employed to evaluate the capability of DESMAOA to solve practical problems, which were the pressure vessel design problem, three-bar truss design problem, and tension/compression spring design problem. In the same way, 30 search agents and 500 iterations were utilized in the design procedure of engineering problems for a fair comparison. Meanwhile, other related results of optimization algorithms proposed by scholars are also given and compared with proposed algorithm here. Detailed descriptions are shown below.

5.1. Pressure Vessel Design

The design of the pressure vessel is an optimization problem with four variables and four constraints in the industrial field [43]. The lowest cost of pressure vessel was the ultimate goal. The structure of pressure vessel is shown in Figure 5. The four design variables were the thickness of the shell (T_s), thickness of the head (T_h), inner radius (R), and length of the cylindrical section (L). Table 14 lists the comparison between DESMAOA and other competitor algorithms. From Table 14, we can see that DESMAOA was capable of finding the optimal solution with the lowest cost.

Table 14. Optimal results for comparative algorithms on the pressure vessel design problem.

Algorithm	Optimal Values for Variables				Optimal Cost
	T_s	T_h	R	L	
DESMAOA	7.943124×10^{-1}	3.927124×10^{-1}	4.288001×10^{1}	1.671866×10^{2}	5.8363262×10^{3}
SMA [33]	7.931×10^{-1}	3.932×10^{-1}	4.06711×10^{1}	1.962178×10^{2}	5.9941857×10^{3}
AOA [34]	8.303737×10^{-1}	4.162057×10^{-1}	4.275127×10^{1}	1.693454×10^{2}	6.0487844×10^{3}
MVO [5]	8.125×10^{-1}	4.375×10^{-1}	4.2090738×10^{1}	1.7673869×10^{2}	6.0608066×10^{3}
WOA [39]	8.12500×10^{-1}	4.37500×10^{-1}	4.2098209×10^{1}	1.76638998×10^{2}	6.0597410×10^{3}
MFO [44]	8.125×10^{-1}	4.375×10^{-1}	4.2098445×10^{1}	1.76636596×10^{2}	6.0597143×10^{3}
GWO [38]	8.125×10^{-1}	4.345×10^{-1}	4.20892×10^{1}	1.767587×10^{2}	6.0515639×10^{3}
MOSCA [45]	7.781909×10^{-1}	3.830476×10^{-1}	4.03207539×10^{1}	$1.999841994 \times 10^{2}$	5.88071150×10^{3}
LWOA [46]	7.78858×10^{-1}	3.85321×10^{-1}	4.032609×10^{1}	2.00×10^{2}	5.893339×10^{3}
IMFO [47]	7.781948×10^{-1}	3.846621×10^{-1}	4.032097×10^{1}	1.999812×10^{2}	5.8853778×10^{3}

5.2. Three-Bar Truss Design

The aim of the three-bar truss design is to achieve the lowest weight of three-bar truss with the constraints of stress, deflection, and buckling, which belongs to the field of civil engineering [48]. In this design problem, two parameters x_1 (or A1) and x_2 (or A2) were involved, as shown in Figure 6. The solutions obtained by the DESMAOA and other representative algorithms are listed in Table 15. It can be seen that the proposed hybrid method apparently outperformed other approaches. Moreover, 30 repeated tests were also performed to evaluate the robustness of the proposed algorithm. The worst value, mean value, best value, and stand deviation were 2.639079×10^{2}, 2.638562×10^{2}, 2.638523×10^{2}, and 1.0451×10^{-2}. Hence, the statistical results revealed that the proposed algorithm had very stable and superior performance in solving this design problem.

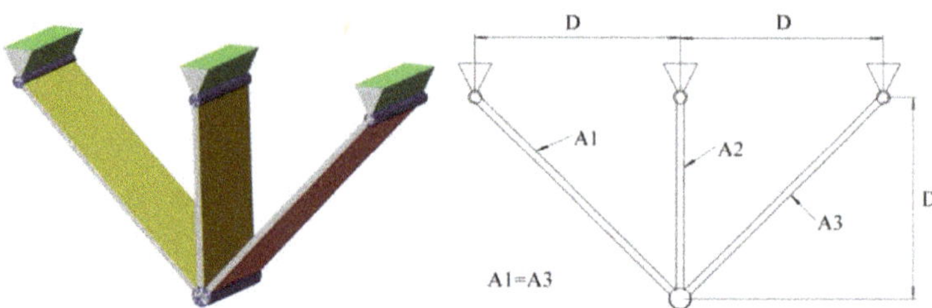

Figure 6. Three-bar truss design problem: model diagram (**left**) and structure parameters (**right**).

Table 15. Optimal results for comparative algorithms on the three-bar truss design problem.

Algorithm	Optimal Values for Variables		Optimal Weight
	x_1	x_2	
DESMAOA	7.882549×10^{-1}	4.085642×10^{-1}	$2.638523657 \times 10^{2}$
SMA [33]	7.729316×10^{-1}	4.718874×10^{-1}	$2.658067955 \times 10^{2}$
AOA [34]	7.9369×10^{-1}	3.9426×10^{-1}	2.639154×10^{2}
MBA [48]	7.885650×10^{-1}	4.085597×10^{-1}	$2.638958522 \times 10^{2}$
SSA [40]	$7.88665414 \times 10^{-1}$	$4.08275784 \times 10^{-1}$	$2.638958434 \times 10^{2}$
MFO [44]	$7.88244771 \times 10^{-1}$	$4.09466906 \times 10^{-1}$	$2.638959797 \times 10^{2}$
PSO-DE [49]	7.886751×10^{-1}	4.082482×10^{-1}	$2.638958433 \times 10^{2}$
HSCAHS [50]	7.885721×10^{-1}	4.084012×10^{-1}	2.63881992×10^{2}

5.3. Tension/Compression Spring Design

In the design of a tension/compression spring [51], the objective is to obtain the minimum optimal weight under three constraints: (1) shear stress, (2) surge frequency, and (3) deflection. As shown in Figure 7, there were three variables that needed to be considered. They were the wire diameter (d), mean coil diameter (D), and the number of active coils (N). The results of DESMAOA and other comparative algorithms are listed in Table 16. By comparison, the proposed DESMAOA achieved the best solution for this problem, which was 5.44827×10^{-2}, 4.83109×10^{-1}, and 5.746128×10^{0} for d, D, and N, respectively. Moreover, the optimal weight was 1.11083×10^{-2}.

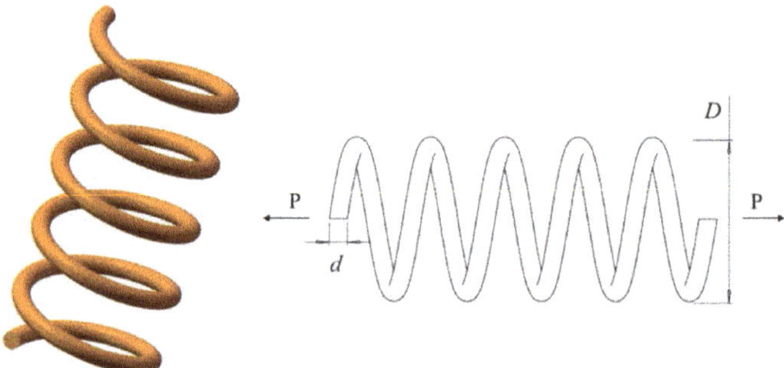

Figure 7. Tension/compression spring design problem: model diagram (**left**) and structure parameters (**right**).

Table 16. Optimal results for comparative algorithms on the tension/compression spring design problem.

Algorithm	Optimal Values for Variables			Optimal Weight
	d	D	p	
DESMAOA	5.44827×10^{-2}	4.83109×10^{-1}	5.746128×10^{0}	1.11083×10^{-2}
SMA [33]	5.8992×10^{-2}	6.23402×10^{-1}	3.590304×10^{0}	1.2128×10^{-2}
AOA [34]	5.00×10^{-2}	3.49809×10^{-1}	1.18637×10^{1}	1.2124×10^{-2}
MVO [5]	5.251×10^{-2}	3.7602×10^{-1}	1.033513×10^{1}	1.2790×10^{-2}
AO [14]	5.02439×10^{-2}	3.5262×10^{-1}	1.05425×10^{1}	1.1165×10^{-2}
SSA [40]	5.1207×10^{-2}	3.45215×10^{-1}	1.2004032×10^{1}	1.26763×10^{-2}
GWO [38]	5.169×10^{-2}	3.56737×10^{-1}	1.128885×10^{1}	1.2666×10^{-2}
GSA [6]	5.0276×10^{-2}	3.23680×10^{-1}	1.3525410×10^{1}	1.27022×10^{-2}
WSA [51]	5.168626×10^{-2}	3.5665047×10^{-1}	$1.129291654 \times 10^{1}$	1.267061×10^{-2}

6. Conclusions and Future Works

To overcome the shortcomings of basic meta-heuristic algorithms, this paper presents an effective deep ensemble method of two very new optimization algorithms, i.e., the SMA and AOA. A preliminary hybrid of these two algorithms was firstly conducted to enhance the capability of exploration. Then, two strategies were integrated to the hybridized algorithm to assist it to jump out of the local minima and improve the accuracy of the solution. The performance of proposed DESMAOA was extensively analyzed by using 23 classical test functions.

First, different combinations of SMAOA and two strategies were analyzed and discussed. The results revealed the effectiveness of proposed strategies. Then, the results of DESMAOA were compared with SMA, AOA, and five well-known algorithms. The results showed that the proposed method had the advantages of both SMA and AOA and that it also was evidently better than other comparison algorithms. Afterward, experimental tests in high dimensional environments (50, 200, and 1000) were also investigated among these comparative algorithms, and the results of scalability test also confirmed the superior performance of the proposed method. Finally, the proposed DESMAOA was employed to deal with three engineering design problems. The results show that the proposed method was good at solving these problems, and in particular it was very stable when solving the three-bar truss design problem.

As future perspectives, the DESMAOA can be utilized to solve more optimization problems in other disciplines, such as the feature selection, training of multi-layer perceptron neural network, and image processing. Another investigation is to consider the implementation of this hybrid method on other optimization algorithms for better optimization performance.

Author Contributions: Conceptualization, R.Z. and H.J.; methodology, R.Z. and H.J.; software, R.Z. and S.W.; validation, R.Z., H.J. and L.A.; formal analysis, R.Z. and S.W.; investigation, R.Z. and H.J.; resources, R.Z., H.J. and L.A.; data curation, R.Z.; writing—original draft preparation, R.Z.; writing—review and editing, R.Z. and H.J.; visualization, Q.L.; supervision, H.J. and L.A.; project administration, R.Z. and H.J.; funding acquisition, R.Z. and H.J. All authors have read and agreed to the published version of the manuscript.

Funding: This work was supported by the Sanming University introduces high-level talents to start scientific research funding support project (21YG01, 20YG14), Fujian Natural Science Foundation Project (2021J011128), Guiding science and technology projects in Sanming City (2021-S-8), Educational research projects of young and middle-aged teachers in Fujian Province (JAT200618), Scientific research and development fund of Sanming University (B202009), and Funded By Open Research Fund Program of Fujian Provincial Key Laboratory of Agriculture Internet of Things Application (ZD2101).

Data Availability Statement: Not applicable.

Conflicts of Interest: The authors declare no conflict of interest.

References

1. Abualigah, L.; Diaba, A. Advances in sine cosine algorithm: A comprehensive survey. *Artif. Intell. Rev.* **2021**, *54*, 2567–2608. [CrossRef]
2. Abualigah, L.; Diaba, A. A comprehensive survey of the Grasshopper optimization algorithm: Results, variants, and applications. *Neural Comput. Appl.* **2020**, *32*, 15533–15556. [CrossRef]
3. Jia, H.; Lang, C.; Oliva, D.; Song, W.; Peng, X. Dynamic Harris Hawks Optimization with Mutation Mechanism for Satellite Image Segmentation. *Remote Sens.* **2019**, *11*, 1421. [CrossRef]
4. Jia, H.; Peng, X.; Lang, C. Remora optimization algorithm. *Expert Syst. Appl.* **2021**, *185*, 115665. [CrossRef]
5. Mirjalili, S.; Mirjalili, S.M.; Hatamlou, A. Multi-verse optimizer: A nature-inspired algorithm for global optimization. *Neural Comput. Appl.* **2016**, *27*, 495–513. [CrossRef]
6. Rashedi, E.; Nezamabadi-pour, H.; Saryazdi, S. GSA: A Gravitational Search Algorithm. *Inf. Sci.* **2009**, *179*, 2232–2248. [CrossRef]
7. Kaveh, A.; Dadras, A. A novel meta-heuristic optimization algorithm: Thermal exchange optimization. *Adv. Eng. Softw.* **2017**, *110*, 69–84. [CrossRef]
8. Asef, F.; Majidnezhad, V.; Feizi-Derakhshi, M.R.; Parsa, S. Heat transfer relation-based optimization algorithm (HTOA). *Soft. Comput.* **2021**, *25*, 8129–8158. [CrossRef]
9. Corriveau, G.; Guilbault, R.; Tahan, A.; Sabourin, R. Bayesian network as an adaptive parameter setting approach for genetic algorithms. *Complex Intell. Syst.* **2016**, *2*, 1–22. [CrossRef]
10. Storn, R.; Price, K. Differential evolution-a simple and efficient heuristic for global optimization over continuous spaces. *J. Glob. Optim.* **2010**, *23*, 689–694.
11. Yao, X.; Liu, Y.; Lin, G. Evolutionary Programming Made Faster. *IEEE Trans. Evol. Comput.* **1999**, *3*, 82–102.
12. Chen, G.; Yu, J. Particle swarm optimization algorithm. *Inf. Control* **2005**, *186*, 454–458.
13. Gaurav, D.; Vijay, K. Emperor penguin optimizer: A bio-inspired algorithm for engineering problems. *Knowl. Based Syst.* **2018**, *159*, 20–50.
14. Abualigah, L.; Yousri, D.; Elaziz, M.A.; Ewees, A.A.; Al-qaness, M.A.A.; Gandomi, A.H. Aquila optimizer: A novel meta-heuristic optimization Algorithm. *Comput. Ind. Eng.* **2021**, *157*, 107250. [CrossRef]
15. Faramarzi, A.; Heidarinejad, M.; Mirjalili, S.; Gandomi, A.H. Marine predators algorithm: A nature-inspired metaheuristic. *Expert Syst. Appl.* **2020**, *152*, 113377. [CrossRef]
16. Rao, R.V.; Savsani, V.J.; Vakharia, D.P. Teaching-learning-based optimization: A novel method for constrained mechanical design optimization problems. *Comput. Aided Des.* **2011**, *43*, 303–315. [CrossRef]
17. Satapathy, S.; Naik, A. Social group optimization (SGO): A new population evolutionary optimization technique. *Complex Intell. Syst.* **2016**, *2*, 173–203. [CrossRef]
18. Al-Betar, M.A. β-hill climbing: An exploratory local search. *Neural Comput. Appl.* **2017**, *28*, 153–168. [CrossRef]
19. Martínez-Lvarez, F.; Asencio-Cortés, G.; Torres, J.F.; Gutiérrez-Avilés, D.; Melgar-García, L.; Pérez-Chacón, R.; Rubio-Escudero, C.; Riquelme, J.C.; Troncoso, A. Coronavirus Optimization Algorithm: A bioinspired metaheuristic based on the COVID-19 propagation model. *Big Data* **2020**, *8*, 308–322. [CrossRef] [PubMed]
20. Hussain, A.; Muhammad, Y.S. Trade-off between exploration and exploitation with genetic algorithm using a novel selection operator. *Complex Intell. Syst.* **2019**, *6*, 1–14. [CrossRef]
21. Wolpert, D.H.; Macready, W.G. No Free Lunch Theorems for Optimization. *IEEE Trans. Evol. Comput.* **1997**, *1*, 67–82. [CrossRef]
22. Shehadeh, H.A. A hybrid sperm swarm optimization and gravitational search algorithm (HSSOGSA) for global optimization. *Neural Comput. Appl.* **2021**, *33*, 11739–11752.
23. Kaveh, A.; Rahmani, P.; Eslamlou, A.D. An efficient hybrid approach based on Harris Hawks optimization and imperialist competitive algorithm for structural optimization. *Eng. Comput.* **2021**, *277*, 1–29.
24. Dhiman, G.; Kaur, A. A hybrid algorithm based on particle swarm and spotted hyena optimizer for global optimization. In *Soft Computing for Problem Solving*; Springer: Singapore, 2019; Volume 1, pp. 599–615.
25. Dhiman, G. SSC: A hybrid nature-inspired meta-heuristic optimization algorithm for engineering applications. *Knowl. Based Syst.* **2021**, *222*, 106926. [CrossRef]
26. Banaie-Dezfouli, M.; Nadimi-Shahraki, M.H.; Beheshti, Z. R-GWO: Representative-based grey wolf optimizer for solving engineering problems. *Appl. Soft Comput.* **2021**, *106*, 107328. [CrossRef]
27. Zhang, H.; Wang, Z.; Chen, W.; Heidari, A.A.; Wang, M.; Zhao, X.; Liang, G.; Chen, H.; Zhang, X. Ensemble mutation-driven salp swarm algorithm with restart mechanism: Framework and fundamental analysis. *Expert Syst. Appl.* **2021**, *165*, 113897. [CrossRef]
28. Yu, C.; Heidari, A.A.; Xue, X.; Zhang, L.; Chen, H.; Chen, W. Boosting quantum rotation gate embedded slime mould algorithm. *Expert Syst. Appl.* **2021**, *181*, 115082. [CrossRef]
29. Zhang, H.; Cai, Z.; Ye, X.; Wang, M.; Kuang, F.; Chen, H.; Li, C.; Li, Y. A multi-strategy enhanced salp swarm algorithm for global optimization. *Eng. Comput.* **2020**. [CrossRef]
30. Che, Y.; He, D. A Hybrid Whale Optimization with Seagull Algorithm for Global Optimization Problems. *Math. Probl. Eng.* **2021**, *2021*, 1–31.
31. Hassan, B.A. CSCF: A chaotic sine cosine firefly algorithm for practical application problems. *Neural Comput. Appl.* **2021**, *33*, 7011–7030. [CrossRef]

32. Yue, S.; Zhang, H. A hybrid grasshopper optimization algorithm with bat algorithm for global optimization. *Multimed. Tools Appl.* **2021**, *80*, 3863–3884. [CrossRef]
33. Li, S.; Chen, H.; Wang, M.; Heidari, A.A.; Mirjalili, S. Slime Mould Algorithm: A new method for stochastic optimization. *Future Gener. Comput. Syst.* **2020**, *111*, 300–323. [CrossRef]
34. Abualigah, L.; Diabat, A.; Mirjalili, S.; Abd Elaziz, M.; Gandomi, A.H. The Arithmetic Optimization Algorithm. *Comput. Methods Appl. Mech. Eng.* **2021**, *376*, 113609. [CrossRef]
35. Mirjalili, S. SCA: A Sine Cosine algorithm for solving optimization problems. *Knowl. Based Syst.* **2016**, *96*, 120–133. [CrossRef]
36. Molga, M.; Smutnicki, C. Test Functions for Optimization Needs. 2005. Available online: http://www.zsd.ict.pwr.wroc.pl/files/docs/functions.pdf (accessed on 1 October 2021).
37. Mohamed, A.W.; Hadi, A.A.; Mohamed, A.K.; Agrawal, P.; Kumar, A.; Suganthan, P.N. Problem Definitions and Evaluation Criteria for the CEC 2021 Special Session and Competition on Single Objective Bound Constrained Numerical Optimization. Cairo University. *Tech. Rep.* 2020. Available online: http://home.elka.pw.edu.pl/~{}ewarchul/cec2021-specification.pdf (accessed on 1 October 2021).
38. Mirjalili, S.; Mirjalili, S.M.; Lewis, A. Grey Wolf Optimizer. *Adv. Eng. Softw.* **2014**, *69*, 46–61. [CrossRef]
39. Mirjalili, S.; Lewis, A. The Whale Optimization Algorithm. *Adv. Eng. Softw.* **2016**, *95*, 51–67. [CrossRef]
40. Mirjalili, S.; Gandomi, A.H.; Mirjalili, S.Z.; Saremi, S.; Faris, H.; Mirjalili, S.M. Salp swarm algorithm: A bio-inspired optimizer for engineering design problems. *Adv. Eng. Softw.* **2017**, *114*, 163–191. [CrossRef]
41. Demsar, J. Statistical comparisons of classifiers over multiple data sets. *J. Mach. Learn. Res.* **2006**, *7*, 1–30.
42. Garcia, S.; Fernandez, A.; Luengo, J.; Herrera, F. Advanced nonparametric tests for multiple comparisons in the design of experiments in computational intelligence and data mining: Experimental analysis of power. *Inform. Sci.* **2010**, *180*, 2044–2064. [CrossRef]
43. Kannan, B.; Kramer, S.N. An augmented lagrange multiplier based method for mixed integer discrete continuous optimization and its applications to mechanical design. *J. Mech. Des.* **1994**, *116*, 405–411. [CrossRef]
44. Mirjalili, S. Moth-flame optimization algorithm: A novel nature-inspired heuristic paradigm. *Knowl. Based Syst.* **2015**, *89*, 228–249. [CrossRef]
45. Rizk-Allah, R.M. Hybridizing sine cosine algorithm with multi-orthogonal search strategy for engineering design problems. *J. Comput. Des. Eng.* **2018**, *5*, 249–273. [CrossRef]
46. Zhou, Y.; Ling, Y.; Luo, Q. Lévy flight trajectory-based whale optimization algorithm for engineering optimization. *Eng. Comput.* **2018**, *35*, 2406–2428. [CrossRef]
47. Pelusi, D.; Mascella, R.; Tallini, L.; Nayak, J.; Naik, B.; Deng, Y. An Improved Moth-Flame Optimization algorithm with hybrid search phase. *Knowl. Based Syst.* **2020**, *191*, 105277. [CrossRef]
48. Sadollah, A.; Bahreininejad, A.; Eskandar, H.; Hamdi, M. Mine blast algorithm: A new population based algorithm for solving constrained engineering optimization problems. *Appl. Soft Comput.* **2013**, *13*, 2592–2612. [CrossRef]
49. Liu, H.; Cai, Z.; Wang, Y. Hybridizing particle swarm optimization with differential evolution for constrained numerical and engineering optimization. *Appl. Soft Comput.* **2010**, *10*, 629–640. [CrossRef]
50. Singh, N.; Kaur, J. Hybridizing sine-cosine algorithm with harmony search strategy for optimization design problems. *Soft. Comput.* **2021**. [CrossRef]
51. Baykasoğlu, A.; Akpinar, S. Weighted superposition attraction (WSA): A swarm intelligence algorithm for optimization problems–part2: Constrained optimization. *Appl. Soft Comput.* **2015**, *37*, 396–415. [CrossRef]

processes

MDPI

Article

Migration-Based Moth-Flame Optimization Algorithm

Mohammad H. Nadimi-Shahraki [1,2,*], Ali Fatahi [1,2], Hoda Zamani [1,2], Seyedali Mirjalili [3,4,*], Laith Abualigah [5,6] and Mohamed Abd Elaziz [7,8,9,10]

1. Faculty of Computer Engineering, Najafabad Branch, Islamic Azad University, Najafabad 8514143131, Iran; fatahi.ali.edu@sco.iaun.ac.ir (A.F.); hoda_zamani@sco.iaun.ac.ir (H.Z.)
2. Big Data Research Center, Najafabad Branch, Islamic Azad University, Najafabad 8514143131, Iran
3. Centre for Artificial Intelligence Research and Optimisation, Torrens University Australia, Brisbane 4006, Australia
4. Yonsei Frontier Lab, Yonsei University, Seoul 03722, Korea
5. Faculty of Computer Sciences and Informatics, Amman Arab University, Amman 11953, Jordan; laythdyabat@aau.edu.jo
6. School of Computer Sciences, University Sains Malaysia, Pulau Pinang 11800, Malaysia
7. Department of Mathematics, Faculty of Science, Zagazig University, Zagazig 44519, Egypt; abd_el_aziz_m@yahoo.com
8. Artificial Intelligence Research Center (AIRC), Ajman University, Ajman 346, United Arab Emirates
9. Department of Artificial Intelligence Science & Engineering, Galala University, Suze 435611, Egypt
10. School of Computer Science and Robotics, Tomsk Polytechnic University, 634050 Tomsk, Russia

* Correspondence: nadimi@iaun.ac.ir (M.H.N.-S.); ali.mirjalili@torrens.edu.au (S.M.); Tel.: +98-3142292632 (M.H.N.-S.)

Abstract: Moth–flame optimization (MFO) is a prominent swarm intelligence algorithm that demonstrates sufficient efficiency in tackling various optimization tasks. However, MFO cannot provide competitive results for complex optimization problems. The algorithm sinks into the local optimum due to the rapid dropping of population diversity and poor exploration. Hence, in this article, a migration-based moth–flame optimization (M-MFO) algorithm is proposed to address the mentioned issues. In M-MFO, the main focus is on improving the position of unlucky moths by migrating them stochastically in the early iterations using a random migration (RM) operator, maintaining the solution diversification by storing new qualified solutions separately in a guiding archive, and, finally, exploiting around the positions saved in the guiding archive using a guided migration (GM) operator. The dimensionally aware switch between these two operators guarantees the convergence of the population toward the promising zones. The proposed M-MFO was evaluated on the CEC 2018 benchmark suite on dimension 30 and compared against seven well-known variants of MFO, including LMFO, WCMFO, CMFO, CLSGMFO, LGCMFO, SMFO, and ODSFMFO. Then, the top four latest high-performing variants were considered for the main experiments with different dimensions, 30, 50, and 100. The experimental evaluations proved that the M-MFO provides sufficient exploration ability and population diversity maintenance by employing migration strategy and guiding archive. In addition, the statistical results analyzed by the Friedman test proved that the M-MFO demonstrates competitive performance compared to the contender algorithms used in the experiments.

Keywords: optimization; metaheuristic algorithms; swarm intelligence algorithm; moth-flame optimization (MFO); exploration and exploitation; population diversity

Citation: Nadimi-Shahraki, M.H.; Fatahi, A.; Zamani, H.; Mirjalili, S.; Abualigah, L.; Abd Elaziz, M. Migration-Based Moth-Flame Optimization Algorithm. *Processes* **2021**, *9*, 2276. https://doi.org/10.3390/pr9122276

Academic Editor: Gurutze Arzamendi

Received: 18 November 2021
Accepted: 14 December 2021
Published: 18 December 2021

1. Introduction

During past decades, optimization techniques have been developed widely to solve complex problems that emerged in different fields of science, such as engineering [1–9], clustering [10–18], feature selection [19–28], and task scheduling [29–32]. Such optimization problems mainly involve characteristics such as linear/non-linear constraints, non-differentiable functions, and a substantial number of decision variables. These characteristics make optimization problems almost impossible to solve by exact methods reasonably, and an effective approach is needed to tackle such complexities. Approximate algorithms are recognized as an effective approach for solving issues due to their stochastic techniques and global and local search strategies. Although metaheuristic algorithms cannot guarantee the optimality of their solutions, they can offer near-optimal solutions in a reasonable amount of time, which helps solve real-world problems [33–37].

Metaheuristic algorithms mostly employ stochastic techniques to solve optimization problems by exploring the search space to promote population diversity in the early iterations. In the exploitation phase, the algorithm locally searches the promising areas to enhance the quality of solutions discovered in the exploration phase. Striking a proper balance between these two tendencies leads the algorithm toward the global optimum after a limited number of iterations. The bio-inspired algorithms are the primary approach to solve optimization problems by employing biological concepts. In the literature, some of the bio-inspired algorithms, such as genetic algorithm (GA) [38], differential evolution (DE) [39], particle swarm optimization (PSO) [40], and artificial bee colony (ABC) [41], have been used to find the optimum of optimization problems in polynomial time. Although the mentioned algorithms demonstrate satisfactory results for many problems, no single metaheuristic can solve all optimization issues based on the no free lunch (NFL) theorem [42]. The NFL is the main reason for continuous developments in the field of optimization. As a result, numerous bio-inspired algorithms have been developed by introducing novel methods.

To comprehensively investigate the bio-inspired algorithms, we can classify them based on their source of inspiration to evolutionary and swarm intelligence (SI) [43]. The natural biological evolution, reproduction, mutation, and Darwin's theory of evolution are the most used fundamentals for developing evolutionary optimization algorithms. Genetic algorithm (GA) [44], genetic programming (GP) [45], differential evolution (DE) [39], evolution strategy (ES) [46], and, from recent studies, quantum-based avian navigation optimizer algorithm (QANA) [47] are some evolutionary algorithms. During past years, many variants have been developed to improve the performance of evolutionary algorithms, such as enhanced genetic algorithm (EGA) [48], an ensemble of mutation strategies and control parameters with the DE (EPSDE) [49], the real-coded genetic algorithm using a directional crossover operator (RGA-DX) [50], and an effective multi-trial vector-based differential evolution (MTDE) [51].

Swarm intelligence (SI) algorithms are grounded in the collective behavior of a group of biological organisms. SI algorithms can be divided into four categories: aquatic animals, terrestrial animals, birds, and insects [52]. The natural behavior of aquatic animals, such as prey besieging and foraging, has been mimicked in many SI algorithms, including the krill herd (KH) algorithm [53], whale optimization algorithm (WOA) [54], and salp swarm algorithm (SSA) [55]. Many researchers have simulated the biological behavior of terrestrial animals to propose functional metaheuristic algorithms, such as grey wolf optimizer (GWO) [41], red fox optimization algorithm (RFO) [56], chimp optimization algorithm (ChOA) [57], and horse herd optimization algorithm (HOA) [58]. In the third category, bat algorithm (BA) [59], cuckoo search algorithm (CS) [60], crow search algorithm (CSA) [61], and Aquila optimizer (AO) [62] are among the well-known algorithms inspired by birds' behaviors. Social behaviors of insects, such as self-organization and cooperation, are the main sources of inspiration behind the fourth group of SI algorithms, including ant colony optimization (ACO) [63], artificial bee colony (ABC) [64], ant lion optimization (ALO) [65], dragonfly algorithm (DA) [66], and moth–flame optimization (MFO) [67].

The SI algorithms intrinsically benefit from autonomy, adaptability, and acceptable time complexity. However, loss of population diversity and sinking into the local optimum are common issues among most SI algorithms. Therefore, many variants have been proposed to address these shortcomings and enhance the performance of the algorithms. Karaboga et al. [68] introduced a quick artificial bee colony (qABC) algorithm to improve the exploitation ability of the traditional algorithm. The conscious neighborhood-based crow search algorithm (CCSA) [52] addresses the imbalance between exploration and exploitation. An improved grey wolf optimizer (I-GWO) [69] was proposed to maintain the population diversity. An enhanced chimp optimization algorithm (EChOA) [70] has been introduced to avoid local optimum.

The moth–flame optimization (MFO) is a prominent bio-inspired metaheuristic algorithm inspired by the moths' spiral movement around the light source at night. The MFO algorithm stands out among many metaheuristic algorithms for its simplicity and acceptable time complexity. Therefore, the MFO is used for solving a broad range of real-world problems, such as clustering [71–77], feature selection [78–85], and image processing [86–91]. Although the MFO is applicable for solving real-world problems and many improvements have been developed, it has been observed that the MFO and its variants hereditarily suffer from poor exploration and loss of population diversity before the near-optimal solution is met, which leads the algorithm toward local optima trapping and premature convergence.

In this study, an enhanced MFO algorithm, named migration-based moth–flame optimization (M-MFO) algorithm, is proposed to cope with these weaknesses. The M-MFO introduces a guiding archive to maintain the population diversity and a hybrid simple strategy named migration strategy consists of two random migration (RM) and guided migration (GM) operators which take advantage of an adapted crossover introduced in the GA [44]. The RM operator is introduced to enhance the exploration ability and population diversity by crossing the unlucky moths with a randomly generated moth to migrate to new areas. If the migrated moths obtain better positions, they are updated and added to the guiding archive to guide other unlucky moths. When the guiding archive size reaches the size of the problem variables, the archive is mature enough to guide other unlucky moths using the GM operator. This dimensionally aware switch between operators can guarantee the convergence of the algorithm toward promising areas.

To evaluate the efficiency of the M-MFO, the CEC 2018 benchmark functions were conducted to investigate the characteristics and performance of the proposed algorithm and its competitors in different dimensions, 30, 50, and 100. The convergence curves and population diversity provided in Section 5, show that the M-MFO can maintain population diversity until the optimal solution emerges and effectively facilitates the convergence behavior. Moreover, the Friedman test was conducted to evaluate the obtained results statistically. The experimental and statistical results were first compared with seven well-known variants of MFO, including LMFO [92], WCMFO [93], CMFO [94], CLSGMFO [95], LGCMFO [96], SMFO [97], and ODSFMFO [98] in dimension 30. Then, the top four algorithms and eight other state-of-the-art swarm intelligence algorithms were considered for the main experiments. Hence, the total competitors for the main experiments were KH [53], GWO [41], MFO [67], WOA [54], WCMFO [93], CMFO [94], HGSO [99], RGA-DX [50], ChOA [57], AOA [100], and ODSFMFO [98]. The experimental evaluations and statistical tests revealed that the M-MFO algorithm outperforms other competitor algorithms with overall effectiveness of 91%. The experimental results revealed that the migration strategy enhances the exploration ability and maintains the population diversity to avoid local optimum by stochastically migrating the worst individuals across the search space in the first iterations and exploiting promising areas discovered by the RM operator in the next iterations. The main contributions of this study are summarized as follows.

- Introducing a guiding archive for storing improved moths to guide other unlucky moths.
- Introducing a migration strategy using RM and GM operators to improve unlucky moths.
- The RM operator enhances the exploration ability, while the GM operator converges the population toward the promising areas by exploiting around improved moths.
- The experiments prove that the M-MFO effectively maintains the population diversity by taking advantage of the guiding archive.

- The Friedman test demonstrated that the M-MFO provides the best results compared to competitors and stands out among MFO variants for solving global optimization problems.

The remainder of the paper is organized as follows. A literature overview of the MFO variants is included in Section 2. Section 3 briefly presents the MFO algorithm. Section 4 comprehensively presents the proposed M-MFO algorithm. A rigorous examination of the effectiveness of the proposed algorithm is provided experimentally in Section 5 and statistically in Section 6. Finally, Section 7 summarizes the conclusions.

2. Related Work

The MFO algorithm is known as a prominent problem solver due to its simple framework, fewer control parameters, and ease of implementation. However, the MFO suffers from some issues for solving complex optimization problems. Therefore, since the release of the MFO, many variants have been developed to address MFO's shortcomings and offer improved performance. These variants can be categorized into hybrid improvements and non-hybrid improvements, as illustrated in Figure 1.

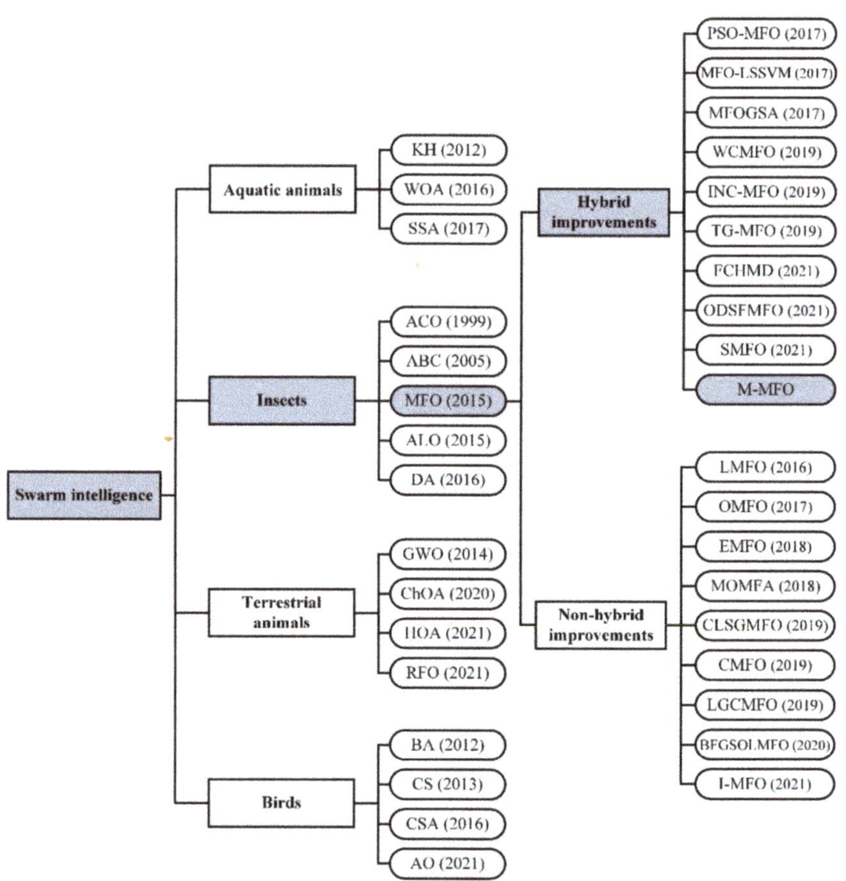

Figure 1. The classification of SI algorithms and variants of MFO.

Since the introduction of MFO, many researchers have proposed hybrid variants to effectively address shortcomings of the canonical MFO by employing operators of other algorithms. Bhesdadiya et al. [101] introduced a hybrid PSO-MFO algorithm by combining particle swarm optimization (PSO) with MFO to boost the exploitation ability of the MFO algorithm. MFO-LSSVM [102] is a hybridization of MFO with least squares support vector machines (LSSVM) to enhance the generalization in the prediction of the MFO algorithm. To boost the exploitation ability of the MFO, Sarma et al. [103] introduced the gravitational search algorithm (GSA) to the canonical MFO and proposed MFOGSA. In WCMFO, Khalilpourazari et al. [93] introduced a combined MFO, water cycle algorithm (WCA) and a random walk to avoid local optimum and enhance the solution quality. Rezk et al. [104] designed a hybrid MPPT method by combining an incremental conductance (INC) approach and MFO, called (INC-MFO), to reach a maximum-power solar PV/thermoelectric system under different environmental conditions.

Ullah et al. [105] introduced a time-constrained genetic moth–flame optimization (TG-MFO) algorithm for an energy management system (EMS) in smart homes and buildings. The FCHMD [106] algorithm combines Harris hawks optimizer (HHO) and MFO to cope with the insufficient exploitation and exploration rate of the HHO and MFO, respectively. Moreover, the method of evolutionary population dynamics (EPD) is employed to address premature convergence and local optima stagnation. ODSFMFO, proposed by Li et al. [98], is a hybridization of MFO with differential evolution (DE) and shuffled frog leaping algorithm (SFLA). In addition, the algorithm is enhanced by the addition of a flame generation strategy and death mechanism. Dang et al. [107] brought up a hybridization of MFO and three different methods, including the Taguchi method, fuzzy logic, and response surface method, to solve the flexure hinge design problem. SMFO has been recently proposed by [97] to enhance the solution quality and convergence speed of the MFO by introducing the sine cosine strategy to the MFO algorithm.

The non-hybrid algorithms are mostly developed to cope with issues such as local optima trapping, premature convergence, the imbalance between search strategies, and poor local and global search abilities. The LMFO algorithm proposed by Li et al. [92] is an enhanced version of MFO, improved by Lévy flight to address premature convergence and local optimum trapping by improving the population diversity. Apinantanakon et al. [108] established an opposition-based moth–flame optimization (OMFO) algorithm to evade local optimum by boosting the exploration ability of the MFO. Xu et al. [109] addressed the MFO's low population diversity and introduced EMFO by taking advantage of the Gaussian mutation (GM). Li et al. [110] presented a multi-objective moth–flame optimization algorithm (MOMFA) to enhance water resource efficiency by maintaining population diversity and accelerating convergence speed by taking advantage of opposition-based learning and indicator-based learning selection-efficient mechanisms.

The CLSGMFO [95] presents an efficient chaotic mutative moth–flame-inspired algorithm by employing Gaussian mutation and chaotic local search to enhance the population diversity and exploitation rate, respectively. Chaos-enhanced moth–flame optimization (CMFO), proposed by Hongwei et al. [94], is an improved MFO algorithm that employs ten chaotic maps. Xu et al. [96] developed LGCMFO to enhance the global and local search ability of the MFO and avoid local optimum by employing new operators, such as Gaussian mutation (GM), Lévy mutation (LM), and Cauchy mutation (CM). In BFGSOLMFO, Zhang et al. [111] introduced orthogonal learning (OL) and Broyden–Fletcher–Goldfarb–Shanno (BFGS) to the MFO to enhance the solution quality of the MFO. Nadimi-Shahraki et al. [112] proposed an improved moth–flame optimization (I-MFO) algorithm to evade the local optima trapping and premature convergence by adding a memory mechanism and taking advantage of the adapted wandering around search (AWAS) strategy. This algorithm is designed to solve the numerical and mechanical engineering problems.

3. Moth–Flame Optimization (MFO) Algorithm

The moth–flame optimization (MFO) is a prominent SI algorithm inspired by the spiral locomotion behavior of moths around a light source at night. This behavior is derived from the navigation mechanism of moths that is used to fly a long distance in a straight line by maintaining a fixed inclination to the moon. However, this principled navigation mechanism turns into a deadly spiral path toward the light source if the light source is relatively close to the moths. According to the brief description, the MFO algorithm consists of moths and flames. As shown in Equation (1), moths are considered search agents, organized in matrix $M(t)$, that explore the D-dimensional search space, and N is the number of moths.

$$M(t) = \begin{bmatrix} m_{1,1} & m_{1,2} & \cdots & m_{1,D} \\ m_{2,1} & m_{2,2} & \cdots & m_{2,D} \\ \vdots & \vdots & \vdots & \vdots \\ m_{N,1} & m_{N,2} & \cdots & m_{N,D} \end{bmatrix} \quad (1)$$

Additionally, the fitness of the corresponding moth is stored in an array $OM(t)$, as shown below.

$$OM(t) = \begin{bmatrix} OM_1(t) \\ OM_2(t) \\ \vdots \\ OM_N(t) \end{bmatrix} \quad (2)$$

On the other hand, flames are the best positions discovered by moths and are stored in a similar matrix $F(t)$, along with their fitness values in an array $OF(t)$. The moths spirally move around their corresponding flames, as shown in Equation (3), where $M_i(t)$ is the position of ith moth in the current iteration, the Dis_i determines the distance between M_i and its corresponding jth flame (F_j) formulated in Equation (4), b indicates the shape of the logarithmic spiral, and k is a random number value between intervals $[-1, 1]$.

$$M_i(t) = Dis_i(t) \times e^{bk} \times Cos(2\pi k) + F_j(t) \quad (3)$$

$$Dis_i(t) = |F_j(t) - M_i(t)| \quad (4)$$

To converge the algorithm and provide more exploitation, the number of flames decreases in the course of iterations based on Equation (5), where t determines the current number of iterations, while N and $MaxIt$ demonstrate the total number of flames and the maximum number of iterations, respectively.

$$Flame_{Num}(t) = round\left(N - t \times \frac{N-1}{MaxIt}\right) \quad (5)$$

4. Proposed Algorithm

The MFO is a prominent population-based algorithm that is successfully applied in different fields. However, based on the conducted analysis reported in Section 5.2 and related studies [113–115], the MFO algorithm suffers from poor exploration and rapid loss of population diversity. While the number of flames converges, the algorithm provides more local searches throughout the course of the iterations. Hence, the algorithm is prone to sink into the local optimum due to its limited simple spiral movement of moths around their corresponding flames which cannot offer further exploration to avoid the local optimum. Therefore, this study proposes a migration-based moth–flame optimization (M-MFO) algorithm, which is a hybridization of the MFO algorithm and the crossover operator introduced in the GA. Moreover, the M-MFO utilizes a guiding archive to maintain population diversity and a migration strategy that uses the crossover operator to boost exploration ability. The migration strategy introduces two operators, RM and GM, by taking advantage of an adapted GA's crossover. The RM operator is introduced to provide sufficient exploration ability during the early iterations, while the GM operator converges the population toward promising areas. Moreover, to maintain the population diversity, a guiding archive is introduced, as outlined in Definition 1, to store lucky moths that have improved using the migration strategy.

Definition 1 (guiding archive)**.** *The guiding archive keeps the position of lucky moths improved by the migration strategy to maintain the population diversity and suppress the premature convergence of the population. Both RM and GM add improved moths to the guiding archive, although only the GM operator exploits the archive. The guiding archive capacity (MaxArc) is formulated in Equation (6), where D and N are dimensions and population size.*

$$MaxArc = D \times [\ln N] \tag{6}$$

To ensure that the guiding archive is mature enough to guide other unlucky moths, the current size of the archive (δ) needs to be greater than the size of the problem variables (D). This limitation provides a dimensionally aware switch to choose the right operator in the migration strategy effectively. In addition, if the value of δ exceeds the *MaxArc*, the next moth is replaced with a random member of the guiding archive.

Migration strategy includes RM and GM operators to ensure that there is high enough exploration capability and convergence toward promising zones. The RM operator provides further exploration ability by changing the position of M_i stochastically. At the same time, the GM operator is introduced to converge the population toward promising zones by exploiting improved moths kept in the guiding archive. Moreover, the migration strategy benefits from a dimensionally aware switch between these two operators as represented in Equation (7), where δ indicates the current size of the guiding archive. The pseudo-code and the flowchart of the M-MFO are presented in Algorithm 1 and Figure 2, respectively.

$$M_i(t+1) = \begin{cases} \text{RM operator} & \delta < D \\ \text{GM operator} & \delta \geq D \end{cases} \tag{7}$$

Random migration (RM) operator: Let unlucky moths (t) = $\{M_1, M_2, \ldots, M_i, \ldots\}$, which is a finite set of unlucky moths, such that OM_i (t) > OM_i ($t - 1$). Hence, in this operator, the position of M_i changes by considering a randomly generated moth (M_r) and M_i represents the parents in the crossover formulated in Equations (8) and (9), where α is a random number in [0, 1]. The crossover produces two offspring, and the one with better fitness is selected and compared with other offspring to choose the best one, and the position of OM_i (t + 1) is added to the guiding archive if it can dominate the OM_i (t). The RM operator satisfies the need for exploration by stochastically moving the unlucky moths to discover promising areas in the early iterations.

$$Offspring_1 = \alpha \times M_i + (1 - \alpha) \times M_r \tag{8}$$

$$Offspring_2 = \alpha \times M_r + (1 - \alpha) \times M_i \tag{9}$$

Guided migration (GM) operator: The GM operator is employed to change the position of unlucky moth, M_i, when the size of the GM reaches the size of the problem variables. The GM changes the position of M_i by employing the crossover formulated in Equations (10) and (11), where LM_r is a random lucky moth from the guiding archive. Similarly, to the RM operator, if the new offspring obtains a better position compared to M_i (t), the position of M_i (t + 1) is updated, and it is appended to the guiding archive.

$$Offspring_1 = \alpha \times M_i + (1 - \alpha) \times LM_r \tag{10}$$

$$Offspring_2 = \alpha \times LM_r + (1 - \alpha) \times M_i \tag{11}$$

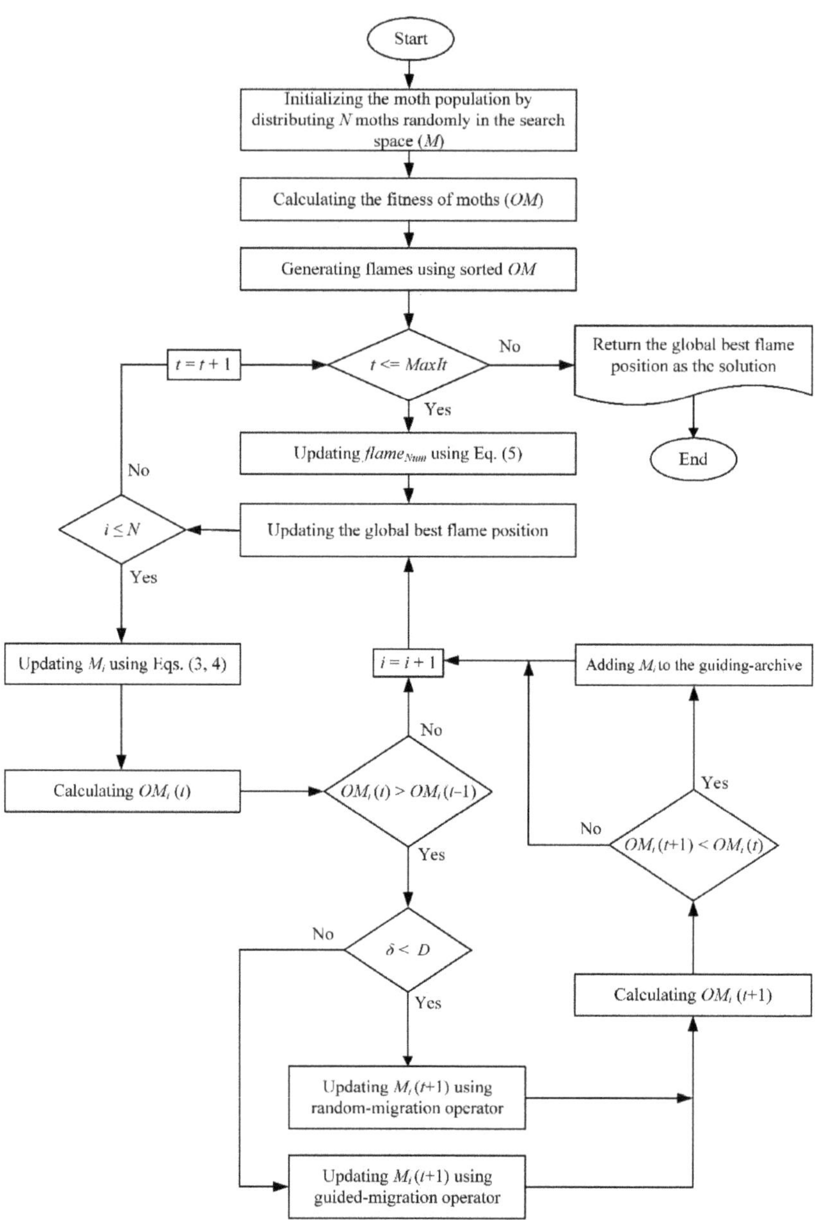

Figure 2. The flowchart of the proposed M-MFO algorithm.

Algorithm 1. The pseudocode of proposed M-MFO algorithm.

Input: Max iterations (*MaxIt*), number of moths (*N*), max size of guiding archive (*MaxArc*), and dimension (*D*).
Output: The best flame position and its fitness value.

```
Begin
  Randomly distributing M moths in the D-dimensional search space.
  Calculating moths' fitness (OM).
  Set t and δ = 1 //δ is the number of guiding archive members.
  OF (t) ← sort OM (t).
  F (t) ← sort M (t).
  While t ≤ MaxIt
    Updating F and OF by the best N moths from F and current M.
    Updating Flame_Num (t) using Equation (5).
    For i = 1: N
      Updating the position of M_i (t) using Equation (3) and computing the OM_i (t).
      If OM_i (t) > OM_i (t − 1)
        τ ← Generating a random number between intervals [1, D].
        For j = 1: τ
          If δ ≤ D (The guiding archive is still immature)
          Generating the next position M_i (t + 1) using RM operator and
Equations (8) and (9).
          Else (The guiding archive is mature)
            Generating next position M_i (t + 1) using GM operator and
Equations (10) and (11).
          End If
        End For
        If OM_i (t + 1) < OM_i (t)
            Updating position M_i (t) and adding M_i (t + 1) to guiding archive
using Definition 1 and MaxArc.
        End If
      End If
    End For
    Updating the position and fitness value of the global best flame.
    t = t +1.
  End while
```

5. Numerical Experiment and Analysis

In this section, the performance of the proposed M-MFO has been evaluated on several benchmark functions. In the first section, the population diversity and convergence behavior of the canonical MFO have been analyzed on some selected functions to provide some useful information about the shortcomings of the MFO algorithm. Then, the MFO variants and the M-MFO have been compared on dimension 30 to determine the top four superior MFO variants to participate in the next experiments. Following that, the performance of M-MFO has been compared with ten of the state-of-the-art swarm intelligence algorithms: KH [53], GWO [41], MFO [67], WOA [54], WCMFO [93], CMFO [94], HGSO [99], RGA-DX [50], ChOA [57], AOA [100], and ODSFMFO [98]. The parameters of the competitor algorithms are reported in Table 1.

Table 1. Parameter values for the optimization algorithms.

Alg.	Parameter Settings
KH	$V_f = 0.02$, $D_{max} = 0.005$, $N_{max} = 0.01$, $Sr = 0$.
GWO	The parameter a is linearly decreased from 2 to 0.
MFO	$b = 1$, a is decreased linearly from −1 to −2.
WOA	α variable decreases linearly from 2 to 0, $b = 1$.
WCMFO	The number of *rivers* and *sea* = 4.
CMFO	$b = 1$, a is decreased linearly from −1 to −2, *chaotic map* = Singer.
HGSO	$l_1 = 5 \times 10^{-3}$, $l_2 = 100$, $l_3 = 1 \times 10^{-2}$, $alpha = 1$, $beta = 1$, $M_1 = 0.1$, $M_2 = 0.2$.
RGA-DX	$p_{cv} = 0.9$, $\alpha = 0.95$, and $p_d = 0.75$.
ChOA	f decreases linearly from 2 to 0.
AOA	$\mu = 0.5$, $\alpha = 5$.
ODSFMFO	$m = 6$, $pc = 0.5$, $\gamma = 5$, $\alpha = 1$, $l = 10$, $b = 1$, $\beta = 1.5$.
M-MFO	τ = random number between 1 and D, $MaxArc = D \times [ln\ N]$.

5.1. Experimental Environment and Benchmark Functions

The M-MFO and competitor algorithms were implemented in Matlab 2020a. All experiments were performed 20 times, independently, on a laptop with Intel Core i7-10750H CPU (2.60 GHz) and 24 GB of memory to ensure fair comparisons. In each run, the maximum number of iterations (MaxIt) was set by $(D \times 10^4)/N$ where D and N were respectively set to the dimensions of the problem and 100. In this study, the CEC 2018 benchmark functions [116] were used to evaluate the effectiveness of the proposed M-MFO. There are 29 test functions in the CEC 2018 benchmark suite, each with its own set of characteristics and different dimensions 30, 50, and 100. These test functions can be classified into unimodal functions F_1 and F_3, multimodal functions F_4–F_{10}, hybrid functions F_{11}–F_{20}, and composition functions F_{21}–F_{30}. The experimental results tabulated in Tables 2–5 and Tables A1–A4 in the Appendix A are based on each algorithm's average and minimum fitness value, where the bold values illustrate the winning algorithm. Moreover, the symbols "W | T | L" in the last row of each table demonstrate the number of wins, ties, and losses for each algorithm.

Table 2. The overall results of MFO variants for dimension 30.

	Metric	LMFO (2016)	WCMFO (2019)	CMFO (2019)	CLSGMFO (2019)	LGCMFO (2019)	SMFO (2021)	ODSFMFO (2021)	M-MFO
Overall results	W | T | L	0 | 0 | 29	0 | 0 | 29	1 | 0 | 28	0 | 0 | 29	0 | 0 | 29	0 | 0 | 29	0 | 0 | 29	**28 | 0 | 1**

Table 3. The overall results of the M-MFO and comparative algorithms on unimodal and multimodal test functions.

	D	Metrics	KH (2012)	GWO (2014)	MFO (2015)	WOA (2016)	WCMFO (2019)	CMFO (2019)	HGSO (2019)	RGA-DX (2019)	ChOA (2020)	AOA (2021)	ODSFMFO (2021)	M-MFO
Overall results	30	W | T | L	0 | 0 | 9	0 | 0 | 9	0 | 0 | 9	0 | 0 | 9	0 | 0 | 9	0 | 0 | 9	0 | 0 | 9	0 | 1 | 8	0 | 0 | 9	0 | 0 | 9	0 | 0 | 9	**8 | 1 | 0**
	50	W | T | L	0 | 0 | 9	0 | 0 | 9	0 | 0 | 9	0 | 0 | 9	0 | 0 | 9	0 | 0 | 9	0 | 0 | 9	0 | 0 | 9	0 | 0 | 9	0 | 0 | 9	0 | 0 | 9	**9 | 0 | 0**
	100	W | T | L	0 | 0 | 9	0 | 0 | 9	0 | 0 | 9	0 | 0 | 9	0 | 0 | 9	0 | 0 | 9	0 | 0 | 9	0 | 0 | 9	0 | 0 | 9	0 | 0 | 9	0 | 0 | 9	**9 | 0 | 0**

Table 4. The overall results of the M-MFO and comparative algorithms on hybrid test functions.

	D	Metrics	KH (2012)	GWO (2014)	MFO (2015)	WOA (2016)	WCMFO (2019)	CMFO (2019)	HGSO (2019)	RGA-DX (2019)	ChOA (2020)	AOA (2021)	ODSFMFO (2021)	M-MFO
Overall results	30	W | T | L	0 | 0 | 10	0 | 0 | 10	0 | 0 | 10	0 | 0 | 10	0 | 0 | 10	1 | 0 | 9	0 | 0 | 10	1 | 0 | 9	0 | 0 | 10	0 | 0 | 10	0 | 0 | 10	**8 | 0 | 2**
	50	W | T | L	0 | 0 | 10	0 | 0 | 10	0 | 0 | 10	0 | 0 | 10	0 | 0 | 10	0 | 0 | 10	0 | 0 | 10	0 | 0 | 10	0 | 0 | 10	0 | 0 | 10	1 | 0 | 9	**9 | 0 | 1**
	100	W | T | L	0 | 0 | 10	0 | 0 | 10	0 | 0 | 10	0 | 0 | 10	0 | 0 | 10	0 | 0 | 10	0 | 0 | 10	0 | 0 | 10	0 | 0 | 10	0 | 0 | 10	0 | 0 | 10	**10 | 0 | 0**

Table 5. The overall results of the M-MFO and comparative algorithms on composition test functions.

<table>
<tr><th></th><th>D</th><th>Metrics</th><th>KH (2012)</th><th>GWO (2014)</th><th>MFO (2015)</th><th>WOA (2016)</th><th>WCMFO (2019)</th><th>CMFO (2019)</th><th>HGSO (2019)</th><th>RGA-DX (2019)</th><th>ChOA (2020)</th><th>AOA (2021)</th><th>ODSFMFO (2021)</th><th>M-MFO</th></tr>
<tr><td rowspan="3">Overall results</td><td>30</td><td>W|T|L</td><td>0|0|10</td><td>0|0|10</td><td>0|0|10</td><td>0|0|10</td><td>0|0|10</td><td>0|0|10</td><td>1|0|9</td><td>0|0|10</td><td>0|0|10</td><td>0|0|10</td><td>0|0|10</td><td>9|0|1</td></tr>
<tr><td>50</td><td>W|T|L</td><td>0|0|10</td><td>0|0|10</td><td>0|0|10</td><td>0|0|10</td><td>1|0|9</td><td>0|0|10</td><td>1|0|9</td><td>0|0|10</td><td>0|0|10</td><td>0|0|10</td><td>0|0|10</td><td>8|0|2</td></tr>
<tr><td>100</td><td>W|T|L</td><td>0|0|10</td><td>0|0|10</td><td>0|0|10</td><td>0|0|10</td><td>0|0|10</td><td>0|0|10</td><td>1|0|9</td><td>1|0|9</td><td>0|0|10</td><td>0|0|10</td><td>0|0|10</td><td>8|0|2</td></tr>
</table>

5.2. Population Diversity Analysis

Maintaining population diversity plays a crucial role in metaheuristic algorithms during the optimization process, as low diversity among search agents can lead the algorithm toward getting stuck at a local optimum. In this experiment, the population diversity and convergence behavior of the MFO algorithm were comprehensively examined to discover the shortcomings of the MFO algorithm. The population diversity curves shown in Figure 3 were measured by a moment of inertia (I_c) [117], where the I_c represents the spreading of each individual from their centroid given by Equation (12) and the centroid c_j for $j = 1, 2 \ldots D$ was calculated using Equation (13).

$$I_c = \sum_{j=1}^{D} \sum_{i=1}^{N} (M_{ji} - c_j)^2 \tag{12}$$

$$c_j = \frac{1}{N} \sum_{i=1}^{N} M_{ji} \tag{13}$$

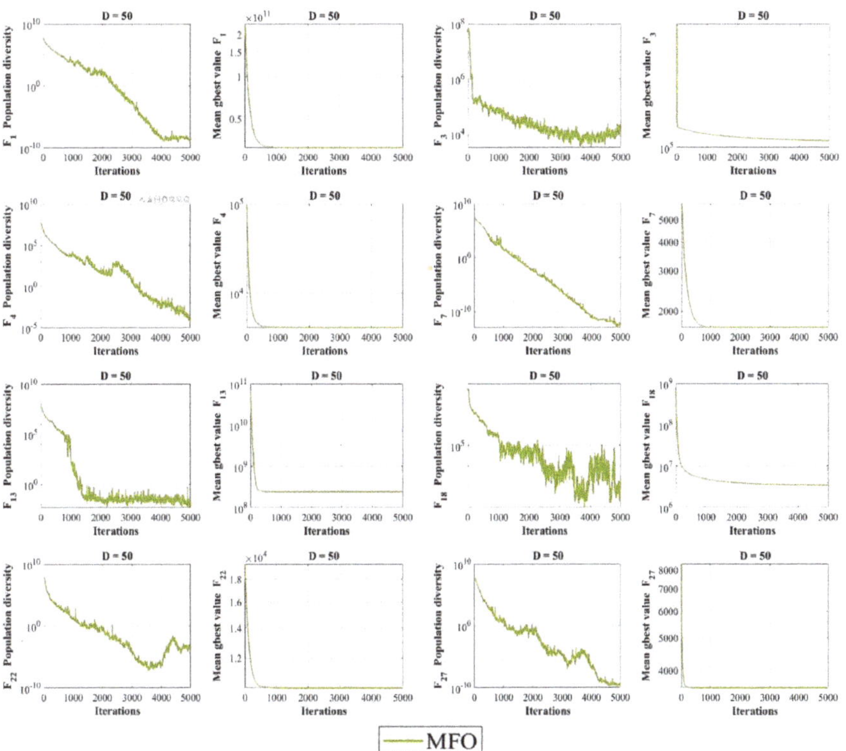

Figure 3. The diversity and convergence curves of the MFO.

In order to perform a fair analysis and develop a better understanding of how population diversity affects the optimization process, Figure 3 illustrates the population diversity and convergence curves of the MFO side by side. For unimodal function F_1, the diversity fell while the optimal solution had not been met yet. Hence, the algorithm sunk into the local optimum until the last iterations. For F_3, the slope of losing diversity was not as sharp as F_1, and the convergence trend continued until the last iterations. However, the population diversity was still low, and convergence speed was too slow to reach the near-optimal solution in the course of iterations. F_4 and F_7 represent the multimodal functions in which the MFO could not maintain the population diversity, and the algorithm experienced premature convergence, beginning at early iterations. F_{13}, F_{18}, F_{22}, and F_{27} were plotted as representatives of hybrid and composition functions with similar diversity and convergence behaviors. The standard behavior was the effort of the algorithm to avoid the local optimum by increasing the population diversity; however, the simple movements of search agents in MFO could not satisfy the needs of exploration to escape the local optimum. To sum up, it can be concluded from the plots that the MFO loses its population diversity before reaching an optimal solution. This behavior was repeated for most of the functions, proving the deficiency of the algorithm in maintaining the population diversity.

5.3. Comparison of M-MFO with MFO Variants

To compare M-MFO with more variants, the results of M-MFO and seven other well-known variants of MFO, including LMFO [92], WCMFO [93], CMFO [94], CLSGMFO [95], LGCMFO [96], SMFO [97], and ODSFMFO [98], were assessed and tabulated in Table 2 and Table A1 in the Appendix A, where the M-MFO outperformed its competitors in dimension 30. Then, the top four algorithms, including the proposed M-MFO, ODSFMFO, WCMFO, and CMFO, were selected for main experiments, including comparison results on dimensions 30, 50, and 100; convergence analysis; population diversity analysis; and the Friedman test.

5.4. Evaluation of Exploitation and Exploration

The exploitation and exploration abilities of the proposed M-MFO have been evaluated by unimodal and multimodal test functions, respectively. As the unimodal functions, F_1 and F_3 have a single global optimum and they are suitable for assessing the exploitation abilities of optimization algorithms. Based on the results of the unimodal functions reported in Table 3 and Table A2 in the Appendix A, the M-MFO outperformed competitors for 30, 50, and 100 dimensions, particularly on test function F_3, where the M-MFO provided the global best solution. The main reason for this exploitation ability is to employ the GM operator, which effectively exploits improved moths kept in the guiding archive. To assess the exploration ability of the M-MFO, the multimodal test functions F_4–F_{10} were considered, as multimodal functions have many local optima. The results of multimodal test functions demonstrated that the M-MFO provides very competitive results compared to other competitors, mainly because of the RM operator and its stochastic movement employed for exploring the landscape effectively in the early iterations.

5.5. Local Optima Avoidance Analysis

In this experiment, the local optima avoidance ability and balance between exploration and exploitation of M-MFO were investigated using hybrid F_{11}–F_{20} and composition F_{21}–F_{30} functions with dimensions 30, 50, and 100. The related results, tabulated in Tables 4 and 5 and Tables A3 and A4 in the Appendix A, proved that the M-MFO is very competitive in comparison to other algorithms used for approximating the global optima values. The main reason is that the algorithm optimally trades off exploration and exploitation by defining two operators—the RM operator, which stochastically moves the unlucky search agents across the search space, and the GM operator, which exploits the promising areas located by successful migrants. Additionally, a guiding archive is introduced to maintain the population diversity by storing new solutions obtained by migration strategy. Moreover, defining a dimensionally aware switching between operators of migration strategy guarantees a proper trade-off between exploration and exploitation.

5.6. The Overall Effectiveness of M-MFO

This study evaluated the overall effectiveness (OE) of the M-MFO and contender algorithms based on the results reported in Tables 3–5 and Tables A2–A4 in Appendix A. The OE results reported in Table 6 were calculated using Equation (14), where N indicates the total number of test functions and L is the number of losing tests for each algorithm. The results prove that the M-MFO, with overall effectiveness of 91%, is the most effective algorithm for the various dimensions 30, 50, and 100.

$$OE\ (\%) = \frac{N - L}{N} \times 100 \tag{14}$$

Table 6. The overall effectiveness of the M-MFO and contender algorithms.

D	Algorithms											
	KH (W/T/L)	GWO (W/T/L)	MFO (W/T/L)	WOA (W/T/L)	WCMFO (W/T/L)	CMFO (W/T/L)	HGSO (W/T/L)	RGA-DX (W/T|L)	ChOA (W|T|L)	AOA (W|T|L)	ODSFMFO (W|T|L)	M-MFO (W|T|L)
30	0/0/29	0/0/29	0/0/29	0/0/29	0/0/29	1/0/28	1/0/28	1/1/27	0/0/29	0/0/29	0/0/29	25/1/3
50	0/0/29	0/0/29	0/0/29	0/0/29	1/0/28	0/0/29	1/0/28	0/0/29	0/0/29	0/0/29	1/0/28	26/0/3
100	0/0/29	0/0/29	0/0/29	0/0/29	0/0/29	0/0/29	1/0/28	1/0/28	0/0/29	0/0/29	0/0/29	27/0/2
Total	0/0/87	0/0/87	0/0/87	0/0/87	1/0/85	1/0/86	3/0/84	2/1/85	0/0/87	0/0/87	1/0/86	78/1/8
OE	0%	0%	0%	0%	1%	1%	4%	2%	0%	0%	1%	**91%**

5.7. Convergence Rate Analysis

In this experiment set, the convergence properties of the M-MFO were examined and the results were compared with contender algorithms for dimensions 30, 50, and 100. Figure 4 illustrates the convergence curves of the average fitness values obtained by each algorithm on unimodal, multimodal, hybrid, and composition test functions. The first row shows the convergence behavior of algorithms on F_3. The M-MFO hit the global optimum solution in the early iterations for all dimensions, which proved the sufficient exploitation ability of the M-MFO. In contrast, the convergence trends of other algorithms were hampered by local minima or demonstrated a prolonged convergence rate. For multimodal function F_{10}, the M-MFO provided the best solution among competitors in the early iterations due to its exploration ability derived from the migration strategy. The third and fourth rows present the convergence of the hybrid functions. The M-MFO bypassed the local optima and continued its gradual trend toward the near-optimum solutions until the final iterations by striking a balance between exploration and exploitation. The convergence curves of the composition function illustrated in the last row demonstrate that the M-MFO obtained the best solution among competitors in early iterations. To sum up, the plots proved that the M-MFO is superior to the other algorithms and provides sufficient exploitation, exploration, and balance between these two tendencies. In addition, it can be noticed that the M-MFO offered more consistent results by increasing the size of the problem variables.

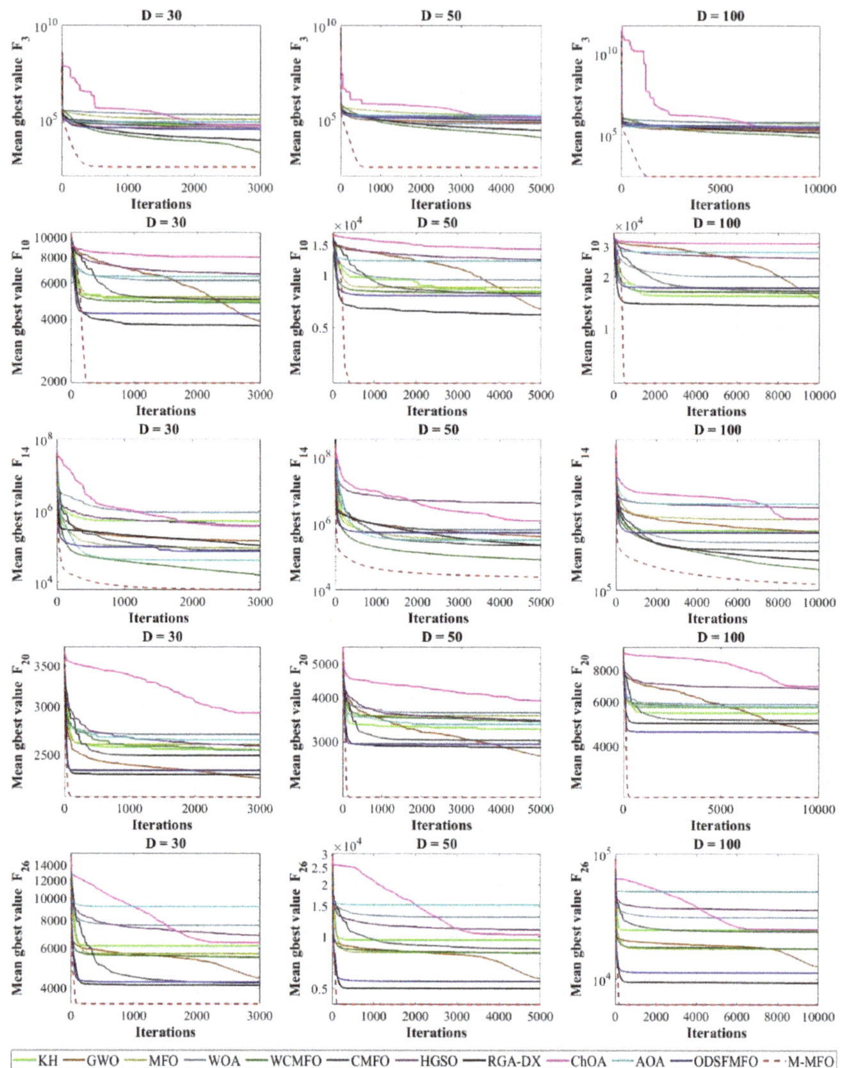

Figure 4. The convergence behavior of the proposed M-MFO and contender algorithms.

5.8. Population Diversity Analysis

As mentioned in Section 5.2, adequate population diversity can suspend the algorithm from local optima trapping. Therefore, in this section, the population diversity of M-MFO and other competitors is provided in Figure 5. Comparing the population diversity with convergence curves provided in the previous section demonstrated that the M-MFO effectively maintained its population diversity until the promising area was met for different test functions with dimensions of 30, 50, and 100. Furthermore, the plots suggest that the M-MFO shows strong robustness for maintaining the population diversity as the size of the problem variables increases, mainly for its dimensionally aware switch between operators.

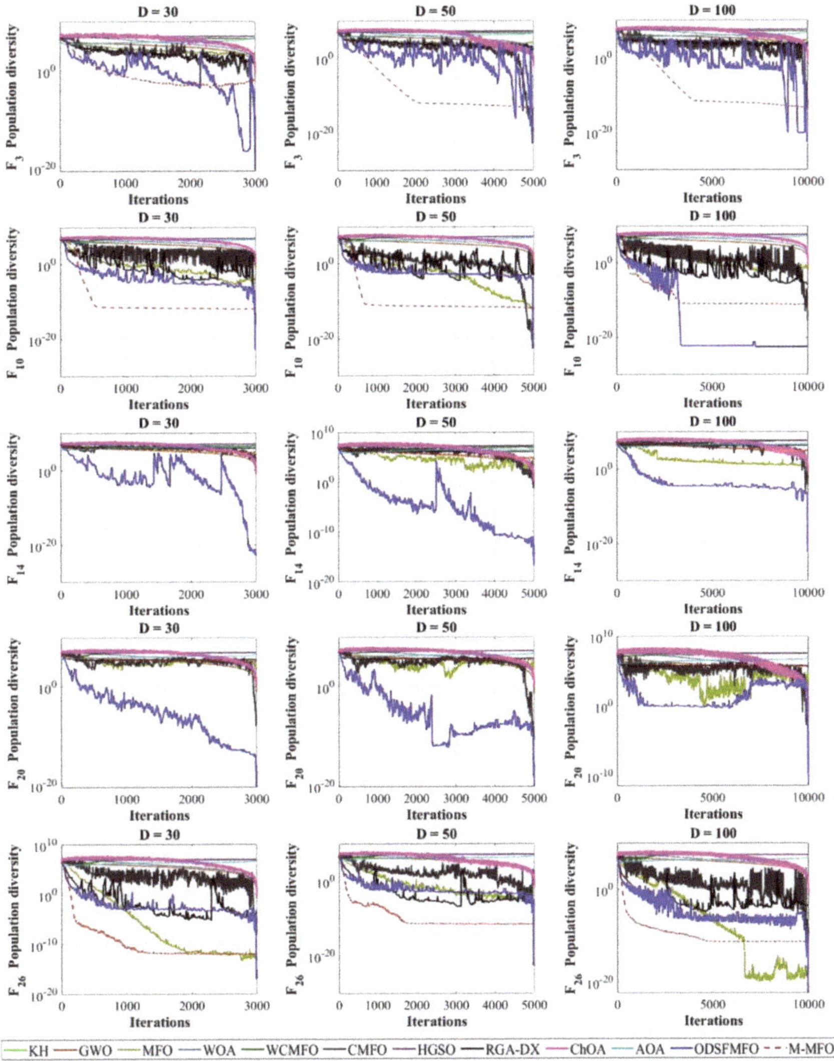

Figure 5. Population diversity of M-MFO and contender algorithms on different dimensions.

5.9. Sensitivity Analysis on the Guiding Archive Maturity Size

As discussed in Definition 1, the M-MFO algorithm switches from RM to GM operator when the guiding archive has matured. Hence, in this experiment, the impact of considering different maturity sizes of the guiding archive was evaluated and is discussed in relation to four different scenarios. Table 7 shows the fitness values gained in each scenario among five functions (i.e., F_1, F_9, F_{17}, and F_{30}) for different dimensions 30, 50, and 100.

Table 7. Results of sensitivity analysis on the guiding archive maturity size.

	Scenario 1 (δ = 1)			Scenario 2 (δ = 3)			Scenario 3 (δ = 5)			Scenario 4 (δ = D) M-MFO		
Dim	D30	D50	D100	D30	D50	D100	D30	D50	D100	D30	D50	D100
F_1 (Unimodal)	2.04×10^3	1.43×10^3	3.48×10^3	2.28×10^3	1.54×10^3	4.24×10^3	2.30×10^3	1.83×10^3	4.69×10^3	1.66×10^3	1.46×10^3	4.46×10^3
F_9 (Multimodal)	9.01×10^2	9.06×10^2	9.58×10^2	9.01×10^2	9.05×10^2	9.57×10^2	9.00×10^2	9.05×10^2	9.49×10^2	9.00×10^2	9.04×10^2	9.45×10^2
F_{17} (Hybrid)	1.75×10^3	2.05×10^3	2.39×10^3	1.74×10^3	1.99×10^3	2.35×10^3	1.74×10^3	1.99×10^3	2.29×10^3	1.74×10^3	1.93×10^3	2.29×10^3
F_{30} (Composition)	6.78×10^3	8.29×10^5	1.30×10^4	6.77×10^3	8.26×10^5	1.14×10^4	6.58×10^3	8.22×10^5	1.05×10^4	6.64×10^3	8.16×10^5	9.59×10^3

The reported results indicate that the fourth scenario ($\delta = D$) among all the tested functions provided better results overall compared to other scenarios. Nevertheless, it can be noticed that other scenarios provided competitive results, especially for unimodal functions. The main reason lies behind the fact that the higher value for maturity provides more population diversity and exploration ability, while the need for exploitation ability is more highlighted for unimodal functions. Moreover, previous studies [47,52] have proved that increasing the dimension has a negative impact on the effectiveness and scalability of metaheuristic algorithms. Hence, the size of the problem variables for maturity condition ($\delta = D$) can provide dimensional robustness for the algorithm. Furthermore, considering D as the maturity size does not add any additional parameters to the algorithm, and it provides an autotune parameter for different dimensions.

6. Statistical Analysis

In this experiment, the Friedman test [118] was conducted to statistically prove the superiority of M-MFO by ranking the algorithms based on their performance on CEC 2018 benchmark functions. Table 8 illustrates the obtained results for unimodal and multimodal test functions. In addition, hybrid and composition functions have been tabulated in Table 9. Inspecting the overall rank of the Friedman test, it is evident that the M-MFO demonstrated superior performance in comparison to contender algorithms for dimensions of 30, 50, and 100.

Table 8. Friedman test for unimodal and multimodal functions of the CEC 2018.

Functions	Unimodal Functions						Multimodal Functions					
Dimensions	30		50		100		30		50		100	
Algorithms	Avg. Rank	Overall Rank	Avg. Rank	Overall Rank	Avg. Rank	Overall Rank	Avg. Rank	Overall Rank	Avg. Rank	Overall Rank	Avg. Rank	Overall Rank
KH	4.9500	3	5.9500	4	6.6750	5	4.4250	2	4.1250	2	4.3250	3
GWO	6.4000	5	6.6750	5	6.1000	4	4.4500	3	4.7500	3	4.9750	4
MFO	9.0500	9	8.8000	8	8.9500	9	7.8750	8	7.8000	8	7.9250	8
WOA	8.4750	7	4.6250	3	8.3250	7	7.5000	7	6.8000	6	6.8000	6
WCMFO	2.8750	2	2.5750	2	2.6000	2	4.8750	4	4.8750	4	4.2250	2
CMFO	7.7000	6	6.9250	7	6.0750	3	7.4250	6	7.4000	7	7.1250	7
HGSO	8.5000	8	10.3000	10	9.0000	10	10.5500	9	10.6250	9	10.3750	9
RGA-DX	2.0250	1	2.1500	1	1.9250	1	2.2250	1	1.8000	1	2.2500	1
ChOA	9.6750	10	9.9500	9	8.7500	8	11.2750	11	11.2250	11	11.1500	10
AOA	11.2000	11	11.8000	11	10.5500	11	11.0750	10	11.1500	10	11.4250	11
ODSFMFO	5.6750	4	6.9000	6	7.5500	6	5.3250	5	6.1250	5	6.4000	5
M-MFO	1.4750	1	1.3500	1	1.5000	1	1.0000	1	1.3250	1	1.0250	1

In Figures 6 and 7, the M-MFO and competitors are visually ranked based on their performance in CEC 2018 benchmark suite for various dimensions. Figure 6 depicts the ranking results of algorithms in solving CEC 2018 benchmark functions, expressed through a radar graph. Meanwhile the clustered bar chart of Friedman's test average results is shown in Figure 7. The radar graph demonstrates that the M-MFO surpassed other algorithms in the majority of test functions for various dimensions. The clustered bar chart shows that the M-MFO achieved the best rank among competitors since it has the shortest bar in the different dimensions of 30, 50, and 100.

Table 9. Friedman test for hybrid and composition functions of the CEC 2018.

Functions	Hybrid Functions						Composition Functions					
Dimensions	30		50		100		30		50		100	
Algorithms	Avg. Rank	Overall Rank	Avg. Rank	Overall Rank	Avg. Rank	Overall Rank	Avg. Rank	Overall Rank	Avg. Rank	Overall Rank	Avg. Rank	Overall Rank
KH	6.0350	6	6.0150	6	6.0400	5	6.4450	6	6.9750	6	6.4550	5
GWO	5.8350	4	5.5000	4	6.0500	6	5.1700	3	4.7950	3	5.0550	4
MFO	6.2950	7	7.2050	7	8.1350	8	6.6900	7	6.5750	5	6.9750	6
WOA	9.2400	9	8.0600	8	6.3300	7	8.5450	8	8.6150	8	8.0200	8
WCMFO	5.9700	5	5.7550	5	5.0400	3	5.7500	4	5.3100	4	5.0100	3
CMFO	5.3550	3	5.1900	3	5.4400	4	6.3850	5	7.0250	7	7.3350	7
HGSO	10.9100	11	10.7300	10	10.8550	10	8.9800	9	9.0400	9	9.7600	9
RGA-DX	3.0000	1	2.8400	1	2.3950	1	2.2500	1	2.2350	1	2.0700	1
ChOA	10.8550	10	10.8750	11	10.3950	9	10.3650	10	10.0700	10	9.9350	10
AOA	8.4900	8	9.7400	9	11.2950	11	11.6400	11	11.6800	11	11.8000	11
ODSFMFO	4.3600	2	4.6450	2	4.8800	2	4.3750	2	4.2750	2	4.2250	2
M-MFO	1.6550	**1**	1.4450	**1**	1.1450	**1**	1.4050	**1**	1.4050	**1**	1.3600	**1**

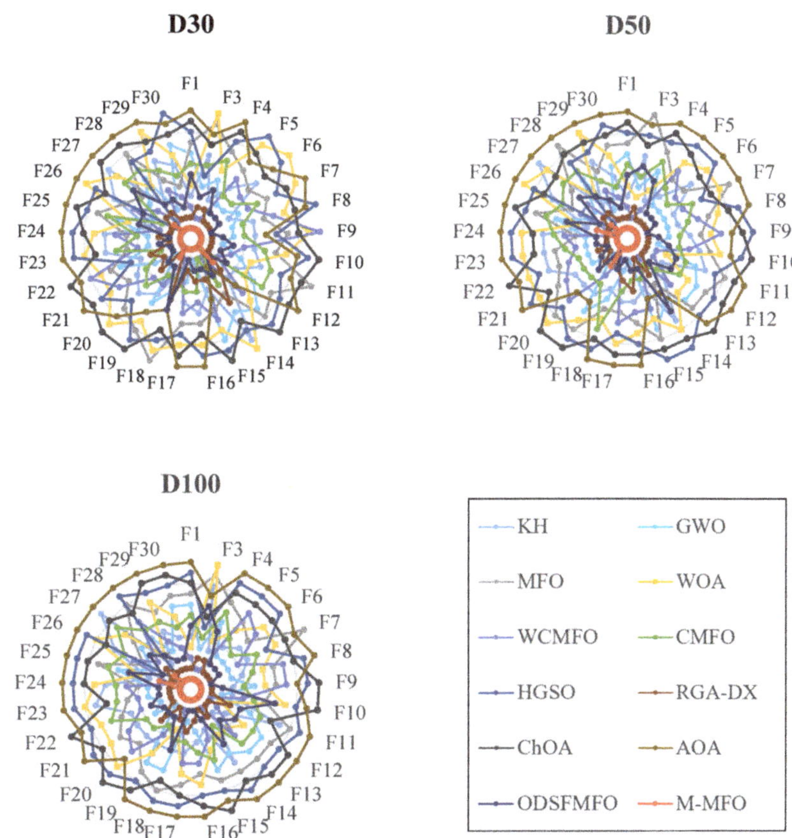

Figure 6. The radar graphs of M-MFO and competitors in different dimensions.

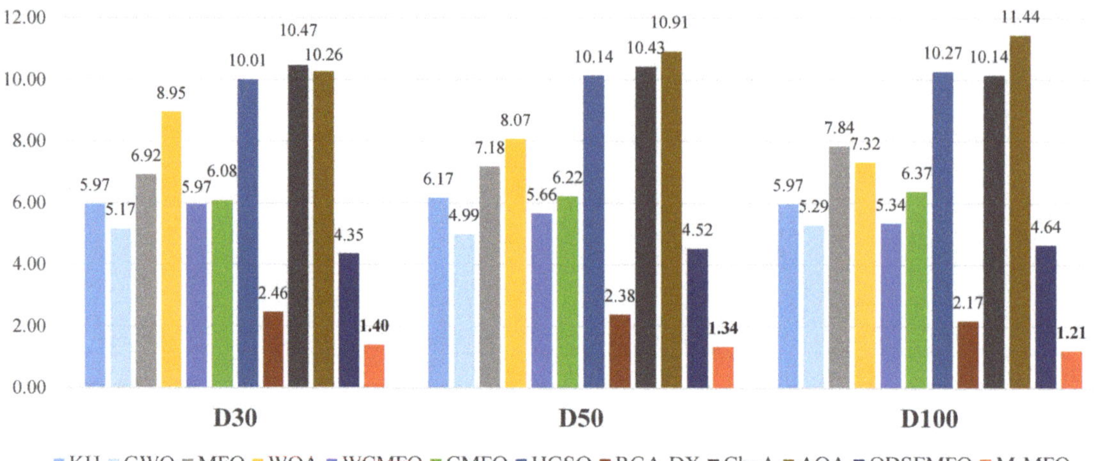

Figure 7. Friedman's test average results in different dimensions.

7. Conclusions

The MFO is a successful metaheuristic algorithm inspired by moths' behavior converging to a light source at night. The MFO has been used in various real-world optimization problems during recent years, mainly due to its simple structure. However, as the experiments revealed, the canonical MFO algorithm experiences local optima trapping and premature convergence due to the rapid dropping of population diversity and poor exploration. Hence, the M-MFO algorithm is proposed to overcome MFO's shortcomings by introducing a migration strategy that includes two new operators to boost exploration ability and maintain the population diversity.

The performance of M-MFO was experimentally evaluated by conducting CEC 2018 benchmark functions on dimension 30 and compared with seven recent variants of MFO, including LMFO, WCMFO, CMFO, CLSGMFO, LGCMFO, SMFO, and ODSFMFO. The top four latest high-performing variants and eight other state-of-the-art swarm intelligence algorithms were considered for experiments on the 30, 50, and 100 dimensions. The M-MFO stood out among competitors by providing highly competitive results and maintaining robustness while the size of the problem variables increased. In addition, to rank the algorithms, M-MFO and competitors were analyzed statistically by the Friedman test, in which the M-MFO obtained the first rank. For future works and studies, the migration strategy and guiding archive could be considered a reference in handling low population diversity and inefficient exploration of other metaheuristic algorithms. Moreover, the M-MFO can be used to solve engineering design problems. It can be converted for solving discrete optimization problems, such as feature selection, data mining, and image segmentation.

Author Contributions: Conceptualization, M.H.N.-S.; methodology, M.H.N.-S. and A.F.; software, M.H.N.-S., A.F. and H.Z.; validation, M.H.N.-S. and H.Z.; formal analysis, M.H.N.-S., A.F. and H.Z.; investigation, M.H.N.-S., A.F. and H.Z.; resources, M.H.N.-S., S.M., L.A. and M.A.E.; data curation, M.H.N.-S., A.F. and H.Z.; writing, M.H.N.-S., A.F. and H.Z.; original draft preparation, M.H.N.-S., A.F. and H.Z.; writing—review and editing, M.H.N.-S., A.F., H.Z., S.M., L.A. and M.A.E.; visualization, M.H.N.-S., A.F. and H.Z.; supervision, M.H.N.-S. and S.M.; project administration, M.H.N.-S. and S.M. All authors have read and agreed to the published version of the manuscript.

Funding: This research received no external funding.

Institutional Review Board Statement: Not applicable.

Informed Consent Statement: Not applicable.

Data Availability Statement: The data and code used in the research may be obtained from the corresponding author upon request.

Conflicts of Interest: The authors declare no conflict of interest.

Appendix A

Table A1 provides the detailed results of the proposed M-MFO algorithm and other variants of the MFO for solving CEC 2018 benchmark functions in dimension 30. Furthermore, the detailed results of the proposed M-MFO and contender algorithms for unimodal, multimodal, hybrid, and composition functions of CEC 2018 benchmark suite in dimensions 30, 50, and 100 are reported in Tables A2–A4.

Table A1. Comparison results of MFO variants.

F	D	Metrics	LMFO (2016)	WCMFO (2019)	CMFO (2019)	CLSGMFO (2019)	LGCMFO (2019)	SMFO (2021)	ODSFMFO (2021)	M-MFO
F_1	30	Avg	2.402×10^{7}	1.328×10^{4}	1.878×10^{8}	9.430×10^{8}	4.532×10^{8}	3.091×10^{10}	7.519×10^{6}	**1.660×10^{3}**
		Min	1.731×10^{7}	1.214×10^{2}	2.464×10^{6}	1.923×10^{8}	9.606×10^{7}	2.010×10^{10}	1.570×10^{6}	1.002×10^{2}
F_3	30	Avg	2.786×10^{3}	1.887×10^{3}	4.825×10^{4}	5.191×10^{4}	4.491×10^{4}	8.189×10^{4}	2.875×10^{4}	**3.006×10^{2}**
		Min	1.424×10^{3}	3.092×10^{2}	2.643×10^{4}	3.418×10^{4}	3.166×10^{4}	7.186×10^{4}	1.359×10^{4}	3.000×10^{2}
F_4	30	Avg	5.335×10^{2}	4.886×10^{2}	6.536×10^{2}	6.009×10^{2}	5.829×10^{2}	5.977×10^{3}	5.335×10^{2}	**4.247×10^{2}**
		Min	4.755×10^{2}	4.239×10^{2}	5.311×10^{2}	5.259×10^{2}	4.890×10^{2}	3.030×10^{3}	4.722×10^{2}	4.001×10^{2}
F_5	30	Avg	6.470×10^{2}	6.744×10^{2}	6.150×10^{2}	6.587×10^{2}	6.468×10^{2}	8.721×10^{2}	5.526×10^{2}	**5.136×10^{2}**
		Min	5.707×10^{2}	6.104×10^{2}	5.709×10^{2}	6.140×10^{2}	6.040×10^{2}	8.041×10^{2}	5.270×10^{2}	5.070×10^{2}
F_6	30	Avg	6.038×10^{2}	6.225×10^{2}	6.179×10^{2}	6.285×10^{2}	6.208×10^{2}	6.830×10^{2}	6.037×10^{2}	**6.000×10^{2}**
		Min	6.017×10^{2}	6.096×10^{2}	6.116×10^{2}	6.117×10^{2}	6.117×10^{2}	6.615×10^{2}	6.010×10^{2}	6.000×10^{2}
F_7	30	Avg	8.986×10^{2}	8.986×10^{2}	9.041×10^{2}	9.291×10^{2}	9.167×10^{2}	1.349×10^{3}	8.151×10^{2}	**7.446×10^{2}**
		Min	8.438×10^{2}	8.402×10^{2}	8.306×10^{2}	8.555×10^{2}	8.532×10^{2}	1.175×10^{3}	7.824×10^{2}	7.363×10^{2}
F_8	30	Avg	9.379×10^{2}	9.841×10^{2}	9.127×10^{2}	9.350×10^{2}	9.273×10^{2}	1.096×10^{3}	8.460×10^{2}	**8.141×10^{2}**
		Min	8.797×10^{2}	9.344×10^{2}	8.679×10^{2}	8.859×10^{2}	8.940×10^{2}	1.058×10^{3}	8.262×10^{2}	8.070×10^{2}
F_9	30	Avg	1.064×10^{3}	8.747×10^{3}	2.277×10^{3}	3.984×10^{3}	3.275×10^{3}	9.591×10^{3}	1.062×10^{3}	**9.005×10^{2}**
		Min	9.456×10^{2}	5.118×10^{3}	1.626×10^{3}	2.051×10^{3}	2.089×10^{3}	7.754×10^{3}	9.647×10^{2}	9.000×10^{2}
F_{10}	30	Avg	4.422×10^{3}	4.808×10^{3}	4.967×10^{3}	5.252×10^{3}	4.970×10^{3}	8.363×10^{3}	4.221×10^{3}	**1.958×10^{3}**
		Min	3.149×10^{3}	3.333×10^{3}	3.912×10^{3}	4.057×10^{3}	3.939×10^{3}	7.473×10^{3}	3.066×10^{3}	1.348×10^{3}
F_{11}	30	Avg	1.292×10^{3}	1.336×10^{3}	2.126×10^{3}	1.784×10^{3}	1.512×10^{3}	5.265×10^{3}	1.285×10^{3}	**1.122×10^{3}**
		Min	1.177×10^{3}	1.255×10^{3}	1.360×10^{3}	1.334×10^{3}	1.275×10^{3}	2.547×10^{3}	1.210×10^{3}	1.105×10^{3}
F_{12}	30	Avg	5.460×10^{6}	1.254×10^{6}	4.296×10^{7}	1.932×10^{7}	1.814×10^{7}	4.462×10^{9}	2.297×10^{6}	**7.118×10^{4}**
		Min	1.046×10^{6}	3.718×10^{4}	8.503×10^{5}	3.033×10^{6}	4.012×10^{6}	1.749×10^{9}	2.010×10^{5}	2.650×10^{4}
F_{13}	30	Avg	4.494×10^{5}	1.047×10^{5}	**6.841×10^{3}**	2.837×10^{5}	1.553×10^{5}	8.738×10^{8}	9.819×10^{3}	1.116×10^{4}
		Min	2.705×10^{5}	1.436×10^{4}	2.860×10^{3}	4.152×10^{4}	2.770×10^{4}	2.189×10^{8}	1.596×10^{3}	5.053×10^{3}
F_{14}	30	Avg	2.614×10^{4}	2.074×10^{4}	7.665×10^{4}	3.506×10^{5}	2.108×10^{5}	1.548×10^{6}	5.267×10^{4}	**6.136×10^{3}**
		Min	2.724×10^{3}	6.252×10^{3}	2.082×10^{3}	8.610×10^{3}	1.008×10^{4}	7.879×10^{4}	4.686×10^{3}	1.533×10^{3}
F_{15}	30	Avg	9.006×10^{4}	3.448×10^{4}	5.245×10^{3}	1.596×10^{4}	1.131×10^{4}	4.469×10^{7}	6.098×10^{3}	**2.252×10^{3}**
		Min	5.003×10^{4}	2.547×10^{3}	1.693×10^{3}	3.298×10^{3}	2.834×10^{3}	2.375×10^{6}	1.703×10^{3}	1.508×10^{3}
F_{16}	30	Avg	2.640×10^{3}	2.807×10^{3}	2.581×10^{3}	2.763×10^{3}	2.736×10^{3}	4.402×10^{3}	2.334×10^{3}	**1.774×10^{3}**
		Min	2.110×10^{3}	2.095×10^{3}	2.009×10^{3}	2.046×10^{3}	2.025×10^{3}	3.607×10^{3}	1.949×10^{3}	1.602×10^{3}
F_{17}	30	Avg	2.203×10^{3}	2.315×10^{3}	2.050×10^{3}	2.125×10^{3}	2.076×10^{3}	2.752×10^{3}	1.937×10^{3}	**1.738×10^{3}**
		Min	1.801×10^{3}	1.942×10^{3}	1.805×10^{3}	1.864×10^{3}	1.762×10^{3}	2.359×10^{3}	1.771×10^{3}	1.703×10^{3}
F_{18}	30	Avg	3.682×10^{5}	1.734×10^{5}	4.761×10^{5}	1.747×10^{6}	1.130×10^{6}	2.844×10^{7}	9.249×10^{5}	**9.790×10^{4}**
		Min	8.629×10^{4}	3.793×10^{4}	3.856×10^{4}	1.058×10^{5}	8.154×10^{4}	2.007×10^{6}	9.952×10^{4}	5.073×10^{4}
F_{19}	30	Avg	6.193×10^{4}	3.223×10^{4}	1.816×10^{4}	1.949×10^{4}	1.084×10^{4}	1.072×10^{8}	9.241×10^{3}	**6.433×10^{3}**
		Min	1.764×10^{4}	2.168×10^{3}	2.460×10^{3}	2.390×10^{3}	2.973×10^{3}	5.192×10^{6}	1.968×10^{3}	1.946×10^{3}
F_{20}	30	Avg	2.498×10^{3}	2.468×10^{3}	2.494×10^{3}	2.496×10^{3}	2.333×10^{3}	2.847×10^{3}	2.306×10^{3}	**2.128×10^{3}**
		Min	2.180×10^{3}	2.073×10^{3}	2.162×10^{3}	2.069×10^{3}	2.132×10^{3}	2.454×10^{3}	2.053×10^{3}	2.028×10^{3}
F_{21}	30	Avg	2.439×10^{3}	2.493×10^{3}	2.396×10^{3}	2.421×10^{3}	2.420×10^{3}	2.653×10^{3}	2.352×10^{3}	**2.312×10^{3}**
		Min	2.378×10^{3}	2.398×10^{3}	2.347×10^{3}	2.386×10^{3}	2.375×10^{3}	2.552×10^{3}	2.331×10^{3}	2.303×10^{3}
F_{22}	30	Avg	5.006×10^{3}	6.637×10^{3}	2.373×10^{3}	2.507×10^{3}	2.492×10^{3}	8.654×10^{3}	2.318×10^{3}	**2.300×10^{3}**
		Min	2.325×10^{3}	5.330×10^{3}	2.315×10^{3}	2.371×10^{3}	2.358×10^{3}	5.677×10^{3}	2.307×10^{3}	2.300×10^{3}
F_{23}	30	Avg	2.786×10^{3}	2.785×10^{3}	2.828×10^{3}	2.795×10^{3}	2.787×10^{3}	3.283×10^{3}	2.718×10^{3}	**2.662×10^{3}**
		Min	2.710×10^{3}	2.721×10^{3}	2.764×10^{3}	2.745×10^{3}	2.707×10^{3}	3.027×10^{3}	2.699×10^{3}	2.647×10^{3}
F_{24}	30	Avg	2.928×10^{3}	2.978×10^{3}	2.928×10^{3}	2.952×10^{3}	2.959×10^{3}	3.433×10^{3}	2.871×10^{3}	**2.827×10^{3}**
		Min	2.898×10^{3}	2.928×10^{3}	2.877×10^{3}	2.878×10^{3}	2.920×10^{3}	3.217×10^{3}	2.848×10^{3}	2.820×10^{3}

Table A1. *Cont.*

F	D	Metrics	LMFO (2016)	WCMFO (2019)	CMFO (2019)	CLSGMFO (2019)	LGCMFO (2019)	SMFO (2021)	ODSFMFO (2021)	M-MFO
F_{25}	30	Avg	2.889×10^{3}	2.894×10^{3}	3.004×10^{3}	2.980×10^{3}	3.005×10^{3}	3.940×10^{3}	2.928×10^{3}	$\mathbf{2.888 \times 10^{3}}$
		Min	2.888×10^{3}	2.884×10^{3}	2.933×10^{3}	2.939×10^{3}	2.920×10^{3}	3.463×10^{3}	2.890×10^{3}	2.887×10^{3}
F_{26}	30	Avg	5.012×10^{3}	5.447×10^{3}	4.227×10^{3}	4.838×10^{3}	4.163×10^{3}	8.871×10^{3}	4.425×10^{3}	$\mathbf{3.408 \times 10^{3}}$
		Min	4.607×10^{3}	4.955×10^{3}	2.936×10^{3}	3.514×10^{3}	3.241×10^{3}	5.057×10^{3}	2.876×10^{3}	2.800×10^{3}
F_{27}	30	Avg	3.241×10^{3}	3.228×10^{3}	3.257×10^{3}	3.286×10^{3}	3.275×10^{3}	3.688×10^{3}	3.230×10^{3}	$\mathbf{3.221 \times 10^{3}}$
		Min	3.200×10^{3}	3.201×10^{3}	3.232×10^{3}	3.224×10^{3}	3.218×10^{3}	3.397×10^{3}	3.222×10^{3}	3.210×10^{3}
F_{28}	30	Avg	3.255×10^{3}	3.194×10^{3}	3.444×10^{3}	3.451×10^{3}	3.290×10^{3}	5.524×10^{3}	3.295×10^{3}	$\mathbf{3.110 \times 10^{3}}$
		Min	3.210×10^{3}	3.100×10^{3}	3.247×10^{3}	3.285×10^{3}	3.268×10^{3}	4.419×10^{3}	3.251×10^{3}	3.100×10^{3}
F_{29}	30	Avg	3.785×10^{3}	3.965×10^{3}	4.050×10^{3}	3.976×10^{3}	3.872×10^{3}	5.698×10^{3}	3.669×10^{3}	$\mathbf{3.319 \times 10^{3}}$
		Min	3.596×10^{3}	3.650×10^{3}	3.631×10^{3}	3.601×10^{3}	3.556×10^{3}	4.728×10^{3}	3.475×10^{3}	3.312×10^{3}
F_{30}	30	Avg	1.579×10^{5}	2.812×10^{4}	8.574×10^{5}	3.989×10^{5}	3.406×10^{5}	3.278×10^{8}	1.629×10^{4}	$\mathbf{6.645 \times 10^{3}}$
		Min	4.934×10^{4}	1.582×10^{4}	7.507×10^{4}	3.747×10^{4}	4.618×10^{4}	3.212×10^{7}	7.769×10^{3}	6.062×10^{3}
Summary	W|T|L		0|0|29	0|0|29	1|0|28	0|0|29	0|0|29	0|0|29	0|0|29	**28|0|1**

Table A2. Results of the comparative algorithms on unimodal and multimodal test functions.

F	D	Metrics	KH (2012)	GWO (2014)	MFO (2015)	WOA (2016)	WCMFO (2019)	CMFO (2019)	HGSO (2019)	RGA-DX (2019)	ChOA (2020)	AOA (2021)	ODSFMFO (2021)	M-MFO
F_1	30	Avg	1.371×10^{4}	8.223×10^{8}	6.952×10^{9}	1.906×10^{6}	1.328×10^{4}	5.824×10^{7}	1.455×10^{10}	2.575×10^{3}	2.395×10^{10}	4.015×10^{10}	7.519×10^{6}	$\mathbf{1.660 \times 10^{3}}$
		Min	3.462×10^{3}	4.405×10^{7}	1.027×10^{9}	5.654×10^{5}	1.214×10^{2}	2.553×10^{6}	7.442×10^{9}	1.272×10^{2}	1.123×10^{10}	3.092×10^{10}	1.570×10^{6}	1.002×10^{2}
	50	Avg	1.954×10^{5}	4.523×10^{9}	3.099×10^{10}	7.172×10^{6}	2.826×10^{4}	1.046×10^{9}	3.844×10^{10}	3.059×10^{3}	4.417×10^{10}	1.003×10^{11}	3.066×10^{8}	$\mathbf{1.466 \times 10^{3}}$
		Min	4.342×10^{4}	1.231×10^{9}	7.095×10^{9}	1.980×10^{6}	6.883×10^{2}	2.822×10^{8}	2.159×10^{10}	1.327×10^{2}	3.506×10^{10}	8.424×10^{10}	9.629×10^{7}	1.001×10^{2}
	100	Avg	5.646×10^{7}	3.207×10^{10}	1.173×10^{11}	3.677×10^{7}	2.017×10^{5}	3.803×10^{9}	1.643×10^{11}	4.575×10^{3}	1.463×10^{11}	2.629×10^{11}	5.316×10^{9}	$\mathbf{4.465 \times 10^{3}}$
		Min	2.550×10^{6}	1.634×10^{10}	6.748×10^{10}	1.409×10^{7}	1.093×10^{4}	1.832×10^{9}	1.299×10^{11}	1.587×10^{2}	1.282×10^{11}	2.350×10^{11}	1.894×10^{9}	1.032×10^{3}
F_3	30	Avg	4.863×10^{4}	2.993×10^{4}	1.009×10^{5}	1.715×10^{5}	1.887×10^{3}	4.248×10^{4}	3.687×10^{4}	7.905×10^{3}	5.178×10^{4}	7.318×10^{4}	2.875×10^{4}	$\mathbf{3.006 \times 10^{2}}$
		Min	2.979×10^{4}	1.576×10^{4}	1.920×10^{3}	8.481×10^{4}	3.092×10^{2}	3.385×10^{4}	2.335×10^{4}	9.933×10^{2}	3.954×10^{4}	5.445×10^{4}	1.359×10^{4}	3.000×10^{2}
	50	Avg	1.216×10^{5}	7.147×10^{4}	1.650×10^{5}	6.180×10^{4}	1.150×10^{4}	9.610×10^{4}	1.363×10^{5}	2.712×10^{4}	1.309×10^{5}	1.625×10^{5}	9.729×10^{4}	$\mathbf{3.000 \times 10^{2}}$
		Min	6.121×10^{4}	3.628×10^{4}	1.176×10^{4}	3.098×10^{4}	7.428×10^{2}	6.899×10^{4}	1.050×10^{5}	1.139×10^{4}	1.006×10^{5}	1.249×10^{5}	6.538×10^{4}	3.000×10^{2}
	100	Avg	3.477×10^{5}	2.023×10^{5}	4.556×10^{5}	5.928×10^{5}	7.361×10^{4}	2.317×10^{5}	2.945×10^{5}	1.387×10^{5}	3.065×10^{5}	3.325×10^{5}	3.252×10^{5}	$\mathbf{3.000 \times 10^{2}}$
		Min	2.569×10^{5}	1.595×10^{5}	1.191×10^{5}	3.355×10^{5}	3.430×10^{4}	2.058×10^{5}	2.601×10^{5}	7.593×10^{4}	2.849×10^{5}	3.027×10^{5}	2.456×10^{5}	3.000×10^{2}
F_4	30	Avg	4.963×10^{2}	5.441×10^{2}	9.082×10^{2}	5.476×10^{2}	4.886×10^{2}	5.631×10^{2}	2.171×10^{3}	4.896×10^{2}	2.545×10^{3}	8.649×10^{3}	5.335×10^{2}	$\mathbf{4.247 \times 10^{2}}$
		Min	4.043×10^{2}	4.963×10^{2}	5.424×10^{2}	4.995×10^{2}	4.239×10^{2}	4.759×10^{2}	1.194×10^{3}	4.040×10^{2}	1.134×10^{3}	3.825×10^{3}	4.722×10^{2}	4.001×10^{2}
	50	Avg	5.683×10^{2}	8.767×10^{2}	4.098×10^{3}	6.676×10^{2}	5.493×10^{2}	1.069×10^{3}	8.889×10^{3}	5.095×10^{2}	9.023×10^{3}	2.568×10^{4}	7.414×10^{2}	$\mathbf{4.872 \times 10^{2}}$
		Min	4.996×10^{2}	6.745×10^{2}	1.216×10^{3}	5.138×10^{2}	4.849×10^{2}	6.394×10^{2}	5.286×10^{3}	4.285×10^{2}	5.017×10^{3}	1.686×10^{4}	6.237×10^{2}	4.092×10^{2}
	100	Avg	7.431×10^{2}	2.813×10^{3}	2.348×10^{4}	9.992×10^{2}	6.423×10^{2}	2.354×10^{3}	2.840×10^{4}	6.436×10^{2}	2.822×10^{4}	7.733×10^{4}	1.400×10^{3}	$\mathbf{5.378 \times 10^{2}}$
		Min	6.443×10^{2}	1.870×10^{3}	6.743×10^{3}	8.615×10^{2}	5.980×10^{2}	1.123×10^{3}	1.677×10^{4}	5.671×10^{2}	2.145×10^{4}	6.186×10^{4}	1.103×10^{3}	4.753×10^{2}
F_5	30	Avg	6.363×10^{2}	5.855×10^{2}	6.894×10^{2}	8.044×10^{2}	6.744×10^{2}	5.957×10^{2}	8.061×10^{2}	5.430×10^{2}	7.905×10^{2}	7.873×10^{2}	5.526×10^{2}	$\mathbf{5.136 \times 10^{2}}$
		Min	5.936×10^{2}	5.508×10^{2}	6.280×10^{2}	7.242×10^{2}	6.104×10^{2}	5.678×10^{2}	7.824×10^{2}	5.259×10^{2}	7.471×10^{2}	7.217×10^{2}	5.270×10^{2}	5.070×10^{2}
	50	Avg	7.659×10^{2}	6.892×10^{2}	8.934×10^{2}	9.209×10^{2}	8.940×10^{2}	8.095×10^{2}	1.049×10^{3}	6.004×10^{2}	1.043×10^{3}	1.074×10^{3}	6.212×10^{2}	$\mathbf{5.296 \times 10^{2}}$
		Min	7.050×10^{2}	6.379×10^{2}	7.731×10^{2}	8.081×10^{2}	7.743×10^{2}	7.138×10^{2}	9.990×10^{2}	5.677×10^{2}	9.853×10^{2}	9.951×10^{2}	5.630×10^{2}	5.179×10^{2}
	100	Avg	1.216×10^{3}	1.058×10^{3}	1.666×10^{3}	1.413×10^{3}	1.726×10^{3}	1.297×10^{3}	1.824×10^{3}	7.916×10^{2}	1.787×10^{3}	1.960×10^{3}	8.658×10^{2}	$\mathbf{5.564 \times 10^{2}}$
		Min	1.054×10^{3}	9.864×10^{2}	1.455×10^{3}	1.329×10^{3}	1.328×10^{3}	1.154×10^{3}	1.701×10^{3}	7.259×10^{2}	1.743×10^{3}	1.842×10^{3}	7.800×10^{2}	5.368×10^{2}
F_6	30	Avg	6.428×10^{2}	6.043×10^{2}	6.267×10^{2}	6.671×10^{2}	6.225×10^{2}	6.166×10^{2}	6.655×10^{2}	$\mathbf{6.000 \times 10^{2}}$	6.603×10^{2}	6.654×10^{2}	6.037×10^{2}	$\mathbf{6.000 \times 10^{2}}$
		Min	6.175×10^{2}	6.011×10^{2}	6.144×10^{2}	6.410×10^{2}	6.096×10^{2}	6.078×10^{2}	6.511×10^{2}	6.000×10^{2}	6.537×10^{2}	6.566×10^{2}	6.010×10^{2}	6.000×10^{2}
	50	Avg	6.515×10^{2}	6.105×10^{2}	6.437×10^{2}	6.760×10^{2}	6.400×10^{2}	6.355×10^{2}	6.813×10^{2}	6.001×10^{2}	6.710×10^{2}	6.837×10^{2}	6.081×10^{2}	$\mathbf{6.000 \times 10^{2}}$
		Min	6.440×10^{2}	6.052×10^{2}	6.270×10^{2}	6.638×10^{2}	6.165×10^{2}	6.257×10^{2}	6.724×10^{2}	6.000×10^{2}	6.608×10^{2}	6.747×10^{2}	6.041×10^{2}	6.000×10^{2}
	100	Avg	6.587×10^{2}	6.276×10^{2}	6.648×10^{2}	6.768×10^{2}	6.664×10^{2}	6.528×10^{2}	6.936×10^{2}	6.001×10^{2}	6.860×10^{2}	7.029×10^{2}	6.184×10^{2}	$\mathbf{6.000 \times 10^{2}}$
		Min	6.527×10^{2}	6.229×10^{2}	6.467×10^{2}	6.676×10^{2}	6.526×10^{2}	6.418×10^{2}	6.867×10^{2}	6.000×10^{2}	6.761×10^{2}	6.970×10^{2}	6.133×10^{2}	6.000×10^{2}

Table A2. *Cont.*

F	D	Metrics	KH (2012)	GWO (2014)	MFO (2015)	WOA (2016)	WCMFO (2019)	CMFO (2019)	HGSO (2019)	RGA-DX (2019)	ChOA (2020)	AOA (2021)	ODSFMFO (2021)	M-MFO
F_7	30	Avg	8.280×10^{2}	8.418×10^{2}	1.011×10^{3}	1.238×10^{3}	8.986×10^{2}	8.989×10^{2}	1.080×10^{3}	7.801×10^{2}	1.187×10^{3}	1.302×10^{3}	8.151×10^{2}	$\mathbf{7.446 \times 10^{2}}$
		Min	7.853×10^{2}	7.801×10^{2}	8.671×10^{2}	1.089×10^{3}	8.402×10^{2}	8.415×10^{2}	1.032×10^{3}	7.586×10^{2}	1.063×10^{3}	1.154×10^{3}	7.824×10^{2}	7.363×10^{2}
	50	Avg	1.070×10^{3}	1.016×10^{3}	1.701×10^{3}	1.684×10^{3}	1.141×10^{3}	1.224×10^{3}	1.530×10^{3}	8.481×10^{2}	1.663×10^{3}	1.862×10^{3}	9.911×10^{2}	$\mathbf{7.684 \times 10^{2}}$
		Min	9.625×10^{2}	9.654×10^{2}	1.113×10^{3}	1.500×10^{3}	1.020×10^{3}	1.021×10^{3}	1.333×10^{3}	8.062×10^{2}	1.464×10^{3}	1.744×10^{3}	9.354×10^{2}	7.588×10^{2}
	100	Avg	2.118×10^{3}	1.710×10^{3}	4.169×10^{3}	3.250×10^{3}	1.988×10^{3}	2.421×10^{3}	3.184×10^{3}	1.129×10^{3}	3.326×10^{3}	3.694×10^{3}	1.619×10^{3}	$\mathbf{8.363 \times 10^{2}}$
		Min	1.819×10^{3}	1.542×10^{3}	2.576×10^{3}	2.814×10^{3}	1.531×10^{3}	2.103×10^{3}	2.813×10^{3}	9.899×10^{2}	3.182×10^{3}	3.580×10^{3}	1.416×10^{3}	8.161×10^{2}
F_8	30	Avg	9.196×10^{2}	8.713×10^{2}	9.790×10^{2}	1.000×10^{3}	9.841×10^{2}	8.945×10^{2}	1.051×10^{3}	8.434×10^{2}	1.031×10^{3}	1.041×10^{3}	8.460×10^{2}	$\mathbf{8.141 \times 10^{2}}$
		Min	8.707×10^{2}	8.435×10^{2}	8.938×10^{2}	9.488×10^{2}	9.344×10^{2}	8.574×10^{2}	1.033×10^{3}	8.249×10^{2}	9.726×10^{2}	1.002×10^{3}	8.262×10^{2}	8.070×10^{2}
	50	Avg	1.065×10^{3}	9.792×10^{2}	1.229×10^{3}	1.249×10^{3}	1.213×10^{3}	1.055×10^{3}	1.369×10^{3}	9.002×10^{2}	1.305×10^{3}	1.425×10^{3}	9.168×10^{2}	$\mathbf{8.315 \times 10^{2}}$
		Min	1.019×10^{3}	9.384×10^{2}	1.118×10^{3}	1.132×10^{3}	1.087×10^{3}	9.831×10^{2}	1.308×10^{3}	8.567×10^{2}	1.251×10^{3}	1.339×10^{3}	8.625×10^{2}	8.199×10^{2}
	100	Avg	1.576×10^{3}	1.397×10^{3}	1.968×10^{3}	1.897×10^{3}	2.026×10^{3}	1.533×10^{3}	2.240×10^{3}	1.063×10^{3}	2.151×10^{3}	2.414×10^{3}	1.193×10^{3}	$\mathbf{8.694 \times 10^{2}}$
		Min	1.465×10^{3}	1.225×10^{3}	1.717×10^{3}	1.716×10^{3}	1.756×10^{3}	1.410×10^{3}	2.093×10^{3}	9.900×10^{2}	2.052×10^{3}	2.248×10^{3}	1.122×10^{3}	8.398×10^{2}
F_9	30	Avg	3.059×10^{3}	1.384×10^{3}	6.278×10^{3}	7.233×10^{3}	8.747×10^{3}	1.893×10^{3}	5.814×10^{3}	9.064×10^{2}	6.551×10^{3}	5.578×10^{3}	1.062×10^{3}	$\mathbf{9.005 \times 10^{2}}$
		Min	1.768×10^{3}	1.025×10^{3}	4.471×10^{3}	4.425×10^{3}	5.118×10^{3}	1.554×10^{3}	3.388×10^{3}	9.009×10^{2}	5.576×10^{3}	4.101×10^{3}	9.647×10^{2}	9.000×10^{2}
	50	Avg	9.536×10^{3}	4.571×10^{3}	1.644×10^{4}	1.783×10^{4}	2.195×10^{4}	7.504×10^{3}	2.616×10^{4}	9.773×10^{2}	2.577×10^{4}	2.294×10^{4}	1.750×10^{3}	$\mathbf{9.045 \times 10^{2}}$
		Min	6.223×10^{3}	2.135×10^{3}	8.748×10^{3}	1.187×10^{4}	1.190×10^{4}	3.715×10^{3}	2.123×10^{4}	9.213×10^{2}	1.969×10^{4}	1.804×10^{4}	1.299×10^{3}	9.007×10^{2}
	100	Avg	2.251×10^{4}	2.638×10^{4}	4.508×10^{4}	3.820×10^{4}	5.208×10^{4}	2.315×10^{4}	6.515×10^{4}	2.428×10^{3}	6.876×10^{4}	5.410×10^{4}	4.877×10^{3}	$\mathbf{9.454 \times 10^{2}}$
		Min	1.965×10^{4}	1.102×10^{4}	3.679×10^{4}	2.557×10^{4}	3.986×10^{4}	1.973×10^{4}	5.587×10^{4}	1.304×10^{3}	5.806×10^{4}	4.674×10^{4}	3.620×10^{3}	9.174×10^{2}
F_{10}	30	Avg	4.876×10^{3}	3.909×10^{3}	5.130×10^{3}	6.156×10^{3}	4.808×10^{3}	4.728×10^{3}	6.636×10^{3}	3.700×10^{3}	7.996×10^{3}	6.444×10^{3}	4.221×10^{3}	$\mathbf{1.958 \times 10^{3}}$
		Min	3.664×10^{3}	2.718×10^{3}	3.575×10^{3}	4.506×10^{3}	3.333×10^{3}	3.522×10^{3}	5.706×10^{3}	2.875×10^{3}	7.199×10^{3}	5.410×10^{3}	3.066×10^{3}	1.348×10^{3}
	50	Avg	8.127×10^{3}	6.428×10^{3}	8.566×10^{3}	9.478×10^{3}	7.956×10^{3}	7.769×10^{3}	1.242×10^{4}	5.949×10^{3}	1.427×10^{4}	1.216×10^{4}	7.662×10^{3}	$\mathbf{2.391 \times 10^{3}}$
		Min	6.288×10^{3}	4.582×10^{3}	6.288×10^{3}	6.969×10^{3}	6.204×10^{3}	6.035×10^{3}	1.100×10^{4}	4.819×10^{3}	1.301×10^{4}	1.073×10^{4}	5.766×10^{3}	1.246×10^{3}
	100	Avg	1.549×10^{4}	1.498×10^{4}	1.728×10^{4}	2.012×10^{4}	1.618×10^{4}	1.578×10^{4}	2.579×10^{4}	1.361×10^{4}	3.140×10^{4}	2.787×10^{4}	1.733×10^{4}	$\mathbf{4.814 \times 10^{3}}$
		Min	1.267×10^{4}	1.141×10^{4}	1.417×10^{4}	1.687×10^{4}	1.147×10^{4}	1.335×10^{4}	2.440×10^{4}	1.115×10^{4}	3.051×10^{4}	2.582×10^{4}	1.468×10^{4}	2.834×10^{3}
Summary	30	W|T|L	0|0|19	0|0|19	0|0|19	0|0|19	0|0|19	0|0|19	0|0|19	0|1|18	0|0|19	0|0|19	0|0|19	**8|1|10**
	50	W|T|L	0|0|19	0|0|19	0|0|19	0|0|19	0|0|19	0|0|19	0|0|19	0|0|19	0|0|19	0|0|19	0|0|19	**9|0|10**
	100	W|T|L	0|0|19	0|0|19	0|0|19	0|0|19	0|0|19	0|0|19	0|0|19	0|0|19	0|0|19	0|0|19	0|0|19	**9|0|10**

Table A3. Results of the comparative algorithms on hybrid test functions.

F	D	Metrics	KH (2012)	GWO (2014)	MFO (2015)	WOA (2016)	WCMFO (2019)	CMFO (2019)	HGSO (2019)	RGA-DX (2019)	ChOA (2020)	AOA (2021)	ODSFMFO (2021)	M-MFO
F_{11}	30	Avg	1.514×10^{3}	1.406×10^{3}	3.749×10^{3}	1.462×10^{3}	1.336×10^{3}	2.126×10^{3}	2.811×10^{3}	1.138×10^{3}	3.267×10^{3}	3.249×10^{3}	1.285×10^{3}	$\mathbf{1.122 \times 10^{3}}$
		Min	1.262×10^{3}	1.271×10^{3}	1.363×10^{3}	1.282×10^{3}	1.255×10^{3}	1.360×10^{3}	1.707×10^{3}	1.109×10^{3}	1.731×10^{3}	1.797×10^{3}	1.210×10^{3}	1.105×10^{3}
	50	Avg	4.926×10^{3}	3.078×10^{3}	7.297×10^{3}	1.591×10^{3}	1.491×10^{3}	1.984×10^{3}	5.800×10^{3}	1.251×10^{3}	8.848×10^{3}	1.587×10^{4}	1.725×10^{3}	$\mathbf{1.128 \times 10^{3}}$
		Min	2.518×10^{3}	1.480×10^{3}	1.574×10^{3}	1.421×10^{3}	1.344×10^{3}	1.278×10^{3}	3.881×10^{3}	1.129×10^{3}	6.441×10^{3}	9.287×10^{3}	1.394×10^{3}	1.123×10^{3}
	100	Avg	7.658×10^{4}	3.531×10^{4}	1.257×10^{5}	7.762×10^{3}	2.191×10^{3}	4.024×10^{4}	1.281×10^{5}	1.033×10^{4}	7.093×10^{4}	1.631×10^{5}	3.043×10^{4}	$\mathbf{1.195 \times 10^{3}}$
		Min	3.912×10^{4}	1.647×10^{4}	2.137×10^{4}	4.463×10^{3}	1.841×10^{3}	2.012×10^{4}	1.128×10^{5}	2.076×10^{3}	6.100×10^{4}	1.268×10^{5}	1.322×10^{4}	1.127×10^{3}
F_{12}	30	Avg	3.051×10^{6}	3.900×10^{7}	6.158×10^{7}	3.770×10^{7}	1.254×10^{6}	4.296×10^{7}	1.343×10^{9}	7.608×10^{5}	3.594×10^{9}	7.204×10^{9}	2.297×10^{6}	$\mathbf{7.118 \times 10^{4}}$
		Min	1.788×10^{5}	2.109×10^{6}	7.306×10^{4}	2.509×10^{6}	3.718×10^{4}	8.503×10^{5}	6.717×10^{8}	1.016×10^{5}	6.620×10^{8}	3.034×10^{9}	2.010×10^{5}	2.650×10^{4}
	50	Avg	1.249×10^{7}	4.764×10^{8}	2.475×10^{9}	1.861×10^{8}	7.229×10^{6}	1.684×10^{8}	1.550×10^{10}	1.983×10^{6}	1.896×10^{10}	5.311×10^{10}	1.951×10^{7}	$\mathbf{3.292 \times 10^{5}}$
		Min	1.930×10^{6}	7.558×10^{7}	1.646×10^{7}	5.114×10^{7}	1.549×10^{6}	6.965×10^{6}	9.677×10^{9}	5.848×10^{5}	1.045×10^{10}	2.948×10^{10}	7.129×10^{6}	1.293×10^{5}
	100	Avg	6.902×10^{7}	4.919×10^{9}	3.523×10^{10}	6.875×10^{8}	3.428×10^{7}	1.982×10^{9}	5.643×10^{10}	3.019×10^{6}	6.718×10^{10}	1.822×10^{11}	4.254×10^{8}	$\mathbf{2.834 \times 10^{3}}$
		Min	2.478×10^{7}	1.450×10^{9}	1.435×10^{10}	2.918×10^{8}	3.806×10^{6}	9.911×10^{7}	3.888×10^{10}	1.129×10^{6}	4.928×10^{10}	1.296×10^{11}	1.525×10^{8}	2.195×10^{5}

Table A3. *Cont.*

F	D	Metrics	KH (2012)	GWO (2014)	MFO (2015)	WOA (2016)	WCMFO (2019)	CMFO (2019)	HGSO (2019)	RGA-DX (2019)	ChOA (2020)	AOA (2021)	ODSFMFO (2021)	M-MFO
F_{13}	30	Avg	3.536×10^{4}	8.368×10^{5}	7.958×10^{6}	1.463×10^{5}	1.047×10^{5}	**6.841×10^{3}**	4.947×10^{8}	1.173×10^{4}	8.944×10^{8}	4.457×10^{4}	9.819×10^{3}	1.116×10^{4}
		Min	1.619×10^{4}	1.991×10^{4}	1.122×10^{4}	2.283×10^{4}	1.436×10^{4}	2.860×10^{3}	1.781×10^{8}	1.376×10^{3}	5.646×10^{7}	2.158×10^{4}	1.596×10^{3}	5.053×10^{3}
	50	Avg	4.578×10^{4}	1.532×10^{8}	2.428×10^{8}	1.657×10^{5}	8.895×10^{4}	2.085×10^{4}	2.604×10^{9}	4.464×10^{3}	6.036×10^{9}	4.764×10^{9}	1.952×10^{4}	**2.083×10^{3}**
		Min	2.381×10^{4}	1.312×10^{5}	1.136×10^{5}	4.764×10^{4}	2.174×10^{4}	6.621×10^{3}	1.204×10^{9}	1.455×10^{3}	8.730×10^{8}	1.041×10^{7}	9.648×10^{3}	1.317×10^{3}
	100	Avg	3.464×10^{4}	4.163×10^{8}	4.053×10^{9}	8.423×10^{4}	1.378×10^{5}	1.269×10^{6}	9.302×10^{9}	5.906×10^{3}	1.894×10^{10}	3.573×10^{10}	5.455×10^{4}	**3.236×10^{3}**
		Min	2.377×10^{4}	1.579×10^{6}	2.629×10^{8}	3.701×10^{4}	3.658×10^{4}	1.538×10^{4}	5.247×10^{9}	1.409×10^{3}	1.203×10^{10}	2.155×10^{10}	1.136×10^{4}	1.611×10^{3}
F_{14}	30	Avg	5.166×10^{5}	1.438×10^{5}	8.969×10^{4}	9.075×10^{5}	2.074×10^{4}	7.665×10^{4}	3.822×10^{5}	1.064×10^{5}	3.622×10^{5}	4.148×10^{4}	5.267×10^{4}	**6.136×10^{3}**
		Min	1.184×10^{4}	3.679×10^{3}	2.197×10^{3}	1.364×10^{5}	6.252×10^{3}	2.082×10^{3}	8.651×10^{4}	5.924×10^{3}	5.125×10^{4}	2.213×10^{3}	4.686×10^{3}	1.533×10^{3}
	50	Avg	5.216×10^{5}	4.016×10^{5}	3.086×10^{5}	6.358×10^{5}	8.151×10^{4}	2.215×10^{5}	4.049×10^{6}	2.251×10^{5}	1.203×10^{6}	3.163×10^{5}	5.207×10^{5}	**2.475×10^{4}**
		Min	1.101×10^{5}	4.749×10^{4}	1.072×10^{4}	9.639×10^{4}	1.194×10^{4}	1.658×10^{4}	1.690×10^{6}	2.594×10^{4}	5.706×10^{5}	4.727×10^{4}	6.023×10^{4}	8.641×10^{3}
	100	Avg	3.785×10^{6}	3.480×10^{6}	7.558×10^{6}	1.876×10^{6}	3.627×10^{5}	1.108×10^{6}	1.595×10^{7}	6.317×10^{5}	8.127×10^{6}	1.993×10^{7}	3.338×10^{6}	**1.466×10^{5}**
		Min	2.148×10^{6}	1.057×10^{6}	3.097×10^{5}	6.461×10^{5}	1.387×10^{5}	3.212×10^{5}	1.187×10^{7}	8.401×10^{4}	5.390×10^{6}	6.674×10^{6}	1.282×10^{6}	1.009×10^{5}
F_{15}	30	Avg	1.744×10^{4}	3.637×10^{5}	3.412×10^{4}	8.683×10^{4}	3.448×10^{4}	5.245×10^{3}	2.685×10^{6}	6.528×10^{3}	5.743×10^{6}	2.428×10^{4}	6.098×10^{3}	**2.252×10^{3}**
		Min	8.598×10^{3}	1.847×10^{4}	3.640×10^{3}	1.368×10^{4}	2.547×10^{3}	1.693×10^{3}	3.185×10^{5}	1.537×10^{3}	1.019×10^{6}	1.478×10^{4}	1.703×10^{3}	1.508×10^{3}
	50	Avg	2.004×10^{4}	9.315×10^{6}	2.145×10^{7}	7.839×10^{4}	7.164×10^{4}	9.137×10^{3}	2.119×10^{8}	7.314×10^{3}	1.006×10^{8}	3.197×10^{4}	7.557×10^{3}	**5.907×10^{3}**
		Min	1.128×10^{4}	1.565×10^{4}	4.235×10^{4}	2.225×10^{4}	1.422×10^{4}	2.035×10^{3}	1.213×10^{8}	1.598×10^{3}	6.005×10^{7}	1.979×10^{4}	2.315×10^{3}	2.972×10^{3}
	100	Avg	2.449×10^{4}	9.478×10^{7}	1.045×10^{9}	2.527×10^{5}	9.337×10^{4}	2.651×10^{6}	2.409×10^{9}	2.975×10^{3}	5.122×10^{9}	4.998×10^{9}	6.568×10^{3}	**1.821×10^{3}**
		Min	1.274×10^{4}	5.864×10^{5}	1.058×10^{5}	2.549×10^{4}	1.223×10^{4}	3.473×10^{3}	1.423×10^{9}	1.621×10^{3}	1.096×10^{9}	1.070×10^{9}	3.039×10^{3}	1.522×10^{3}
F_{16}	30	Avg	2.908×10^{3}	2.287×10^{3}	2.995×10^{3}	3.519×10^{3}	2.807×10^{3}	2.581×10^{3}	3.628×10^{3}	2.435×10^{3}	3.456×10^{3}	3.700×10^{3}	2.334×10^{3}	**1.774×10^{3}**
		Min	2.538×10^{3}	1.744×10^{3}	2.487×10^{3}	2.728×10^{3}	2.095×10^{3}	2.009×10^{3}	3.221×10^{3}	1.854×10^{3}	2.949×10^{3}	2.867×10^{3}	1.949×10^{3}	1.602×10^{3}
	50	Avg	3.336×10^{3}	2.791×10^{3}	4.150×10^{3}	4.689×10^{3}	3.778×10^{3}	3.302×10^{3}	4.712×10^{3}	3.313×10^{3}	5.278×10^{3}	6.365×10^{3}	2.936×10^{3}	**2.003×10^{3}**
		Min	2.736×10^{3}	2.209×10^{3}	3.133×10^{3}	3.895×10^{3}	3.014×10^{3}	2.761×10^{3}	3.890×10^{3}	2.592×10^{3}	4.488×10^{3}	3.693×10^{3}	2.394×10^{3}	1.845×10^{3}
	100	Avg	6.038×10^{3}	5.610×10^{3}	8.085×10^{3}	9.811×10^{3}	6.869×10^{3}	6.627×10^{3}	1.213×10^{4}	5.397×10^{3}	1.224×10^{4}	1.873×10^{4}	5.155×10^{3}	**2.566×10^{3}**
		Min	5.126×10^{3}	4.748×10^{3}	6.389×10^{3}	7.513×10^{3}	4.978×10^{3}	4.601×10^{3}	9.757×10^{3}	3.740×10^{3}	1.047×10^{4}	1.409×10^{4}	4.009×10^{3}	1.851×10^{3}
F_{17}	30	Avg	2.253×10^{3}	1.956×10^{3}	2.411×10^{3}	2.520×10^{3}	2.315×10^{3}	2.050×10^{3}	2.488×10^{3}	1.941×10^{3}	2.595×10^{3}	2.601×10^{3}	1.937×10^{3}	**1.738×10^{3}**
		Min	1.884×10^{3}	1.777×10^{3}	1.975×10^{3}	1.931×10^{3}	1.942×10^{3}	1.805×10^{3}	2.223×10^{3}	1.718×10^{3}	2.277×10^{3}	2.085×10^{3}	1.771×10^{3}	1.703×10^{3}
	50	Avg	3.405×10^{3}	2.676×10^{3}	3.708×10^{3}	3.892×10^{3}	3.758×10^{3}	3.115×10^{3}	3.827×10^{3}	2.846×10^{3}	4.046×10^{3}	4.165×10^{3}	2.635×10^{3}	**1.931×10^{3}**
		Min	2.871×10^{3}	2.257×10^{3}	2.866×10^{3}	3.106×10^{3}	2.932×10^{3}	2.590×10^{3}	3.518×10^{3}	2.326×10^{3}	3.304×10^{3}	3.228×10^{3}	2.084×10^{3}	1.858×10^{3}
	100	Avg	5.589×10^{3}	4.439×10^{3}	7.668×10^{3}	7.212×10^{3}	6.345×10^{3}	5.366×10^{3}	1.919×10^{4}	4.515×10^{3}	1.341×10^{4}	3.461×10^{5}	4.401×10^{3}	**2.292×10^{3}**
		Min	4.266×10^{3}	3.338×10^{3}	5.623×10^{3}	5.421×10^{3}	4.935×10^{3}	3.832×10^{3}	9.150×10^{3}	3.706×10^{3}	9.608×10^{3}	1.666×10^{4}	3.410×10^{3}	1.868×10^{3}
F_{18}	30	Avg	4.488×10^{5}	6.631×10^{5}	3.177×10^{6}	2.408×10^{6}	1.734×10^{5}	4.761×10^{5}	2.212×10^{6}	6.722×10^{5}	1.276×10^{6}	6.751×10^{5}	9.249×10^{5}	**9.790×10^{4}**
		Min	5.229×10^{4}	8.000×10^{4}	3.737×10^{4}	1.933×10^{5}	3.793×10^{4}	3.856×10^{4}	3.218×10^{5}	5.547×10^{4}	4.340×10^{5}	1.206×10^{5}	9.952×10^{4}	5.073×10^{4}
	50	Avg	2.760×10^{6}	3.300×10^{6}	3.443×10^{6}	4.272×10^{6}	4.064×10^{5}	5.021×10^{6}	8.705×10^{6}	2.036×10^{6}	8.529×10^{6}	2.406×10^{7}	2.111×10^{6}	**1.126×10^{5}**
		Min	3.941×10^{5}	2.968×10^{5}	1.807×10^{5}	1.009×10^{6}	1.509×10^{5}	8.293×10^{5}	3.908×10^{6}	2.080×10^{5}	3.639×10^{6}	8.365×10^{5}	6.113×10^{5}	5.963×10^{4}
	100	Avg	2.777×10^{6}	4.158×10^{6}	1.162×10^{7}	2.020×10^{6}	8.326×10^{5}	2.307×10^{6}	2.039×10^{7}	1.049×10^{6}	1.063×10^{7}	3.147×10^{7}	3.743×10^{6}	**1.629×10^{5}**
		Min	1.197×10^{6}	7.431×10^{5}	4.881×10^{5}	8.476×10^{5}	3.782×10^{5}	6.200×10^{5}	1.197×10^{7}	2.595×10^{5}	5.745×10^{6}	9.728×10^{6}	1.203×10^{6}	1.177×10^{5}
F_{19}	30	Avg	1.127×10^{5}	2.913×10^{5}	4.071×10^{6}	2.647×10^{6}	3.223×10^{4}	1.816×10^{4}	8.962×10^{6}	4.640×10^{3}	4.874×10^{7}	1.068×10^{6}	9.241×10^{3}	**6.433×10^{3}**
		Min	5.515×10^{3}	9.466×10^{3}	2.093×10^{3}	1.744×10^{5}	2.168×10^{3}	2.460×10^{3}	4.803×10^{6}	2.110×10^{3}	2.507×10^{6}	8.696×10^{5}	1.968×10^{3}	1.946×10^{3}
	50	Avg	2.440×10^{5}	2.362×10^{6}	6.151×10^{6}	2.457×10^{6}	2.362×10^{4}	8.702×10^{4}	1.443×10^{8}	1.438×10^{4}	3.026×10^{8}	4.636×10^{5}	**1.433×10^{4}**	1.566×10^{4}
		Min	2.445×10^{4}	6.908×10^{4}	5.031×10^{3}	1.534×10^{5}	2.700×10^{3}	4.883×10^{3}	7.728×10^{7}	3.740×10^{3}	3.919×10^{7}	4.438×10^{5}	2.057×10^{3}	9.683×10^{3}
	100	Avg	5.676×10^{5}	1.003×10^{8}	3.561×10^{8}	1.529×10^{7}	7.032×10^{4}	5.510×10^{4}	2.661×10^{9}	2.871×10^{3}	2.968×10^{9}	4.646×10^{9}	9.029×10^{3}	**2.852×10^{3}**
		Min	8.460×10^{4}	2.250×10^{6}	2.761×10^{6}	5.273×10^{6}	1.223×10^{4}	2.334×10^{3}	1.245×10^{9}	2.008×10^{3}	7.255×10^{8}	1.529×10^{9}	2.774×10^{3}	1.974×10^{3}

70. Jia, H.; Sun, K.; Zhang, W.; Leng, X. An enhanced chimp optimization algorithm for continuous optimization domains. *Complex Intell. Syst.* **2021**, *7*, 1–18. [CrossRef]
71. Singh, T.; Saxena, N.; Khurana, M.; Singh, D.; Abdalla, M.; Alshazly, H. Data Clustering Using Moth-Flame Optimization Algorithm. *Sensors* **2021**, *21*, 4086. [CrossRef] [PubMed]
72. Shah, Y.A.; Habib, H.A.; Aadil, F.; Khan, M.F.; Maqsood, M.; Nawaz, T. CAMONET: Moth-Flame Optimization (MFO) Based Clustering Algorithm for VANETs. *IEEE Access* **2018**, *6*, 48611–48624. [CrossRef]
73. Kotary, D.K.; Nanda, S.J. Distributed robust data clustering in wireless sensor networks using diffusion moth flame optimization. *Eng. Appl. Artif. Intell.* **2020**, *87*, 103342. [CrossRef]
74. Fei, W.; Hexiang, B.; Deyu, L.; Jianjun, W. Energy-Efficient Clustering Algorithm in Underwater Sensor Networks Based on Fuzzy C Means and Moth-Flame Optimization Method. *IEEE Access* **2020**, *8*, 97474–97484. [CrossRef]
75. Ishtiaq, A.; Ahmed, S.; Khan, M.F.; Aadil, F.; Maqsood, M.; Khan, S. Intelligent clustering using moth flame optimizer for vehicular ad hoc networks. *Int. J. Distrib. Sens. Netw.* **2019**, *15*, 1550147718824460. [CrossRef]
76. Mittal, N. Moth Flame Optimization Based Energy Efficient Stable Clustered Routing Approach for Wireless Sensor Networks. *Wirel. Pers. Commun.* **2018**, *104*, 677–694. [CrossRef]
77. Nadimi-Shahraki, M.H.; Moeini, E.; Taghian, S.; Mirjalili, S. DMFO-CD: A Discrete Moth-Flame Optimization Algorithm for Community Detection. *Algorithms* **2021**, *14*, 314. [CrossRef]
78. Zawbaa, H.M.; Emary, E.; Parv, B.; Sharawi, M. Feature selection approach based on moth-flame optimization algorithm. In Proceedings of the 2016 IEEE Congress on Evolutionary Computation (CEC), Vancouver, BC, Canada, 24–29 July 2016; pp. 4612–4617.
79. Khurma, R.A.; Aljarah, I.; Sharieh, A. An Efficient Moth Flame Optimization Algorithm using Chaotic Maps for Feature Selection in the Medical Applications. In Proceedings of the ICPRAM, Valletta, Malta, 22–24 February 2020; pp. 175–182.
80. Wang, M.; Chen, H.; Yang, B.; Zhao, X.; Hu, L.; Cai, Z.; Huang, H.; Tong, C. Toward an optimal kernel extreme learning machine using a chaotic moth-flame optimization strategy with applications in medical diagnoses. *Neurocomputing* **2017**, *267*, 69–84. [CrossRef]
81. Hassanien, A.E.; Gaber, T.; Mokhtar, U.; Hefny, H. An improved moth flame optimization algorithm based on rough sets for tomato diseases detection. *Comput. Electron. Agric.* **2017**, *136*, 86–96. [CrossRef]
82. Ewees, A.A.; Sahlol, A.T.; Amasha, M.A. A bio-inspired moth-flame optimization algorithm for arabic handwritten letter recognition. In Proceedings of the 2017 International Conference on Control, Artificial Intelligence, Robotics & Optimization (ICCAIRO), Prague, Czech Republic, 20–22 May 2017; pp. 154–159.
83. Gupta, D.; Ahlawat, A.K.; Sharma, A.; Rodrigues, J.J.P.C. Feature selection and evaluation for software usability model using modified moth-flame optimization. *Computing* **2020**, *102*, 1503–1520. [CrossRef]
84. Elaziz, M.A.; Ewees, A.A.; Ibrahim, R.A.; Lu, S. Opposition-based moth-flame optimization improved by differential evolution for feature selection. *Math. Comput. Simul.* **2020**, *168*, 48–75. [CrossRef]
85. Nadimi-Shahraki, M.H.; Banaie-Dezfouli, M.; Zamani, H.; Taghian, S.; Mirjalili, S. B-MFO: A Binary Moth-Flame Optimization for Feature Selection from Medical Datasets. *Computers* **2021**, *10*, 136. [CrossRef]
86. Khan, M.A.; Sharif, M.; Akram, T.; Damaševičius, R.; Maskeliūnas, R. Skin Lesion Segmentation and Multiclass Classification Using Deep Learning Features and Improved Moth Flame Optimization. *Diagnostics* **2021**, *11*, 811. [CrossRef]
87. Nguyen, T.-T.; Wang, H.-J.; Dao, T.-K.; Pan, J.-S.; Ngo, T.-G.; Yu, J. A Scheme of Color Image Multithreshold Segmentation Based on Improved Moth-Flame Algorithm. *IEEE Access* **2020**, *8*, 174142–174159. [CrossRef]
88. Jaiswal, V.; Sharma, V.; Varma, S. MMFO: Modified moth flame optimization algorithm for region based RGB color image segmentation. *Int. J. Electr. Comput. Eng.* **2020**, *10*, 196. [CrossRef]
89. Said, S.; Mostafa, A.; Houssein, E.H.; Hassanien, A.E.; Hefny, H. Moth-flame Optimization Based Segmentation for MRI Liver Images. In Proceedings of the International Conference on Advanced Intelligent Systems and Informatics 2017, Cairo, Egypt, 9–11 September 2017; Advances in Intelligent Systems and Computing; Springer: Cham, Switzerland, 2018; pp. 320–330.
90. Jia, H.; Ma, J.; Song, W. Multilevel thresholding segmentation for color image using modified moth-flame optimization. *IEEE Access* **2019**, *7*, 44097–44134. [CrossRef]
91. Aziz, M.A.E.; Ewees, A.A.; Hassanien, A.E. Whale Optimization Algorithm and Moth-Flame Optimization for multilevel thresholding image segmentation. *Expert Syst. Appl.* **2017**, *83*, 242–256. [CrossRef]
92. Li, Z.; Zhou, Y.; Zhang, S.; Song, J. Lévy-flight moth-flame algorithm for function optimization and engineering design problems. *Math. Probl. Eng.* **2016**, *2016*, 1423930. [CrossRef]
93. Khalilpourazari, S.; Khalilpourazary, S. An efficient hybrid algorithm based on Water Cycle and Moth-Flame Optimization algorithms for solving numerical and constrained engineering optimization problems. *Soft Comput.* **2019**, *23*, 1699–1722. [CrossRef]
94. Hongwei, L.; Jianyong, L.; Liang, C.; Jingbo, B.; Yangyang, S.; Kai, L. Chaos-enhanced moth-flame optimization algorithm for global optimization. *J. Syst. Eng. Electron.* **2019**, *30*, 1144–1159.
95. Xu, Y.; Chen, H.; Heidari, A.A.; Luo, J.; Zhang, Q.; Zhao, X.; Li, C. An efficient chaotic mutative moth-flame-inspired optimizer for global optimization tasks. *Expert Syst. Appl.* **2019**, *129*, 135–155. [CrossRef]
96. Xu, Y.; Chen, H.; Luo, J.; Zhang, Q.; Jiao, S.; Zhang, X. Enhanced Moth-flame optimizer with mutation strategy for global optimization. *Inf. Sci.* **2019**, *492*, 181–203. [CrossRef]

97. Chen, C.; Wang, X.; Yu, H.; Wang, M.; Chen, H. Dealing with multi-modality using synthesis of Moth-flame optimizer with sine cosine mechanisms. *Math. Comput. Simul.* **2021**, *188*, 291–318. [CrossRef]
98. Li, Z.; Zeng, J.; Chen, Y.; Ma, G.; Liu, G. Death mechanism-based moth–flame optimization with improved flame generation mechanism for global optimization tasks. *Expert Syst. Appl.* **2021**, *183*, 115436. [CrossRef]
99. Hashim, F.A.; Houssein, E.H.; Mabrouk, M.S.; Al-Atabany, W.; Mirjalili, S. Henry gas solubility optimization: A novel physics-based algorithm. *Future Gener. Comput. Syst.* **2019**, *101*, 646–667. [CrossRef]
100. Abualigah, L.; Diabat, A.; Mirjalili, S.; Abd Elaziz, M.; Gandomi, A.H. The arithmetic optimization algorithm. *Comput. Methods Appl. Mech. Eng.* **2021**, *376*, 113609. [CrossRef]
101. Bhesdadiya, R.; Trivedi, I.N.; Jangir, P.; Kumar, A.; Jangir, N.; Totlani, R. A novel hybrid approach particle swarm optimizer with moth-flame optimizer algorithm. In *Advances in Computer and Computational Sciences*; Springer: Singapore, 2017; pp. 569–577.
102. Mustaffa, Z.; Sulaiman, M.H.; Ernawan, F.; Kamarulzaman, S.F. Hybrid least squares support vector machines for short term predictive analysis. In Proceedings of the 2017 3rd International Conference on Control, Automation and Robotics (ICCAR), Nagoya, Japan, 24–26 April 2017; pp. 571–574.
103. Sarma, A.; Bhutani, A.; Goel, L. Hybridization of moth flame optimization and gravitational search algorithm and its application to detection of food quality. In Proceedings of the 2017 Intelligent Systems Conference (IntelliSys), London, UK, 7–8 September 2017; pp. 52–60.
104. Rezk, H.; Ali, Z.M.; Abdalla, O.; Younis, O.; Gomaa, M.R.; Hashim, M. Hybrid moth-flame optimization algorithm and incremental conductance for tracking maximum power of solar PV/thermoelectric system under different conditions. *Mathematics* **2019**, *7*, 875. [CrossRef]
105. Ullah, I.; Hussain, S. Time-Constrained Nature-Inspired Optimization Algorithms for an Efficient Energy Management System in Smart Homes and Buildings. *Appl. Sci.* **2019**, *9*, 792. [CrossRef]
106. Abd Elaziz, M.; Yousri, D.; Mirjalili, S. A hybrid Harris hawks-moth-flame optimization algorithm including fractional-order chaos maps and evolutionary population dynamics. *Adv. Eng. Softw.* **2021**, *154*, 102973. [CrossRef]
107. Dang, M.P.; Le, H.G.; Chau, N.L.; Dao, T.-P. Optimization for a flexure hinge using an effective hybrid approach of fuzzy logic and moth-flame optimization algorithm. *Math. Probl. Eng.* **2021**, *2021*, 6622655. [CrossRef]
108. Apinantanakon, W.; Sunat, K. Omfo: A new opposition-based moth-flame optimization algorithm for solving unconstrained optimization problems. In Proceedings of the International Conference on Computing and Information Technology, Singapore, 27–29 December 2017; Springer: Cham, Switzerland, 2017; pp. 22–31.
109. Xu, L.; Li, Y.; Li, K.; Beng, G.H.; Jiang, Z.; Wang, C.; Liu, N. Enhanced moth-flame optimization based on cultural learning and Gaussian mutation. *J. Bionic Eng.* **2018**, *15*, 751–763. [CrossRef]
110. Li, W.K.; Wang, W.L.; Li, L. Optimization of water resources utilization by multi-objective moth-flame algorithm. *Water Resour. Manag.* **2018**, *32*, 3303–3316. [CrossRef]
111. Zhang, H.; Li, R.; Cai, Z.; Gu, Z.; Heidari, A.A.; Wang, M.; Chen, H.; Chen, M. Advanced orthogonal moth flame optimization with Broyden–Fletcher–Goldfarb–Shanno algorithm: Framework and real-world problems. *Expert Syst. Appl.* **2020**, *159*, 113617. [CrossRef]
112. Nadimi-Shahraki, M.H.; Fatahi, A.; Zamani, H.; Mirjalili, S.; Abualigah, L. An Improved Moth-Flame Optimization Algorithm with Adaptation Mechanism to Solve Numerical and Mechanical Engineering Problems. *Entropy* **2021**, *23*, 1637. [CrossRef]
113. Kaur, K.; Singh, U.; Salgotra, R. An enhanced moth flame optimization. *Neural Comput. Appl.* **2020**, *32*, 2315–2349. [CrossRef]
114. Pelusi, D.; Mascella, R.; Tallini, L.; Nayak, J.; Naik, B.; Deng, Y. An Improved Moth-Flame Optimization algorithm with hybrid search phase. *Knowl.-Based Syst.* **2020**, *191*, 105277. [CrossRef]
115. Li, Y.; Zhu, X.; Liu, J. An improved moth-flame optimization algorithm for engineering problems. *Symmetry* **2020**, *12*, 1234. [CrossRef]
116. Awad, N.; Ali, M.; Liang, J.; Qu, B.; Suganthan, P. *Problem Definitions and Evaluation Criteria for the cec 2017 Special Session and Competition on Single Objective Bound Constrained Real-Parameter Numerical Optimization*; Technical Report; Nanyang Technological University: Singapore, 2016.
117. Morrison, R.W. *Designing Evolutionary Algorithms for Dynamic Environments*; Springer Science & Business Media: Berlin/Heidelberg, Germany, 2004.
118. Derrac, J.; García, S.; Molina, D.; Herrera, F. A practical tutorial on the use of nonparametric statistical tests as a methodology for comparing evolutionary and swarm intelligence algorithms. *Swarm Evol. Comput.* **2011**, *1*, 3–18. [CrossRef]

Review

Scheduling by NSGA-II: Review and Bibliometric Analysis

Iman Rahimi [1], Amir H. Gandomi [1,*], Kalyanmoy Deb [2], Fang Chen [1] and Mohammad Reza Nikoo [3]

[1] Faculty of Engineering & Information Technology, University of Technology Sydney, Sydney, NSW 2007, Australia; iman83@gmail.com (I.R.); Fang.Chen@uts.edu.au (F.C.)
[2] Computional Optimization and Innovation (COIN) Laboratory, Michigan State University, East Lansing, MI 48824, USA; kdeb@egr.msu.edu
[3] Department of Civil and Architectural Engineering, Sultan Qaboos University, Muscat P.O. Box 50, Oman; nikoo@squ.edu.om
* Correspondence: Gandomi@uts.edu.au

Abstract: NSGA-II is an evolutionary multi-objective optimization algorithm that has been applied to a wide variety of search and optimization problems since its publication in 2000. This study presents a review and bibliometric analysis of numerous NSGA-II adaptations in addressing scheduling problems. This paper is divided into two parts. The first part discusses the main ideas of scheduling and different evolutionary computation methods for scheduling and provides a review of different scheduling problems, such as production and personnel scheduling. Moreover, a brief comparison of different evolutionary multi-objective optimization algorithms is provided, followed by a summary of state-of-the-art works on the application of NSGA-II in scheduling. The next part presents a detailed bibliometric analysis focusing on NSGA-II for scheduling applications obtained from the Scopus and Web of Science (WoS) databases based on keyword and network analyses that were conducted to identify the most interesting subject fields. Additionally, several criteria are recognized which may advise scholars to find key gaps in the field and develop new approaches in future works. The final sections present a summary and aims for future studies, along with conclusions and a discussion.

Keywords: NSGA-II; scheduling; multi-objective optimization; review; scientometric analysis

Citation: Rahimi, I.; Gandomi, A.H.; Deb, K.; Chen, F.; Nikoo, M.R. Scheduling by NSGA-II: Review and Bibliometric Analysis. *Processes* **2022**, *10*, 98. https://doi.org/10.3390/pr10010098

Academic Editors: Luis Puigjaner and Jie Zhang

Received: 22 October 2021
Accepted: 29 December 2021
Published: 4 January 2022

Publisher's Note: MDPI stays neutral with regard to jurisdictional claims in published maps and institutional affiliations.

1. Introduction

Non-Dominated Sorting Genetic Algorithm II (NSGA-II) [1] has been proposed as a powerful decision space exploration engine based on a genetic algorithm for solving multi-objective optimization problems. The NSGA-II algorithm has been applied to a wide variety of search and optimization problems since its publication in 2000.

Scheduling problems are dedicated to allocating tasks to resources. Two major schools of thought in relation to schedule generation are algorithmic and knowledge-based approaches [2]. The first approach is based on a mathematical formulation that includes objective function(s) and constraints, while the second approach is not easy to explain in an analytical format and is often used in cases where a feasible solution is sufficient. In addition, scheduling problems are generally known to be complex, large-scale, challenging, NP-hard, and involve several constraints [3,4].

Therefore, discovering efficient and low-cost procedures for use of the scheduling systems is significantly essential. Although numerous techniques have been proposed to solve the optimization problem mentioned above, there is still a crucial need for more suitable techniques. A viable method to manage these issues is to employ global optimization algorithms, including exact optimization methods (e.g., branch-and-bound and branch-and-cut) and, in some cases, evolutionary computation (EC) techniques [5–9]. EC techniques have been employed for large, complex real-world problems that cannot be solved using classical methods [10–12].

Another serious problem is that numerous objectives could be identified to optimize systems simultaneously. Hence, several objectives must usually be identified for optimizing a real-world scheduling problem. Furthermore, multi-objective optimization problems arise naturally in most disciplines, and solving them has been a challenging issue for researchers. Although a variety of techniques have been developed in operations research and other fields to address these problems, alternative approaches are urgently needed because of the complexities of their solutions [13–15]. Since EC methods are identified as the more effective methods to handle this limitation, they are suitable for solving multi-objective optimization problems (MOOPs) [12,16]. EC approaches repeatedly modify a population of individual solutions to find the optimal set of solutions to a problem. Additionally, multi-objective evolutionary algorithms are able to find a set of non-dominated solutions, known as Pareto solutions, in a single run within an ideal time [12,17,18]. Among the EC approaches, genetic algorithm (GA)-based solution methods are quickly gaining popularity due to their dependence on the population and, therefore, are suitable for solving MOOPs.

Non-dominated sorting is a technique used to assign solutions in a population to different Pareto fronts according to their dominance relationships. Because individuals of the population in the first front have the maximum fitness value, they can obtain more copies [1,12,19]. The NGA-II [1,20] is a well-known evolutionary computation technique that has been used widely by researchers, with more than 40,000 citations as of April 2021. Owing to its lower computational complexity, elitism, and parameterless nature [20–23], it has been applied to a wide variety of search and optimization problems since its introduction. The NSGA-II algorithm creates a population of individuals, ranks and sorts each individual based on the nondomination level, and then performs crowding distance sorting to keep the population diverse [1]. This paper presents a review of the application of NSGA-II in scheduling problems.

To better understand the research field in this study and provide new insights from publications, the information provided in this work attempts to answer the following questions:

- What is the basic concept of scheduling, and why is NSGA-II important (Section 2)?
- What is the contribution of NSGA-II in scheduling (Section 3.2)?
- What are the different types of scheduling? Which fields of scheduling are the most important (Section 2)?
- What are the most important problems in scheduling, how do researchers tackle them, and what do researchers find from their experiments (Sections 2 and 3.2)?
- What are the main topics and keywords regarding NSGA-II and scheduling problems (Section 4)?
- Which journals have the most contributions in the field? Who are the best researchers in the area, and what are their respective countries of origin (Section 4)?
- What are the current gaps and future trajectories in scheduling (Sections 5 and 6)?

After a brief introduction of different scheduling problems, scheduling algorithms are introduced. A comparison of algorithms in both single-objective and multi-objective scheduling problems is addressed, followed by introducing the application of NSGA-II in scheduling problems. Moreover, scientometric analysis is conducted in the field. The last section provides a summary and future studies.

The research procedure in this work was divided into five stages (Figure 1). In the first stage, documents were gathered from the Scopus (https://www.scopus.com/ 31 December 2020) and WoS (https://clarivate.com/products/web-of-science/, accessed on 31 December 2020) databases. Before initiating the search in the databases, special keywords, namely "NSGA-II" AND "scheduling", were searched for in titles, abstracts, and keywords to identify related articles. First, the authors filtered the documents with the special keywords to find the results, such as the type of objective function, problem statement, and solution approaches. Second, in some special cases where the research methodology using the title, abstract, and keywords did not help, the content of the papers was reviewed.

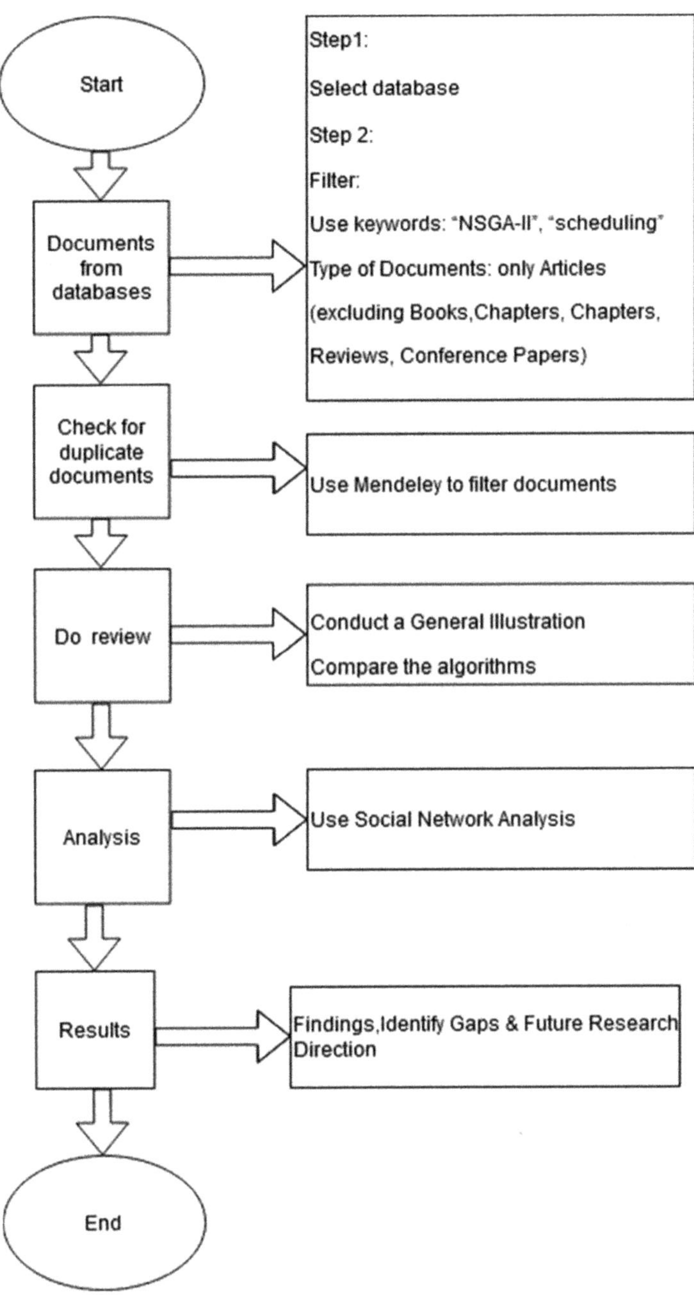

Figure 1. Research methodology.

It is worthy to note that this work reviewed only the research article type, excluding books, book chapters, reviews, conference papers, and short letters, and 683 and 875 published articles (between 2000 and 2020) were extracted from WOS (Supplementary Material A) and Scopus (Supplementary Material B), respectively. Since some of the articles were duplicates, they were identified and removed from the library in stage 2 using Mendeley as a powerful reference manager. In addition, some research questions for this study were designed in stage 2. A comprehensive review was initiated in stage 3 with a general illustration of the basic concepts of scheduling and comparison of the algorithms. In stage 4, social network analysis was performed to provide a scientometric analysis of the documents using VOSviewer 1.6.17 and CitNetExplorer 1.0.0 [24,25], which have been identified as powerful tools for scientometric analysis. Stage 4 required several steps, including co-occurrence, co-authorship, citation, bibliographic coupling, and citation network analyses. In the last stage, the results were obtained to formulate a discussion to answer the proposed research questions. In stage 5, the findings were prepared, important gaps were identified, and future research directions were determined.

The remainder of the paper is organized as follows. Section 2 gives an overview of the scheduling. Section 3 discusses the scheduling algorithms, the solution methods based on the genetic algorithm in scheduling, and state-of-the-art works on the application of NSGA-II in scheduling. Section 4 presents a detailed scientometric analysis in the field. Finally, a summary and suggested future studies are given in Section 5, followed by concluding remarks and a discussion in Section 6.

2. Overview of Scheduling

The following subsections provide an overview of the different aspects of scheduling in manufacturing and services.

2.1. Scheduling

Scheduling and sequencing are the processes of arranging and optimizing the manufacturing and service activities that play an important role in industries [3,26]. Firms use backward and forward scheduling to allocate plants and resources, plan production processes, and purchase materials [27–29]. In addition, the benefits of production scheduling include the following: inventory reduction, leveling [30–32], increased production efficiency [33–35], accurate delivery date quotes [36–38], and real-time information [39–43]. "Manufacturing model" specifies the machine(s) or resource configuration used in the production process. Classification of scheduling in manufacturing was built over the last few decades, and it is proven and applied in defining the complexity of a scheduling problem. Since the mathematical model is related to the machine configuration, the system uses the machine configuration instead of the industry type for categorizing problems [44,45]. Table 1 presents a classification of different models.

2.1.1. Scheduling in Manufacturing

In industry, each order should be converted into a list of operations that the organization must carry out. These operations should be handled by different machines and are based on certain sequences. It is pertinent to note that the provided schedule of the organizations helps to optimize the strategic usage of resources, forecasting of demands, and resource requirements. Single-machine scheduling or single-resource scheduling is an optimization problem in computer science and operations research. We are given n jobs of varying processing times, which need to be scheduled on a single machine in a way that optimizes a certain objective. Parallel machine scheduling (PMS) is for scheduling jobs processed on a series of machines with the same function with the optimized objective.

In a general job scheduling problem, we are given n jobs of varying processing times, which need to be scheduled on m machines with varying processing power, while trying to minimize the makespan (i.e., the total length of the schedule). Flow-shop scheduling is a special case of job-shop scheduling where there is a strict order of all operations to be performed on all jobs. Flow-shop scheduling may apply to production facilities for computing designs as well.

Table 1. Classification of different scheduling models.

Manufacturing Model	Model Type
Single Parallel Machines Job-Shop Flow- or Open-Shop Flexible Manufacturing Lot Scheduling System Project Scheduling	Linear Programing Mixed-Integer Programming Mixed-Integer Quadratic Programming Mixed-Integer Non-Linear Programming Queuing Techniques and Simulation
Objective Function	**Constraints**
Economic-Related Objective	**Economic-Related Constraints**
Minimize Makespan Minimize or Maximize Tardiness Minimize Electricity Cost Minimize Labor Cost Minimize Inventory Cost, etc.	Makespan Equation Makespan Value Limitation Tardiness Equation Tardiness Value Limitation Amount of Demand
Environment-Related Objective	Total Energy Cost
Minimize Total Energy Consumption Minimize Peak Power Minimize Carbon Emissions Minimize Squatted Deviation	Energy Cost in Specific Mode Electricity Price Revenue from Power Sold Labor Cost Equation, etc.
Maximize Utilization	**Environment-Related Constraints**
Minimize Water Consumption Maximize Total Availability System, etc.	Power's Peak Constraint Total Energy Consumption
Social-Related Objective	Energy Consumption in Specific Mode
Minimize Noise Level	Total Power Supply Capacity Limitation Duration of Initiatives Carbon Emissions Value Limitation Carbon Emissions Equation Amount of Water Water Quality Class Function Cleaning Cost Amount of Water Discharge Amount of Contaminant Waste Water and Effluent Limitation, etc.
	Social-Related Constraints
	Recovery Time Ergonomic Time Value Limitation, etc.

Machine Scheduling

This type of scheduling includes single-machine, parallel-machine, multi-stage flow-shop, multi-stage flexible (hybrid) flow-shop, multi-stage assembly flow-shop, job-shop, flexible job-shop, or open-shop flow-shop, job-shop, and open-shop [46–50]. For example, there are several objectives pertaining to job-shop scheduling problems, including maximizing completion time (C_{max}), total flow time (C_{total}), machine workload (W_{max}), total machine workload (W_{total}), and minimizing earliness or tardiness (E/T).

Flexible Manufacturing

A flexible manufacturing system (FMS) is a production approach which is designed to easily adjust to changes in the type and quantity of the product being produced. As a result, flexible manufacturing can be an important element of a make-to-order strategy that allows customers to customize the products [51–54].

Lot Scheduling System

This type of scheduling is suitable for tactical and strategic processes. Unlike the previous three classes, the production and demand processes are continuous. The objective functions of lot scheduling include minimizing inventory and cost [55–57].

2.1.2. Personnel Scheduling

In personnel scheduling, a good schedule should satisfy management and increase the time an employee stays with an employer [1]. All problems are originally divided into static and dynamic categories. Static scheduling has a structure that does not change over time. An example could be a 3-month flight schedule chart at an airport. Dynamic scheduling often has a variable schedule structure. There are several scheduling classifications in the literature. As an example, Figure 2 presents a classification of personnel scheduling in service system scheduling. An example is given in the resource schedules [58–101].

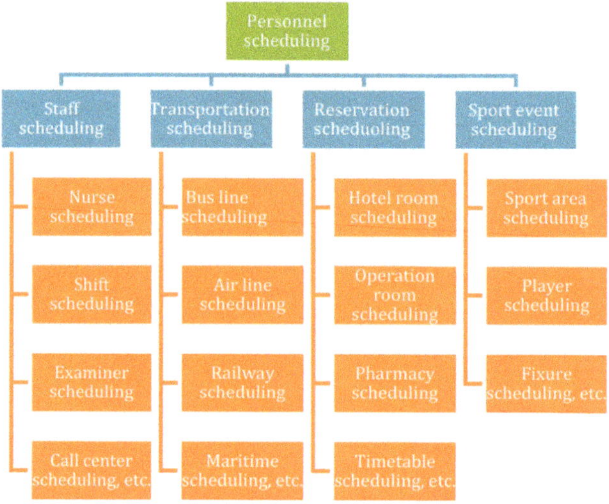

Figure 2. A classification of personnel scheduling problems.

3. Scheduling Algorithms

This section and the following subsections aim to justify the importance of NSGA-II and compare it with other evolutionary algorithms statistically and briefly.

Several optimization methods have been addressed in the literature to solve scheduling problems in addition to different classifications for solving optimization problems, namely the exact and approximate approaches (Figure 3). Exact methods include the efficient rule approach [49], mathematical programming approach [102], and branch-and-bound method [103,104]. Approximate methods pertain to constructive methods [105–107], artificial intelligence methods [108], local search methods [109], and metaheuristic approaches [110,111]. While exact methods are typically expensive in terms of computing time and often result in poor quality solutions, metaheuristic approaches produce alternative optimal solutions in a single run [112]. Most exact solution approaches convert MOOPs into a single optimization problem, while metaheuristic methods solve MOO problems without this conversion. Some metaheuristics incorporate certain mathematical methods [113], and others are suitable for solving global optimization problems [114].

Exact optimization methods

Enumerative methods
(Mixed)(Integer) Linear programming
Decomposition methods
Branch and bound
(Augmented)Lagrangian relaxation, etc.

Approximation algorithms

Priority dispatching rule
Insertion algorithm
Bottleneck based heuristics
Constraint satisfaction
Metaheuristics, etc.

Figure 3. Proposed solution approaches used in scheduling problems.

For example, the authors of [115] proposed an idea and scheduling for a flexible job-shop (FJS) based on a hierarchical approach considering multiple performance objectives. A genetic algorithm for generating robust solutions for flexible job-shop schedules was introduced in [116].

The authors of [117] presented a two-job-shop scheduling problem with unrelated machines and solved it using the classical geometric approach. A hybrid algorithm based on swarm optimization and simulated annealing for solving multi-objective flexible job-shop scheduling problems was introduced in [118]. The authors of [119] presented a mixed-integer nonlinear program for solving common cycle economic lot scheduling in flexible job-shops. The latter study considered a combination of job-shops and parallel machines, and the authors suggested an efficient enumeration method for solving the mentioned problem. An integer linear programming model for flexible job-shop scheduling for the jobs that were on a make-to-order basis was proposed in [120]. The authors of [121] used a genetic algorithm approach for solving FJS scheduling under resource constraints. A Tabu search approach for flexible job-shop scheduling by minimizing C_{max} was proposed in [122]. The authors of [123] established evolving dispatching rules for solving FJS scheduling using applied genetic programming. In total, more than 29,688 articles have been published in the area of optimization of scheduling problems (since 2000). Among the published articles, the genetic algorithm owns the most contributions (just above 26%) for solving scheduling problems, followed by particle swarm optimization (just above 9%), simulated annealing (6.4%), ant colony optimization (4.09%), and then tau search (4.47%) (Figure 4).

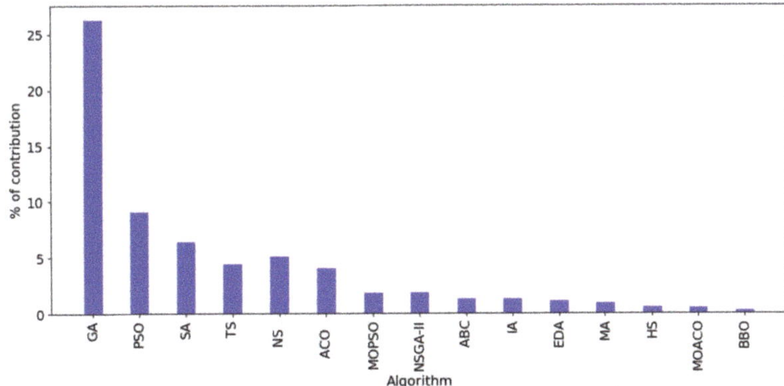

Figure 4. Contribution of different metaheuristic algorithms applied to scheduling problems.

3.1. Genetic Algorithm (GA)-Based Solution Methods

Since genetic algorithms are based on the population, they are suitable for solving MOO problems. The most famous MOO algorithms based on genetic algorithms are as follows:

- VEGA (Vector-Evaluated Genetic Algorithm) [124]
- MOGA (Multi-Objective Genetic Algorithm) [125]
- WBGA (Weighted Based Genetic Algorithm) [126]
- RWGA (Random Weighted Genetic Algorithm) [127]
- NSGA (Non-Dominated Sorted Genetic Algorithm) [128]
- NSGA-II (Fast Non-Dominated Sorted Genetic Algorithm) [20]
- RDGA (Rank Density-Based Genetic Algorithm) [129]
- NPGA (Niched Pareto Genetic Algorithm) [130,131]
- DMOEA (Dynamic Multi-Objective Evolutionary Algorithm) [132]

Table 2 presents a comparison of the above-mentioned genetic algorithms based on three criteria: elitism, diversity, and fitness function.

Table 2. Comparison of different GA-based solution methods.

Algorithm	Fitness Function	Diversity	Elitism	Strengths	Weakness
VEGA [133–136]	Select subpopulation using an objective function	No	No	Easy to code	Fast convergence to an objective function
MOGA [137–139]	Pareto ranking	Using fitness function	No	Extension of single objective	Slow convergence and dependency on niche size parameter
WBGA [140]	Average normalized weighted objective function	Identifying weights	No	Extension of single objective	Difficulty in nonconvex space
RWGA [141,142]	Average normalized weighted objective function	Assign weight randomly	Yes	Easy to code	Difficulty in nonconvex space
RDGA [143]	Ranking based and reducing problem	Non-concentration based on cells	Yes	Updated cells	Difficulty in run
NPGA [144–148]	No	Niche count	No	Easy tournament selection	Dependency on niche size parameter
DMOEA [149]	Ranking based on cells	Adjusting density of cells	Yes	Updated cells	Difficulty in run
NSGA [150–154]	Ranking based on non-dominated solutions	Using fitness function	No	Fast convergence	Dependency on niche size parameter
NSGA-II [22,155–159]	Ranking based on non-dominated solutions	Crowding distance	Yes	Uses non-dominated sorting, crowding distance, and elitist techniques	Crowding distance performs only in objective functions

3.2. NSGA-II

Most evolutionary multi-objective optimization (EMO) algorithms possess the following difficulties:

- Computational cost in non-dominated sorting increases significantly when the population increases;
- Lack of elitism reduces the algorithm's performance and inhibits individuals with good fitness values in different generations;

- Difficulty in the parameter settings largely affects the performance of the majority of evolutionary algorithms.

To alleviate these difficulties, NSGA-II was proposed in 2000 [19] and has become one of the most popular EMO algorithms in use to date, along with multi-objective particle swarm optimization (MOPSO) and multi-objective ant colony optimization (MOACO). Figure 5 shows the trend of published articles considering the contributions of these algorithms. Since 2014, NSGA-II has been the most studied algorithm in scheduling, followed by MOPSO then MOACO.

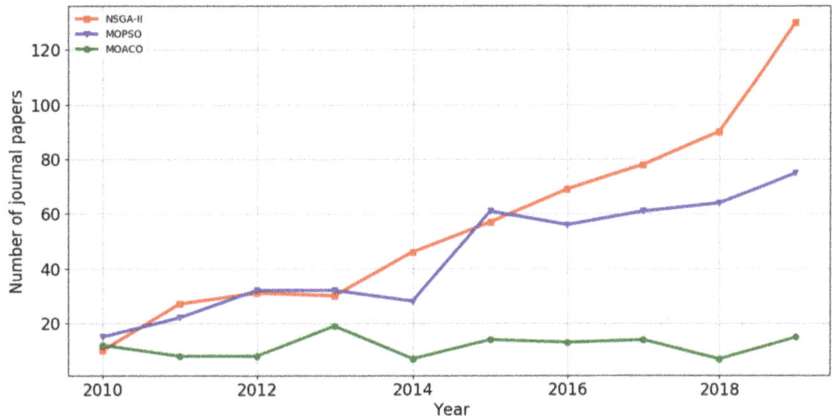

Figure 5. Publication counts of the three most-popular EMO algorithms from 2010 to 2019.

The authors of [20] proposed NSGA-II as a revised version of the NSGA [128] that has lower computational complexity, is parameterless, and possesses elitism [20]. NSGA-II has been applied in many different fields of study by numerous researchers [22,160–164]. Figure 6 presents the total citations of original NSGA-II papers over the years and indicates that NSGA-II has been considered by numerous researchers.

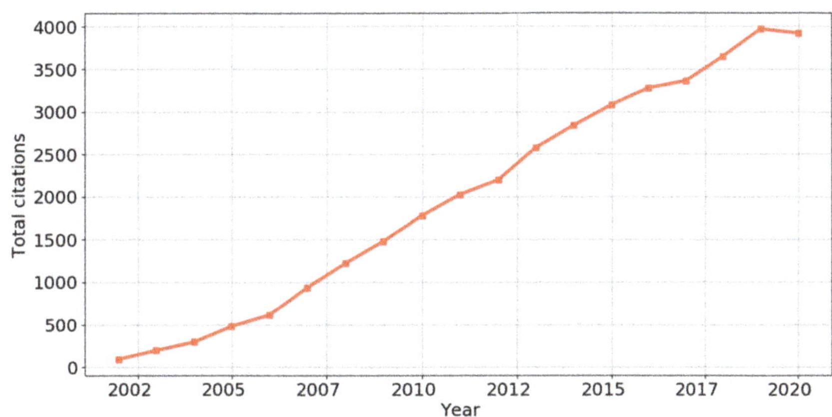

Figure 6. Total citations of original NSGA-II papers over the years.

Tables 3–8 present a summary of the literature review in the scheduling field, where the predominant areas include scheduling problems for job-shop scheduling, routing, satellites, projects, weapon selection, forest planning, and machinery. It is noteworthy to mention that some studies have compared NSGA-II with other well-known evolutionary algorithms, such as MOPSO, Tabu search, and the memetic algorithm. In addition, some authors have tried to improve the original version of NSGA-II and expand it for a specific problem. Optimizing the makespan, machining cost, and idle time are among the top objective functions in the published documents. Multi-objective constrained optimization was also found to be an interesting area for researchers.

Table 3. Summary of the literature review on NSGA-II applications in different fields, along with methodology and results.

Source	Problem	Objective	Methodology	Results and Findings
[165]	Weapon selection and planning problem	Optimizing net present value (NPV) and effectiveness	An MOEA based on NSGA-II is employed	The proposed measures are able to adapt to dynamic changes.
[166]	Allocation problem	Integrating MOMO process and Monte Carlo simulation technique.	Integrating MOMO, NSGA-II, and Tabu search	The MOMO technique possesses a better performance of seeking global optimum than other proposed methods.
[167]	Satellite scheduling problem	Proposing a multi-objective optimization method to solve the mentioned problem	Designing a decomposition method. Expressing a multi-objective integer-programming model. Designing multi-objective genetic algorithm NSGA-II.	The applicability of the proposed method under different situations has been proven.
[168]	High-dose rate brachytherapy planning	Determining an appropriate schedule of a radiation source	Four different MOEAs have been employed	Results present that MO-RV-GOMEA is the best performing MOEA.
[169]	FJS problem under mixed work calendars	Proposing two key technologies, namely time reckoning and sequential scheduling	Designing NSGA-II with an elite strategy	The suggested technique can gain an effective Pareto set within an acceptable time.

Table 4. Summary of the literature review on NSGA-II applications in different fields, along with methodology and results.

Source	Problem	Objective	Methodology	Results and Findings
[170]	Reliability in Cyber-Physical Systems (CPS) components	Designing and verifying CPS using multi-objective evolutionary optimization.	Using three scheduling methods: fixed priority, earliest deadline first, and deadline monotonic.	The results show that the proposed approach can be used to design and validate CPS for performance and verify timing guarantees.
[171]	Job-shop scheduling problem	Minimizing the mean weighted completion time and the sum of the weighted tardiness costs.	Proposing a new integer linear programming. Modifying PSO and comparing with NSGA-II.	The results depict that the proposed PSO outperforms NSGA-II.
[172]	Multi-objective unreliable unbalanced production lines	Maximizes the throughput rate and minimizes the total buffer capacities and cost.	Proposing DOE and RSM along with NRGA and NSGA-II.	The proposed system could be applied to a large-scale production line.

Table 5. Summary of the literature review on NSGA-II applications in different fields, along with methodology and results.

Source	Problem	Objective	Methodology	Results and Findings
[173]	Multi-objective traveling salesman problem	Improving a GA-based algorithm, namely Physarum-inspired computational model (PCM).	Using the hill-climbing algorithm to improve the proposed method.	Findings show that the proposed method has a better performance compared with the other MOTSP.
[174]	Project scheduling problem	Proposing a robust project scheduling.	Two-stage multi-objective buffer allocation approach.	The results indicate that the obtained buffered schedule reduces the cost of disruptions.
[175]	Process planning and FJS scheduling.	Makespan, critical machine workload, and machine total workload.	Integration of WGA and NSGA-II.	The proposed algorithm outperforms the exact solutions.

Table 6. Summary of the literature review on NSGA-II application in different fields, along with methodology and results.

Source	Problem	Objective	Methodology	Results and Findings
[176]	Generator scheduling considering environmental and economic issues.	Optimal generation scheduling.	Two-phase approach (hourly and 24-h scheduling)	Effectiveness of the proposed approach has been approved.
[177]	Multi-objective spatial forest planning.	Maximizing timber volume and minimizing sediment level.	Spatial NSGA-II approach	The results show that the proposed method has better performance for both constrained and unconstrained problems.
[178]	Resource allocation problem in a hospital.	Daily scheduling for residents or patients in a hospital.	Using variable neighborhood search, scatter search, and NSGA-II	Able to find efficient solutions.
[59]	Nurse scheduling problem considering human factors.	Minimizing the total cost of staffing as well as the sum of incompatibility and maximizing the satisfaction.	Keshtel algorithm, NSGA-II, and Tabu search.	Effectiveness of the proposed methods is approved.

Table 7. Summary of the literature review on NSGA-II applications in different fields, along with methodology and results.

Source	Problem	Objective	Methodology	Results and Findings
[179]	Process planning and scheduling	Optimizing the makespan, machine workload, and the total workload of machines.	Multi-objective memetic algorithm.	The results compared with NSGA-II show that the proposed algorithm has better performance.
[180]	Scheduling of locks and transshipment problem	Optimizing water–land transshipment co-scheduling.	Hybrid heuristic method using binary NSGA-II.	The feasibility and the superiority of the model have been verified.
[181]	Integration of process planning and scheduling	Minimizing of makespan, machining cost, and idle time.	Improved version of NSGA-II.	Results provide optimal and robust solutions.
[182]	Sudden drinking water contamination incident	Minimizing the volume of contaminated water and the operational costs.	Integration of NSGA-II and EPANET simulation model.	The validity of the model has been approved by two water distribution networks.

Table 8. Summary of the literature review on NSGA-II applications in different fields, along with methodology and results.

Source	Problem	Objective	Methodology	Results and Findings
[183]	Single machine scheduling with controllable processing times.	Developing a new multi-objective discrete backtracking search algorithm.	Through adaptive selection scheme and total cost reduction strategy.	The performance of the proposed method compared with other algorithms was validated.
[184]	Reentrant hybrid flow-shop scheduling.	Optimizing of makespan and total tardiness.	Genetic algorithms with Minkowski distance-based crossover operator.	The results show that NSGA-II outperformed in terms of convergence, diversity, and the dominance of solution.
[185]	Sustainable ship routing and scheduling.	Estimating the total fuel consumed and carbon emission from each vessel as well as improving the service level of the port.	Mixed-integer nonlinear programming using NSGA-II and MOPSO.	The robustness of the model has been approved by experimental results and comparative, and sensitive analysis.

Less than 10% (9.51%) of the papers published on NSGA-II in scheduling have addressed uncertainty. Among the above-mentioned papers, the power system owns the most contributions in the field at 32%, followed by project scheduling (13%), resource allocation (8%), and then job-shop scheduling (8%) (shown in Figure 7).

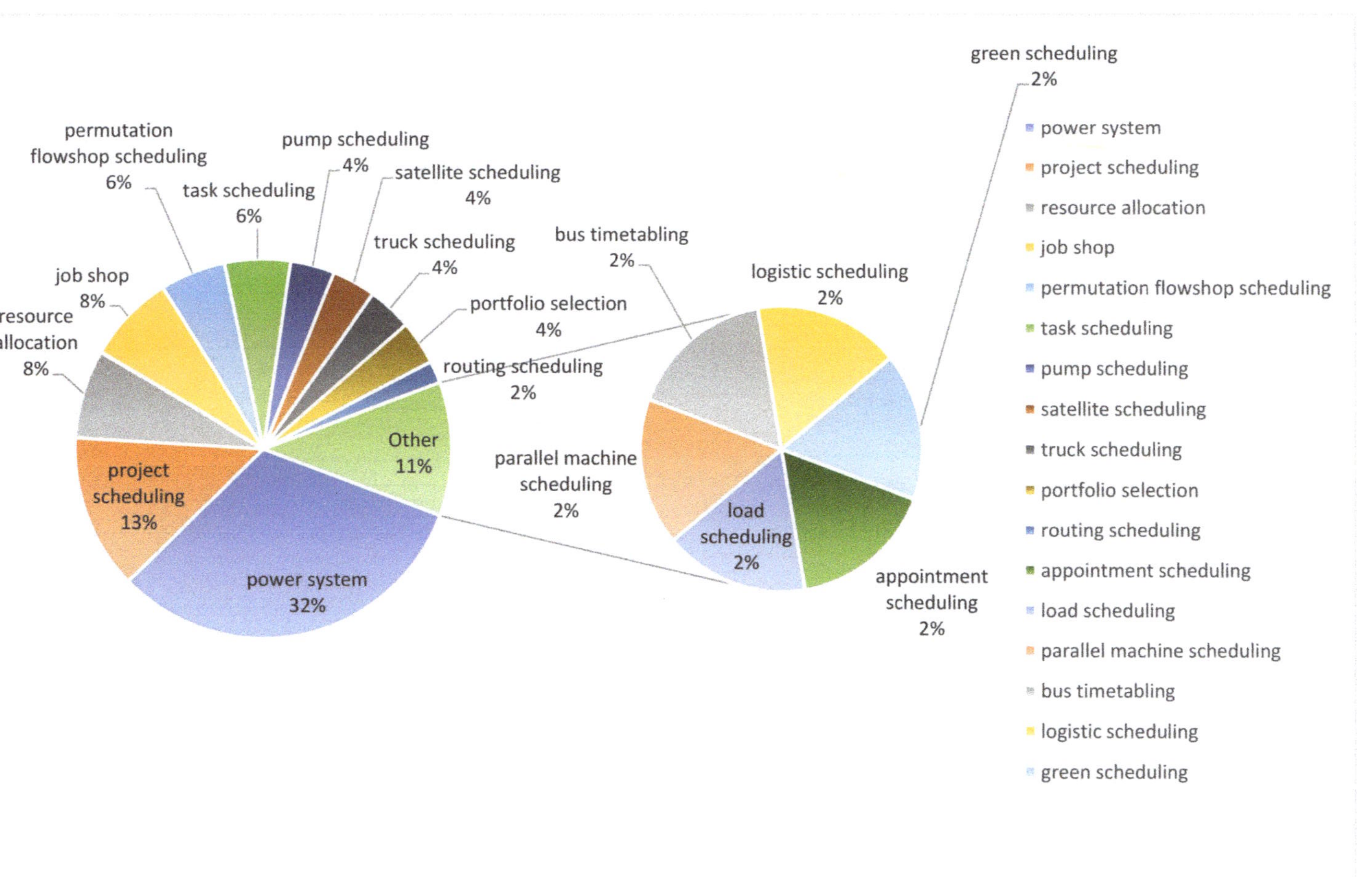

Figure 7. Different uncertainty scheduling problems that have been solved by NSGA-II.

4. Scientometric Analysis

Scientometric analysis is the field of study that scientifically measures and analyzes the literature [186]. Bibliometrics is the most famous field of scientometrics that uses statistics to analyze and measure the impacts of books, research articles, conference papers, etc. [187]. Recently, this field of analysis has attracted much attention from researchers and has been used in various literature review fields [4,188–192]. To achieve this aim, VOSviewer 1.6.17 [24] and CitNetExplorer 1.0.0 [25] were employed in this work. The following subsections provide new insights into scientometric analysis in the field of scheduling.

4.1. Statistics Based on Document Types

Among the document types, including articles, proceedings papers, reviews, and other items indexed by WoS, a total of 683 publications on scheduling and NSGA-II were found (Table 9). From the search, articles were the most popular document type with a total of 462 (67.64% of 683 documents) and 2.77 authors per publication (APP). Additionally, reviews as a document type had the highest CPP 2020 of 31, followed by articles (18.97). Moreover, there was a significant difference between the TC 2020 article and that of the proceedings paper. Figure 8 presents the distribution of documents based on different types, according to WoS. It is clear from the figure that proceedings papers had the greatest contributions before 2010, followed by articles. However, since 2010, articles had the most contributions in the field.

Table 9. Citation analysis based on document type.

Document type	TP	%	AU	APP	TC2020	CPP2020
Article	462	67.64	1282	2.77	8766	18.97
Proceedings paper	231	33.82	652	2.82	1126	4.87
Review	5	0.73	15	3.0	155	31
Other items	15	2.19	154	10.26	269	17.93

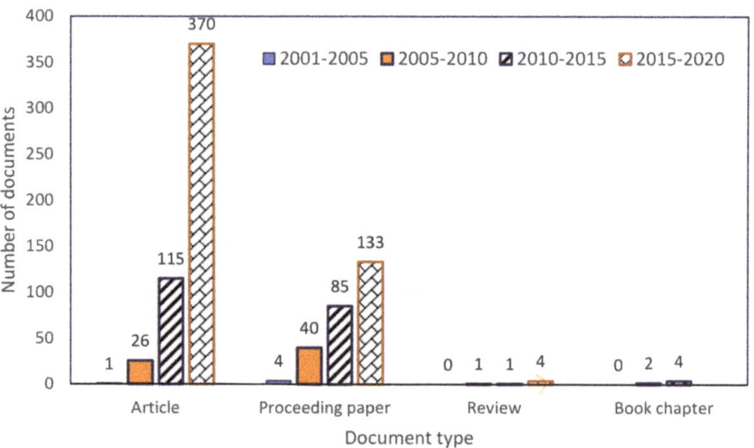

Figure 8. Type of research outputs.

TP, AU, APP, TC2020, and CPP2020 refer to the total number of articles, total number of authors, total number of authors for each publication, total citations from WoS since the publication year to the end of 2020, and total citations for each paper, respectively. Other items include early access and letters.

4.2. Keyword Analysis

Figure 9 presents the total citations per year for the published documents (NSGA-II in scheduling). As an overall trend, it is clear that the sum of the number of times articles were cited increased gradually until the end of 2012, and then the trend increased sharply up to 2020. Figure 10 presents a treemap visualization of the different categories found by WoS. Accordingly, computer science artificial intelligence (#86), operation research management science (#86), computer science interdiscipline applications (#82), industrial engineering (#71), and manufacturing engineering (#65) were among the top categories, while computer science information systems (#28), computer science theory methods (#29), and automation control systems (#33) contributed the least in the field.

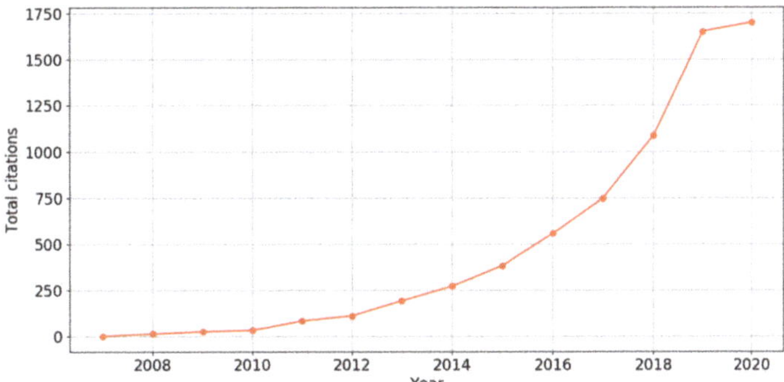

Figure 9. Total citations per year including percentage change (NSGA-II in scheduling).

Figure 10. Treemap visualization of different categories (database: WoS https://clarivate.com/products/web-of-science/, accessed on 31 December 2020).

4.3. Network Visualization

The keywords indicate the basic parts of a certain field of research and can offer insight into the organization and knowledge provided in the articles. Figures 10 and 11 depict the overlay visualization co-occurrence analyses via a network map based on the Scopus and WoS databases, respectively. In Figure 11, "scheduling", "optimization", "NSGA-II", "multi-objective optimization", "multi-objective genetic algorithm", "Pareto-optimal", "makespan", and "stochastic models" were identified as the top keywords in

Scopus. Figure 12 reveals that "genetic algorithm", "algorithm", "multi-objective genetic algorithm", "design", "cost", "parallel machines", "task analysis", and "operations" were the most important keywords in WoS. The color of each circle represents the identified cluster, and the size of each circle illustrates the importance of the keywords; in other words, the keywords with larger circles were used more than others. The green and yellow colors show the keywords that were used recently, while the dark blue color indicates those that were used earlier (around 2012).

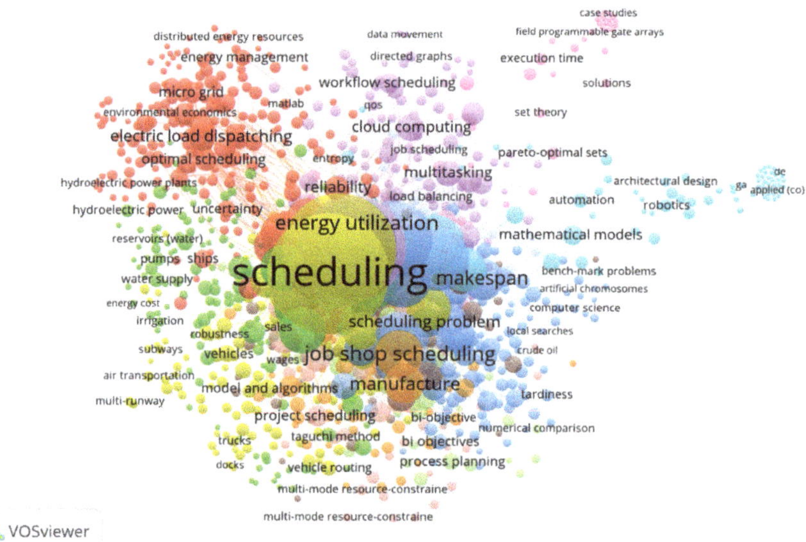

Figure 11. Overlay visualization occurrences (database: Scopus www.scopus.com, accessed on 31 December 2020).

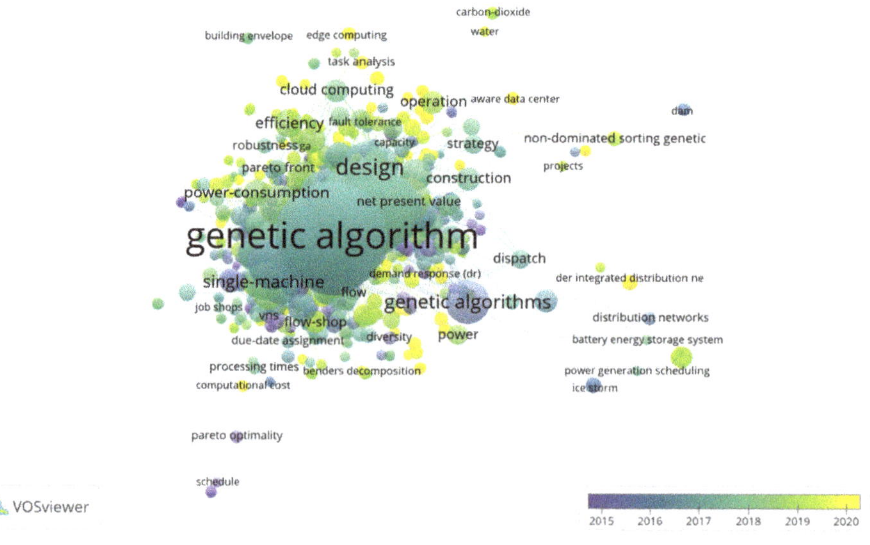

Figure 12. Overlay visualization occurrences (database: WoS https://clarivate.com accessed on 31 December 2020).

Tables 10 and 11 present the top 10 keywords of 1-word, 2-word, and 3-word lengths extracted from WoS and Scopus, respectively. NSGA-II, scheduling, and makespan were the top three one-word-long keywords for both WoS and Scopus. Multi-objective optimization, genetic algorithm, and multi-objective were the top three two-word-long keywords in WoS, while preventive maintenance, NSGA-II algorithm, and project scheduling were the top three two-word-long keywords in Scopus. In WoS, multi-objective genetic algorithm and particle swarm optimization were the top two three-word-long keywords, while EMO algorithm and non-dominated sorting were the top two three-word-long keywords in Scopus.

Table 10. The top 10 keywords of 1-word, 2-word, and 3-word lengths (WoS https://clarivate.com, accessed on 31 December 2020).

1-Word		2-Word		3-Word	
Keyword	**Frequency**	**Keyword**	**Frequency**	**Keyword**	**Frequency**
NSGA-II	76	Multi-objective optimization	86	Multi-objective genetic algorithm	9
Scheduling	38	Multi-objective	19	Particle swarm optimization	6
Makespan	22	Genetic algorithms	17	Unrelated parallel machine	3
Optimization	10	Energy consumption	13	Differential evolution algorithm	3
Reliability	9	Production scheduling	10	Single machine scheduling	3
Uncertainty	8	Cloud computing	9	Flexible job-shop	3
Microgrid	6	Project scheduling	8	Grey wolf optimizer	3
Metaheuristics	5	Preventive maintenance	8	Job-shop scheduling	3
Tardiness	4	Memetic algorithm	7	Just-in-time	3
Heuristic	3	Dynamic scheduling	6	Charge-discharge scheduling	1

Table 11. The top 10 keywords of 1-word, 2-word, and 3-word lengths (Scopus www.scopus.com, accessed on 31 December 2020).

1-Word		2-Word		3-Word	
Keyword	**Frequency**	**Keyword**	**Frequency**	**Keyword**	**Frequency**
NSGA-II	100	Preventive maintenance	10	Multi-objective evolutionary algorithm	13
Scheduling	54	NSGA-II algorithm	10	Particle swarm optimization	7
Multi-objective	32	Project scheduling	10	Non-dominated sorting	7
Makespan	28	Evolutionary algorithm	9	Ant colony optimization	6
Reliability	9	Multi-objective scheduling	8	Variable neighborhood search	6
Optimization	9	Optimal scheduling	7	Energy efficient scheduling	6
Microgrid	9	Task scheduling	7	Hybrid flow-shop	4
Metaheuristics	7	Memetic algorithm	6	Controllable processing times	4
Rescheduling	7	Generation scheduling	5	Demand side management	3
Uncertainty	6	Demand response	5	Single-machine scheduling	3

4.4. Bibliographic Coupling

When two documents reference other common documents, bibliographic coupling occurs [147,193]. Figure 13a–d shows the bibliographic coupling in documents from the WOS database. Specifically, Figure 13a,b presents the network visualization and overlay bibliographic visualization coupling, revealing that most bibliographic coupling [194–197] occurred prior to 2016 (dark blue), while the yellow color represents recent studies [198–201].

Figure 13c,d displays the network and overlay visualization bibliographic coupling organization over the studied time period, revealing that the Islamic Azad University (Iran), Capital University of Economics and Business (China), and Hong Kong University of Science and Technology (Hong Kong) were the three top universities in 2016, 2018, and 2020, respectively. Figure 14 shows the density visualization of bibliographic coupling based on item density sources. It is apparent that *Computers & Industrial Engineering, International Journal of Advanced Manufacturing Technology*, and *Applied Soft Computing* were three major sources, while *Science of the Total Environment, Advanced Science Letters*, and the *IEEE Internet of Things Journal* were three minor sources.

4.5. Publication Statistics Based on the Journal

Table 12 presents the top 10 journals that published the greatest number of related papers based on Scopus. Accordingly, *Lecture Notes in Computer Science* (#30), *Computers and Industrial Engineering* (#20), and *Computer Integrated Manufacturing Systems* (#20) predominated in the field of optimization and evolutionary computations.

Table 12. The top 10 productive Scopus categories.

	Scopus	ISSN	Number of Documents
1	*Lecture Notes in Computer Science*	1611-3349	30
2	*Computers and Industrial Engineering*	0360-8352	20
3	*Robotics and Computer-Integrated Manufacturing*	0736-5845	20
4	*International Journal of Advanced Manufacturing Technology*	1433-3015	19
5	*Applied Soft Computing Journal*	1568-4946	18
6	*International Journal of Production Research*	0020-7543	15
7	*Advances in Intelligent Systems and Computing*	2194-5365	13
8	*IEEE Access*	2169-3536	13
9	*China Mechanical Engineering*	2192-8258	13
10	*Computers and Operations Research*	0305-0548	11

A total of 683 articles were published in 432 journals, which were classified among the 46 WoS categories in Sci-Expanded. Table 13 lists the 10 most productive WoS categories. A total of 175 articles (25.62% of 683 articles) were published in the first category (computer science artificial intelligence), followed by computer science theory methods (7.17%) and engineering electrical electronic (5.85%). When comparing the top 10 categories, the highest CPP 2020 of articles published in the computer science cybernetics category was 28.14, followed by engineering manufacturing (18.55). The highest APP for articles published in the computer science information systems category was 3.66.

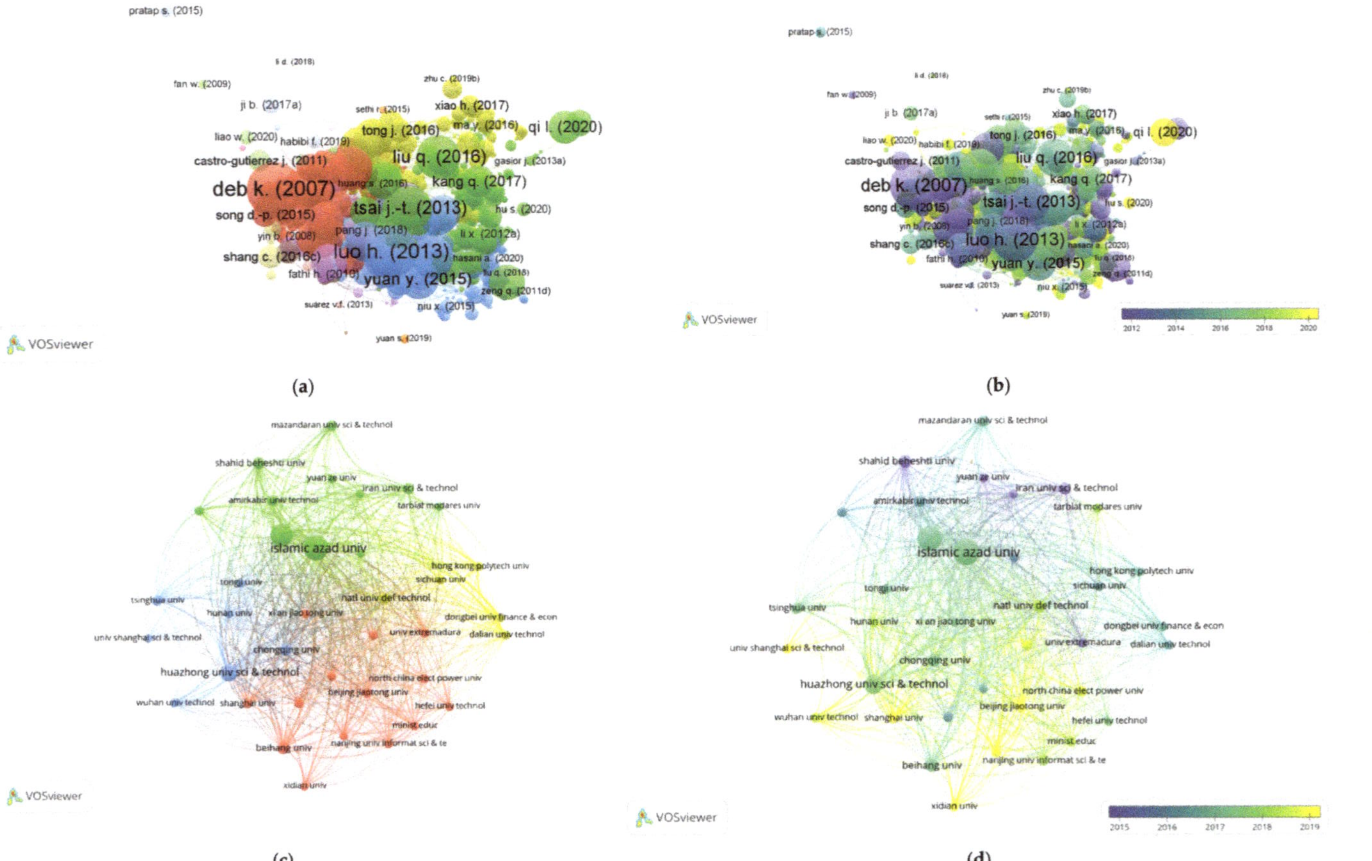

Figure 13. Bibliographic coupling (network and overlay visualization). **(a)** Network visualization bibliographic coupling document **(b)** Overlay visualization bibliographic coupling document **(c)** Network visualization bibliographic coupling organization **(d)** Overlay visualization bibliographic coupling organization.

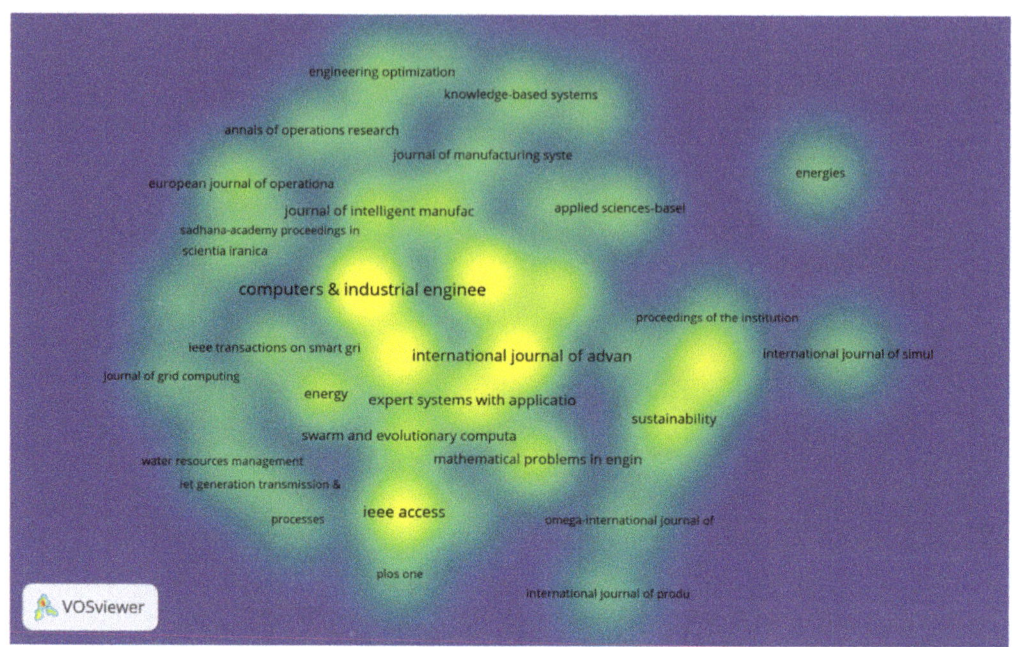

Figure 14. Density visualization bibliographic coupling (item density sources).

Table 13. The top 10 productive WoS categories.

	Web of Science Category	TP	AU	APP	TC 2020	CPP 2020
1	Computer Science Artificial Intelligence	175	523	2.98	2023	11.56
2	Computer Science Theory Methods	49	159	3.24	290	5.91
3	Engineering Electrical Electronic	40	125	3.12	541	13.52
4	Computer Science Interdisciplinary Applications	37	118	3.18	664	17.94
5	Operations Research Management Science	24	82	3.41	418	17.41
6	Automation Control Systems	16	44	2.75	201	12.56
7	Computer Science Information Systems	15	55	3.66	46	3.60
8	Engineering Manufacturing	9	31	3.44	167	18.55
9	Robotics	8	25	3.12	14	1.75
10	Computer Science Cybernetics	7	19	2.71	197	28.14

TP, AU, APP, TC 2020, and CPP 2020 present the total number of articles, total number of authors, total number of authors for each publication, total citations from WoS from the publication year to the end of 2020, and total citations for each paper, respectively. Other items: early access and letters.

4.6. Statistics Based on Authors

Figure 15 shows the top authors with the most publications according to Scopus. Reza Tavakkoli-Moghaddam from the University of Tehran (Tehran, Iran), Farouk Yalaoui from

Université de Technologie de Troyes (Troyes, France), and Mostafa Zabdieh from Shahid Beheshti University (Tehran, Iran) were the top 3 authors in the field, as indexed by Scopus, with 22, 18, and 14 publications, respectively.

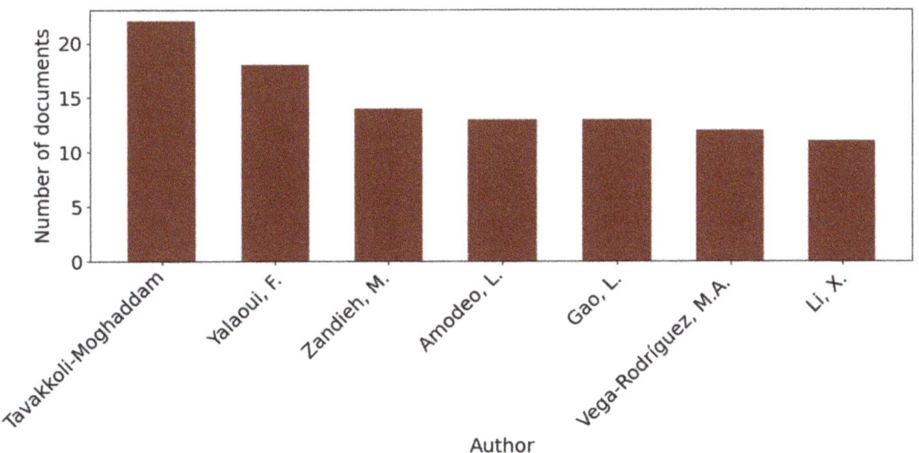

Figure 15. The most active authors in the field (Scopus https://www.scopus.com).

4.7. Publication Statistics by Country

Figure 16 presents the distribution of documents by the most active countries in the database (Scopus). It is apparent that China, Iran, and India were the top three most active countries in the field. Additionally, it can be seen that there was a significant difference between the first rank (China) and second rank (Iran) based on the number of publications indexed by Scopus. Although Iran was identified as the second-ranked country in the field, when comparing the populations of China and Iran, it is noteworthy to mention that Iran performed well in this area.

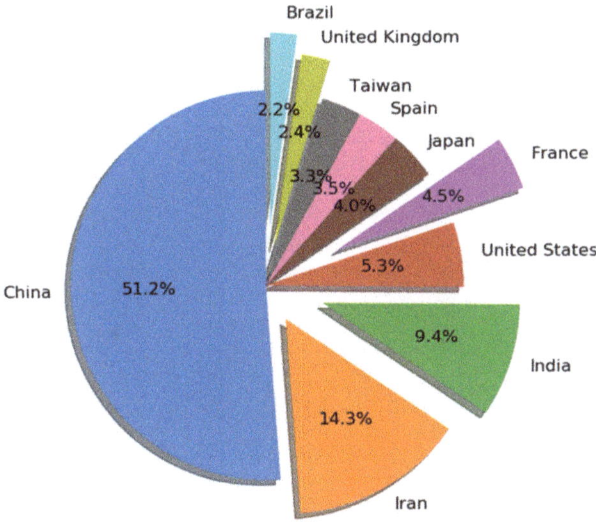

Figure 16. Research output of top 10 most productive countries across the database.

Figure 17 displays the growth rate of the top five active countries. While China, Iran, and France had smooth trends between 2000 and 2020, India and the United States showed some fluctuations. Between 2009 and 2015, the US presented the highest growth rate (positive and negative rate), and then the trend continued smoothly until the end of 2020.

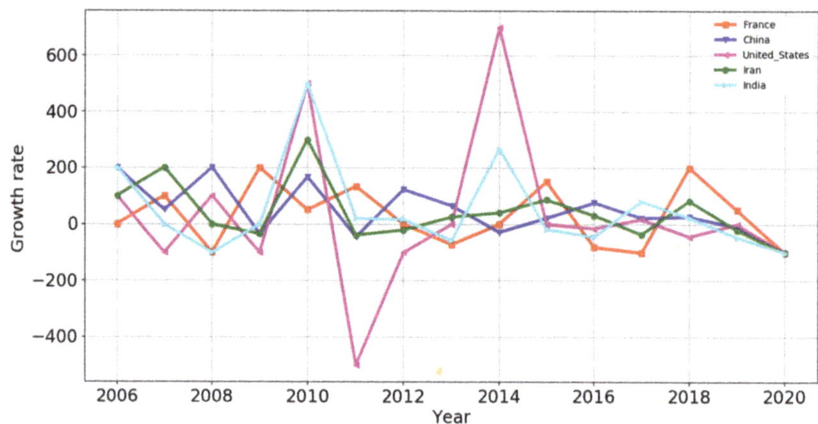

Figure 17. Growth rate of published documents for top 5 countries.

5. Summary

This paper presents a comprehensive review of NSGA-II applied to different scheduling problems. In the first part of the paper, the main idea of scheduling was defined, and the second part described the scientometric analysis in the field in detail.

It is noteworthy to mention that the *European Journal of Operational Research* owns the most contributions (19%) of published documents in scheduling, which is in the area of operations research.

This paper also reviewed different aspects of scheduling, namely production scheduling and personnel scheduling. It should be noted that about 9.51% of the published articles in the field considered uncertainty, while the majority of the mentioned articles addressed scheduling in power systems (32%), followed by project scheduling (13%), resource allocation (8%), and job-shop scheduling (8%). Among the different objective functions pertaining to job-shop scheduling, maximum completion time (C_{max}) possessed the most contributions (32%), followed by maximum machine workload (19%).

Although there are several optimization algorithms, metaheuristics are among the top solution approaches that have been used by researchers. Since genetic algorithms are based on populations, researchers have widely used genetic algorithms for scheduling problems (about 26%), followed by simulated annealing (6.4%), ant colony optimization (4.09%), and tau search (4.47%). The other GA-based solution methods in the field include VEGA, MOGA, WBGA, RWGA, NSGA, NSGA-II, RDGA, NPGA, and DMOEA. While most of the evolutionary algorithms possess difficulties, such as high computational cost, lack of elitism, and difficulty in parameter settings, NSGA-II, proposed in 2002, has attempted to alleviate all of the above difficulties.

Furthermore, the scientometric analysis indicated that computer science artificial intelligence (#86), operation research management science (#86), and computer science interdicipline applications (#82) were among the top categories. In addition, network visualization identified that scheduling, optimization, NSGA-II, multi-objective optimization, multi-objective genetic algorithm, Pareto-optimal, makespan, stochastic models, design, cost, parallel machines, task analysis, and operations were the top keywords. Moreover, the authors of this paper found that NSGA-II, scheduling, and makespan were the top three one-word-long keywords for both WoS and Scopus. Additionally, two-word- and three-word-long keywords were identified. Additional analyses, namely citation network, bibliographic coupling, and journal mapping, were conducted in this work.

Future Studies

In this paper, we discussed the benefits of NSGA-II and its application in different fields of study. Since NSGA-II was specifically designed to solve two- and three-objective problems, less than 1% of NSGA-II articles have considered many-objective scheduling problems (with more than three objectives) [202]. NSGA-III [203,204], its successor, was designed to solve problems with more than three objectives. Hence, it is suggested to review the application of NSGA-III in the field while considering many-objective scheduling problems. Furthermore, the majority of the studies used deterministic approaches, and there is an urgent need to provide more robust approaches for tackling uncertainties in scheduling problems. Additionally, a comprehensive review in other fields of solution methods applied to scheduling problems is encouraged for future studies. As the authors presented in the paper, MOPSO and MOACO are two other famous EMO algorithms, and thus, a comprehensive review in the area using the above-mentioned solution approaches is suggested. Moreover, the application of scheduling for energy conservation is an interesting area for research.

6. Conclusions and Discussions

Since exact methods are expensive in terms of computing time and often possess poor-quality solutions, researchers have become more interested in applying metaheuristics in scheduling problems, which can produce alternative optimal solutions in a single run. This study reviewed the most important scheduling problems that have been solved by the NSGA-II method and provides a bibliometric analysis of the published literature. In terms of MOO problems, most of the exact solution approaches convert MOO problems into a single optimization problem, while metaheuristic methods obtain solutions without this conversion.

This study addressed the most important subject fields based on keywords and network analysis. Moreover, a detailed scientometric analysis was employed as an influential tool in the bibliometric analyses and reviews.

According to the analyses performed in the work, several key arguments that are worthy of further discussion are offered below:

- In terms of keyword analysis, scheduling, optimization, NSGA-II, makespan, design, cost, genetic algorithm, and decision making are the most prevalent keywords for scholars;
- Among the current scheduling problems, machine scheduling (specifically job-shop scheduling), routing, satellite scheduling, project scheduling, weapon selection, and forest planning are most predominant in the reviewed articles;
- Among the proposed solution methods for solving scheduling problems, the genetic algorithm possessed the greatest contribution of (26%), followed by PSO (9%), SA (6.4%), ACO (4.09%), and then tabu search (4.47%);
- Since 2014, NSGA-II has been the most studied algorithm, followed by MOPSO and then MOACO;
- Despite the increasing complexity of scheduling problems, metaheuristic algorithms (specifically NSGA-II) are more suitable for finding efficient solutions or near-optimal solutions.

Supplementary Materials: The following supporting information can be downloaded at: https://www.mdpi.com/article/10.3390/pr10010098/s1, The supplementary materials are available for research purposes in Supplementary Files A and B.

Author Contributions: Conceptualization, I.R., K.D. and A.H.G.; methodology, I.R.; software, I.R.; validation, I.R., K.D., A.H.G., M.R.N. and F.C.; formal analysis, I.R.; investigation, I.R.; resources, I.R.; data curation, I.R.; writing—original draft preparation, I.R.; writing—review and editing, A.H.G., M.R.N., F.C. and K.D.; visualization, I.R.; supervision, A.H.G. and K.D.; project administration, A.H.G.; funding acquisition, A.H.G. All authors have read and agreed to the published version of the manuscript.

Funding: This research received no external funding.

Institutional Review Board Statement: Not applicable.

Informed Consent Statement: Not applicable.

Data Availability Statement: Not applicable.

Conflicts of Interest: The authors declare no conflict of interest.

References

1. Deb, K.; Agrawal, S.; Pratap, A.; Meyarivan, T. A fast elitist non-dominated sorting genetic algorithm for multi-objective optimization: NSGA-II. In *International Conference on Parallel Problem Solving from Nature*; Springer: Berlin/Heidelberg, Germany, 2000; pp. 849–858.
2. Salvendy, G. *Handbook of Industrial Engineering: Technology and Operations Management*; John Wiley & Sons: Hoboken, NJ, USA, 2001.
3. Lenstra, J.K.; Kan, A.H.G.R. Complexity of vehicle routing and scheduling problems. *Networks* **1981**, *11*, 221–227. [CrossRef]
4. Pinedo, M.; Hadavi, K. Scheduling: Theory, Algorithms and Systems Development. In *Operations Research Proceedings 1991*; Springer: Berlin/Heidelberg, Germany, 1992; pp. 35–42.
5. Gandomi, A.H.; Emrouznejad, A.; Rahimi, I. Evolutionary Computation in Scheduling: A Scientometric Analysis. In *Evolutionary Computation in Scheduling*; John Wiley & Sons: Hoboken, NJ, USA, 2020; pp. 1–10.
6. Wang, Y.; Dang, C. An Evolutionary Algorithm for Global Optimization Based on Level-Set Evolution and Latin Squares. *IEEE Trans. Evol. Comput.* **2007**, *11*, 579–595. [CrossRef]
7. Sun, J.; Zhang, Q.; Tsang, E.P.K. DE/EDA: A new evolutionary algorithm for global optimization. *Inf. Sci.* **2005**, *169*, 249–262. [CrossRef]
8. Wang, Y.-J.; Zhang, J.-S. Global optimization by an improved differential evolutionary algorithm. *Appl. Math. Comput.* **2007**, *188*, 669–680. [CrossRef]
9. Guo, D.; Wang, J.; Huang, J.; Han, R.; Song, M. Chaotic-NSGA-II: An effective algorithm to solve multi-objective optimization problems. In Proceedings of the 2010 International Conference on Intelligent Computing and Integrated Systems, Guilin, China, 22–24 October 2010; pp. 20–23.
10. Liu, J.; Abbass, H.A.; Tan, K.C. Evolutionary Computation and Complex Networks. In *Evolutionary Computation and Complex Networks*; Springer: Singapore, 2019; pp. 3–22.
11. Simon, D. *Evolutionary Optimization Algorithms*; John Wiley & Sons: Hoboken, NJ, USA, 2013.
12. Coello, C.A.C.; Lamont, G.B.; Van Veldhuizen, D.A. *Evolutionary Algorithms for Solving Multi-Objective Problems*; Springer: Berlin/Heidelberg, Germany, 2007; Volume 5.
13. Behmanesh, R.; Rahimi, I.; Gandomi, A.H. Evolutionary Many-Objective Algorithms for Combinatorial Optimization Problems: A Comparative Study. *Arch. Comput. Methods Eng.* **2021**, *28*, 673–688. [CrossRef]
14. Deb, K. Multi-objective optimization. In *Search Methodologies*; Springer: New York, NY, USA, 2014.
15. Marler, R.; Arora, J. Survey of multi-objective optimization methods for engineering. *Struct. Multidiscip. Optim.* **2004**, *26*, 369–395. [CrossRef]
16. Deb, K. Multi-objective Optimisation Using Evolutionary Algorithms: An Introduction. In *Multi-Objective Evolutionary Optimisation for Product Design and Manufacturing*; Springer: Berlin/Heidelberg, Germany, 2011; pp. 3–34.
17. Liu, L.; Gu, S.; Fu, D.; Zhang, M.; Buyya, R. A New Multi-objective Evolutionary Algorithm for Inter-Cloud Service Composition. *KSII Trans. Internet Inf. Syst.* **2018**, *12*, 1–20.
18. Yuan, S.; Deng, G.; Feng, Q.; Zheng, P.; Song, T. Multi-Objective Evolutionary Algorithm Based on Decomposition for Energy-aware Scheduling in Heterogeneous Computing Systems. *J. Univers. Comput. Sci.* **2017**, *23*, 636–651.
19. Long, Q.; Wu, X.; Wu, C. Non-dominated sorting methods for multi-objective optimization: Review and numerical comparison. *J. Ind. Manag. Optim.* **2021**, *17*, 1001–1023. [CrossRef]
20. Deb, K.; Pratap, A.; Agarwal, S.; Meyarivan, T. A fast and elitist multiobjective genetic algorithm: NSGA-II. *IEEE Trans. Evol. Comput.* **2002**, *6*, 182–197. [CrossRef]
21. Yusoff, Y.; Ngadiman, M.S.; Zain, A.M. Overview of NSGA-II for Optimizing Machining Process Parameters. *Procedia Eng.* **2011**, *15*, 3978–3983. [CrossRef]
22. Deb, K.; Rao, U.B.N.; Karthik, S. Dynamic multi-objective optimization and decision-making using modified NSGA-II: A case study on hydro-thermal power scheduling. In Proceedings of the International Conference on Evolutionary Multi-Criterion Optimization, Matsushima, Japan, 5–8 March 2007; pp. 803–817.
23. Bekele, E.G.; Nicklow, J.W. Multi-objective automatic calibration of SWAT using NSGA-II. *J. Hydrol.* **2007**, *341*, 165–176. [CrossRef]
24. Van Eck, N.J.; Waltman, L. *VOSviewer Manual*; Univeristeit Leiden: Leiden, The Netherlands, 2013; Volume 1, pp. 1–53.
25. Van Eck, N.J.; Waltman, L. CitNetExplorer: A new software tool for analyzing and visualizing citation networks. *J. Inf.* **2014**, *8*, 802–823. [CrossRef]
26. Pinedo, M. *Planning and Scheduling in Manufacturing and Services*; Springer: Berlin/Heidelberg, Germany, 2005.
27. Özdamar, L.; Ulusoy, G. A note on an iterative forward/backward scheduling technique with reference to a procedure by Li and Willis. *Eur. J. Oper. Res.* **1996**, *89*, 400–407. [CrossRef]

28. Li, K.; Willis, R. An iterative scheduling technique for resource-constrained project scheduling. *Eur. J. Oper. Res.* **1992**, *56*, 370–379. [CrossRef]
29. Gonçalves, J.F.; Resende, M.G.C.; Mendes, J.J.M. A biased random-key genetic algorithm with forward-backward improvement for the resource constrained project scheduling problem. *J. Heuristics* **2010**, *17*, 467–486. [CrossRef]
30. Qi, X. A logistics scheduling model: Inventory cost reduction by batching. *Nav. Res. Logist.* **2005**, *52*, 312–320. [CrossRef]
31. Tiemessen, H.; van Houtum, G. Reducing costs of repairable inventory supply systems via dynamic scheduling. *Int. J. Prod. Econ.* **2013**, *143*, 478–488. [CrossRef]
32. Liu, W.; Ke, G.Y.; Chen, J.; Zhang, L. Scheduling the distribution of blood products: A vendor-managed inventory routing approach. *Transp. Res. Part E Logist. Transp. Rev.* **2020**, *140*, 101964. [CrossRef]
33. Wang, S.; Wang, X.; Chu, F.; Yu, J. An energy-efficient two-stage hybrid flow shop scheduling problem in a glass production. *Int. J. Prod. Res.* **2020**, *58*, 2283–2314. [CrossRef]
34. Xu, Z.; Zheng, Z.; Gao, X. Energy-efficient steelmaking-continuous casting scheduling problem with temperature constraints and its solution using a multi-objective hybrid genetic algorithm with local search. *Appl. Soft Comput.* **2020**, *95*, 106554. [CrossRef]
35. Li, X.; Jin, X.; Lu, S.; Li, Z.; Wang, Y.; Cao, J. Carbon-Efficient Production Scheduling of a Bioethanol Plant Considering Diversified Feedstock Pelletization Density: A Case Study. *Processes* **2020**, *8*, 1189. [CrossRef]
36. Wang, S.; Che, Y.; Zhao, H.; Lim, A. Accurate Tracking, Collision Detection, and Optimal Scheduling of Airport Ground Support Equipment. *IEEE Internet Things J.* **2020**, *8*, 572–584. [CrossRef]
37. Zhou, B.; Zhu, Z. Optimally scheduling and loading tow trains of in-plant milk-run delivery for mixed-model assembly lines. *Assem. Autom.* **2020**, *40*, 511–530. [CrossRef]
38. Torabbeigi, M.; Lim, G.J.; Kim, S.J. Drone delivery scheduling optimization considering payload-induced battery consumption rates. *J. Intell. Robot. Syst.* **2020**, *97*, 471–487. [CrossRef]
39. Sheikh, S.Z.; Pasha, M.A. Energy-efficient real-time scheduling on multicores: A novel approach to model cache contention. *ACM Trans. Embed. Comput. Syst.* **2020**, *19*, 1–25. [CrossRef]
40. Wang, J.; Yang, J.; Zhang, Y.; Ren, S.; Liu, Y. Infinitely repeated game based real-time scheduling for low-carbon flexible job shop considering multi-time periods. *J. Clean. Prod.* **2020**, *247*, 119093. [CrossRef]
41. Kim, E.; Lee, Y.; He, L.; Shin, K.G.; Lee, J. Power Guarantee for Electric Systems Using Real-Time Scheduling. *IEEE Trans. Parallel Distrib. Syst.* **2020**, *31*, 1783–1798. [CrossRef]
42. Fathollahi-Fard, A.M.; Ranjbar-Bourani, M.; Cheikhrouhou, N.; Hajiaghaei-Keshteli, M. Novel modifications of social engineering optimizer to solve a truck scheduling problem in a cross-docking system. *Comput. Ind. Eng.* **2019**, *137*, 106103. [CrossRef]
43. Bossche, T.V.D.; Çalık, H.; Jacobs, E.-J.; Toffolo, T.A.; Berghe, G.V. Truck scheduling in tank terminals. *EURO J. Transp. Logist.* **2020**, *9*, 100001. [CrossRef]
44. Demeulemeester, E.L.; Herroelen, W.S. *Project Scheduling: A Research Handbook*; Springer Science & Business Media: Berlin, Germany, 2006; Volume 49.
45. Brucker, P.; Drexl, A.; Möhring, R.; Neumann, K.; Pesch, E. Resource-constrained project scheduling: Notation, classification, models, and methods. *Eur. J. Oper. Res.* **1999**, *112*, 3–41. [CrossRef]
46. Biskup, D. Single-machine scheduling with learning considerations. *Eur. J. Oper. Res.* **1999**, *115*, 173–178. [CrossRef]
47. Mosheiov, G. Parallel machine scheduling with a learning effect. *J. Oper. Res. Soc.* **2001**, *52*, 1165–1169. [CrossRef]
48. Cheng, T.C.E.; Sin, C.C.S. A state-of-the-art review of parallel-machine scheduling research. *Eur. J. Oper. Res.* **1990**, *47*, 271–292. [CrossRef]
49. Garey, M.R.; Johnson, D.S.; Sethi, R. The Complexity of Flowshop and Jobshop Scheduling. *Math. Oper. Res.* **1976**, *1*, 117–129. [CrossRef]
50. Fu, Y.; Tian, G.; Fathollahi-Fard, A.M.; Ahmadi, A.; Zhang, C. Stochastic multi-objective modelling and optimization of an energy-conscious distributed permutation flow shop scheduling problem with the total tardiness constraint. *J. Clean. Prod.* **2019**, *226*, 515–525. [CrossRef]
51. Denzler, D.R.; Boe, W.J. Experimental investigation of flexible manufacturing system scheduling decision rules. *Int. J. Prod. Res.* **1987**, *25*, 979–994. [CrossRef]
52. Zhang, W.; Freiheit, T.; Yang, H. Dynamic scheduling in flexible assembly system based on timed Petri nets model. *Robot. Comput. Manuf.* **2005**, *21*, 550–558. [CrossRef]
53. Sawik, T. Loading and scheduling of a flexible assembly system by mixed integer programming. *Eur. J. Oper. Res.* **2004**, *154*, 1–19. [CrossRef]
54. Valckenaers, P.; Van Brussel, H.; Bongaerts, L.; Bonneville, F. Programming, scheduling, and control of flexible assembly systems. *Comput. Ind.* **1995**, *26*, 209–218. [CrossRef]
55. Elmaghraby, S.E. The Economic Lot Scheduling Problem (ELSP): Review and Extensions. *Manag. Sci.* **1978**, *24*, 587–598. [CrossRef]
56. Dobson, G. The Economic Lot-Scheduling Problem: Achieving Feasibility Using Time-Varying Lot Sizes. *Oper. Res.* **1987**, *35*, 764–771. [CrossRef]
57. Rogers, J. A Computational Approach to the Economic Lot Scheduling Problem. *Manag. Sci.* **1958**, *4*, 264–291. [CrossRef]
58. Schoenfelder, J.; Bretthauer, K.M.; Wright, P.D.; Coe, E. Nurse scheduling with quick-response methods: Improving hospital performance, nurse workload, and patient experience. *Eur. J. Oper. Res.* **2020**, *283*, 390–403. [CrossRef]

59. Hamid, M.; Tavakkoli-Moghaddam, R.; Golpaygani, F.; Vahedi-Nouri, B. A multi-objective model for a nurse scheduling problem by emphasizing human factors. *Proc. Inst. Mech. Eng. Part H J. Eng. Med.* **2019**, *234*, 179–199. [CrossRef]
60. Simić, S.; Milutinović, D.; Sekulić, S.; Simić, D.; Simić, S.D.; Đorđević, J. A hybrid case-based reasoning approach to detecting the optimal solution in nurse scheduling problem. *Log. J. IGPL* **2020**, *28*, 226–238. [CrossRef]
61. Legrain, A.; Omer, J.; Rosat, S. An online stochastic algorithm for a dynamic nurse scheduling problem. *Eur. J. Oper. Res.* **2020**, *285*, 196–210. [CrossRef]
62. Jiang, J.; Xiong, X.; Ou, Y.; Wang, H. An Improved Bacterial Foraging Optimization with Differential and Poisson Distribution Strategy and its Application to Nurse Scheduling Problem. In *International Conference on Swarm Intelligence, Belgrade, Serbia, 14–20 July 2020*; Springer: Berlin/Heidelberg, Germany, 2020; pp. 312–324.
63. Rerkjirattikal, P.; Huynh, V.-N.; Olapiriyakul, S.; Supnithi, T. A Framework for a Practical Nurse Scheduling Approach: A Case of Operating Room of a Hospital in Thailand. In *International Conference on Applied Human Factors and Ergonomics, San Diego, CA, USA, 16–20 July 2020*; Springer: Singapore, 2020; pp. 259–264.
64. İnanç, Ş.; Şenaras, A.E. Solving Nurse Scheduling Problem via Genetic Algorithm in Home Healthcare. In *Transportation, Logistics, and Supply Chain Management in Home Healthcare: Emerging Research and Opportunities*; IGI Global: Hershey, PA, USA, 2020; pp. 20–28.
65. Aydas, O.T.; Ross, A.D.; Scanlon, M.C.; Aydas, B. New results on integrated nurse staffing and scheduling: The medium-term context for intensive care units. *J. Oper. Res. Soc.* **2021**, *72*, 2631–2648. [CrossRef]
66. Batun, S.; Karpuz, E. Nurse Scheduling and Rescheduling Under Uncertainty. *Hacettepe Univ. J. Econ. Adm. Sci. Üniversitesi Iktis. ve Idari Bilim. Fakültesi Derg.* **2020**, *38*, 75–95. [CrossRef]
67. Roshanaei, V.; Luong, C.; Aleman, D.M.; Urbach, D.R. Reformulation, linearization, and decomposition techniques for balanced distributed operating room scheduling. *Omega* **2020**, *93*, 102043. [CrossRef]
68. Schiele, J.; Koperna, T.; Brunner, J.O. Predicting intensive care unit bed occupancy for integrated operating room scheduling via neural networks. *Nav. Res. Logist.* **2020**, *68*, 65–88. [CrossRef]
69. Zhu, S.; Fan, W.; Liu, T.; Yang, S.; Pardalos, P.M. Dynamic three-stage operating room scheduling considering patient waiting time and surgical overtime costs. *J. Comb. Optim.* **2019**, *39*, 185–215. [CrossRef]
70. Ahmed, A.; Ali, H. Modeling patient preference in an operating room scheduling problem. *Oper. Res. Health Care* **2020**, *25*, 100257. [CrossRef]
71. Najjarbashi, A.; Lim, G.J. A Decomposition Algorithm for the Two-Stage Chance-Constrained Operating Room Scheduling Problem. *IEEE Access* **2020**, *8*, 80160–80172. [CrossRef]
72. Varmazyar, M.; Akhavan-Tabatabaei, R.; Salmasi, N.; Modarres, M. Operating room scheduling problem under uncertainty: Application of continuous phase-type distributions. *IISE Trans.* **2020**, *52*, 216–235. [CrossRef]
73. Barrera, J.; Carrasco, R.A.; Mondschein, S.; Canessa, G.; Rojas-Zalazar, D. Operating room scheduling under waiting time constraints: The Chilean GES plan. *Ann. Oper. Res.* **2020**, *286*, 501–527. [CrossRef]
74. Abdeljaouad, M.A.; Bahroun, Z.; Saadani, N.E.H.; Zouari, B. A simulated annealing for a daily operating room scheduling problem under constraints of uncertainty and setup. *INFOR Inf. Syst. Oper. Res.* **2020**, *58*, 456–477. [CrossRef]
75. Divsalar, A.; Jokar, A.; Emami, S. Operating Room Scheduling considering Patient Priority: Case of Shomal Hospital in Amol. *Int. J. Ind. Eng. Manag. Sci.* **2020**, *7*, 57–68.
76. Roshanaei, V.; Booth, K.E.; Aleman, D.M.; Urbach, D.R.; Beck, J.C. Branch-and-check methods for multi-level operating room planning and scheduling. *Int. J. Prod. Econ.* **2020**, *220*, 107433. [CrossRef]
77. Akbarzadeh, B.; Moslehi, G.; Reisi-Nafchi, M.; Maenhout, B. A diving heuristic for planning and scheduling surgical cases in the operating room department with nurse re-rostering. *J. Sched.* **2020**, *23*, 265–288. [CrossRef]
78. Rahimi, I.; Gandomi, A.H. A Comprehensive Review and Analysis of Operating Room and Surgery Scheduling. *Arch. Comput. Methods Eng.* **2021**, *28*, 1667–1688. [CrossRef]
79. Bandi, C.; Gupta, D. Operating Room Staffing and Scheduling. *Manuf. Serv. Oper. Manag.* **2020**, *22*, 958–974. [CrossRef]
80. Oliveira, M.; Bélanger, V.; Marques, I.; Ruiz, A. Assessing the impact of patient prioritization on operating room schedules. *Oper. Res. Health Care* **2020**, *24*, 100232. [CrossRef]
81. Moosavi, A.; Ebrahimnejad, S. Robust operating room planning considering upstream and downstream units: A new two-stage heuristic algorithm. *Comput. Ind. Eng.* **2020**, *143*, 106387. [CrossRef]
82. Bovim, T.R.; Christiansen, M.; Gullhav, A.N.; Range, T.M.; Hellemo, L. Stochastic master surgery scheduling. *Eur. J. Oper. Res.* **2020**, *285*, 695–711. [CrossRef]
83. Gegg, D.L. The Impact of Middle School Scheduling Practices on Adolescent Math Achievement in Louisiana Public Schools. Ph.D. Thesis, 2020.
84. Khan, K.; Sahai, A. Tabu Ant Colony Optimisation for School Timetable Scheduling Problem. *Int. J. Eng. Res. Appl.* **2020**, *10*, 1–9.
85. Hao, X.; Liu, J.; Zhang, Y.; Sanga, G. Mathematical model and simulated annealing algorithm for Chinese high school timetabling problems under the new curriculum innovation. *Front. Comput. Sci.* **2020**, *15*, 1–11. [CrossRef]
86. Tan, J.S.; Goh, S.L.; Sura, S.; Kendall, G.; Sabar, N.R. Hybrid particle swarm optimization with particle elimination for the high school timetabling problem. *Evol. Intell.* **2020**, *14*, 1915–1930. [CrossRef]

87. Hoshino, R.; Fabris, I. Optimizing Student Course Preferences in School Timetabling. In Proceedings of the International Conference on Integration of Constraint Programming, Artificial Intelligence, and Operations Research, Vienna, Austria, 5–8 July 2020; pp. 283–299.
88. Tassopoulos, I.X.; Iliopoulou, C.A.; Beligiannis, G.N. Solving the Greek school timetabling problem by a mixed integer programming model. *J. Oper. Res. Soc.* **2019**, *71*, 117–132. [CrossRef]
89. Chen, P.S.; Huang, W.T.; Peng, N.C.; Chen, G.Y.H. Modularising school timetabling problems in different types of classes for Taiwanese elementary and junior high schools. *Int. J. Math. Oper. Res.* **2020**, *17*, 110. [CrossRef]
90. Saviniec, L.; Santos, M.O.; Costa, A.M.; Santos, L.M.R.d. Pattern-based models and a cooperative parallel metaheuristic for high school timetabling problems. *Eur. J. Oper. Res.* **2020**, *280*, 1064–1081. [CrossRef]
91. Cacchiani, V.; Salazar-González, J.-J. Heuristic approaches for flight retiming in an integrated airline scheduling problem of a regional carrier. *Omega* **2020**, *91*, 102028. [CrossRef]
92. Zhou, L.; Liang, Z.; Chou, C.-A.; Chaovalitwongse, W.A. Airline planning and scheduling: Models and solution methodologies. *Front. Eng. Manag.* **2020**, *7*, 1–26. [CrossRef]
93. Khanmirza, E.; Nazarahari, M.; Haghbeigi, M. A heuristic approach for optimal integrated airline schedule design and fleet assignment with demand recapture. *Appl. Soft Comput.* **2020**, *96*, 106681. [CrossRef]
94. Bayliss, C.; De Maere, G.; Atkin, J.A.D.; Paelinck, M. Scheduling airline reserve crew using a probabilistic crew absence and recovery model. *J. Oper. Res. Soc.* **2019**, *71*, 543–565. [CrossRef]
95. Sanchez, D.T. *Optimising Airline Maintenance Scheduling Decisions*; Lancaster University: Lancaster, UK, 2020.
96. Fairbrother, J.; Zografos, K.G.; Glazebrook, K.D. A Slot-Scheduling Mechanism at Congested Airports That Incorporates Efficiency, Fairness, and Airline Preferences. *Transp. Sci.* **2019**, *54*, 115–138. [CrossRef]
97. Kerkemezos, Y.; Karreman, B. *On the Benefits of Being Alone: Scheduling Changes, Intensity of Competition and Dynamic Airline Pricing*; Tinbergen Institute Discussion Paper 2020-042/VII; Tinbergen Institute: Amsterdam, The Netherlands, 2020.
98. Shiau, J.-Y.; Huang, M.-K.; Huang, C.-Y. A Hybrid Personnel Scheduling Model for Staff Rostering Problems. *Mathematics* **2020**, *8*, 1702. [CrossRef]
99. Chutima, P.; Arayikanon, K. Many-objective low-cost airline cockpit crew rostering optimisation. *Comput. Ind. Eng.* **2020**, *150*, 106844. [CrossRef]
100. Sun, J.Y. Airport curfew and scheduling differentiation: Domestic versus international competition. *J. Air Transp. Manag.* **2020**, *87*, 101839. [CrossRef]
101. Nenem, S.; Graham, A.; Dennis, N. Airline schedule and network competitiveness: A consumer-centric approach for business travel. *Ann. Tour. Res.* **2020**, *80*, 102822. [CrossRef]
102. Wagner, H.M. An integer linear-programming model for machine scheduling. *Nav. Res. Logist. Q.* **1959**, *6*, 131–140. [CrossRef]
103. Brooks, G.H. An algorithm for finding optimal or near optimal solutions to the production scheduling problem. *J. Ind. Eng.* **1969**, *16*, 34–40.
104. Lomnicki, Z.A. A "Branch-and-Bound" Algorithm for the Exact Solution of the Three-Machine Scheduling Problem. *J. Oper. Res. Soc.* **1965**, *16*, 89–100. [CrossRef]
105. Barker, J.R.; McMahon, G.B. Scheduling the General Job-Shop. *Manag. Sci.* **1985**, *31*, 594–598. [CrossRef]
106. French, S. Sequencing and scheduling. In *An Introduction to the Mathematics of the Job-Shop*; Wiley: Hoboken, NJ, USA, 1982.
107. Morton, T.; Pentico, D.W. *Heuristic Scheduling Systems: With Applications to Production Systems and Project Management*; John Wiley & Sons: Hoboken, NJ, USA, 1993; Volume 3.
108. Fonseca, D.J.; Navaresse, D. Artificial neural networks for job shop simulation. *Adv. Eng. Inform.* **2002**, *16*, 241–246. [CrossRef]
109. Aarts, E.; Aarts, E.H.L.; Lenstra, J.K. *Local Search in Combinatorial Optimization*; Princeton University Press: Princeton, NJ, USA, 2003.
110. Reeves, C.R. *Modern Heuristic Techniques for Combinatorial Problems*; John Wiley & Sons, Inc.: Hoboken, NJ, USA, 1993.
111. Jones, D.F.; Mirrazavi, S.; Tamiz, M. Multi-objective meta-heuristics: An overview of the current state-of-the-art. *Eur. J. Oper. Res.* **2002**, *137*, 1–9. [CrossRef]
112. Sarker, R.; Ray, T. An improved evolutionary algorithm for solving multi-objective crop planning models. *Comput. Electron. Agric.* **2009**, *68*, 191–199. [CrossRef]
113. Poojari, C.; Beasley, J. Improving benders decomposition using a genetic algorithm. *Eur. J. Oper. Res.* **2009**, *199*, 89–97. [CrossRef]
114. Herrmann, J.W.; Lee, C.-Y.; Hinchman, J. Global job shop scheduling with a genetic algorithm. *Prod. Oper. Manag.* **1995**, *4*, 30–45. [CrossRef]
115. Tung, L.-F.; Lin, L.; Nagi, R. Multiple-objective scheduling for the hierarchical control of flexible manufacturing systems. *Int. J. Flex. Manuf. Syst.* **1999**, *11*, 379–409. [CrossRef]
116. Jensen, M.T. Generating robust and flexible job shop schedules using genetic algorithms. *IEEE Trans. Evol. Comput.* **2003**, *7*, 275–288. [CrossRef]
117. Mati, Y.; Xie, X. The complexity of two-job shop problems with multi-purpose unrelated machines. *Eur. J. Oper. Res.* **2004**, *152*, 159–169. [CrossRef]
118. Xia, W.; Wu, Z. An effective hybrid optimization approach for multi-objective flexible job-shop scheduling problems. *Comput. Ind. Eng.* **2005**, *48*, 409–425. [CrossRef]
119. Torabi, S.A.; Karimi, B.; Ghomi, S.F. The common cycle economic lot scheduling in flexible job shops: The finite horizon case. *Int. J. Prod. Econ.* **2005**, *97*, 52–65. [CrossRef]

120. Gomes, M.C.; Barbosa-Povoa, A.P.; Novais, A.Q. Optimal scheduling for flexible job-shop operation. *Int. J. Prod. Res.* **2005**, *43*, 2323–2353. [CrossRef]
121. Chan, F.T.; Wong, T.C.; Chan, L.Y. Flexible job-shop scheduling problem under resource constraints. *Int. J. Prod. Res.* **2006**, *44*, 2071–2089. [CrossRef]
122. Fattahi, P.; Mehrabad, M.S.; Jolai, F. Mathematical modeling and heuristic approaches to flexible job shop scheduling problems. *J. Intell. Manuf.* **2007**, *18*, 331–342. [CrossRef]
123. Tay, J.C.; Ho, N.B. Evolving dispatching rules using genetic programming for solving multi-objective flexible job-shop problems. *Comput. Ind. Eng.* **2008**, *54*, 453–473. [CrossRef]
124. Schaffer, J.D. Multiple objective optimization with vector evaluated genetic algorithms. In Proceedings of the 1st International Conference on Genetic Algorithms, Pittsburgh, PA, USA, 24–26 July 1985.
125. Konak, A.; Coit, D.W.; Smith, A.E. Multi-objective optimization using genetic algorithms: A tutorial. *Reliab. Eng. Syst. Saf.* **2006**, *91*, 992–1007. [CrossRef]
126. Fang, Z. A Weight-Based Multiobjective Genetic Algorithm for Flowshop Scheduling. In Proceedings of the 2009 International Conference on Artificial Intelligence and Computational Intelligence, Shanghai, China, 7–8 November 2009; Volume 1, pp. 373–377.
127. Zhou, H.; Cheung, W.; Leung, L.C. Minimizing weighted tardiness of job-shop scheduling using a hybrid genetic algorithm. *Eur. J. Oper. Res.* **2009**, *194*, 637–649. [CrossRef]
128. Srinivas, N.; Deb, K. Muiltiobjective Optimization Using Nondominated Sorting in Genetic Algorithms. *Evol. Comput.* **1994**, *2*, 221–248. [CrossRef]
129. Lu, H.; Yen, G. Rank-density-based multiobjective genetic algorithm and benchmark test function study. *IEEE Trans. Evol. Comput.* **2003**, *7*, 325–343.
130. Horn, J.; Nafpliotis, N.; Goldberg, D.E. *Multiobjective Optimization using the Niched Pareto Genetic Algorithm*; IlliGAL Report, No. 93005; Illinois Genetic Algorithms Laboratory, University of Illinois at Urbana-Champaign: Urbana-Champaign, IL, USA, 1993; pp. 1–32.
131. Horn, J.; Nafpliotis, N.; Goldberg, D.E. A niched Pareto genetic algorithm for multiobjective optimization. In Proceedings of the first IEEE conference on evolutionary computation. IEEE world congress on computational intelligence, Orlando, FL, USA, 27–29 June 1994; pp. 82–87.
132. Wang, Y.; Dang, C. An evolutionary algorithm for dynamic multi-objective optimization. *Appl. Math. Comput.* **2008**, *205*, 6–18. [CrossRef]
133. Mao, T.; Xu, Z.; Hou, R.; Peng, M. Efficient Satellite Scheduling Based on Improved Vector Evaluated Genetic Algorithm. *J. Netw.* **2012**, *7*, 517. [CrossRef]
134. Zhang, W.; Fujimura, S. Multiobjective process planning and scheduling using improved vector evaluated genetic algorithm with archive. *IEEJ Trans. Electr. Electron. Eng.* **2012**, *7*, 258–267. [CrossRef]
135. Zhang, W.; Fujimura, S. Improved vector evaluated genetic algorithm with archive for solving multiobjective pps problem. In Proceedings of the 2010 International Conference on E-Product E-Service and E-Entertainment, Henan, China, 7–9 November 2010; pp. 1–4.
136. Zhang, W.; Gen, M.; Jo, J. Hybrid sampling strategy-based multiobjective evolutionary algorithm for process planning and scheduling problem. *J. Intell. Manuf.* **2014**, *25*, 881–897. [CrossRef]
137. Wang, X.; Gao, L.; Zhang, C.; Shao, X. A multi-objective genetic algorithm based on immune and entropy principle for flexible job-shop scheduling problem. *Int. J. Adv. Manuf. Technol.* **2010**, *51*, 757–767. [CrossRef]
138. Lee, L.H.; Lee, C.U.; Tan, Y.P. A multi-objective genetic algorithm for robust flight scheduling using simulation. *Eur. J. Oper. Res.* **2007**, *177*, 1948–1968. [CrossRef]
139. Chang, F.-S.; Wu, J.-S.; Lee, C.-N.; Shen, H.-C. Greedy-search-based multi-objective genetic algorithm for emergency logistics scheduling. *Expert Syst. Appl.* **2014**, *41*, 2947–2956. [CrossRef]
140. Balasubramanian, H.; Mönch, L.; Fowler, J.; Pfund, M. Genetic algorithm based scheduling of parallel batch machines with incompatible job families to minimize total weighted tardiness. *Int. J. Prod. Res.* **2004**, *42*, 1621–1638. [CrossRef]
141. Kar, C.; Rakesh, V.K.; Samanta, T.; Banerjee, S. A New Approach to Grid Scheduling using Random Weighted Genetic Algorithm with Fault Tolerance Strategy. *Int. J. Comput. Appl.* **2012**, *48*, 42–47. [CrossRef]
142. Qian, B.; Wang, L.; Huang, D.-X.; Wang, X. Multi-objective flow shop scheduling using differential evolution. In *Intelligent Computing in Signal Processing and Pattern Recognition*; Springer: Berlin/Heidelberg, Germany, 2006; pp. 1125–1136.
143. Elloumi, S.; Fortemps, P. A hybrid rank-based evolutionary algorithm applied to multi-mode resource-constrained project scheduling problem. *Eur. J. Oper. Res.* **2010**, *205*, 31–41. [CrossRef]
144. Kim, K.; Walewski, J.; Cho, Y.K. Multiobjective Construction Schedule Optimization Using Modified Niched Pareto Genetic Algorithm. *J. Manag. Eng.* **2016**, *32*, 04015038. [CrossRef]
145. Benedict, S.; Vasudevan, V. Scheduling of scientific workflows using Niched Pareto GA for Grids. In Proceedings of the 2006 IEEE International Conference on Service Operations and Logistics, and Informatics, Shanghai, China, 21–23 June 2006; pp. 908–912.
146. Benedict, S.; Vasudevan, V. A Niched Pareto GA Approach for Scheduling Scientific Workflows in Wireless Grids. *J. Comput. Inf. Technol.* **2008**, *16*, 101–108. [CrossRef]

147. Azevedo, S.G.; Santos, M.; Antón, J.R. Supply chain of renewable energy: A bibliometric review approach. *Biomass Bioenergy* **2019**, *126*, 70–83. [CrossRef]
148. Sankar, S.S.; Ponnambalam, S.G.; Rathinavel, V.; Gurumarimuthu, M. A pareto based multi-objective genetic algorithm for scheduling of FMS. *IEEE Conf. Cybern. Intell. Syst.* **2004**, *2*, 700–705.
149. Lu, H.; Xu, X.; Zhang, M.; Yin, L. Dynamic multi-objective evolutionary algorithm based on decomposition for test task scheduling problem. In Proceedings of the 2015 Sixth International Conference on Intelligent Control and Information Processing (ICICIP), Wuhan, China, 26–28 November 2015; pp. 11–18.
150. Bagchi, T.P.; Jayaram, K.; Srinivas, T.D. Pareto optimal production scheduling by meta-heuristic methods. In Proceedings of the PICMET'99: Portland International Conference on Management of Engineering and Technology, Proceedings Vol-1: Book of Summaries (IEEE Cat. No. 99CH36310), Portland, OR, USA, 29 July 1999; Volume 1, p. 448.
151. Bagchi, T.P. A Comparison of Multiobjective Flowshop Sequencing by NSGA and ENGA. In *Multiobjective Scheduling by Genetic Algorithms*; Springer: Singapore, 1999; pp. 245–255.
152. Bagchi, T.P. Multiobjective Job Shop Scheduling. In *Multiobjective Scheduling by Genetic Algorithms*; Springer: Singapore, 1999; pp. 256–266.
153. Bagchi, T.P. Multiobjective Open Shop Scheduling. In *Multiobjective Scheduling by Genetic Algorithms*; Springer: Singapore, 1999; pp. 267–276.
154. Bagchi, T.P. Multiobjective Flowshop Scheduling. In *Multiobjective Scheduling by Genetic Algorithms*; Springer: Singapore, 1999; pp. 203–215.
155. Bandyopadhyay, S.; Bhattacharya, R. Solving multi-objective parallel machine scheduling problem by a modified NSGA-II. *Appl. Math. Model.* **2013**, *37*, 6718–6729. [CrossRef]
156. Ciro, G.C.; Dugardin, F.; Yalaoui, F.; Kelly, R. A NSGA-II and NSGA-III comparison for solving an open shop scheduling problem with resource constraints. *IFAC PapersOnLine* **2016**, *49*, 1272–1277. [CrossRef]
157. Rabiee, M.; Zandieh, M.; Ramezani, P. Bi-objective partial flexible job shop scheduling problem: NSGA-II, NRGA, MOGA and PAES approaches. *Int. J. Prod. Res.* **2012**, *50*, 7327–7342. [CrossRef]
158. Han, Y.-Y.; Gong, D.-W.; Sun, X.-Y.; Pan, Q.-K. An improved NSGA-II algorithm for multi-objective lot-streaming flow shop scheduling problem. *Int. J. Prod. Res.* **2013**, *52*, 2211–2231. [CrossRef]
159. Makaremi, Y.; Haghighi, A.; Ghafouri, H.R. Optimization of Pump Scheduling Program in Water Supply Systems Using a Self-Adaptive NSGA-II; a Review of Theory to Real Application. *Water Resour. Manag.* **2017**, *31*, 1283–1304. [CrossRef]
160. Huang, B.; Buckley, B.; Kechadi, M.T. Multi-objective feature selection by using NSGA-II for customer churn prediction in telecommunications. *Expert Syst. Appl.* **2010**, *37*, 3638–3646. [CrossRef]
161. Xu, W.; Xu, J.; He, D.; Tan, K.C. A combined differential evolution and NSGA-II approach for parametric optimization of a cancer immunotherapy model. In Proceedings of the 2017 IEEE Symposium Series on Computational Intelligence (SSCI), Honolulu, HI, USA, 27 November–1 December 2017; pp. 1–8.
162. Atiquzzaman, M.; Liong, S.-Y.; Yu, X. Alternative Decision Making in Water Distribution Network with NSGA-II. *J. Water Resour. Plan. Manag.* **2006**, *132*, 122–126. [CrossRef]
163. Wang, S.; Zhao, D.; Yuan, J.; Li, H.; Gao, Y. Application of NSGA-II Algorithm for fault diagnosis in power system. *Electr. Power Syst. Res.* **2019**, *175*, 105893. [CrossRef]
164. Sadeghi, J.; Sadeghi, S.; Niaki, S.T.A. A hybrid vendor managed inventory and redundancy allocation optimization problem in supply chain management: An NSGA-II with tuned parameters. *Comput. Oper. Res.* **2014**, *41*, 53–64. [CrossRef]
165. Xiong, J.; Zhou, Z.; Tian, K.; Liao, T.; Shi, J. A multi-objective approach for weapon selection and planning problems in dynamic environments. *J. Ind. Manag. Optim.* **2017**, *13*, 1189–1211. [CrossRef]
166. Guo, Z.; Wong, W.K.; Leung, S. A hybrid intelligent model for order allocation planning in make-to-order manufacturing. *Appl. Soft Comput.* **2013**, *13*, 1376–1390. [CrossRef]
167. Niu, X.; Tang, H.; Wu, L. Satellite scheduling of large areal tasks for rapid response to natural disaster using a multi-objective genetic algorithm. *Int. J. Disaster Risk Reduct.* **2018**, *28*, 813–825. [CrossRef]
168. Luong, N.H.; Alderliesten, T.; Bel, A.; Niatsetski, Y.; Bosman, P.A. Application and benchmarking of multi-objective evolutionary algorithms on high-dose-rate brachytherapy planning for prostate cancer treatment. *Swarm Evol. Comput.* **2018**, *40*, 37–52. [CrossRef]
169. Zeng, Q.; Wang, M.; Shen, L.; Song, H. Sequential Scheduling Method for FJSP with Multi-Objective under Mixed Work Calendars. *Processes* **2019**, *7*, 888. [CrossRef]
170. Balasubramaniyan, S.; Srinivasan, S.; Buonopane, F.; Subathra, B.; Vain, J.; Ramaswamy, S. Design and verification of Cyber-Physical Systems using TrueTime, evolutionary optimization and UPPAAL. *Microprocess. Microsyst.* **2016**, *42*, 37–48. [CrossRef]
171. Tavakkoli-Moghaddam, R.; Azarkish, M.; Sadeghnejad-Barkousaraie, A. Solving a multi-objective job shop scheduling problem with sequence-dependent setup times by a Pareto archive PSO combined with genetic operators and VNS. *Int. J. Adv. Manuf. Technol.* **2011**, *53*, 733–750. [CrossRef]
172. Motlagh, M.M.; Azimi, P.; Amiri, M.; Madraki, G. An efficient simulation optimization methodology to solve a multi-objective problem in unreliable unbalanced production lines. *Expert Syst. Appl.* **2019**, *138*, 112836. [CrossRef]
173. Chen, X.; Liu, Y.; Li, X.; Wang, Z.; Wanga, S.; Gao, C. A New Evolutionary Multiobjective Model for Traveling Salesman Problem. *IEEE Access* **2019**, *7*, 66964–66979. [CrossRef]

174. Ghoddousi, P.; Ansari, R.; Makui, A. An improved robust buffer allocation method for the project scheduling problem. *Eng. Optim.* **2016**, *49*, 718–731. [CrossRef]
175. Shokouhi, E. Integrated multi-objective process planning and flexible job shop scheduling considering precedence constraints. *Prod. Manuf. Res.* **2017**, *6*, 61–89. [CrossRef]
176. Li, D.; Das, S.; Pahwa, A.; Deb, K. A multi-objective evolutionary approach for generator scheduling. *Expert Syst. Appl.* **2013**, *40*, 7647–7655. [CrossRef]
177. Fotakis, D.G.; Sidiropoulos, E.; Myronidis, D.; Ioannou, K. Spatial genetic algorithm for multi-objective forest planning. *For. Policy Econ.* **2012**, *21*, 12–19. [CrossRef]
178. Jerić, S.V.; Figueira, J.R. Multi-objective scheduling and a resource allocation problem in hospitals. *J. Sched.* **2012**, *15*, 513–535. [CrossRef]
179. Jin, L.; Zhang, C.; Shao, X.; Yang, X.; Tian, G. A multi-objective memetic algorithm for integrated process planning and scheduling. *Int. J. Adv. Manuf. Technol.* **2016**, *85*, 1513–1528. [CrossRef]
180. Ji, B.; Sun, H.; Yuan, X.; Yuan, Y.; Wang, X. Coordinated optimized scheduling of locks and transshipment in inland waterway transportation using binary NSGA-II. *Int. Trans. Oper. Res.* **2019**, *27*, 1501–1525. [CrossRef]
181. Mohapatra, P.; Nayak, A.; Kumar, S.; Tiwari, M. Multi-objective process planning and scheduling using controlled elitist non-dominated sorting genetic algorithm. *Int. J. Prod. Res.* **2014**, *53*, 1712–1735. [CrossRef]
182. Hu, C.; Yan, X.; Gong, W.; Liu, X.; Wang, L.; Gao, L. Multi-objective based scheduling algorithm for sudden drinking water contamination incident. *Swarm Evol. Comput.* **2020**, *55*, 100674. [CrossRef]
183. Lu, C.; Gao, L.; Li, X.; Wang, Q.; Liao, W.; Zhao, Q. An Efficient Multiobjective Backtracking Search Algorithm for Single Machine Scheduling with Controllable Processing Times. *Math. Probl. Eng.* **2017**, *2017*, 1–24. [CrossRef]
184. Cho, H.-M.; Bae, S.-J.; Kim, J.; Jeong, I.-J. Bi-objective scheduling for reentrant hybrid flow shop using Pareto genetic algorithm. *Comput. Ind. Eng.* **2011**, *61*, 529–541. [CrossRef]
185. De, A.; Choudhary, A.; Tiwari, M.K. Multiobjective Approach for Sustainable Ship Routing and Scheduling with Draft Restrictions. *IEEE Trans. Eng. Manag.* **2019**, *66*, 35–51. [CrossRef]
186. Leydesdorff, L.; Milojević, S. Scientometrics. *arXiv* **2012**, arXiv:1208.4566.
187. Childress, D. Citation tools in academic libraries: Best practices for reference and instruction. *Ref. User Serv. Q.* **2011**, *51*, 143. [CrossRef]
188. Estabrooks, C.A.; Derksen, L.; Winther, C.; Lavis, J.N.; Scott, S.D.; Wallin, L.; Profetto-McGrath, J. The intellectual structure and substance of the knowledge utilization field: A longitudinal author co-citation analysis, 1945 to 2004. *Implement. Sci.* **2008**, *3*, 49. [CrossRef]
189. Emrouznejad, A.; Marra, M. Big Data: Who, What and Where? Social, Cognitive and Journals Map of Big Data Publications with Focus on Optimization. In *Big Data Optimization: Recent Developments and Challenges*; Springer: Berlin/Heidelberg, Germany, 2016; pp. 1–16.
190. Rahimi, I.; Ahmadi, A.; Zobaa, A.F.; Emrouznejad, A.; Aleem, S.H.E.A. *Big Data Optimization in Electric Power Systems: A Review*; CRC Press: Boca Raton, FL, USA, 2017.
191. Musigmann, B.; Von Der Gracht, H.; Hartmann, E. Blockchain Technology in Logistics and Supply Chain Management—A Bibliometric Literature Review from 2016 to January 2020. *IEEE Trans. Eng. Manag.* **2020**, *67*, 988–1007. [CrossRef]
192. Neelam, S.; Sood, S.K. A Scientometric Review of Global Research on Smart Disaster Management. *IEEE Trans. Eng. Manag.* **2021**, *68*, 317–329. [CrossRef]
193. Weinberg, B.H. Bibliographic coupling: A review. *Inf. Storage Retr.* **1974**, *10*, 189–196. [CrossRef]
194. Luo, H.; Du, B.; Huang, G.Q.; Chen, H.; Li, X. Hybrid flow shop scheduling considering machine electricity consumption cost. *Int. J. Prod. Econ.* **2013**, *146*, 423–439. [CrossRef]
195. Liu, P.; Guo, S.; Xu, X.; Chen, J. Derivation of Aggregation-Based Joint Operating Rule Curves for Cascade Hydropower Reservoirs. *Water Resour. Manag.* **2011**, *25*, 3177–3200. [CrossRef]
196. Sengupta, S.; Das, S.; Nasir, M.; Vasilakos, A.V.; Pedrycz, W. An Evolutionary Multiobjective Sleep-Scheduling Scheme for Differentiated Coverage in Wireless Sensor Networks. *IEEE Trans. Syst. Man Cybern. Part C Appl. Rev.* **2012**, *42*, 1093–1102. [CrossRef]
197. Langdon, W.B.; Harman, M.; Jia, Y. Efficient multi-objective higher order mutation testing with genetic programming. *J. Syst. Softw.* **2010**, *83*, 2416–2430. [CrossRef]
198. Wang, W.; Tian, G.; Chen, M.; Tao, F.; Zhang, C.; Ai-Ahmari, A.; Li, Z.; Jiang, Z. Dual-objective program and improved artificial bee colony for the optimization of energy-conscious milling parameters subject to multiple constraints. *J. Clean. Prod.* **2020**, *245*, 118714. [CrossRef]
199. Wu, X.; Cao, Y.; Xiao, Y.; Guo, J. Finding of urban rainstorm and waterlogging disasters based on microblogging data and the location-routing problem model of urban emergency logistics. *Ann. Oper. Res.* **2020**, *290*, 865–896. [CrossRef]
200. Xu, X.; Mo, R.; Dai, F.; Lin, W.; Wan, S.; Dou, W. Dynamic Resource Provisioning with Fault Tolerance for Data-Intensive Meteorological Workflows in Cloud. *IEEE Trans. Ind. Inform.* **2020**, *16*, 6172–6181. [CrossRef]
201. Salkuti, S.R. Day-ahead thermal and renewable power generation scheduling considering uncertainty. *Renew. Energy* **2019**, *131*, 956–965. [CrossRef]

202. Verma, S.; Pant, M.; Snassel, V. A Comprehensive Review on NSGA-II for Multi-Objective Combinatorial Optimization Problems. *IEEE Access* **2021**, *9*, 57757–57791. [CrossRef]
203. Deb, K.; Jain, H. An Evolutionary Many-Objective Optimization Algorithm Using Reference-Point-Based Nondominated Sorting Approach, Part I: Solving Problems with Box Constraints. *IEEE Trans. Evol. Comput.* **2013**, *18*, 577–601. [CrossRef]
204. Zhang, H.; Wang, G.-G.; Dong, J.; Gandomi, A. Improved NSGA-III with Second-Order Difference Random Strategy for Dynamic Multi-Objective Optimization. *Processes* **2021**, *9*, 911. [CrossRef]

Article

Enhance Teaching-Learning-Based Optimization for Tsallis-Entropy-Based Feature Selection Classification Approach

Di Wu [1], Heming Jia [2,*], Laith Abualigah [3,4], Zhikai Xing [5,*], Rong Zheng [2], Hongyu Wang [2] and Maryam Altalhi [6]

[1] School of Education and Music, Sanming University, Sanming 365004, China; wudi@fjsmu.edu.cn
[2] School of Information Engineering, Sanming University, Sanming 365004, China; zhengr@fjsmu.edu.cn (R.Z.); 19890432@fjsmu.edu.cn (H.W.)
[3] Faculty of Computer Sciences and Informatics, Amman Arab University, Amman 11953, Jordan; aligah.2020@gmail.com
[4] School of Computer Science, Universiti Sains Malaysia, Penang, Gelungor 11800, Malaysia
[5] School of Electrical Engineering and Automation, Wuhan University, Wuhan 430072, China
[6] Department of Management Information System, College of Business Administration, Taif University, P.O. BOX 11099, Taif 21944, Saudi Arabia; marem.m@tu.edu.sa
* Correspondence: jiaheming@fjsmu.edu.cn (H.J.); xingzk@whu.edu.cn (Z.X.)

Abstract: Feature selection is an effective method to reduce the number of data features, which boosts classification performance in machine learning. This paper uses the Tsallis-entropy-based feature selection to detect the significant feature. Support Vector Machine (SVM) is adopted as the classifier for classification purposes in this paper. We proposed an enhanced Teaching-Learning-Based Optimization (ETLBO) to optimize the SVM and Tsallis entropy parameters to improve classification accuracy. The adaptive weight strategy and Kent chaotic map are used to enhance the optimal ability of the traditional TLBO. The proposed method aims to avoid the main weaknesses of the original TLBO, which is trapped in local optimal and unbalance between the search mechanisms. Experiments based on 16 classical datasets are selected to test the performance of the ETLBO, and the results are compared with other well-established optimization algorithms. The obtained results illustrate that the proposed method has better performance in classification accuracy.

Keywords: feature selection; optimization algorithm; Tsallis-entropy; teaching and learning; adaptive weight strategy; Kent chaotic map

Citation: Wu, D.; Jia, H.; Abualigah, L.; Xing, Z.; Zheng, R.; Wang, H.; Altalhi, M. Enhance Teaching-Learning-Based Optimization for Tsallis-Entropy-Based Feature Selection Classification Approach. *Processes* **2022**, *10*, 360. https://doi.org/10.3390/pr10020360

Academic Editor: Jean-Pierre Corriou

Received: 9 January 2022
Accepted: 11 February 2022
Published: 14 February 2022

Publisher's Note: MDPI stays neutral with regard to jurisdictional claims in published maps and institutional affiliations.

1. Introduction

Machine learning has been widely used in many practical applications such as data mining, text processing, pattern recognition, and medical image analysis, which often rely on large data sets [1,2]. From utilizing label information, feature selection algorithms are mainly categorized as filters or wrapper approaches [3,4]. The wrapper-based methods are commonly used to finish the classification task [5]. The main step includes classifiers, evaluation criteria of features, and finding the optimal features [6].

The SVM algorithm is one of the most popular supervised models and is regarded as one of the most robust methods in the machine learning field [7,8]. SVM has some robust characteristics compared to other methods, such as excellent generalization performance, which is able to generate high-quality decision boundaries based on a small subset of training data points [9]. The largest problems encountered in setting up the SVM model are how to select the kernel function and its parameter values. Inappropriate parameter settings will lead to poor classification results [10].

Swarm intelligence algorithms can solve complex engineering problems, but different optimization algorithms solve different engineering problems with different effects [11,12]. The optimization algorithms can reduce the time and improve the segmentation accuracy.

There many optimization algorithms are proposed, such as Genetic Algorithm (GA) [13], Particle Swarm Optimization (PSO) [14], Differential Evolution (DE) [15], Ant Colony Optimization (ACO) [16], Artificial Bee Colony (ABC) algorithm [17], Grey Wolf Optimizer (GWO) [18], Ant Lion Optimizer (ALO) [19], Moth-flame Optimization (MFO) [20], Whale Optimization Algorithm (WOA) [21], Invasive weed optimization algorithm [22], Flower Pollination Algorithm [23]. Although all algorithms have advantages, no-free lunch (NFL) [24] has proved that no algorithm can solve all optimization problems.

There is no perfect optimization algorithm, and the optimization algorithm should be improved to solve engineering problems better. Many scholars study the strategies for improving optimization algorithm. The strategies commonly used by scholars are as follows adaptive weight strategy and chaotic map. Zhang Y. proposed an improved particle swarm optimization algorithm with an adaptive learning strategy [25]. The adaptive learning strategy increased the population diversity of PSO. Dong Z. proposed a self-adaptive weight vector adjustment strategy based on a chain segmentation strategy [26]. The self-adaptive solved the shape of the true Pareto front (PF) of the multi-objective problem. Li E. proposed a multi-objective decomposition algorithm based on adaptive weight vector and matching strategy [27]. The adaptive weight vector solved the degradation of the performance of the solution set. The chaotic map is also a general nonlinear phenomenon, and its behavior is complex and semi-random. It is mathematically defined as the randomness generated by a simple deterministic system [28]. Xu C. proposed an improved boundary bird swarm algorithm [29]. The algorithm combined the good global convergence and robustness of the birds' swarm algorithm. Tran, N. T. presented a method for fatigue life prediction of 2-DOF compliant mechanism which combined the differential evolution algorithm and the adaptive neuro-fuzzy inference system [30]. The experiment result shows that the accuracy of the proposed method is high.

Teaching-Learning-Based Optimization (TLBO) is proposed by R. V. Rao, which solves the global problem of continuous nonlinear functions [31]. The TLBO approach works on the philosophy of teaching and learning. Many scholars study the strategies to improve the optimization ability for a different problem. Gunji A. B. proposed improved TLBO for solving assembly sequence problems [32]. Zhang H. proposed a hybridizing TLBO [33]. The approach can enable better tracking accuracy and efficiency. Ho, N.L. presented a hybrid Taguchi-teaching learning-based optimization algorithm (HTLBO) [34]. The proposed method had good agreement with the predicted results. The strategies can improve the optimal ability of TLBO. In this paper, for solving the problem of learning efficiency and initial parameter setting, we use several strategies to enhance the optimal ability of the TLBO.

The main contribution of our work includes:

(1) The enhanced Teaching-Learning-Based Optimization (ETLBO) is proposed to improve optimal ability. The adaptive weight and Kent chaotic map are used to enhance the TLBO. These two strategies can improve the searching ability of the students and teachers in TLBO.

(2) We adopt the Tsaliis entropy-based feature selection method for finding the crucial feature. The selected feature x and the parameter α of Tsallis entropy are optimized by ETLBO.

(3) The parameter c of the SVM classifier is optimized by ETLBO for obtaining high classification accuracy. The core idea of this method is to automatically determine the parameter α of Tsallis entropy and parameter c of the SVM under different data.

The proposed method is tested on several feature selection and classification problems in terms of several comment evaluation measures. The results are compared with other well-established optimization methods. The obtained results showed that the proposed ETLBO got better and promising results in almost all the tested problems compared to other methods.

The rest of the paper is described as follows: Section 2 introduces Tsallis's entropy-based feature selection formula. Section 3, Enhance Teaching-learning-based optimization, and the ETLBO optimizes the feature selection design is introduced. In Sections 4 and 5, the feature selection results and the algorithm analysis are given. Finally, the conclusions are summarized in Section 6.

2. Related work

2.1. Tsallis Entropy-Based Feature Selection (TEFS)

TEFS estimates the importance of a feature by calculating its information gain (IG) with respect to the target feature. The IG is calculated by subtracting the Tsallis entropy of features concerning target from the total entropy of the target feature. The Tsallis entropy and IG are defined as follow:

$$H(m) = \frac{1}{1-\alpha}\log\sum_{i=1}^{n} p_i^{\alpha} \tag{1}$$

$$IG(m|n) = H(m) - H(m|n) \tag{2}$$

where, $H(m)$ represents the Tsallis entropy of a feature m, $IG(m|n)$ represents the Tsallis entropy of a target in terms of a feature n, m is the number of target feature, n is the total number of the feature.

IG measures the significance of a feature by calculating how much information a feature obtains us about the target.

2.2. SVM Classifier

SVM finds the optimal separation of hyperplanes between classes by focusing on the training cases of the edges of effectively discarded classes. For training samples $F = \{(x_1, y_1), \ldots, (x_n, y_n)\}$ in different dimensional spaces, a classifier can be accurately summarized. The main core of SVM is finding a suitable kernel function $k(x_i, x_j) = \phi(x_i)\cdot\phi(x_j)$, where $\phi(x_i)$ is a nonlinear function, and the function is used to transfer the nonlinear space of the sample input to two hyperplanes. The formula can be written as:

$$f(x) = w\cdot\phi(x) + b \tag{3}$$

where, w is the weight vector, b is the threshold value, and $(\cdot)$ represents the inner product operation. The objective of SVM is to determine the w, and b when minimizing the $w^T w/2$, it can be seen below:

$$\min\frac{1}{2}\|w\| + C\sum_{i=1}^{n}\xi_i \tag{4}$$

where, ξ_i is the slack variable, C is the penalty parameter.

The most commonly used kernel is the Gaussian kernel, used for data conversion in SVM. The Gaussian kernel is defined as:

$$K(x_i, x_j) = \exp(-\frac{\|x_i - x_j\|^2}{2\delta^2}) \tag{5}$$

where, $\delta > 0$ denotes the width parameter, and δ controls the mapping results.

The strategy of reducing multi-class problems to a set of dichotomies enables support vector machines to be used more appropriately with fewer computational requirements, that is, to consider all classes at once and thus to obtain a multi-class support vector machine. One way to do this is by solving a single optimization problem, similar to the "one for all" approach on a fundamental basis. There are n decision functions or hyperplanes, and the problems can convert to one problem as:

$$\min\frac{1}{2}\sum_{i=1}^{n} w_i^T w_i + C\sum_{j=1}^{m}\sum_{i\neq y_j}\xi_j^i w_{y_j}^T\varphi(x_j) + b_{y_j} \tag{6}$$

where, $\xi_j^i \geq 0$. The resulting decision function can be represented as:

$$\text{argmax}_i(w_i^T \varphi(x_i) + b_i) \tag{7}$$

2.3. Fitness Function Design

The main indexes influencing FS are the classification error accuracy and the number of features. So, how to balance the number of features and the classification is the essential key for the FS problem. Whereas, f_1 is the Normalized Mutual Information (NMI) [13]. The formula can be seen as follow:

$$f_1(x) = NMI(X, S) = \frac{MI(X; S)}{[G(X) + G(S)]/2} \tag{8}$$

where, X is the set of clusters and S is the set of classes. The MI is the mutual information between X and S [35]. It can be defined as follow:

$$\begin{aligned} MI(X; S) &= \sum_k \sum_j P(X_k \cap S_j) \log \frac{P(X_k \cap S_j)}{P(X_k)P(S_j)} \\ &= \sum_k \sum_j \frac{|X_k \cap S_j|}{N} \log \frac{N|X_k \cap S_j|}{|X_k||S_j|} \end{aligned} \tag{9}$$

where, $P(X_k)$, $P(S_j)$, $P(X_k \cap S_j)$ is the probability of the X_k, S_j, and $X_k \cap S_j$. The $G(X)$ comes from the maximum likelihood estimation of probability.

$$\begin{aligned} G(X) &= -\sum_k P(X_k) \log P(X_k) \\ &= -\sum_k \frac{|X_k|}{N} \log \frac{|X_k|}{N} \end{aligned} \tag{10}$$

3. Enhance Teaching-Learning-Based Optimization (ETLBO)

In this section, we introduce the proposed method in detail. Firstly, we introduce the TLBO and the strategies used in the proposed method. And then, the ETLBO is introduced. Finally, the flowchart of the proposed method is described.

3.1. Teacher Phase

It is the first part of the algorithm where the learner with the highest marks acts as a teacher, and the teacher's task is to increase the mean marks of the class. The update process of i-th learner in teacher phase is formulated as:

$$X_{i,new} = X_i + rand \times (X_{teacher} - T_F \times X_{ave}) \tag{11}$$

where, X_i is the solution of the i-th learner, $X_{teacher}$ represents the teacher's solution, X_{ave} means the average of all learners, rand is a random number in (0,1), and T_F is the teaching factor that decides the value of mean to be changed. The value can be either 1 or 2, which is again a heuristic step and decided randomly with equal probability $T_F = round[1 + rand(0,1)\{2 - 1\}]$.

In addition, the new solution $X_{i,new}$ is accepted only if it is better than the previous solution, it can be formulated as:

$$X_i = \begin{cases} X_{i,new} & f(X_{i,new}) > f(X_i) \\ X_i & otherwise \end{cases} \tag{12}$$

where, f means the fitness function.

3.2. Learner Phase

The second part of the algorithm is where the learner updates its knowledge through interaction with other learners. In each iteration, two learners interact with X_m and X_n, in which the more innovative learner improves the marks of other learners. In the learner phase, one learner learns new things if the other learner has more knowledge than himself. The phenomenon is described as follows:

$$X_{m,new} = \begin{cases} X_m + rand \times (X_m - X_n); f(X_m) > f(X_n) \\ X_m + rand \times (X_n - X_m); f(X_n) > f(X_m) \end{cases} \quad (13)$$

The temporary solution is accepted only if it is better than the previous solution; it can be formulated as:

$$X_m = \begin{cases} X_{m,new}; f(X_{m,new}) > f(X_m) \\ X_m; otherwise \end{cases} \quad (14)$$

3.3. Adaptive Weight Strategy

The adaptive weight strategy is easier to jump out of local minima, facilitating global optimization. While the TLBO solves the problem of the complex optimized function, the algorithm will easily fall into the local optimum. And a smaller inertia factor is beneficial for precise local search for the current search domain. We design a new weight strategy t which can be written as follows:

$$t = (1 - \frac{iter}{Max\_iter})^{1-\sin(\pi \frac{iter}{Max\_iter})} \quad (15)$$

where, *iter* is the current number of the iteration; *Max_iter* is the max number of the iteration.

3.4. Kent Chaotic Map (KCM)

Chaotic mapping is one kind of nonlinear mapping that can generate a random number sequence. It is sensitive to initial values, which ensures that the encoder can generate an unrelated encoding sequence. There are many kinds of chaotic maps, such as Logistic map, Kent map, etc. In this paper, we use the Kent map as the improved strategy. The formula of the Kent map can be seen as follow:

$$f(x) = \begin{cases} \frac{x}{a} & 0 < x \leq a \\ \frac{1-x}{1-a} & a < x < 1 \end{cases} \quad (16)$$

where, a is a variable value, x is the initial value of the $x(0)$. In this paper, $a = 0.5$.

3.5. Proposed Method

There are two phases in the basic TLBO search process to update the individual's position. In the teacher phase, we use the Kent chaotic map to improve the original state of the teacher. The teacher can be endowed with different abilities to teach the different students. This strategy allows the abilities of different teachers to be demonstrated. In the learner phase, we design a learning efficiency to improve the students' learning state. The adaptive weight strategy can improve itself with the iteration increases. The students will learn more knowledge at the beginning phase of the iteration. The students can obtain enough knowledge at the end of the iteration, and the adaptive weight gets small. The students can learn the different knowledge at the different phases. The formula can be represented as follow:

$$X_{m,new} = \begin{cases} X_m \times t + rand \times (X_m - X_n); f(X_m) > f(X_n) \\ X_m \times t + rand \times (X_n - X_m); f(X_n) > f(X_m) \end{cases} \quad (17)$$

where, t is the adaptive weight.

The proposed classification method can be divided into two parts: feature selection and the parameter selection of the SVM. At first, the Tsallis entropy of the target is calculated using Equation (1). Then the entropy of each feature concerning the target is calculated and subtracted from the target's entropy using Equation (2). In this process, the selected feature

x and the parameter α of Tsallis entropy are optimized by the ETLBO. The parameter α can decide the ability of the Tsallis entropy.

In the second part, we use the ETLBO to optimize parameter c of SVM. The penalty coefficient c is the compromise between the smoothness of the fitting function and the classification accuracy. When c is too large, the training accuracy is high, and the generalization ability is poor; while c is too small, errors will be increased. Therefore, a reasonable selection of parameter c can obviously improve the model's classification accuracy and generalization ability.

Finally, the selected feature x, the parameter α of Tsallis entropy, and the parameter c of SVM are optimized by ETLBO. We use the parameter optimized by the ETLBO and the SVM to classify the test dataset. The SVM classifier output the classification result. The flowchart of the proposed method is shown in Figure 1.

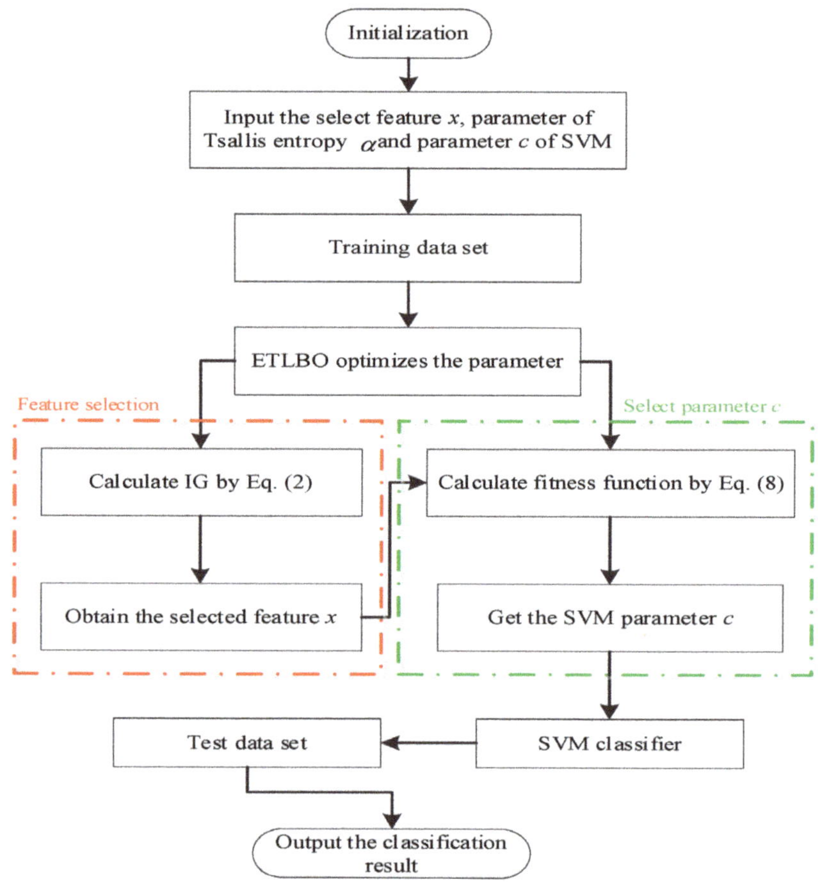

Figure 1. The flowchart of the proposed method.

4. Experiment and Result

To analysis the effectiveness of the proposed method, five optimization algorithms are used for comparison, such as PSO [13], WOA [20], HHO [36], TLBO [29], HSOA [37], and HTLBO [34]. The PSO, WOA, HHO, and TLBO are the original optimization algorithms. These optimization algorithms have the strong ability to find the optimal value of the mathematical function. While these algorithms optimize the engineering problems, the optimization performance is not well. Many schoolers study the strategies to improve the optimization algorithms. The HSOA and HTLBO are improved methods. These two algorithms use the hybrid way to enhance the optimization ability of the SOA and TLBO. The improved methods have the excellent performance to solve the problems which mentioned in the reference [34,37]. However, these algorithms may not solve all problems. Therefore, we select these algorithms as compared algorithms to test the performance of the proposed method.

The set of parameters is the same as the reference. All the methods are coded and implemented in MATLAB 2018B. To keep the fairness of the compared algorithms, each algorithm runs 30 times independently. To test the performance of the comparison algorithm, we set the number of populations to 30 and the maximum iteration to 500. The proposed ETLBO is training in MATLAB2018B. Experiments are managed on a computer with an i7-11800H central processing unit.

The results of the proposed method are described in this section. First, the fitness values obtained by the different optimization algorithms are compared to show the performances of these approaches. Then, we analyze the classification result of the compared algorithms. Finally, the discussion of the proposed method is described.

4.1. Datasets and Evaluation Index

The benchmark datasets used in the evaluations are introduced. The dataset selects 16 standard datasets from the University of California (UCI) data repository [38]. Table 1 records the primary information of these selected datasets.

Table 1. The datasets used in the experiments.

	Datasets	Samples	Features
1	Iris	150	4
2	Wine	178	13
3	Sonar	208	60
4	Vehicle	846	18
5	Balancescale	625	4
6	CMC	1473	9
7	Cancer	683	9
8	Vowel	871	3
9	Thyroid	215	5
10	WDBC	569	30
11	HeartEW	270	13
12	Lymphography	148	18
13	SonarEW	208	60
14	IonosphereEW	351	34
15	Vote	300	16
16	WaveformEW	5000	40

To evaluate the result of the health index diagnosis, we use the F-score, the accuracy of the classification and the CPU time as the metric index.

The function of F-score can be defined as follow:

$$F - score = (1 + \beta^2) \cdot \frac{Precision \cdot Recall}{\beta^2 \cdot Precision + Recall} \tag{18}$$

$$Precision = \frac{Tp}{Tp + Fp} \tag{19}$$

$$Recall = \frac{Tp}{Tp + Fn} \tag{20}$$

where, Tn is the number of negative classes, Fn is the number of negative classes, Tp is the number of positive classes, and Fp is the number of positive classes.

4.2. Experiment 1: Feature Selection

Table 2 shows the fitness value of the compared algorithms. The table shows that when the number of features is small, the compared algorithms can reduce the number of features. When the number of features increases, it takes a huge challenge for the optimization algorithms. The ETLBO obtains better performance than compared algorithms. Table 3 shows the std of the fitness values. It can be known from the given table that the ETLBO has strong robustness.

Table 2. The fitness values of compared algorithms.

	PSO	WOA	HHO	TLBO	HSOA	HTLBO	ETLBO
Iris	0.0271	0.0271	0.0271	0.0271	0.0271	0.0271	0.0271
Wine	0.1846	0.1657	0.1594	0.1709	0.1585	0.1576	0.1521
Sonar	0.2655	0.2147	0.2356	0.2386	0.2268	0.2258	0.2145
Vehicle	0.2972	0.2575	0.2746	0.2528	0.2432	0.2741	0.2716
Balancescale	0.0619	0.0519	0.0540	0.0613	0.0521	0.0556	0.0516
CMC	0.1937	0.1855	0.1837	0.1777	0.1701	0.1758	0.1621
Cancer	0.0629	0.0710	0.0660	0.0645	0.0634	0.0645	0.0612
Vowel	0.1251	0.1251	0.1251	0.1251	0.1251	0.1251	0.1251
Thyroid	0.0934	0.0853	0.0995	0.0922	0.0925	0.0921	0.0851
WDBC	0.3828	0.3155	0.3678	0.3258	0.3234	0.3221	0.3154
HeartEW	0.2942	0.2993	0.3325	0.2869	0.3077	0.2989	0.2814
Lymphography	0.2208	0.2040	0.2127	0.1965	0.1999	0.1989	0.1951
SonarEW	0.3650	0.4065	0.3892	0.3721	0.3610	0.3589	0.3514
IonosphereEW	0.3390	0.3425	0.3718	0.3339	0.3426	0.3411	0.3226
Vote	0.2274	0.2567	0.2460	0.2542	0.2215	0.2218	0.2158
WaveformEW	0.4348	0.4151	0.3998	0.4436	0.4278	0.4025	0.3915

Table 3. The std of fitness values.

	PSO	WOA	HHO	TLBO	HSOA	HTLBO	ETLBO
Iris	4.56×10^{-4}	4.69×10^{-4}	5.40×10^{-4}	4.60×10^{-4}	4.86×10^{-4}	4.88×10^{-4}	4.50×10^{-4}
Wine	6.84×10^{-4}	6.53×10^{-4}	6.07×10^{-4}	6.22×10^{-4}	5.79×10^{-4}	5.82×10^{-4}	5.60×10^{-4}
Sonar	1.27×10^{-4}	1.12×10^{-5}	1.00×10^{-5}	1.08×10^{-5}	1.10×10^{-5}	1.11×10^{-5}	1.00×10^{-5}
Vehicle	6.54×10^{-4}	6.16×10^{-4}	6.37×10^{-4}	6.95×10^{-4}	6.13×10^{-4}	6.15×10^{-4}	6.10×10^{-4}
Balancescale	4.25×10^{-4}	4.49×10^{-4}	4.50×10^{-4}	4.89×10^{-4}	4.17×10^{-4}	4.20×10^{-4}	4.10×10^{-4}
CMC	4.13×10^{-4}	3.47×10^{-4}	3.96×10^{-4}	3.56×10^{-4}	3.40×10^{-4}	3.41×10^{-4}	3.30×10^{-4}
Cancer	1.15×10^{-3}	1.13×10^{-3}	1.03×10^{-3}	1.03×10^{-3}	1.02×10^{-3}	1.02×10^{-4}	9.60×10^{-4}
Vowel	7.80×10^{-4}	7.68×10^{-4}	8.44×10^{-4}	7.94×10^{-4}	7.74×10^{-4}	7.78×10^{-4}	7.50×10^{-4}
Thyroid	6.82×10^{-4}	6.79×10^{-4}	7.33×10^{-4}	7.22×10^{-4}	7.26×10^{-4}	7.26×10^{-4}	6.70×10^{-4}
WDBC	1.05×10^{-3}	1.01×10^{-3}	1.02×10^{-3}	1.01×10^{-3}	9.74×10^{-4}	9.78×10^{-4}	9.40×10^{-4}
HeartEW	9.67×10^{-4}	9.34×10^{-4}	9.08×10^{-4}	1.00×10^{-3}	8.91×10^{-4}	8.93×10^{-4}	8.80×10^{-4}
Lymphography	8.62×10^{-4}	7.11×10^{-4}	8.00×10^{-4}	8.13×10^{-4}	7.44×10^{-4}	7.47×10^{-4}	6.90×10^{-4}
SonarEW	4.54×10^{-4}	3.81×10^{-4}	3.91×10^{-4}	4.50×10^{-4}	4.00×10^{-4}	4.02×10^{-4}	3.80×10^{-4}
IonosphereEW	9.56×10^{-4}	8.85×10^{-4}	8.22×10^{-4}	7.72×10^{-4}	7.82×10^{-4}	7.88×10^{-4}	7.40×10^{-4}
Vote	8.81×10^{-4}	9.30×10^{-4}	9.87×10^{-4}	9.64×10^{-4}	9.23×10^{-4}	9.25×10^{-1}	8.50×10^{-4}
WaveformEW	4.34×10^{-4}	4.51×10^{-4}	4.28×10^{-4}	4.34×10^{-4}	3.92×10^{-4}	3.93×10^{-4}	3.90×10^{-4}

Table 4 shows the number of the selected attributes. The compared algorithms can reduce the number of features. The attributes are little, and the compared algorithms obtain the same result. The ETLBO gets the least attributes among the compared algorithms when the attributes are large. The total attributes of the dataset, the ETLBO also obtain the least attributes than other algorithms. It means that the ETLBO can reduce the number of features. However, reducing the number of features does not mean the classification accuracy is high.

Table 4. The average number of selected attributes.

	Attributes	PSO	WOA	HHO	TLBO	HSOA	HTLBO	ETLBO
Iris	4	3	3	3	3	3	3	3
Wine	13	6	6	5	10	6	6	**5**
Sonar	60	31	32	32	48	29	29	**28**
Vehicle	18	4	4	4	4	4	4	4
Balancescale	4	4	3	4	4	4	3	**3**
CMC	9	7	6	6	8	7	8	**6**
Cancer	9	5	6	5	7	6	6	**5**
Vowel	3	3	3	3	3	3	3	3
Thyroid	5	4	4	3	4	4	4	**3**
WDBC	30	9	8	7	10	7	8	**6**
HeartEW	13	5	5	4	6	5	5	**4**
Lymphography	18	6	7	7	6	8	6	**5**
SonarEW	60	19	18	19	22	15	13	**12**
IonosphereEW	34	25	23	20	17	15	14	**12**
Vote	16	8	7	9	8	6	6	**6**
WaveformEW	40	21	26	24	23	20	18	**16**
Total	336	160	161	155	183	142	136	**121**

Table 5 shows the parameter obtained by ETLBO. It can be seen from the table that the ETBLO obtains the different values under the diverse dataset. The ETBLO not only reduce the number of features but also acquires the parameter α of Tsallis entropy and the parameter c of SVM. We will test the performance of the compared algorithms in the next section.

Table 5. The parameter obtained by ETLBO.

	α	c
Iris	0.52	1.14
Wine	0.64	1.20
Sonar	0.80	1.24
Vehicle	0.22	1.83
Balancescale	0.77	1.17
CMC	0.26	1.22
Cancer	0.15	0.65
Vowel	0.07	1.82
Thyroid	0.61	1.01
WDBC	0.81	1.17
HeartEW	0.78	0.95
Lymphography	0.47	1.64
SonarEW	0.08	1.82
IonosphereEW	0.55	0.99
Vote	0.73	1.58
WaveformEW	0.55	0.73
Total	0.73	1.83

4.3. Experiment 2: Classification

Table 6 shows the classification results of compared algorithms. Table 7 shows the f-score of the compared methods. The table result shows that the ETLBO is better than the original TLBO. The strategies improve the optimal ability of the TLBO. At the same time, the HSOA and ETLBO are better than the other algorithms. It means that the strategies significantly boost the original optimization algorithms. It can be known that the methods can be ordered as follows in terms of them F-score result: ETLBO > HTLBO > HSOA > HHO > PSO > WOA > TLBO. To sum up, the ETLBO obtains the high f-score values.

Table 6. The classification accuracy of compared algorithms.

	PSO	WOA	HHO	TLBO	HSOA	HTLBO	ETLBO
Iris	0.9545	0.9481	0.9320	0.9167	0.9548	0.9546	**0.9579**
Wine	0.9413	0.9303	0.9232	0.9369	0.9411	0.9479	**0.9488**
Sonar	0.9160	0.9375	0.9242	0.9025	0.9347	0.9359	**0.9435**
Vehicle	0.9012	0.9306	0.9147	0.9358	0.9345	0.9339	**0.9414**
Balancescale	0.9338	0.9241	0.9308	0.9287	0.9450	0.9442	**0.9465**
CMC	0.9440	0.9271	0.9259	0.9253	0.9387	0.9371	**0.9454**
Cancer	0.9060	0.9180	0.9403	0.9356	0.9349	0.9408	**0.9438**
Vowel	0.9600	0.9477	0.9429	0.9559	0.9628	0.9581	**0.9651**
Thyroid	0.9031	0.9249	0.9048	0.8849	0.9202	0.9282	**0.9285**
WDBC	0.9329	0.9351	0.9177	0.9057	0.9345	0.9358	**0.9385**
HeartEW	0.9230	0.9118	0.9143	0.8995	0.9323	0.9271	**0.9358**
Lymphography	0.9022	0.9256	0.9001	0.9141	0.9210	0.9180	**0.9268**
SonarEW	0.9360	0.9115	0.9116	0.9164	0.9336	0.9357	**0.9361**
IonosphereEW	0.9361	0.9448	0.9276	0.9233	0.9384	0.9441	**0.9458**
Vote	0.9136	0.9170	0.9217	0.9008	0.9381	0.9378	**0.9451**
WaveformEW	0.9276	0.9354	0.9229	0.9029	0.9292	0.9284	**0.9365**

Table 7. The f-score of compared algorithms.

	PSO	WOA	HHO	TLBO	HSOA	HTLBO	ETLBO
Iris	0.9362	0.9038	0.9193	0.9007	0.9262	0.9334	**0.9498**
ine	0.9118	0.9116	0.9175	0.9219	0.9194	0.9288	**0.9433**
Sonar	0.9043	0.913	0.9206	0.8754	0.9223	0.9091	**0.9359**
Vehicle	0.8782	0.8895	0.9038	0.9142	0.918	0.9164	**0.9387**
Balancescale	0.9169	0.9055	0.9139	0.9139	0.9309	0.9276	**0.9422**
CMC	0.9295	0.9	0.8988	0.8977	0.9217	0.9109	**0.9429**
Cancer	0.8861	0.9119	0.8945	0.9104	0.9133	0.9223	**0.9396**
Vowel	0.9399	0.914	0.9336	0.9345	0.9464	0.9506	**0.9601**
Thyroid	0.8812	0.8759	0.8954	0.8585	0.908	0.904	**0.9211**
WDBC	0.9102	0.9039	0.9071	0.8913	0.9128	0.9143	**0.9329**
HeartEW	0.8936	0.8887	0.8962	0.8892	0.9045	0.9186	**0.9275**
Lymphography	0.8894	0.8725	0.9117	0.8891	0.8883	0.9109	**0.9184**
SonarEW	0.9199	0.8988	0.897	0.8888	0.9066	0.9092	**0.9360**
IonosphereEW	0.9171	0.9088	0.9212	0.9090	0.9304	0.9168	**0.9411**
Vote	0.8890	0.9108	0.9057	0.8776	0.9225	0.9204	**0.9449**
WaveformEW	0.9115	0.8948	0.9106	0.8888	0.9089	0.9102	**0.9346**

To sum up, the ETLBO obtained the best result in compared algorithms. The ETLBO not only reduces the number of features but also obtains high classification accuracy. Table 8 shows that the std of classification accuracy. The ETLBO has a better stable ability than other algorithms. The proposed method has strong robustness to finish the classification task.

A statistical test is an essential and vital measure to evaluate and prove the performance of the tested methods. Parameter statistical test is based on various assumptions. This section uses well-known non-parametric statistical test types, Wilcoxon's rank-sum test [39]. Table 9 shows the results of the Wilcoxon rank-sum test. It can be found that the ETLBO is significantly different from other methods.

The CPU time is also an important index for the practical engineering testing problem. The CPU time results of the compared algorithms can be seen in Table 10. The CPU time ordering of each algorithm is: TLBO < PSO < WOA < HHO < ETLBO < HTLBO < HSOA. Although the ETLBO costs considerable CPU time, the classification accuracy has good performance. At the same time, the ETLBO uses less CPU time than HSOA. It means that the strategies have good adaptive effectiveness for the TLBO. The strategies enhance the TLBO under less CPU time than the improved method.

Table 8. The std of classification accuracy.

	PSO	WOA	HHO	TLBO	HSOA	HTLBO	ETLBO
Iris	4.57×10^{-5}	4.69×10^{-5}	5.45×10^{-5}	4.63×10^{-5}	4.90×10^{-5}	4.60×10^{-5}	4.54×10^{-5}
Wine	6.87×10^{-5}	6.56×10^{-5}	6.08×10^{-5}	6.22×10^{-5}	5.84×10^{-5}	6.90×10^{-5}	5.65×10^{-5}
Sonar	1.28×10^{-6}	1.12×10^{-5}	1.01×10^{-6}	1.08×10^{-5}	1.10×10^{-6}	1.28×10^{-6}	1.00×10^{-5}
Vehicle	6.60×10^{-5}	6.19×10^{-5}	6.39×10^{-5}	6.98×10^{-5}	6.14×10^{-5}	6.63×10^{-5}	6.12×10^{-5}
Balancescale	4.28×10^{-5}	4.50×10^{-5}	4.52×10^{-5}	4.92×10^{-5}	4.18×10^{-5}	4.31×10^{-5}	4.13×10^{-5}
CMC	4.14×10^{-5}	3.50×10^{-5}	3.97×10^{-5}	3.58×10^{-5}	3.41×10^{-5}	4.18×10^{-5}	3.32×10^{-5}
Cancer	1.15×10^{-4}	1.14×10^{-4}	1.03×10^{-4}	1.04×10^{-4}	1.02×10^{-4}	1.16×10^{-5}	9.62×10^{-5}
Vowel	7.80×10^{-5}	7.69×10^{-5}	8.48×10^{-5}	7.98×10^{-5}	7.78×10^{-5}	7.88×10^{-5}	7.55×10^{-5}
Thyroid	6.86×10^{-5}	6.83×10^{-5}	7.39×10^{-5}	7.28×10^{-5}	7.33×10^{-5}	6.88×10^{-5}	6.74×10^{-5}
WDBC	1.06×10^{-4}	1.01×10^{-4}	1.02×10^{-4}	1.02×10^{-4}	9.84×10^{-5}	1.06×10^{-5}	9.42×10^{-5}
HeartEW	9.76×10^{-5}	9.43×10^{-5}	9.11×10^{-5}	1.00×10^{-4}	8.95×10^{-5}	9.86×10^{-5}	8.86×10^{-5}
Lymphography	8.70×10^{-5}	7.18×10^{-5}	8.07×10^{-5}	8.16×10^{-5}	7.50×10^{-5}	8.73×10^{-5}	6.94×10^{-5}
SonarEW	4.55×10^{-5}	3.84×10^{-5}	3.92×10^{-5}	4.51×10^{-5}	4.02×10^{-5}	4.56×10^{-5}	3.81×10^{-5}
IonosphereEW	9.61×10^{-5}	8.88×10^{-5}	8.24×10^{-5}	7.74×10^{-5}	7.87×10^{-5}	9.66×10^{-5}	7.45×10^{-5}
Vote	8.82×10^{-5}	9.35×10^{-5}	9.97×10^{-5}	9.64×10^{-5}	9.31×10^{-5}	8.89×10^{-5}	8.56×10^{-5}
WaveformEW	4.38×10^{-5}	4.55×10^{-5}	4.29×10^{-5}	4.35×10^{-5}	3.94×10^{-5}	4.41×10^{-5}	3.92×10^{-5}

Table 9. Wilcoxon's rank-sum test of classification accuracy.

	PSO		WOA		HHO		TLBO		HSOA		HTLBO	
	p-Value	h	*p*-Value	h	*p*-Value	h	*p*-Value	h	*p*-Value	h	*p*-Value	h
Iris	<0.05	1	<0.05	1	<0.05	1	<0.05	1	<0.05	1	<0.05	1
Wine	<0.05	1	<0.05	1	<0.05	1	<0.05	1	<0.05	1	<0.05	1
Sonar	<0.05	1	<0.05	1	<0.05	1	<0.05	1	<0.05	1	<0.05	1
Vehicle	<0.05	1	<0.05	1	<0.05	1	<0.05	1	<0.05	1	<0.05	1
Balancescale	<0.05	1	<0.05	1	<0.05	1	<0.05	1	<0.05	1	<0.05	1
CMC	<0.05	1	<0.05	1	<0.05	1	<0.05	1	<0.05	1	<0.05	1
Cancer	<0.05	1	<0.05	1	<0.05	1	<0.05	1	<0.05	1	<0.05	1
Vowel	<0.05	1	<0.05	1	<0.05	1	<0.05	1	<0.05	1	<0.05	1
Thyroid	<0.05	1	<0.05	1	<0.05	1	<0.05	1	<0.05	1	<0.05	1
WDBC	<0.05	1	<0.05	1	<0.05	1	<0.05	1	<0.05	1	<0.05	1
HeartEW	<0.05	1	<0.05	1	<0.05	1	<0.05	1	<0.05	1	<0.05	1
Lymphography	<0.05	1	<0.05	1	<0.05	1	<0.05	1	<0.05	1	<0.05	1
SonarEW	<0.05	1	<0.05	1	<0.05	1	<0.05	1	<0.05	1	<0.05	1
IonosphereEW	<0.05	1	<0.05	1	<0.05	1	<0.05	1	<0.05	1	<0.05	1
Vote	<0.05	1	<0.05	1	<0.05	1	<0.05	1	<0.05	1	<0.05	1
WaveformEW	<0.05	1	<0.05	1	<0.05	1	<0.05	1	<0.05	1	<0.05	1

4.4. Experiment 3: Compared with Different Classifiers

In this section, we compare with the different classifiers. The compared classifiers contain K-NearestNeighbor (KNN), original SVM, and random forest (RF) [40]. Table 11 shows the configuration parameters and characteristics of the classifier models.

Table 12 demonstrates the evaluation index of the compare algorithsm. The BTLBO obtains the best result than other compared classifiers in all index. The BTLBO outperforms KNN, SVM, and RF by yielding an improvement of 3.45%, 2.94%, and 1.62% in F-score index. To sum up, the optimization algorithms obtain the optimal parameter of the SVM. The classification accuracy is higher than other compared classifiers.

Table 10. The CPU time of the compared algorithms.

	ETLBO	PSO	WOA	HHO	TLBO	HSOA	HTLBO
Iris	2.3973	1.8441	2.0285	2.1794	**1.6764**	2.637	2.4128
Wine	4.1247	3.1729	3.4902	3.7498	**2.8844**	4.5372	4.1315
Sonar	6.65	5.1154	5.6269	6.0454	**4.6503**	7.315	6.6542
Vehicle	4.1722	3.2094	3.5303	3.7929	**2.9176**	4.5894	4.1765
Balancescale	2.4302	1.8694	2.0563	2.2093	**1.6994**	2.6732	2.4462
CMC	3.0792	2.3686	2.6055	2.7993	**2.1533**	3.3872	3.0803
Cancer	1.5509	1.193	1.3123	1.41	**1.0846**	1.706	1.5669
Vowel	3.6988	2.8452	3.1298	3.3626	**2.5866**	4.0687	3.7183
Thyroid	3.6676	2.8212	3.1034	3.3342	**2.5648**	4.0344	3.6827
WDBC	6.4025	4.925	5.4175	5.8204	**4.4772**	7.0427	6.4075
HeartEW	4.5164	3.4741	3.8215	4.1058	**3.1583**	4.968	4.5271
Lymphography	6.4656	4.9735	5.4709	5.8778	**4.5214**	7.1122	6.4850
SonarEW	6.9375	5.3365	5.8702	6.3068	**4.8514**	7.6313	6.9569
IonosphereEW	7.552	5.8092	6.3901	6.8654	**5.2811**	8.3072	7.5664
Vote	5.5987	4.3067	4.7374	5.0898	**3.9152**	6.1586	5.6065
WaveformEW	3.3821	2.6016	2.8618	3.0746	**2.3651**	3.7203	3.3848

Table 11. Configuration parameters and characteristics of the classifier models.

Classifier	Caret Method Value	R Package	Tuning Parameters	Characteristics
KNN	knn		k-5	Unique classifier. The number of neighbors is directly compared to the test data using the KNN function in the Caret package.
SVM	svmRadial	E1071	$\Sigma{-7} \times 10^{-2}$ c-1	Radial basic function outperformed linear SVM.
RF	rf	randomForest	mtry-8 ntree-150	Overcomes the disadvantage of simple DT using a large number of DT's to classify by majority vote. Use the randomForest function.

Table 12. The evaluation index of compared algorithms.

Classifier	Ac	Pc	R	F-Score
KNN	0.9275	0.9135	0.9123	0.9097
SVM	0.9241	0.9178	0.9167	0.9142
RF	0.9352	0.9189	0.9197	0.9261
BTLBO	**0.9478**	**0.9455**	**0.9427**	**0.9411**

5. Discussion

The proposed method has an optimal ability to solve the Tsallis-entropy-based feature selection problem in the feature selection domain. The ETLBO selects the suitable parameter of the Tsallis-entropy. At the same time, the proposed method reduces the number of features successfully. The optimization algorithms have a robust optimal ability; however, they do not adapt to solve the different optimized problems. So some adaptive strategies are very effective for improving themselves.

The proposed method obtains better classification accuracy than the compared algorithms in the classification field. The proposed method finds the proper parameter α of the SVM classifier. The proposed method has a higher classification accuracy and strong robustness than the compared algorithms. At the same time, the proposed method is better than orther compared classifiers. So, the ETLBO algorithms can be used in the classification task field.

The proposed method's limitation is that the optimization algorithm needs iteration to find the optimal solution, which is time-consuming. Improving the optimization capability and reducing the number of iterations can solve this problem. Therefore, it is necessary to search for powerful optimization algorithms and new strategies in future work.

6. Conclusions

In this paper, an enhanced teaching-learning-based optimization is proposed. The adaptive weight strategy and Kent chaotic map are used to enhance the TLBO. The ETLBO optimizes the selected feature x, the parameter α of Tsallis entropy, and the parameter c of SVM. The proposed method reduces the number of features through the UCI data experiment and finds the critical features for classification. Finally, the classification accuracy of the proposed method is better than compared algorithms.

We will design an effective and useful function to reduce the number of features in future work. We will focus on solving the randomness of the TLBO and obtaining more stability parameters of the fitness function. At the same time, we will also test the novel strategies to boost the TLBO.

Author Contributions: Conceptualization, D.W. and H.J.; methodology, D.W. and H.J.; software, D.W. and Z.X.; validation, H.J. and Z.X.; formal analysis, D.W., R.Z. and H.W.; investigation, D.W. and H.J.; writing—original draft preparation, D.W. and M.A.; writing—review and editing, D.W., L.A., M.A. and H.J.; visualization, D.W., H.W., M.A. and H.J.; funding acquisition. All authors have read and agreed to the published version of the manuscript.

Funding: This work was supported by National fund cultivation project of Sanming University (PYS2107), the Sanming University Introduces High-level Talents to Start Scientific Research Funding Support Project (21YG01S), The 14th five year plan of Educational Science in Fujian Province (FJJKBK21-149), Bidding project for higher education research of Sanming University (SHE2101), Research project on education and teaching reform of undergraduate colleges and universities in Fujian Province (FBJG20210338), Fujian innovation strategy research joint project (2020R0135). This study was financially supported via a funding grant by Deanship of Scientific Research, Taif University Researchers Supporting Project number (TURSP-2020/300), Taif University, Taif, Saudi Arabia.

Institutional Review Board Statement: Not applicable.

Informed Consent Statement: Not applicable.

Data Availability Statement: Not applicable.

Conflicts of Interest: The authors declare no conflict of interest.

References

1. Ji, B.; Lu, X.; Sun, G.; Zhang, W.; Li, J.; Xiao, Y. Bio-inspired feature selection: An improved binary particle swarm optimization approach. *IEEE Access* **2020**, *8*, 85989–86002. [CrossRef]
2. Kumar, S.; Tejani, G.G.; Pholdee, N.; Bureerat, S. Multiobjecitve structural optimization using improved heat transfer search. *Knowl.-Based Syst.* **2021**, *219*, 106811. [CrossRef]
3. Sun, L.; Wang, L.; Ding, W.; Qian, Y.; Xu, J. Feature selection using fuzzy neighborhood entropy-based uncertainty measures for fuzzy neighborhood multigranulation rough sets. *IEEE Trans. Fuzzy Syst.* **2020**, *99*, 1–14. [CrossRef]
4. Zhao, J.; Liang, J.; Dong, Z.; Tang, D.; Liu, Z. NEC: A nested equivalence class-based dependency calculation approach for fast feature selection using rough set theory. *Inform. Sci.* **2020**, *536*, 431–453. [CrossRef]
5. Liu, H.; Zhao, Z. Manipulating data and dimension reduction methods: Feature selection. In *Encyclopedia Complexity Systems Science*; Springer: New York, NY, USA, 2009; pp. 5348–5359.
6. Al-Tashi, Q.; Said, J.A.; Helmi, M.R.; Seyedali, M.; Hitham, A. Binary optimization using hybrid grey wolf optimization for feature selection. *IEEE Access* **2019**, *7*, 39496–39508. [CrossRef]
7. Homayoun, H.; Mahdi, J.; Xinghuo, Y. An opinion formation based binary optimization approach for feature selection. *Phys. A Stat. Mech. Its Appl.* **2018**, *491*, 142–152.
8. Mafarja, M.M.; Mirjalili, S. Hybrid whale optimization algorithm with simulated annealing for feature selection. *Neurocomputing* **2017**, *260*, 302–312. [CrossRef]
9. Aljarah, L.; Ai-zoubl, A.M.; Faris, H.; Hassonah, M.A.; Mirjalili, S.; Saadeh, H. Simultaneous feature selection and support vector machine optimization using the grasshopper optimization algorithm. *Cogn. Comput.* **2018**, *2*, 1–18. [CrossRef]
10. Lin, S.; Ying, K.; Chen, S.; Lee, Z. Particle swarm optimization for parameter determination and feature selection of support vector machines. *Expert Syst. Appl.* **2008**, *35*, 1817–1824. [CrossRef]
11. Sherpa, S.R.; Wolfe, D.W.; Van Es, H.M. Sampling and data analysis optimization for estimating soil organic carbon stocks in agroecosystems. *Soil Sci. Soc. Am. J.* **2016**, *80*, 1377. [CrossRef]
12. Lee, H.M.; Yoo, D.G.; Sadollah, A.; Kim, J.H. Optimal cost design of water distribution networks using a decomposition approach. *Eng. Optim.* **2016**, *48*, 16. [CrossRef]

13. Roberge, V.; Tarbouchi, M.; Okou, F. Strategies to accelerate harmonic minimization in multilevel inverters using a parallel genetic algorithm on graphical processing unit. *IEEE Trans. Power Electron.* **2014**, *29*, 5087–5090. [CrossRef]
14. Russell, E.; James, K. A new optimizer using particle swarm theory. In Proceedings of the 6th International Symposium on Micro Machine and Human Science, MHS'95, Nagoya, Japan, 4–6 October 1995; pp. 39–43.
15. Rahnamayan, S.; Tizhoosh, H.; Salama, M. Opposition-based differential evolution. *IEEE Trans. Evolut. Comput.* **2008**, *12*, 64–79. [CrossRef]
16. Liao, T.; Socha, K.; Marco, A.; Stutzle, T.; Dorigo, M. Ant colony optimization for mixed-variable optimization problems. *IEEE Trans. Evolut. Comput.* **2013**, *18*, 53–518. [CrossRef]
17. Taran, S.; Bajaj, V. Sleep apnea detection using artificial bee colony optimize hermite basis functions for eeg signals. *IEEE Trans. Instrum. Meas.* **2019**, *69*, 608–616. [CrossRef]
18. Precup, R.; David, R.; Petriu, E.M. Grey wolf optimizer algorithm-based tuning of fuzzy control systems with reduced parametric sensitivity. *IEEE Trans. Ind. Electron.* **2017**, *64*, 527–534. [CrossRef]
19. Hatata, A.Y.; Lafi, A. Ant lion optimizer for optimal coordination of doc relays in distribution systems containing dgs. *IEEE Access* **2018**, *6*, 72241–72252. [CrossRef]
20. Mirjalili, S. Moth-flame optimization algorithm: A novel natureinspired heuristic paradigm. *Knowl.-Based Syst.* **2015**, *89*, 228–249. [CrossRef]
21. Mohamed, A.; Ahmed, A.; Aboul, E. Whale optimization algorithm and moth-flame optimization for multilevel thresholding image segmentation. *Expert Syst. Appl.* **2017**, *83*, 51–67.
22. Sang, H.; Pan, Q.; Li, J.; Wang, P.; Han, Y.; Gao, K.; Duan, P. Effective invasive weed optimization algorithms for distributed assembly permutation flowshop problem with total flowtime criterion. *Swarm Evolut. Comput.* **2019**, *444*, 64–73. [CrossRef]
23. Zhou, Y.; Luo, Q.; Chen, H.; He, A.; Wu, J. A discrete invasive weed optimization algorithm for solving traveling salesman problem. *Neurocomputing* **2015**, *151*, 1227–1236. [CrossRef]
24. Wolpert, D.H.; Macready, W.G. No free lunch theorems for optimization. *IEEE Trans. Evolut. Comput.* **1997**, *1*, 67–82. [CrossRef]
25. Zhang, Y.; Liu, X.; Bao, F.; Chi, J.; Zhang, C.; Liu, P. Particle swarm optimization with adaptive learning strategy. *Knowl.-Based Syst.* **2020**, *196*, 105789. [CrossRef]
26. Dong, Z.; Wang, X.; Tang, L. Moea/d with a self-adaptive weight vector adjustment strategy based on chain segmentation. *Inform. Sci.* **2020**, *521*, 209–230. [CrossRef]
27. Li, E.; Chen, R. Multi-objective decomposition optimization algorithm based on adaptive weight vector and matching strategy. *Appl. Intell.* **2020**, *6*, 1–17. [CrossRef]
28. Feng, J.; Zhang, J.; Zhu, X.; Lian, W. A novel chaos optimization algorithm. *Multimedia Tools Appl.* **2016**, *76*, 1–32. [CrossRef]
29. Xu, C.B.; Yang, R. Parameter estimation for chaotic systems using improved bird swarm algorithm. *Mod. Phys. Lett. B* **2017**, *1*, 1750346. [CrossRef]
30. Tran, N.T.; Dao, T.-P.; Nguyen-Trang, T.; Ha, C.-N. Prediction of Fatigue Life for a New 2-DOF Compliant Mechanism by Clustering-Based ANFIS Approach. *Math. Probl. Eng.* **2021**, *2021*, 1–14. [CrossRef]
31. Rao, R.V.; Savsani, V.J.; Vakharia, D.P. Teaching learning based optimization: A novel method for constrained mechanical design optimization problems. *Comput. Aided Des.* **2011**, *43*, 303–315. [CrossRef]
32. Gunji, A.B.; Deepak, B.; Bahubalendruni, C.; Biswal, D. An optimal robotic assembly sequence planning by assembly subsets detection method using teaching learning-based optimization algorithm. *IEEE Trans. Autom. Sci. Eng.* **2018**, *1*, 1–17. [CrossRef]
33. Zhang, H.; Gao, Z.; Ma, X.; Jie, Z.; Zhang, J. Hybridizing teaching-learning-based optimization with adaptive grasshopper optimization algorithm for abrupt motion tracking. *IEEE Access* **2019**, *7*, 168575–168592. [CrossRef]
34. Ho, N.L.; Dao, T.-P.; Le Chau, N.; Huang, S.-C. Multi-objective optimization design of a compliant microgripper based on hybrid teaching learning-based optimization algorithm. *Microsyst. Technol.* **2018**, *25*, 2067–2083. [CrossRef]
35. Estevez, P.A.; Tesmer, M.; Perez, C.; Zurada, J. Normalized mutual information feature selection. *IEEE Trans. Neural Netw.* **2009**, *20*, 189–201. [CrossRef]
36. Heidari, A.A.; Mirjalili, S.; Faris, H.; Aljarah, I.; Mafarja, M.; Chen, H. Harris hawks optimization: Algorithm and applications. *Future Gener. Comput. Syst.* **2019**, *97*, 849–872. [CrossRef]
37. Jia, H.; Xing, Z.; Song, W. A new hybrid seagull optimization algorithm for feature selection. *IEEE Access* **2019**, *12*, 49614–49631. [CrossRef]
38. Newman, D.J.; Hettich, S.; Blake, C.L.; Merz, C.J. UCI Repository of Machine Learning Databases. Available online: http://www.ics.uci.edu/~{}mlearn/MLRepository.html (accessed on 1 June 2016).
39. Derrac, J.S.; Garcia, D.; Molina, F.; Herrera, A. Practical tutorial on the use of non-parametric statistical test as a methodology for comparing evolutionary and swarm intelligence algorithms. *Swarm Evolut. Comput.* **2011**, *1*, 13–18. [CrossRef]
40. Machaka, R. Machine learning-based prediction of phases in high-entropy alloys. *Comput. Mater. Sci.* **2020**, *188*, 110244. [CrossRef]

Article

Multilayer Reversible Data Hiding Based on the Difference Expansion Method Using Multilevel Thresholding of Host Images Based on the Slime Mould Algorithm

Abolfazl Mehbodniya [1], Behnaz karimi Douraki [2], Julian L. Webber [1], Hamzah Ali Alkhazaleh [3,*], Ersin Elbasi [4], Mohammad Dameshghi [5], Raed Abu Zitar [6] and Laith Abualigah [7]

1 Department of Electronics and Communication Engineering, Kuwait College of Science and Technology (KCST), Kuwait City 7207, Kuwait; a.niya@kcst.edu.kw (A.M.); jwebber@ieee.org (J.L.W.)
2 Department of Mathematics, University of Isfahan, Isfahan 81431-33871, Iran; bkarimidouraki@gmail.com
3 IT Department, College of Engineering and IT, University of Dubai, Academic City, United Arab Emirates
4 College of Engineering and Technology, American University of the Middle East, Kust Kuwait 15453, Kuwait; ersin.elbasi@aum.edu.kw
5 Department of Computer Engineering, Faculty of Electrical and Computer Engineering, University of Tabriz, Tabriz 5166-15731, Iran; dameshghi@aol.com
6 Sorbonne Center of Artificial Intelligence, Sorbonne University-Abu Dhabi, Abu Dhabi, United Arab Emirates; raed.zitar@sorbonne.ae
7 Faculty of Computer Sciences and Informatics, Amman Arab University, Amman 11953, Jordan; aligah.2020@gmail.com
* Correspondence: halkhazaleh@ud.ac.ae

Citation: Mehbodniya, A.; Douraki, B.k.; Webber, J.L.; Alkhazaleh, H.A.; Elbasi, E.; Dameshghi, M.; Abu Zitar, R.; Abualigah, L. Multilayer Reversible Data Hiding Based on the Difference Expansion Method Using Multilevel Thresholding of Host Images Based on the Slime Mould Algorithm. *Processes* **2022**, *10*, 858. https://doi.org/10.3390/pr10050858

Academic Editor: Mengchu Zhou

Received: 16 March 2022
Accepted: 24 April 2022
Published: 26 April 2022

Publisher's Note: MDPI stays neutral with regard to jurisdictional claims in published maps and institutional affiliations.

Abstract: Researchers have scrutinized data hiding schemes in recent years. Data hiding in standard images works well, but does not provide satisfactory results in distortion-sensitive medical, military, or forensic images. This is because placing data in an image can cause permanent distortion after data mining. Therefore, a reversible data hiding (RDH) technique is required. One of the well-known designs of RDH is the difference expansion (DE) method. In the DE-based RDH method, finding spaces that create less distortion in the marked image is a significant challenge, and has a high insertion capacity. Therefore, the smaller the difference between the selected pixels and the more correlation between two consecutive pixels, the less distortion can be achieved in the image after embedding the secret data. This paper proposes a multilayer RDH method using the multilevel thresholding technique to reduce the difference value in pixels and increase the visual quality and the embedding capacity. Optimization algorithms are one of the most popular methods for solving NP-hard problems. The slime mould algorithm (SMA) gives good results in finding the best solutions to optimization problems. In the proposed method, the SMA is applied to the host image for optimal multilevel thresholding of the image pixels. Moreover, the image pixels in different and more similar areas of the image are located next to one another in a group and classified using the specified thresholds. As a result, the embedding capacity in each class can increase by reducing the value of the difference between two consecutive pixels, and the distortion of the marked image can decrease after inserting the personal data using the DE method. Experimental results show that the proposed method is better than comparable methods regarding the degree of distortion, quality of the marked image, and insertion capacity.

Keywords: reversible data hiding (RDH); slime mould algorithm (SMA); difference expansion (DE)

1. Introduction

Data hiding (DH) is a way of secretly sending information or data to others. In this way, data, including text, images, etc., can be inserted into a medium such as an image, video, audio, etc., using a DH algorithm to make them invisible to others. DH is performed in two domains of space and frequency. The related data are inserted directly into the host image pixels in the space domain, which is often reversible. Reversibility means that the inserted data and the original host image are entirely recovered in the extraction phase. In the frequency domain, first, a frequency transform such as discrete wavelet transform (DWT), discrete cosine transform (DCT), etc., is applied to the host image. Then, using a special algorithm, the data are inserted into frequency coefficients that are often irreversible. The data are then extracted, while the original host image is not fully and accurately recovered.

Reversible data hiding (RDH) techniques in the space field are divided into several categories: difference expansion (DE), histogram shifting (HS), prediction-error expansion (PEE), pixel value grouping (PVG), and pixel value ordering (PVO). Over the years, various RDH methods have been proposed by researchers in these fields, all of which aim to reduce distortion and increase the quality of the marked image or increase the capacity to insert the data, and these issues continue. As the main challenge, this has occupied the minds of researchers. The following is a brief description of the papers available in each field:

An RDH method was proposed in [1] to increase the capacity based on the DE method. In this method, the host image is divided into 1×2 blocks, and spaces of -1, 0, and +1 are selected to insert the data. A lossless RDH method based on the DE histogram (DEH) was proposed to increase the capacity in [2], where the difference histogram (DH) peak point was selected to insert the data. In [3], a multilayer RDH method based on the DEH was proposed to increase the capacity and reduce the distortion. However, this method was not able to improve the capacity sufficiently. In [4], the RDH method was presented for gray images using the DEH and the module function. The position matrix and the marked image were sent separately to the receiver with low capacity.

In [5], to improve the capacity performance and increase security, a guided filter predictor and an adaptive PEE design were used to insert the data in color images using intrachannel correlation. A two-layer DH method based on the expansion and displacement of a PE pair in a two-dimensional histogram was proposed to increase the capacity by extracting the correlations between the consecutive PEs in [6], where their work was both low-capacity and low-quality. In [7], the authors presented an RDH method based on the multiple histogram correction and PEE to increase the capacity of the gray images, using a rhombus predictor to predict the host image pixels. In PE histogram (PEH) methods [8], to extract the redundancy between adjacent pixels, the correction path detection strategy is used to obtain a two-dimensional PEH that has a high distortion. In an RDH method [9], based on a two-layer insertion and PEH, the pixel pairs are selected based on pixel density and the spatial distance between two pixels. The distortion is much less than in the previous methods. In [10], an RDH method based on multiple histogram shifting was proposed that used a genetic algorithm (GA) to control the substantial image distortion. In [11], an RDH technique was proposed based on pairwise prediction-error expansion (PPEE) or two-dimensional PEH (2D-PEH) to increase capacity. In [12], an RDH technique based on the host image texture analysis was proposed by blocking the image and inserting the data into image texture blocks [13–20].

In another RDH method [21], the PVG method was used by multilayer insertion to increase the capacity. The PVG was applied on each block. The zero point of the histogram of the difference was selected to insert the data bits. The pixel-based PVG (PPVG) technique in [22] was proposed to increase the capacity applied to each block after image blocking. Each time a pixel is located in the smooth areas of each block for inserting the data, the PE value is the difference between the reference pixel and the particular pixel. This method has a low capacity and moderate distortion. In [23], a technique was presented to increase the reliability of the image for inserting the data. Using PEE based on PVO, the authors inserted the data in the pins of -1 and +1 from the PEH. The problem was that the quality of the image was still low, while the capacity was not high. The authors of [24] presented a robust reversible data hiding (RRDH) scheme based on two-layer embedding with a reduced capacity-distortion tradeoff, where it first decomposes the image into two planes—namely, the higher significant bit (HSB) and least important bit (LSB) planes—and then employs prediction-error expansion (PEE) to embed the secret data into the HSB plane.

In [25], the authors proposed an AMBTC-based RDH method in which the hamming distance and PVD methods were used to insert information. In [26], a PVD-based RDH method was used. In [27], the RDH method was based on PVD and LSB reversible insert information. The authors of [28] proposed an RDH in encrypted images (RDHEI) method with hierarchical embedding based on PE. PEs are divided into three categories: small-magnitude, medium-magnitude, and large-magnitude. In their approach, pixels with small-magnitude/large-magnitude PEs were used to insert data. Their method had a high capacity, but the image quality was still low.

The authors of [29] proposed an RDH method for color images using HS-based double-layer embedding. The authors used the image interpolation method to generate PE matrices for HS in the first-layer embedding, and local pixel similarity to calculate the difference matrices for HS in the second-layer embedding. In their process, the embedding capacity was low. In [30], the authors proposed a dual-image RDH method based on PE shift. Moreover, in their work, a bidirectional-shift strategy was used to extend the shiftable positions in the central zone of the allowable coordinates. In their work, the embedding capacity was low. Today, optimization algorithms such as the grasshopper optimization algorithm (GOA), whale optimization algorithm (WOA), moth–flame optimization (MFO) [31], Harris hawk optimization (HHO) [32], and artificial bee colony (ABC) [33] are used in many papers [34–38].

This paper proposes a DE-based multilayer image RDH method using the multilevel thresholding technique. Optimization algorithms are one of the most popular methods for solving NP-hard problems. Due to this, the slime mould algorithm (SMA) gives good results in finding the best solutions to solve optimization problems. First, the SMA is applied to the host image to find the two thresholds, and image pixels are based on the thresholds located in three different classes. Using the multilevel thresholding creates more similarities between the pixels of each class. Therefore, due to the reduction in the difference between each class' pixels, the quality of the marked image does not decrease with the insertion of data via DE. At the same time, less distortion is created in the picture, and the insertion capacity increases. Table 1 shows the critical acronyms and their meanings.

Table 1. Key acronyms and their meanings.

Key Acronyms	Their Meanings
DH	Data hiding
DWT	Discrete wavelet transform
DCT	Discrete cosine transform
RDH	Reversible data hiding
DE	Difference expansion
HS	Histogram shifting
PEE	Prediction-error expansion
PVG	Pixel value grouping
PVO	Pixel value ordering
DH	Difference histogram
PEH	PE histogram
GA	Genetic algorithm
PPEE	Pairwise prediction-error expansion
2D-PEH	Two-dimensional PEH

Table 1. Cont.

Key Acronyms	Their Meanings
RRDH	Robust reversible data hiding
HSB	Higher significant bit
LSB	Least important bit
PEE	Prediction-error expansion
RDHEI	RDH in encrypted images
MSE	Mean squared error
PSNR	Peak signal-to-noise ratio
bpp	Bits per pixel
SSIM	Structural similarity index measure

The method of Arham et al. [3] is the basis of the proposed method. Arham et al. [3] proposed a multilayer RDH method to reduce the distortion in the image. First, the host image is divided into 2 × 2 blocks in their work. Then, to calculate the value of differences for both consecutive pixels, the pixel vector is created for each block in each layer, as shown in Figure 1. A threshold ($2 \leq \text{Th} \leq 30$) is considered, and only the blocks in which the value of differences for both consecutive pixels is positive and less than the threshold value ($2 \leq \text{Th} \leq 30$) are selected to insert the data.

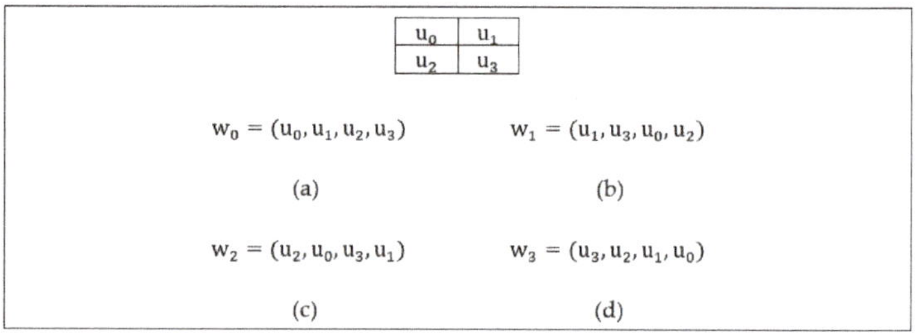

u_0	u_1
u_2	u_3

$w_0 = (u_0, u_1, u_2, u_3)$ (a)

$w_1 = (u_1, u_3, u_0, u_2)$ (b)

$w_2 = (u_2, u_0, u_3, u_1)$ (c)

$w_3 = (u_3, u_2, u_1, u_0)$ (d)

Figure 1. Production of vectors for four layers [3]: (**a**) Layer-1; (**b**) Layer-2; (**c**) Layer-3; (**d**) Layer-4.

For each the vectors w_s where $s = \{0, 1, 2, 3\}$ in the s-th insertion layer, in order to ensure reversibility and extract the data in the extraction phase, the s-th pixel from each block is not inserted, there is a reference pixel in each insertion layer, and three bits of data are inserted in a block of size 2 × 2. In the first insertion layer, in Figure 1, the pixel vector w_s is created to calculate the values of v_1, v_2, v_3,. The value of differences is calculated using Equation (1) for both consecutive pixels:

$$v_1 = u_1 - u_0,\ v_2 = u_2 - u_1,\ v_3 = u_3 - u_2 \tag{1}$$

Furthermore, to reduce the distortion and control of overflow/underflow, Equation (2) is used to reduce the value of v_k $(k = 1, 2, 3)$:

$$v'_k = \begin{cases} v_k - 2^{|v_k|-1} & \text{if } 2 \times 2^{n-1} \leq |v_k| \leq 3 \times 2^{n-1} - 1 \\ v_k - 2^{|v_k|} & \text{if } 3 \times 2^{n-1} \leq |v_k| \leq 4 \times 2^{n-1} - 1 \end{cases} \tag{2}$$

where n is calculated using Equation (3):

$$n = \lfloor \log_2 (|v_k|) \rfloor \tag{3}$$

Then, the value of v''_k is expanded to Equation (4) by collecting the kth bits of the three-bit sequence of confidential data $b = (b_1, b_2, b_3)$, as follows:

$$v''_k = 2 \times v'_k + b_k \quad (4)$$

A location map (LM) is used to separate the range $|v_k|$, as shown in Equation (5), to recover the original differences and the original pixels of the host image in the extraction phase:

$$LM = \begin{cases} 0, & \text{if} \quad 2 \times 2^{n-1} \le |v_k| \le 3 \times 2^{n-1} - 1 \\ 1, & \text{if} \quad 3 \times 2^{n-1} \le |v_k| \le 4 \times 2^{n-1} - 1 \end{cases} \quad (5)$$

Each insertion layer creates a location map (M) to determine the location of blocks containing data. If the block is inserted with the data bits, the value of M is equal to 1. Otherwise, the value of M is 0. Finally, Equation (6) is used to insert the data in the vector u_0:

$$\begin{gathered} v_0 = \frac{u_0 + u_1 + u_2 + u_3}{2} \\ u_0 = u_0,\ u_1 = v''_1 + u_0,\ u_2 = v''_2 + u_2,\ u_3 = v''_3 + u_3 \end{gathered} \quad (6)$$

In the extraction phase, the secret data and the v_k are recovered using location maps LM and M and Equation (7):

$$v_k = \begin{cases} v'_k + (|v'_k|) + 1, & \text{if} \quad LM = 0 \\ v'_k + (|v'_k|), & \text{if} \quad LM = 1 \end{cases} \quad (7)$$

The remainder of this paper is organized as follows: Section 2 introduces the proposed plan that uses the Otsu thresholding method and SMA. In Section 3, the proposed methods are evaluated and compared with other works, and Section 4 contains the conclusions.

2. Proposed Method

The proposed method consists of three phases: In the first phase, the multilevel thresholding technique using a combination of the SMA and the Otsu evaluation function is applied to the host image to classify the host image pixels based on the relevant thresholds. The second phase inserts the data in the pixels of each class using DE, and in the third phase the extracted inserted data and the original image are recovered. In the following section, each of these three phases is described in turn:

2.1. Multilevel Thresholding Using the Slime Mould Algorithm

In the DE-based DH methods, the extraction of the best and most extra space for inserting the data and achieving a high capacity will also create less distortion in the host image—a significant issue. In other words, the smaller the difference between the two consecutive pixels and the more similarity between the pixels, the less distortion created in the image after inserting the data using the DE.

In this paper, we use the multilevel thresholding technique using the combination of the SMA and the Otsu evaluation function [34] to determine the two optimal thresholds (T_1 and T_2) for classifying the image pixels. Thus, increases in the similarity and correlation between the pixels of each class are used to reduce the value of differences for two consecutive pixels in each class, and to increase capacity and reduce distortion in the marked image.

The Otsu method is an automatic thresholding method obtained according to the image histogram, and defines the boundaries of objects in the image with high accuracy [34]. In gray images, the pixel intensity is between 0 and 255. Therefore, the Otsu method selects a threshold from 0 to 255, with the highest interclass variance in gray images, or minimizes intraclass variance [34]. All image pixels are grouped using multilevel thresholding based on correlation and similarity [35].

SMA is a new population-based metaheuristic algorithm [36] inspired by the intelligent behavior of a type of mould called slime mould. Slime mould also behaves intelligently, and can navigate very quickly and without error. In this paper, the input and output of the

SMA are the image histogram and the thresholds, respectively. The related thresholds in the SMA are obtained by $\vec{X}^*$. The vector represents each slime mould $\vec{X}_{kt}$, in which kt = 1, 2:

$$\vec{X}_{kt} = (T_1, T_2) \quad \text{for} \quad 0 < T_1 < T_2 < H \tag{8}$$

In Equation (8), T_1 and T_2 represent the search thresholds. In this paper, we seek to find the two optimal thresholds T_1 and T_2, and utilize the SMA to use these two thresholds to classify image pixels.

The location of each slime mould that represents a threshold value is in the range [lb, ub], where lb and ub represent the lowest and highest brightness of the pixels in the host image, respectively. The initial location of the slime moulds is determined randomly, according to Equation (9):

$$x_{i1} = lb + rand(0,1) \times (ub - lb) \tag{9}$$

where x_{i1} represents each threshold of the threshold vector. The value of the evaluation function is then calculated for all slime moulds, and the slime mould for which the evaluation function obtained is the smallest value is taken as the criterion, while the corresponding location, which is defined as $\vec{X}^*$, is set as the related threshold. Slime mould uses the bait smell released into the air to approach and guide the prey. Equation (10) describes the routing behavior of slime moulds based on bait smell:

$$\vec{X(t+1)} = \begin{cases} \vec{X_b(t)} + \vec{vb}\cdot\left(\vec{W}\cdot\vec{X_A(t)} - \vec{X_B(t)}\right), & r < p \\ \vec{vc}\cdot\vec{X(t)}, & r \geq p \end{cases} \tag{10}$$

where $\vec{vb}$ is a parameter in the range $[-a, a]$, $\vec{vc}$ is a parameter that decreases linearly from 1 to 0, and t represents the current iteration. $\vec{X_b(t)}$ Indicates the location of the slime mould in the tth iteration with the highest smell concentration in the environment. $\vec{X(t)}$ indicates the location of the slime mould in the tth iteration. $\vec{X(t+1)}$ is the next location of the slime mould in the tth iteration. $\vec{X_A(t)}$ and $\vec{X_B(t)}$ represent the two randomly selected locations of the slime mould, and $\vec{W}$ represents the slime mould's weight. The value of p is obtained using Equation (11):

$$p = \tanh|S(i)\text{-}DF| \tag{11}$$

where i = 1, 2, ..., n represents the number of cells in the slime mould, DF represents the best evaluation obtained in all iterations, S(i) represents the value of the evaluation function, $\vec{X(t)}$ DF represents the best value of the evaluation function obtained during all iterations, and $\vec{vb}$ is obtained from Equation (12):

$$\vec{vb} = [-a, a] \tag{12}$$

$$a = \operatorname{arctanh}\left(-\left(\frac{t}{max\_t}\right) + 1\right) \tag{13}$$

$\vec{W}$ is obtained using Equation (14):

$$\overline{W(SmellIndex(i)} = \begin{cases} 1 + r.\log\left(\frac{bF - S(i)}{bF - wF} + 1\right), & \text{condition} \\ 1 - r.\log\left(\frac{bF - S(i)}{bF - wF} + 1\right), & \text{others} \end{cases} \tag{14}$$

$$SmellIndex = sort(S) \tag{15}$$

The condition indicates that S(i) is in the first half of the population. The r represents a random value [0, 1]. The bF represents the value of the obtained optimal evaluation function in the current iteration. The wF represents the worst evaluation function received in the current iteration. SmellIndex represents the sequence of values of the evaluated function (in ascending order). The location of the slime mould is also updated using Equation (16) [36]:

$$\vec{X^*} = \begin{cases} \text{rand}.(\text{UB} - \text{LB}) + \text{LB}, & \text{rand} < z \\ \vec{X_b}(t) + \vec{vb}.\left(W.\vec{X_A}(t) - \vec{X_B}(t)\right), & r < p \\ \vec{vc}.\vec{X}(t), & r \geq p \end{cases} \tag{16}$$

where LB and UB are the lower limit and the upper limit, respectively, which in this case are equal to 0 and 255, respectively; rand and r are randomly determined in the range [0, 1], and the value of z will be discussed in the parameter setting test. Therefore, this process is repeated until the stop condition is met. We set the stop condition to reach 100 iterations. Then, we obtain the output $\vec{X^*}$, which represents the optimal threshold vector. Finally, the host gray image A is divided into three separate classes C_1, C_2, and C_3 using the optimal thresholds T_{kt} (kt = 1, 2), as shown in Equation (17):

$$\begin{cases} C_1 = \{g(i, j) \in A \mid 0 \leq g(i, j) \leq T_1 - 1\} \\ C_2 = \{g(i, j) \in A \mid T_1 \leq g(i, j) \leq T_2 - 1\} \\ C_3 = \{g(i, j) \in A \mid T_2 \leq g(i, j) \leq H\} \end{cases} \tag{17}$$

where g(i, j) represents each pixel in row i and column j of image A, and H represents the gray area of gray image A. Thresholds T_{kt} are obtained by maximizing the evaluation function F, as shown in Equation (18):

$$T_{kt} = \text{MAX}_{T_{kt}}\, F(T_{kt}) \tag{18}$$

where $F(T_{kt})$ is the same evaluation function for the SMA or the Otsu evaluation function. The Otsu evaluation function is calculated using Equation (19) [28]:

$$F = \sum\nolimits_{i=0}^{2} \text{SUM}_i (\mu_i - \mu_1)^2 \tag{19}$$

$$\text{SUM}_i = \sum\nolimits_{j=T_i}^{T_{i+1}-1} P_j \tag{20}$$

$$\mu_i = \sum\nolimits_{j=T_i}^{T_{i+1}-1} i \frac{P_j}{\text{SUM}_i}, \quad P_j = \text{fer}(j)/\text{Num}_p \tag{21}$$

In Equation (19), μ_1 is the average density of the host image A for T_1 = 0 and T_2 = H. The μ_i is the average density of the C_y class for T_1 and T_2, and SUM_i is the sum of the probabilities. In Equations (20) and (21), P_j shows the probability of the gray level j^{th}, fer(j) is the frequency of the jth gray level, and Num_p represents the total number of pixels in the host image A [36].

2.2. Data Insertion Process

In this paper, the value of 0 difference is also expanded for inserting the data ($0 \leq \text{Th} \leq 32$) to increase the capacity, in addition to the fewer equal to 0 and fewer equal to 32 differences [3]. In this paper, two optimal thresholds T_1 and T_2 are obtained for the host image, and the host image pixels are located together in three separate classes based on the two corresponding thresholds. Therefore, there is more similarity and correlation between the pixels of each class, and the difference between two consecutive pixels related to each class is less.

By reducing the value of differences between consecutive pixels in each class, less distortion is created in the image due to data insertion based on the expansion of the value of differences. Furthermore, the capacity increases when increasing the number of pixels with less difference between them.

After determining the optimal thresholds by the SMA, based on the two thresholds of T_1 and T_2, starting from the pixel on the first row and the first column of the host image, pixels belonging to the first class—whose value is smaller than the threshold of T_1—are located in the liner matrix C_1. The second-class pixels, whose value is larger than that of T_1 and smaller than that of T_2, are located in the liner matrix C_2. Eventually, the pixels belonging to the third class, whose value is larger than that of T_2—are located in the liner matrix C_3. Figure 2 shows the status of the three matrices of C_1, C_2, and C_3.

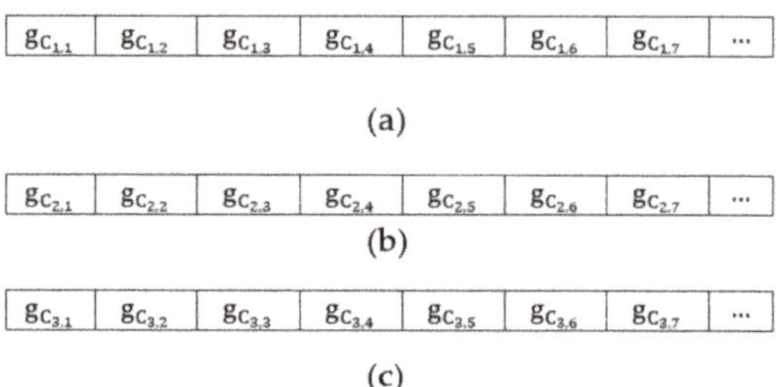

Figure 2. Production of the matrices for three classes: (**a**) Class C_1; (**b**) Class C_2; (**c**) Class C_3.

To insert the data, after classifications of host image pixels, matrix pixels C_y (y = 1, 2, 3) are divided into non-overlapping blocks with a size of 1 × 5. The vector P is then created for pixels of each block to calculate the different values for both consecutive pixels, according to Figure 3.

$p_{C_{y,1}}$	$p_{C_{y,2}}$	$p_{C_{y,3}}$	$p_{C_{y,4}}$	$p_{C_{y,5}}$

$$p=(p_{C_{y,1}}, p_{C_{y,2}}, p_{C_{y,3}}, p_{C_{y,4}}, p_{C_{y,5}})$$

Figure 3. Production of the vector P.

According to Equation (22), differences are calculated for both consecutive pixels in the vector P:

$$\begin{cases} v_1 = p_{C_{y,2}} - p_{C_{y,1}} \\ v_2 = p_{C_{y,3}} - p_{C_{y,2}} \\ v_3 = p_{C_{y,4}} - p_{C_{y,3}} \\ v_4 = p_{C_{y,5}} - p_{C_{y,4}} \end{cases} \tag{22}$$

Given that the proposed method inserts the data by expanding the value of differences to reduce distortion in the marked image, the blocks are selected to insert the data. The difference between two consecutive pixels is between 0 and +32 ($0 \leq Th \leq 32$). Moreover, to reduce distortion in the marked image and the overflow/underflow control, the range of each difference $v_{k'}$ ($k' = 1, 2, 3, 4$) is diminished. A location map M_q (q = 1, 2, 3, 4) is created in each insertion layer to determine the locations of blocks inserted with data bits. If the block is inserted with data bits, it M_q will be equal to 1. Otherwise, it is equal to 0. Furthermore, a location map LM_y to separate each of the differences as $2 \leq v_{k'} \leq 32$ or $0 \leq v_{k'} \leq 1$ is considered, as shown in Equation (23):

$$LM_y = \begin{cases} 0, & \text{if} \quad 0 \leq v_{k'} \leq 1 \\ 1, & \text{if} \quad 2 \leq v_{k'} \leq 32 \end{cases} \tag{23}$$

If $0 \leq v_{k'} \leq 1$, Equation (24) can be used to change the range of $v_{k'}$:

$$v'_{k'} = |v_{k'} + 2| - 2^{\lfloor \log_2 (|v_{k'}|) \rfloor} \tag{24}$$

If $2 \leq v_{k'} \leq 32$, Equation (25) can be used to change the range of $v_{k'}$:

$$v'_{k'} = \begin{cases} |v_{k'}| - 2^{\lfloor \log_2 (|v_{k'}|) \rfloor - 1}, \text{ if } 2 \times 2^{n-1} \leq |v_{k'}| \leq 3 \times 2^{n-1} - 1 \\ |v_{k'}| - 2^{\lfloor \log_2 (|v_{k'}|) \rfloor}, \text{ if } 3 \times 2^{n-1} \leq |v_{k'}| \leq 4 \times 2^{n-1} - 1 \end{cases} \tag{25}$$

where n is calculated using Equation (26):

$$n = \lfloor \log_2(|v_{k'}|) \rfloor \tag{26}$$

In each insertion layer of each y class, a location map $LM'_{y,f}$—where f = 1, 2, as shown in Equations (27) and (28)—is used to recover the original differences. $LM'_{y,1}$ and $LM'_{y,2}$ are used to separate the differences, as $0 \leq v_{k'} \leq 1$ and $2 \leq v_{k'} \leq 32$, respectively.

$$LM'_{y,1} = \begin{cases} 0, & \text{if} \quad |v_{k'}| = 0 \\ 1, & \text{if} \quad |v_{k'}| = 1 \end{cases} \tag{27}$$

$$LM'_{y,2} = \begin{cases} 0, & \text{if} \quad 2 \times 2^{n-1} \leq |v_{k'}| \leq 3 \times 2^{n-1} - 1 \\ 1, & \text{if} \quad 3 \times 2^{n-1} \leq |v_{k'}| \leq 4 \times 2^{n-1} - 1 \end{cases} \tag{28}$$

The matrices M_q 'LM_y, and $LM'_{y,f}$ are used to recover the original differences and original image pixels. Given that in the proposed method, in the sth insertion layer (s = 1, 2, 3, 4, 5), the sth pixel is not inserted for reversibility, and in each layer, four - bits of data are inserted in four pixels of each block, the data bits' sequence is divided into 4-bit subsequences, where $b_{k'} = (b_1, b_2, b_3, b_4)_2$ that $k' = 1, 2, 3, 4$. Then, using Equation (29), $v'_{k'}$ expands with a datum:

$$v''_{k'} = 2 \times v'_{k'} + b_{k'} \tag{29}$$

The first pixel is not inserted from each block in the first layer. Therefore, in the first insertion layer, insertion of the 4-bit sequence b in pixels of each block is carried out as shown in Equation (30). In the second insertion layer, the second pixels of each class block are not selected to insert the data. In this layer, the process of inserting the 4-bit sequence b in pixels of each block is as shown in Equation (31). In the third insertion layer, the third pixel of each class block is not selected to insert the data, according to Equation (32). In the 4th insertion layer, the 4th pixel of each class block is not selected to insert the data, according to Equation (33).

$$\begin{cases} {P_{C_{y,1}}}' = P_{C_{y,1}} \\ {P_{C_{y,2}}}' = v''_1 + P_{C_{y,1}} \\ {P_{C_{y,3}}}' = v''_2 + P_{C_{y,2}} \\ {P_{C_{y,4}}}' = v''_3 + P_{C_{y,3}} \\ {P_{C_{y,5}}}' = v''_4 + P_{C_{y,4}} \end{cases} \tag{30}$$

$$\begin{cases} {P_{C_{y,1}}}' = v''_1 + P_{C_{y,1}} \\ {P_{C_{y,2}}}' = P_{C_{y,2}} \\ {P_{C_{y,3}}}' = v''_2 + P_{C_{y,2}} \\ {P_{C_{y,4}}}' = v''_3 + P_{C_{y,3}} \\ {P_{C_{y,5}}}' = v''_4 + P_{C_{y,4}} \end{cases} \tag{31}$$

$$\begin{cases} P_{C_{y,1}}{}' = v''_1 + P_{C_{y,1}} \\ P_{C_{y,2}}{}' = v''_2 + P_{C_{y,2}} \\ P_{C_{y,3}}{}' = P_{C_{y,2}} \\ P_{C_{y,4}}{}' = v''_3 + P_{C_{y,3}} \\ P_{C_{y,5}}{}' = v''_4 + P_{C_{y,4}} \end{cases} \quad (32)$$

$$\begin{cases} P_{C_{y,1}}{}' = v''_1 + P_{C_{y,1}} \\ P_{C_{y,2}}{}' = v''_2 + P_{C_{y,2}} \\ P_{C_{y,3}}{}' = v''_3 + P_{C_{y,2}} \\ P_{C_{y,4}}{}' = P_{C_{y,3}} \\ P_{C_{y,5}}{}' = v''_4 + P_{C_{y,4}} \end{cases} \quad (33)$$

Therefore, due to the data insertion process of pixels $p_{C_{y,1}}, p_{C_{y,2}}, p_{C_{y,3}}, p_{C_{y,4}}$ and $p_{C_{y,5}}$ from each block, five marked pixels $p'_{C_{y,1}}, p'_{C_{y,2}}, p'_{C_{y,3}}, p'_{C_{y,4}}$, and $p'_{C_{y,5}}$ are obtained, as is the marked image A'.

The steps of the data insertion process in each insertion layer are as follows:

Step 1: Dividing the data sequence into 4-bit b. Step 2: Apply SMA on the host image to determine the optimal thresholds T_1 and T_2. Step 3: Classification of image pixels based on thresholds T_1 and T_2. Step 4: Dividing the class pixels C_y into non-overlapping blocks of size 1×5. Step 5: Production of a location map M_q to determine the block locations with insertion conditions. Step 6: Produce the vector P for each selected block to calculate the value of differences. Step 7: Calculate the differences $v_{k'}$ for both consecutive pixels, and then calculate $v'_{k'}$. Step 8: Calculate the value of $v''_{k'}$ to reduce image distortion and prevent overflow/underflow. Step 9: Produce the location map LM_y. Step 10: Calculation of the marked pixels $p'_{C_{y,1}}, p'_{C_{y,2}}, p'_{C_{y,3}}, p'_{C_{y,4}}$, and $p'_{C_{y,5}}$. Step 11: Produce the marked image A' and save the location maps LM_y, $LM'_{y,f}$, and M_q, and thresholds T_1 and T_2 in order to extract the original data and recover the original image.

2.3. Data Extraction Process

In the extraction phase, the marked image A' is first classified using the thresholds T_1 and T_2, and the pixels belonging to each class are located in the matrix C'_y. Then, the pixels belonging to the matrix C'_y are divided into blocks of size 1×5, and using location maps the LM_y and $LM'_{y,f}$ the secret data are extracted from the corresponding blocks, and the original image pixels are recovered. The extraction process is performed from the first insertion to the last. The vector p' is created for the pixels of each block after dividing the matrix C'_y into blocks of size 1×5, according to Equation (34):

$$p' = \left(p'_{C_{y,1}}, p'_{C_{y,2}}, p'_{C_{y,3}}, p'_{C_{y,4}}, p'_{C_{y,5}}\right) \quad (34)$$

In each insertion layer, the binary value $v''_{k'}$ is obtained. Then, the least significant bit (LSB) of $v''_{k'}$ as the k'^{th} bit is extracted from the secret data. In the 4th insertion layer, extracting the data bits $b_{k'}$ and recovering the original pixels from each vector p' is performed as follows (see Equations (35)–(38)):

$$v''_1 = P'_{C_{y,k'}} - P'_{C_{y,k'}} \quad (35)$$

$$b_{k'} = LSB(v''_{k'}) \quad (36)$$

$$v_{k'}{}' = \left\lfloor \frac{v''_{k'}}{2} \right\rfloor \quad (37)$$

$$P'_{C_{y,k'}} = P'_{C_{y,k'-1}} + v_{k'-1} \quad (38)$$

If the $LM'_{y,1}$ is equal to 0 or 1, Equation (39) can be used to obtain $v_{k'}$. If the $LM'_{y,2}$ is equal to 0 or 1, Equation (40) can be used to obtain $v_{k'}$.

$$v_{k'} = \left|v'_{k'}\right| - 2^{\lfloor \log_2 (|v_{k'}|) \rfloor} \tag{39}$$

$$v_{k'} = \begin{cases} \left|v'_{k'}\right| + 2^{\lfloor \log_2 (|v_{k'}|) \rfloor}, \text{ if } & LM'_{y,2} = 0 \\ \left|v'_{k'}\right| + 2^{\lfloor \log_2 (|v_{k'}|) \rfloor}, \text{ if } & LM'_{y,2} = 1 \end{cases} \tag{40}$$

The values of $p_{C_{y,1}}$ $p_{C_{y,2}}$, b_1, and $v_1{}'$ are obtained using Equations (41)–(57).

$$P_{c_{y,1}} = P'_{c_{y,1}} \tag{41}$$

$$v''_1 = P'_{C_{y,2}} - P_{c_{y,1}} \tag{42}$$

$$b_1 = LSB(v''_1) \tag{43}$$

$$v_1{}' = \left\lfloor \frac{v''_1}{2} \right\rfloor \tag{44}$$

$$P_{c_{y,2}} = P_{c_{y,1}} + v_1 \tag{45}$$

Therefore, the value v_1 is obtained using Equation (39) or Equation (40). Then, $p_{C_{y,3}}$, b_2, and $v_2{}'$ are obtained using the following equations:

$$v''_2 = P'_{C_{y,3}} - P_{c_{y,2}} \tag{46}$$

$$b_2 = LSB(v''_2) \tag{47}$$

$$v_2{}' = \left\lfloor \frac{v''_2}{2} \right\rfloor \tag{48}$$

$$P_{c_{y,3}} = P_{c_{y,2}} + v_2 \tag{49}$$

Therefore, the value v_2 is obtained using Equation (39) or Equation (40).
Then, $p_{C_{y,4}}$, b_3, and $v_3{}'$ are obtained using the following equations:

$$v''_3 = P'_{C_{y,4}} - P_{c_{y,3}} \tag{50}$$

$$b_3 = LSB(v''_3) \tag{51}$$

$$v_3{}' = \left\lfloor \frac{v''_3}{2} \right\rfloor \tag{52}$$

$$P_{c_{y,4}} = P_{c_{y,3}} + v_3 \tag{53}$$

Therefore, the value v_3 is obtained using Equations (39) and (40).
Then, $p_{C_{y,5}}$, b_4, and $v_4{}'$ are obtained using the following equations:

$$v''_4 = P'_{C_{y,5}} - P_{c_{y,4}} \tag{54}$$

$$b_4 = LSB(v''_4) \tag{55}$$

$$v_4{}' = \left\lfloor \frac{v''_4}{2} \right\rfloor \tag{56}$$

$$P_{c_{y,5}} = P_{c_{y,4}} + v_4 \tag{57}$$

Therefore, the value v_4 is obtained using Equations (39) and (40).

The steps of the data extraction process and the original host image pixel recovery in each layer are as follows:

Step 1: Classify the pixels of the image A′ to get three classes C_y.

Step 2: Dividing the class pixels C_y into non-overlapping blocks of size 1 × 5.

Step 3: Identify marked blocks using the location map M_q.

Step 4: Production of the vector p′ for each block.

Step 5: Calculate the value $v''_{k'}$.

Step 6: Extract the inserted data $b_{k'}$ using the first LSB $v''_{k'}$.

Step 7: Calculate the value $v_{k'}{}'$.

Step 8: Calculate the value of the $v_{k'}$ difference using LM_y and $LM'_{y,f}$.

Step 9: Calculate the original pixels of each block.

Step 10: Recover the original image A.

The marked pixels in the respective block are shown in Figure 4. An example of the embedding and extraction process:

106	121	148	158	158

(a)

106	120	144	161	160

(b)

106	121	148	158	158

(c)

Figure 4. An example of quad-pixel reversible data embedding: (**a**) before embedding; (**b**) after embedding; (**c**) after extraction.

Embedding process:

$$P = (106,\ 121,\ 148,\ 158,\ 155)$$

$$\begin{cases} v_1 = 121 - 106 = 15 \\ v_2 = 148 - 121 = 27 \\ v_3 = 158 - 148 = 10 \\ v_4 = 158 - 158 = 0 \end{cases}$$

$$v_{k'} = (15,\ 27,\ 10,\ 0),\ \mathrm{b} = (0,\ 1,\ 1,\ 0)$$

$$v_1 = 15, n = 3,\ v'_{1'} = 7,\ LM'_{y,2} = 1, b_1 = 0, v''_1 = 2 \times 7 + 0 = 14$$

$$v_2 = 27, \mathrm{n} = 4,\ v'_{2'} = 11,\ LM'_{y,2} = 1, b_2 = 1, v''_2 = 2 \times 11 + 1 = 23$$

$$v_3 = 10, \mathrm{n} = 4,\ v'_{3'} = 6,\ LM'_{y,2} = 0, b_3 = 1, v''_3 = 2 \times 6 + 1 = 13$$

$$v_4 = 0, \mathrm{n} = 4,\ v'_{0'} = 1,\ LM'_{y,2} = 1, b_4 = 0, v''_4 = 2 \times 1 + 0 = 2$$

$$v'_{k'} = (7,\ 11,\ 6,\ 1),\ v''_{k'} = (14,\ 55,\ 13,\ 2), LM'_{y,2} = (1,\ 1,\ 0)_2,\ \mathrm{M} = 1$$

$$\begin{cases} p_1{}' = 106 \\ p_2{}' = 106 + 14 = 120 \\ p_3{}' = 121 + 23 = 144 \\ p_4{}' = 148 + 13 = 161 \\ p_5{}' = 158 + 2 = 160 \end{cases}$$

$$p'(106,\ 120,\ 144,\ 161,\ 160)$$

Extraction process:

$$
\begin{aligned}
&p'(106,\ 120,\ 144,\ 161,\ 160)\\
&v_1'' = 120 - 106 = 14\\
&b_1 = LSB(14) = LSB(1110)_2 = 0\\
&v_1{}' = \left\lfloor \tfrac{14}{2} \right\rfloor = 7\\
&LM'_{y,1} = 1, v_1 = 15\\
&\begin{cases} p_{C_{y,1}} = 106 \\ p_{C_{y,2}} = 106 + 15 = 121 \end{cases}\\
&v_2'' = 144 - 121 = 23\\
&b_2 = LSB(23) = LSB(10111)_2 = 2\\
&v_2{}' = \left\lfloor \tfrac{23}{2} \right\rfloor = 11\\
&LM'_{y,2} = 1, v_2 = 27\\
&p_{C_{y,3}} = 27 + 121 = 148\\
&v_3'' = 161 - 148 = 13\\
&b_3 = LSB(13) = LSB(1101\,)_2 = 1\\
&v_3{}' = \left\lfloor \tfrac{13}{2} \right\rfloor = 6\\
&LM'_{y,2} = 0, v_3 = 10\\
&p_{C_{y,4}} = 148 + 10 = 158\\
&v_4'' = 160 - 158 = 2\\
&b_4 = LSB(2) = LSB(10\,)_2 = 0\\
&v_3{}' = \left\lfloor \tfrac{2}{2} \right\rfloor = 1\\
&LM'_{y,1} = 1, v_3 = 0\\
&p_{C_{y,5}} = 158 + 0 = 158
\end{aligned}
$$

Finally, the initial pixels of the corresponding block are retrieved as follows.

P = (106, 121, 148, 158, 155)

3. Results

In this paper, the gray images of Lena, Peppers, Airplane, Baboon, Ship, Lake, Bridge, Cameraman, and Barbara are used as the host image of size 512 × 512, taken from the USC-SIPI database. Figure 5 shows the host images used in this paper. The proposed method is simulated using MATLAB 2018b and a Windows 64-bit operating system with a Core i5 CPU.

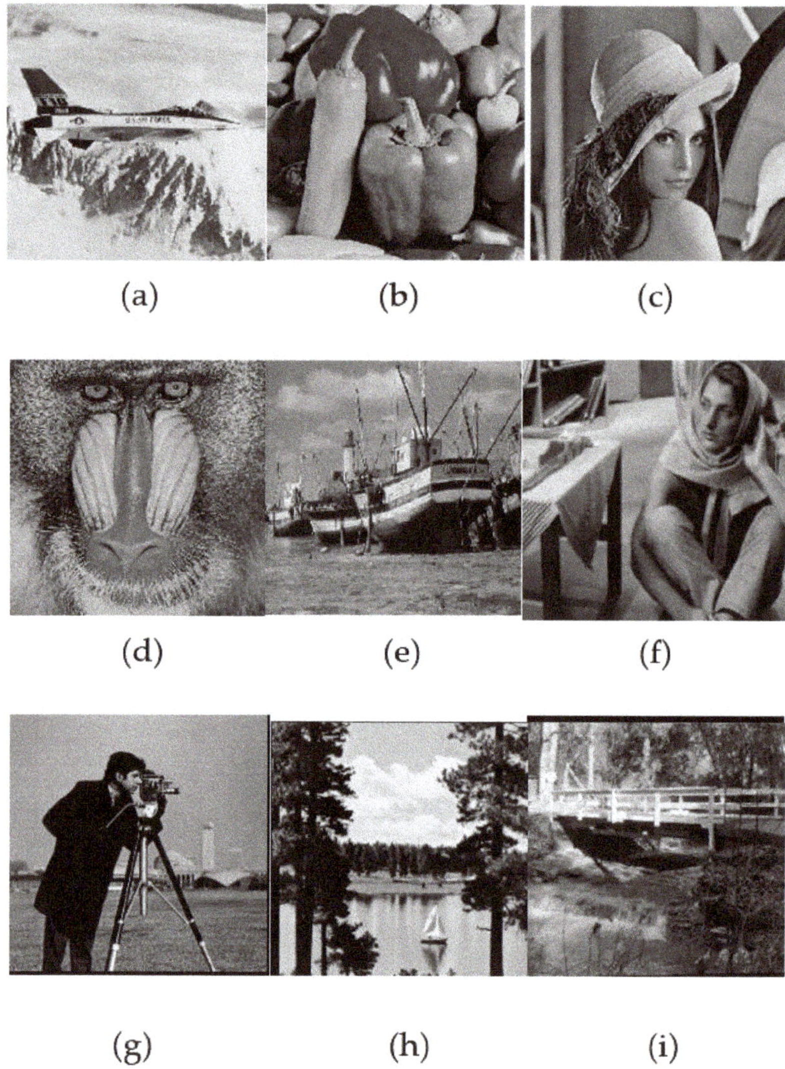

Figure 5. Used host images in this paper: (**a**) Lena; (**b**) Peppers; (**c**) Airplane; (**d**) Baboon; (**e**) Ship; (**f**) Barbara; (**g**) Cameraman; (**h**) Lake; (**i**) Bridge.

3.1. Evaluation Metrics

To evaluate the proposed method and compare it with the methods of Arham et al. [3] and Kumar et al. [24], the peak signal-to-noise ratio (PSNR), insertion capacity, structural similarity index measure (SSIM), and processing time metrics were used.

PSNR: To measure the quality of the marked image and the similarity measurement between the marked image and the original host image, which also specifies the distortion ratio, the PSNR was used. The PSNR value was calculated using the mean squared error (MSE), and has a reverse ratio with MSE. MSE and PSNR were calculated using Equations (59) and (60), respectively [21,22].

$$\mathrm{MSE} = \frac{1}{m \times n} \sum_{i=1}^{m} \sum_{j=1}^{n} (A(i,\ j) - A'(i,\ j))^2 \tag{58}$$

$$\text{PSNR} = 10Log_{10}(\frac{255^2}{MSE}) \tag{59}$$

For the image with a size of $m \times n$, A (i, j) represents the host image pixels, and $A'(i, j)$ represents the marked image pixels.

Insertion capacity: This paper used a random bit sequence as secret data. For a gray host image with a size of $m \times n$, the maximum insertion capacity was calculated according to the number of bits per pixel (bpp), using Equation (61) [3]:

$$\text{Insertion capacity(bpp)} = \frac{Length\ of\ sequence\ bit}{m \times n} \tag{60}$$

In the proposed method, the multilayer insertion technique enhances the capacity. Arham et al. [3] considered the maximum number of insertion layers to be eight. Therefore, the data insertion process was performed two times, and at each time, data bits were inserted under four layers—the first time from the first layer to the fourth layer, and the second time from the fifth layer to the eighth layer.

SSIM: The SSIM metric is a famous metric used to measure the amount of structural similarity between the host image A and image A$'$. This metric can be obtained using Equation (62) [34]:

$$\text{SSIM}(\text{A}, \text{A}') = \frac{(2\mu_1\mu_{\text{A}'} + c_1)(2\sigma_{1,\text{A}'} + c_2)}{(\mu_1^2 + \mu_{\text{A}'}^2 + c_1)(\sigma_1^1 + \sigma_{\text{A}'}^2 + c_2)} \tag{61}$$

where μ_1 and $\mu_{\text{A}'}$ are the mean brightness intensity of A and A$'$, respectively, $\sigma_{\text{A}'}$ represents the standard deviation of images A and A$'$, respectively, $\sigma_{1,\text{A}'}$ represents covariance between images A and A$'$, respectively, and c_1 and c_2 are two constant values of 6.50 and 58.52, respectively. The higher the SSIM value in data hiding methods and the closer it is to 1, the more effective the corresponding process [34].

Processing time: processing time is one of the essential parameters for comparison to DH methods. Therefore, the total processing time is equal to the total insertion and extraction time values to compare the proposed method and other methods.

3.2. Comparison with the Other Methods

In this paper, the proposed method is compared with the methods of Arham et al. [3], Yao et al. [22], and Kumar et al. [16]. Similarity and correlation between pixels belonging to each class are increased using multilevel thresholding, and the difference between the pixels decreases. As a result, inserting a few layers of data creates less distortion in the image.

Because, in the proposed method, all of the values of zero and positive differences are expanded to insert the data, in the first insertion layer and higher insertion layers, it has a higher insertion capacity and PSNR than the methods of Arham et al. [3], Yao et al. [30], and Kumar et al. [24]. As a result, the resulting distortion in the marked image for the proposed method is less than that in the methods of Arham et al. [3], Yao et al. [30], and Kumar et al. [24]. Table 2 shows the values of insertion capacity maxima for different images per thresholds T_1 and T_2 at $0 \leq \text{Th} \leq +32$. The values of T_1 and T_2 are set by the SMA. Table 3 shows the values of processing time (seconds) for different images per thresholds T_1 and T_2 at $0 \leq \text{Th} \leq +32$.

As can be seen from Tables 2 and 3, for the two thresholds and T_2 in the first insertion layer, the proposed method has more insertion capacity and more PSNR for all images than the methods of Arham et al. [3] and Kumar et al. [24].

According to Table 2, the insertion capacity of the first insertion layer in the proposed method for the Airplane, Baboon, Barbara, Ship, Lena, Lake, Bridge, Cameraman, and Peppers images is 834 bits (0.0032 bpp), 1689 bits (0.0064 bpp), 766 bits (0.0029 bpp), 1561 bits (0.0059 bpp), 2241 bits (0.0085 bpp), 1604 bits (0.0061 bpp), 4402 bits (0.0168 bpp), 4794 bits (0.0183 bpp), and 1571 bits (0.0061 bpp), respectively—more than in the method of Arham et al. [3].

Table 2. A comparison of the actual embedding capacities in common images.

	Proposed		Insertion Capacity (bits)			Insertion Capacity (bpp)		
Image	**T1**	**T2**	[3]	[24]	**PM**	[3]	[24]	**PM**
Airplane	111	182	194,250	210,000	215,000	0.7449	0.8010	0.8201
Baboon	78	149	193,314	110,000	195,003	0.7374	0.4196	0.7438
Barbara	55	131	191,466	158,000	192,232	0.7304	0.6027	0.7333
Ship	79	156	194,583	160,000	196,144	0.7423	0.6103	0.7482
Lena	85	152	196,768	200,000	209,000	0.7480	0.7629	0.7972
Lake	76	145	191,854	111,000	193,458	0.7318	0.4234	0.7514
Bridge	84	151	192,472	185,000	196,874	0.7342	0.7186	0.7379
Cameraman	91	174	194,968	159,000	199,762	0.7437	0.6039	0.7510
Peppers	61	136	195,414	178,000	196,985	0.7454	0.6790	0.7514

PM = proposed method.

Table 3. A comparison of the processing times in common images.

	Proposed		Processing Time (s)		
Image	T_1	T_2	[3]	[24]	**PM**
Airplane	111	182	154.09	1.29	173.94
Baboon	78	149	153.72	1.81	173.60
Barbara	55	131	151.44	1.37	175.35
Ship	79	156	138.11	1.44	160.99
Lena	85	152	153.58	1.67	181.56
Lake	76	145	152.41	1.37	174.56
Bridge	84	151	153.25	1.41	175.92
Cameraman	91	174	139.48	1.62	179.45
Peppers	61	136	154.11	1.15	185.98

PM = proposed method.

In the proposed method, the Lena image has the highest increase in capacity (0.0085 bpp), while the Barbara image has the lowest increase in capacity (0.0029 bpp), compared to the method of Arham et al. [3]. Therefore, on average, the capacity of the first insertion layer in the proposed method is 0.0058 bpp more than that of the method of Arham et al. [3]. The average processing time for the proposed method is 173.2367 s, while that for the method of Arham et al. [3] is 150.8417 s. Therefore, the proposed method is slower than the method of Arham et al. [3], due to the use of the SMA and its repetitions to obtain optimal thresholds.

Moreover, the proposed method is better than the method of Kumar et al. [24] in terms of PSNR and embedding capacity values, as can be seen in Tables 2 and 3, but in terms of execution time it is slower compared to the methods of Kumar et al. [24] and Arham et al. [3]. Furthermore, the proposed method, compared to the method of Kumar et al. [24] for the Lake, Bridge, Cameraman, Airplane, Baboon, Barbara, Ship, Lena, and Peppers images, yields 82,485 bits (3145 bpp), 11,874 bits (0.0324 bpp), 40,762 bits (0.1581 bpp), 5000 bits (0.0191 bpp), 85,000 bits (0.3242 bpp), 34,232 bits (0.1306 bpp), 36,144 bits (0.1379 bpp), 9000 bits (0.0343 bpp) and 18,985 bits (0.0724 bpp), respectively, showing greater capacity.

Table 4 also compares the PSNR values of the proposed method with the method of Arham et al. [3] for different capacities of 0.1 bpp, 0.2 bpp, 0.3 bpp, 0.4 bpp, 0.5 bpp, 0.6 bpp, and 0.7 bpp in the first insertion layer. As can be seen from Table 4, the average PSNR of the proposed method for different capacities is higher than that of the method of Arham et al. [3].

Table 4. Comparison of the PSNR (dB) value single-layer embedding in terms of visual quality test images.

		PSNR (dB)						
		0.1 bpp	**0.2 bpp**	**0.3 bpp**	**0.4 bpp**	**0.5 bpp**	**0.6 bpp**	**0.7 bpp**
Airplane	[3]	54.35	51.88	48.85	44.36	42.45	41.44	40.92
	Proposed	55.17	52.73	49.99	45.40	44.07	43.11	41.86
Baboon	[3]	40.95	38.97	37.98	36.95	36.48	35.97	35.40
	Proposed	42.01	40.03	38.88	37.34	37.11	36.54	36.89
Barbara	[3]	49.29	46.73	43.84	41.12	38.98	37.50	36.57
	Proposed	50.74	47.43	44.45	42.31	39.12	38.61	37.63
Ship	[3]	50.71	46.86	43.85	41.95	40.92	40.46	40.19
	Proposed	51.85	48.46	45.48	43.01	42.15	41.64	41.78
Lena	[3]	54.70	50.31	47.74	45.77	44.19	43.03	42.16
	Proposed	55.18	52.12	48.61	46.17	45.46	44.61	43.44
Lake	[3]	48.32	44.16	40.35	38.36	36.56	34.79	33.21
	Proposed	49.97	45.98	41.68	39.86	37.15	35.96	34.56
Bridge	[3]	47.76	43.32	39.86	36.75	34.85	33.45	31.99
	Proposed	49.23	44.53	40.96	37.82	35.98	34.59	32.68
Cameraman	[3]	46.12	42.36	38.25	34.91	32.56	31.20	30.05
	Proposed	48.06	43.74	39.94	36.20	34.11	32.31	31.11
Peppers	[3]	49.66	47.12	45.37	43.92	43.15	42.61	42.05
	Proposed	51.11	48.94	46.33	44.61	45.20	43.45	43.11

Figure 6 shows the PSNR comparison diagram of the proposed method with the method of Arham et al. [3] for different images under the same capacities drawn using the data shown in Tables 4 and 5. As can be seen from the diagrams in Figure 6, the proposed method has a higher PSNR value for all images than the method of Arham et al. [3]. For the first insertion layer, the Airplane image for the capacities of 0.1 bpp, 0.2 bpp, 0.3 bpp, 0.4 bpp, 0.5 bpp, 0.6 bpp, and 0.7 bpp shows increases in quality compared to the method of Arham et al. [3] of 0.8200 dB, 0.8500 dB, 1.1400 dB, 1.04 dB, 1.6200 dB, 1.6700 dB, and 0.94 dB, respectively.

Table 5. A comparison of the PSNR (dB) value multiple-layer embedding in terms of visual quality test images.

		PSNR (dB)							
		0.7 bpp	**1.5 bpp**	**2.2 bpp**	**3 bpp**	**3.7 bpp**	**4.5 bpp**	**5.2 bpp**	**6 bpp**
Airplane	[3]	40.5	36.59	35.1	34	32.5	31.9	31	30
	Proposed	41.12	38.02	36.25	35.01	33.84	32.56	32.97	31.01
Baboon	[3]	35.1	32	30	28.5	27.1	26.2	25.3	25
	Proposed	36.98	33.24	31.12	29.95	28.87	27.97	26.78	26.25
Barbara	[3]	36.2	33.9	32	30.7	29.1	28.83	27.8	27
	Proposed	37.45	34.25	33.12	31.99	30.47	30.02	28.11	28
Ship	[3]	40.58	36.9	35	33.2	32	31.57	30	29.1
	Proposed	42.01	37.2	36.11	34.42	33	33.03	31	30.25
Lena	[3]	42	38.2	37	35.1	34	33	32.1	31.1
	Proposed	43.52	39.44	38	36.55	35	34.11	33	32
Lake	[3]	37.26	34.73	32.82	31.74	30.82	29	27.76	26
	Proposed	38.99	35.47	34.15	33.26	32.42	30.76	29.12	27.22
Bridge	[3]	37.85	33.72	32.25	31.74	29.18	28	26.87	25.10
	Proposed	39.23	35.28	33.76	32.06	30.46	29	27.75	26
Cameraman	[3]	40.51	36.97	34.98	32.76	30.56	28.75	27	25.34
	Proposed	42.07	38.13	36.46	35.36	32.89	31.46	29.42	26.12
Peppers	[3]	41.9	38.3	36.3	35	34	33	32	31.1
	Proposed	43	39.85	37	36.55	35.89	34	33	32.34

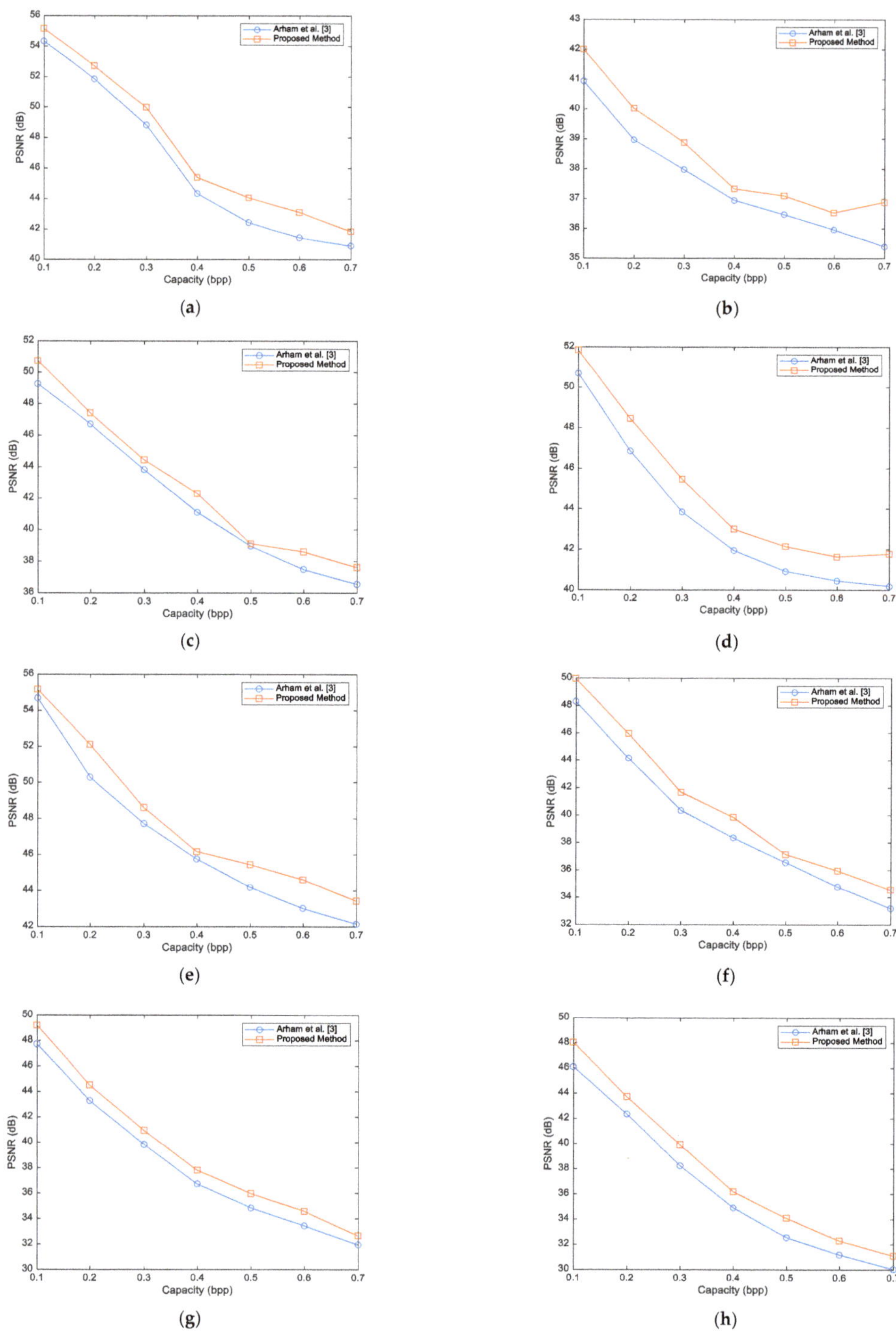

Figure 6. Comparison of image quality for single-layer capacity: (**a**) Airplane; (**b**) Baboon; (**c**) Barbara; (**d**) Ship; (**e**) Lena; (**f**) Peppers; (**g**) Lake; (**h**) Bridge [3].

Moreover, the Baboon image for the capacities of 0.1 bpp, 0.2 bpp, 0.3 bpp, 0.4 bpp, 0.5 bpp, 0.6 bpp, and 0.7 bpp, compared to the method of Arham et al. [3], shows increases in quality of 1.0600 dB, 0.9600 dB, 0.9000 dB, 0.3900 dB, 0.6300 dB, 0.5700 dB and 1.4900 dB, respectively. Meanwhile, the Barbara image for the capacities of 0.1 bpp, 0.2 bpp, 0.3 bpp, 0.4 bpp, 0.5 bpp, 0.6 bpp, and 0.7 bpp increases in quality by 1.4500 dB, 0.7000 dB, 0.610.0 dB, 1.1900 dB, 0.1400 dB, 1.1100 dB, and 1.0600 dB, respectively, compared to the method of Arham et al. [3].

The Ship image for the capacities of 0.1 bpp, 0.2 bpp, 0.3 bpp, 0.4 bpp, 0.5 bpp, 0.6 bpp, and 0.7 bpp increases in quality by 1.1400 dB, 1.6000 dB, 1.6300 dB, 1.0600 dB, 1.2300 dB, 1.1800 dB, and 1.5900 dB, respectively, compared to the method of Arham et al. [3].

The Lena image for capacities of 0.1 bpp, 0.2 bpp, 0.3 bpp, 0.4 bpp, 0.5 bpp, 0.6 bpp and 0.7 bpp, compared to the method of Arham et al. [3], has an increase in quality of 0.4800 dB, 1.8100 dB, 0.8700 dB, 0.4000 dB, 1.2700 dB, 1.5800 dB, and 1.2800 dB, respectively. The Peppers image for capacities of 0.1 bpp, 0.2 bpp, 0.3 bpp, 0.4 bpp, 0.5 bpp, 0.6 bpp, and 0.7 bpp, compared to the method of Arham et al. [3], has an increase in quality of 1.6500 dB, 1.8200 dB, 1.3300 dB, 1.50 dB, 0.5900 dB, 1.1700 dB, and 1.3500 dB, respectively. The Lake image for the capacities of 0.1 bpp, 0.2 bpp, 0.3 bpp, 0.4 bpp, 0.5 bpp, 0.6 bpp, and 0.7 bpp increases in quality by 1.4700 dB, 1.2100 dB, 1.1000 dB, 1.0700 dB, 1.1300 dB, 1.1400 dB, and 0.6900 dB, respectively, compared to the method of Arham et al. [3].

The Cameraman image for capacities of 0.1 bpp, 0.2 bpp, 0.3 bpp, 0.4 bpp, 0.5 bpp, 0.6 bpp and 0.7 bpp, compared to the method of Arham et al. [3], has an increase in quality of 1.9400 dB, 1.3800 dB, 1.6900 dB, 1.2900 dB, 1.5500 dB, 1.1100 dB, and 1.0600 dB, respectively. The Bridge image for capacities of 0.1 bpp, 0.2 bpp, 0.3 bpp, 0.4 bpp, 0.5 bpp, 0.6 bpp, and 0.7 bpp, compared to the method of Arham et al. [3], has an increase in quality of 1.4500 dB, 1.8200 dB, 0.9600 dB, 1.2400 dB, 1.5000 dB, 0.8400 dB, and 1.0600 dB, respectively.

The proposed method has more insertion capacity than that of Arham et al. [3] for the first and higher insertion layers. Table 5 shows the PSNR comparison of the proposed method and the method of Arham et al. [3] for the eight insertion layers. As the number of insertion layers increases, the total insertion capacity increases. As shown in Table 5, the values of insertion capacity and PSNR for the proposed method are higher than those in the method of Arham et al. [3]. According to Table 5, the average insertion capacity increase in the eight insertion layers in the proposed method is 0.625 bpp, and in the method of Arham et al. [3] it is 0.572 bpp. In insertion layer eight, the Peppers, Lena, Barbara, and Baboon images have the highest increase in capacity, while the Airplane, Ship, Lake, Bridge, and Cameraman images have the slightest increase in capacity, compared to the method of Arham et al. [3].

In insertion layer eight, the Baboon image has the highest increase in PSNR (1.25 dB), while the Barbara image has the lowest increase in PSNR (1 dB), compared to the method of Arham et al. [3]. In insertion layer eight, the average insertion capacity of the proposed method and the method of Arham et al. [3] is 6.33 bpp and 6 bpp, respectively, while the average PSNR in the proposed method and the method of Arham et al. [3] is 29.75 dB and 28.88 dB, respectively. In insertion layer eight, the proposed method has an average capacity increase of 0.33 bpp and an average quality increase of 1.63 dB compared to the method of Arham et al. [3]. Figure 7 shows a comparison diagram of the capacity and PSNR values of the proposed method and the method of Arham et al. [3] for eight insertion layers drawn using the data shown in Table 5.

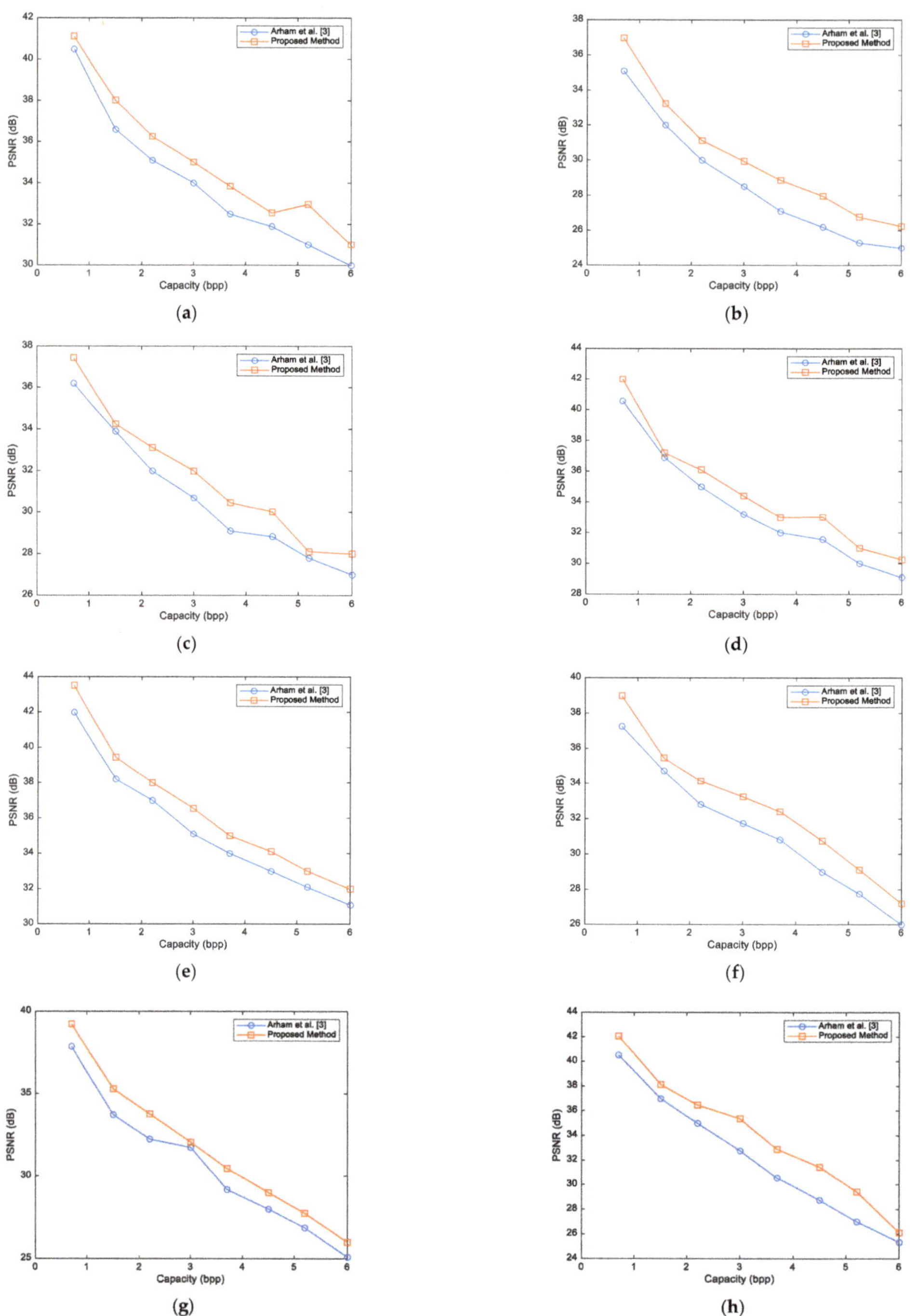

Figure 7. Comparing the quality values for multilayer capacities in different images: (**a**) Airplane; (**b**) Baboon; (**c**) Barbara; (**d**) Ship; (**e**) Lena; (**f**) Peppers; (**g**) Lake; (**h**) Bridge [3].

Table 6 shows the SSIM values of the proposed method and the methods of Arham et al. [3] and Kumar et al. [16] for maximum embedding capacity (bits). According to Table 6, due to its nature, the proposed method has a higher SSIM than the compared methods, while the capacity of the proposed method is also more than that of the compared methods.

Table 6. Comparison of the SSIM in common images for different capacities.

	Proposed		Insertion Capacity (bits)			SSIM		
Image	**T1**	**T2**	[3]	[24]	**PM**	[3]	[24]	**PM**
Airplane	111	182	194,250	210,000	215,000	0.9128	0.9131	0.9261
Baboon	78	149	193,314	110,000	195,003	0.9125	0.9155	0.9298
Barbara	55	131	191,466	158,000	192,232	0.9135	0.9146	0.9283
Ship	79	156	194,583	160,000	196,144	0.9134	0.9124	0.9282
Lena	85	152	196,768	200,000	209,000	0.9132	0.9138	0.9272
Lake	76	145	191,854	111,000	193,458	0.9157	0.9115	0.9284
Bridge	84	151	192,472	185,000	196,874	0.9125	0.9114	0.9279
Cameraman	91	174	194,968	159,000	199,762	0.9148	0.9118	0.9256
Peppers	61	136	195,414	178,000	196,985	0.9136	0.9145	0.9274

PM = proposed method.

Table 7 shows the method proposed by Arham et al. [3], using the SSIM evaluation metric, which compares one embedding layer and several embedding layers. As can be seen in Table 7, the proposed method has superior performance for different capacities (bpp) compared to the method of Arham et al. [3], with a higher SSIM value. As the capacity value increases in bpp, the SSIM value for the proposed method and the method of Arham et al. [3] decreases. Still, in any case, the proposed method compared is superior to the method of Arham et al. [3] in terms of SSIM value.

Table 7. A comparison of the SSIM for single-layer embedding.

		SSIM						
		0.1 bpp	**0.2 bpp**	**0.3 bpp**	**0.4 bpp**	**0.5 bpp**	**0.6 bpp**	**0.7 bpp**
Airplane	[3]	0.9736	0.9658	0.9531	0.9462	0.9312	0.9243	0.9134
	Proposed	0.9865	0.9736	0.9638	0.9582	0.9462	0.9365	0.9245
Baboon	[3]	0.9735	0.9685	0.9538	0.9425	0.9365	0.9235	0.9141
	Proposed	0.9846	0.9734	0.9648	0.9536	0.9468	0.9328	0.9267
Barbara	[3]	0.9762	0.9694	0.9539	0.9462	0.9361	0.9217	0.9119
	Proposed	0.9873	0.9725	0.9647	0.9543	0.9486	0.9369	0.9278
Ship	[3]	0.9748	0.9657	0.9567	0.9474	0.9313	0.9236	0.9167
	Proposed	0.9839	0.9746	0.9625	0.9512	0.9452	0.9385	0.9286
Lena	[3]	0.9743	0.9651	0.9512	0.9436	0.9321	0.9238	0.9118
	Proposed	0.9849	0.9783	0.9674	0.9572	0.9445	0.9368	0.9264
Lake	[3]	0.9739	0.9645	0.9536	0.9412	0.9336	0.9225	0.9136
	Proposed	0.9818	0.9762	0.9652	0.9548	0.9462	0.9368	0.9275
Bridge	[3]	0.9786	0.9638	0.9536	0.9438	0.9356	0.9283	0.9169
	Proposed	0.9863	0.9782	0.9671	0.9582	0.9468	0.9397	0.9247
Cameraman	[3]	0.9768	0.9637	0.9532	0.9413	0.9367	0.9214	0.9179
	Proposed	0.9864	0.9726	0.9682	0.9551	0.9439	0.9369	0.9264
Peppers	[3]	0.9739	0.9632	0.9539	0.9462	0.9363	0.9258	0.9179
	Proposed	0.9827	0.9746	0.9648	0.9583	0.9486	0.9378	0.9248

PM = proposed method.

Table 8 shows the PSNR and SSIM values for the proposed method and the method of Yao et al. [30] for capacities of 30,000 and 50,000 bits. As can be seen from Table 8, the proposed method has higher SSIM and PSNR values compared to the method of Yao et al. [30].

Table 8. Compares the PSNR value for the PM and yao et al. [22] method.

	30,000 (bits)				50,000 (bits)			
	psnr		ssim		psnr		ssim	
Image	**Yao et al. [30]**		**PM**		**Yao et al. [30]**		**PM**	
Airplane	65.56	0.9736	66.12	0.9846	64.34	0.9616	65.22	0.9765
Baboon	65.56	0.9747	66.68	0.9874	64.34	0.9623	65.44	0.9716
Barbara	65.56	0.9732	66.23	0.9834	64.34	0.9675	65.65	0.9736
Ship	65.56	0.9747	66.42	0.9845	64.34	0.9646	65.61	0.9765
Lena	65.56	0.9761	66.29	0.9856	64.34	0.9654	65.12	0.9748
Lake	65.42	0.9719	66.18	0.9885	64.12	0.9638	65.83	0.9716
Bridge	65.56	0.9764	66.47	0.9849	64.34	0.9676	65.79	0.9734
Cameraman	65.55	0.9773	66.42	0.9859	64.34	0.9649	65.68	0.9756
Peppers	65.56	0.9772	66.82	0.9817	64.34	0.9647	65.36	0.9748

PM = proposed method.

In this paper, a multilayer RDH method based on the multilevel thresholding technique is proposed, aiming to increase the insertion capacity and reduce the distortion after data embedding by improving the correlation between consecutive pixels of the image. Firstly, the SMA is applied to find the optimal thresholds of host image segmentation. Next, according to the specified threshold, image pixels located in different image areas are classified into different categories. Finally, the difference between two consecutive pixels is reduced in each class, and then the data are embedded via DE.

4. Conclusions

In this paper, a new multilayer RDH technique was proposed using multilevel thresholding where, first, the SMA was used to determine the optimal thresholds on the host image. By using the SMA, two optimal threshold values were determined. Image pixels were then located in their specific classes based on those thresholds. The similar and more correlated pixels within a group were classified. Therefore, the difference between pixels related to each class decreased. It was proven that insertion and distortion in the marked image decreased after inserting the data by reducing the differences between each class' pixels. The proposed method has a simple implementation. The results showed that the proposed method, in comparison with the methods of Arham et al. [3] and Kumar et al. [24], has more capacity and more PSNR, but is slightly slower in comparison with the methods of Arham et al. [3] and Kumar et al. [24]. In our subsequent work, we intend to use deep learning methods to extract features and image thresholds in order to classify image features and maximize the capacity and quality of the marked image. Therefore, using the multilevel thresholding technique in the proposed method improved the multilayer RDH method compared to the other methods. This technique reduced the difference between the pixels, and the pixels in the same class were very similar to one another.

Author Contributions: Conceptualization, A.M., B.k.D., J.L.W., H.A.A., E.E., M.D., R.A.Z. and L.A.; methodology, A.M. and L.A.; software, A.M., J.L.W., H.A.A., M.D. and L.A.; validation, A.M., B.k.D., J.L.W., H.A.A., E.E., M.D., R.A.Z. and L.A.; formal analysis, A.M., B.k.D., J.L.W., H.A.A., E.E., M.D., R.A.Z. and L.A.; investigation, E.E., M.D., R.A.Z. and L.A.; resources, A.M., B.k.D., J.L.W., H.A.A. and L.A.; data curation, A.M., B.k.D., J.L.W., H.A.A., E.E., M.D., R.A.Z. and L.A.; writing—original draft preparation, A.M., B.k.D., J.L.W., H.A.A., E.E., M.D., R.A.Z. and L.A.; writing—review and editing, A.M., B.k.D., J.L.W.; visualization, A.M. All authors have read and agreed to the published version of the manuscript.

Funding: This research received no external funding.

Institutional Review Board Statement: Not applicable.

Informed Consent Statement: The study did not require ethical approval.

Data Availability Statement: Data is available upon the request.

Conflicts of Interest: The authors declare no conflict of interest.

References

1. Abdullah, S.M.; Manaf, A.A. *Multiple Layer Reversible Images Watermarking Using Enhancement of Difference Expansion Techniques*; Springer: Berlin/Heidelberg, Germany, 2010; Volume 87, pp. 333–342.
2. Zeng, X.; Li, Z.; Ping, L. Reversible data hiding scheme using reference pixel and multi-layer embedding. *Int. J. Electron. Commun. (AEÜ)* **2012**, *66*, 532–539. [CrossRef]
3. Arham, A.; Nugroho, H.A.; Adji, T.B. Multiple Layer Data Hiding Scheme Based on Difference Expansion of Quad. *Signal Process.* **2017**, *137*, 52–62. [CrossRef]
4. Maniriho, P.; Ahmad, T. Information Hiding Scheme for Digital Images Using Difference Expansion and Modulus Function. *J. King Saud Univ. Comput. Inf. Sci.* **2018**, *31*, 335–347. [CrossRef]
5. Yao, H.; Qin, C.; Tang, Z.; Tian, Y. Guided filtering based color image reversible data hiding. *J. Vis. Commun. Image R* **2013**, *43*, 152–163. [CrossRef]
6. Ou, B.; Li, X.; Zhao, Y. Pairwise Prediction-Error Expansion for Efficient Reversible Data Hiding. *IEEE Trans. Image Process.* **2013**, *22*, 5010–5021. [CrossRef] [PubMed]
7. Li, X.; Zhang, W.; Gui, X.; Yang, B. Efficient Reversible Data Hiding Based on Multiple Histograms Modification. *IEEE Trans. Inf. Forensics Secur.* **2015**, *10*, 2016–2027.
8. Fu, D.; Jing, Z.J.; Zhao, S.G.; Fan, J. Reversible data hiding based on prediction-error histogram shifting and EMD mechanis. *Int. J. Electron. Commun.* **2014**, *68*, 933–943. [CrossRef]
9. Ou, B.; Li, X.; Wang, J.; Peng, F. High-fidelity reversible data hiding based on geodesic path and pairwise prediction-error expansion. *Neurocomputing* **2016**, *68*, 933–943. [CrossRef]
10. Wang, J.; Ni, J.; Zhang, X.; Shi, Y.Q. Rate and Distortion Optimization for Reversible Data Hiding Using Multiple Histogram Shifting. *IEEE Trans. Cybern.* **2016**, *47*, 315–326. [CrossRef]
11. Xiao, M.; Li, X.; Wang, Y.; Zhao, Y.; Ni, R. Reversible data hiding based on pairwise embedding and optimal expansion path. *Signal Process.* **2019**, *19*, 30017–30019. [CrossRef]
12. Zhou, H.; Chen, K.; Zhang, W.; Yu, N. Comments on Steganography Using Reversible Texture Synthesis. *IEEE Trans. Image Process.* **2017**, *26*, 1623–1625. [CrossRef] [PubMed]
13. Shehab, M.; Abualigah, L.; Shambour, Q.; Abu-Hashem, M.A.; Shambour, M.K.Y.; Alsalibi, A.I.; Gandomi, A.H. Machine learning in medical applications: A review of state-of-the-art methods. *Comput. Biol. Med.* **2022**, *145*, 105458. [CrossRef] [PubMed]
14. Zhu, X.; Zhou, M. Multiobjective Optimized Cloudlet Deployment and Task Offloading for Mobile-Edge Computing. *IEEE Internet Things J.* **2021**, *8*, 15582–15595. [CrossRef]
15. Zhu, Q.-H.; Tang, H.; Huang, J.-J.; Hou, Y. Task Scheduling for Multi-Cloud Computing Subject to Security and Reliability Constraints. *IEEE/CAA J. Autom. Sin.* **2021**, *8*, 848–865. [CrossRef]
16. Ezugwu, A.E.; Ikotun, A.M.; Oyelade, O.O.; Abualigah, L.; Agushaka, J.O.; Eke, C.I.; Akinyelu, A.A. A comprehensive survey of clustering algorithms: State-of-the-art machine learning applications, taxonomy, challenges, and future research prospects. *Eng. Appl. Artif. Intell.* **2022**, *110*, 104743. [CrossRef]
17. Otair, M.; Abualigah, L.; Qawaqzeh, M.K. Improved near-lossless technique using the Huffman coding for enhancing the quality of image compression. *Multimed. Tools Appl.* **2022**, 1–21. [CrossRef]
18. Tang, J.; Liu, G.; Pan, Q. A Review on Representative Swarm Intelligence Algorithms for Solving Optimization Problems: Applications and Trends. *IEEE/CAA J. Autom. Sin.* **2021**, *8*, 1627–1643. [CrossRef]
19. Abualigah, L.; Abd Elaziz, M.; Sumari, P.; Geem, Z.W.; Gandomi, A.H. Reptile Search Algorithm (RSA): A nature-inspired meta-heuristic optimizer. *Expert Syst. Appl.* **2022**, *191*, 116158. [CrossRef]
20. He, W.; Cai, J.; Xiong, G.; Zhou, K. Improved reversible data hiding using pixel-based pixel value grouping. *Optik* **2017**, *157*, 68–78. [CrossRef]
21. He, W.; Xiong, G.; Zhou, K.; Cai, J. Reversible data hiding based on multilevel histogram modification and pixel value grouping. *J. Vis. Commun. Image Represent.* **2016**, *40*, 459–469. [CrossRef]
22. Li, X.; Li, J.; Li, B.; Yang, B. High-fidelity reversible data hiding scheme based on pixel-value-ordering and prediction-error expansion. *Signal Process.* **2013**, *93*, 198–205. [CrossRef]
23. Kumar, R.; Jung, K.H. Robust reversible data hiding scheme based on two-layer embedding strategy. *Inf. Sci.* **2020**, *512*, 96–107. [CrossRef]
24. Kumar, R.; Kim, D.S.; Jung, K.H. Enhanced AMBTC based data hiding method using hamming distance and pixel value differencing. *J. Inf. Secur. Appl.* **2019**, *47*, 94–103. [CrossRef]
25. Kim, P.H.; Ryu, K.W.; Jun, K.H. Reversible data hiding scheme based on pixel-value differencing in dual images. *Int. J. Distrib. Sens. Netw.* **2020**, *16*, 1550147720911006. [CrossRef]
26. Hussain, M.; Riaz, Q.; Saleem, S.; Ghafoor, A.; Jung, K.H. Enhanced adaptive data hiding method using LSB and pixel value differencing. *Multimed. Tools Appl.* **2021**, *80*, 20381–20401. [CrossRef]
27. Yu, C.; Zhang, X.; Zhang, X.; Li, G.; Tang, Z. Reversible Data Hiding with Hierarchical Embedding for Encrypted Images. *IEEE Trans. Circuits Syst. Video Technol.* **2021**, *32*, 451–466. [CrossRef]

28. Tang, Z.; Nie, H.; Pun, C.M.; Yao, H.; Yu, C.; Zhang, X. Color Image Reversible Data Hiding with Double-Layer Embedding. *IEEE Access* **2020**, *8*, 6915–6926. [CrossRef]
29. Yao, H.; Mao, F.; Tang, Z.; Qin, C. High-fidelity dual-image reversible data hiding via prediction-error shift. *Signal Process.* **2020**, *170*, 107447. [CrossRef]
30. Salehnia, T.; Izadi, S.; Ahmadi, M. Multilevel image thresholding using GOA, WOA and MFO for image segmentation. In Proceedings of the 8th International Conference on New Strategies in Engineering, Information Science and Technology in the Next Century, Dubai, United Arab Emirates (UAE), 2021; Available online: https://civilica.com/doc/1196572/ (accessed on 16 March 2022).
31. Raziani, S.; Salehnia, T.; Ahmadi, M. Selecting of the best features for the knn classification method by Harris Hawk algorithm. In Proceedings of the 8th International Conference on New Strategies in Engineering, Information Science and Technology in the Next Century, Dubai, United Arab Emirates (UAE), 2021; Available online: https://civilica.com/doc/1196573/ (accessed on 16 March 2022).
32. Salehnia, T.; Fath, A. Fault tolerance in LWT-SVD based image watermarking systems using three module redundancy technique. *Expert Syst. Appl.* **2021**, *179*, 115058. [CrossRef]
33. El Aziz, M.A.; Ewees, A.A.; Hassanien, A.E. Whale Optimization Algorithm and Moth-Flame Optimization for Multilevel Thresholding Image Segmentation. *Expert Syst. Appl.* **2017**, *83*, 242–256. [CrossRef]
34. Akay, B. A study on particle swarm optimization and artificial bee colony algorithms for multilevel thresholding. *Appl. Soft Comput.* **2013**, *13*, 3066–3091. [CrossRef]
35. Li, S.; Chen, H.; Wang, M.; Heidari, A.A.; Mirjalili, S. Slime mould algorithm: A new method for stochastic optimization. *Future Gener. Comput. Syst.* **2020**, *111*, 300–323. [CrossRef]
36. Liu, H.; Zhou, M.; Guo, X.; Zhang, Z.; Ning, B.; Tang, T. Timetable Optimization for Regenerative Energy Utilization in Subway Systems. *IEEE Trans. Intell. Transp. Syst.* **2019**, *20*, 3247–3257. [CrossRef]
37. Abualigah, L.; Yousri, D.; Abd Elaziz, M.; Ewees, A.A.; Al-Qaness, M.A.; Gandomi, A.H. Aquila optimizer: A novel meta-heuristic optimization algorithm. *Comput. Ind. Eng.* **2021**, *157*, 107250. [CrossRef]
38. Agushaka, J.O.; Ezugwu, A.E.; Abualigah, L. Dwarf mongoose optimization algorithm. *Comput. Methods Appl. Mech. Eng.* **2022**, *391*, 114570. [CrossRef]

Article

Data Driven Model Estimation for Aerial Vehicles: A Perspective Analysis

Syeda Kounpal Fatima [1], Manzar Abbas [1], Imran Mir [1,*], Faiza Gul [2], Suleman Mir [3], Nasir Saeed [4,*], Abdullah Alhumaidi Alotaibi [5,6], Turke Althobaiti [7] and Laith Abualigah [8,9]

1 Department of Avionics Engineering, Air University, Aerospace and Aviation Campus Kamra, Islamabad 43600, Pakistan; 195125@aack.au.edu.pk (S.K.F.); manzar.abbas@aack.au.edu.pk (M.A.)
2 Department of Electrical Engineering, Air University, Aerospace and Aviation Campus Kamra, Islamabad 43600, Pakistan; faiza.gul@aack.au.edu.pk
3 Electrical Department, Fast-National University of Computer & Emerging Sciences, Peshawar 25000, Pakistan; suleman.mir@nu.edu.pk
4 Department of Electrical Engineering, Northern Border University, Arar 73222, Saudi Arabia
5 Remote Sensing Unit, Northern Border University, Arar 73222, Saudi Arabia; a.alhumaidi@tu.edu.sa
6 Department of Science and Technology, College of Ranyah, Taif Univeristy, P.O. Box 11099, Taif 21944, Saudi Arabia
7 Department of Computer Science, Faculty of Science, Northern Border University, Arar 73222, Saudi Arabia; turke.althobaiti@nbu.edu.sa
8 Faculty of Computer Sciences and Informatics, Amman Arab University, Amman 11953, Jordan; laythdyabat@aau.edu.jo
9 Faculty of Information Technology, Middle East University, Amman 11831, Jordan
* Correspondence: imranmir56@yahoo.com (I.M.); mr.nasir.saeed@ieee.org (N.S.)

Abstract: Unmanned Aerial Vehicles (UAVs) are important tool for various applications, including enhancing target detection accuracy in various surface-to-air and air-to-air missions. To ensure mission success of these UAVs, a robust control system is needed, which further requires well-characterized dynamic system model. This paper aims to present a consolidated framework for the estimation of an experimental UAV utilizing flight data. An elaborate estimation mechanism is proposed utilizing various model structures, such as Autoregressive Exogenous (ARX), Autoregressive Moving Average exogenous (ARMAX), Box Jenkin's (BJ), Output Error (OE), and state-space and non-linear Autoregressive Exogenous. A perspective analysis and comparison are made to identify the salient aspects of each model structure. Model configuration with best characteristics is then identified based upon model quality parameters such as residual analysis, final prediction error, and fit percentages. Extensive validation to evaluate the performance of the developed model is then performed utilizing the flight dynamics data collected. Results indicate the model's viability as the model can accurately predict the system performance at a wide range of operating conditions. Through this, to the best of our knowledge, we present for the first time a model prediction analysis, which utilizes comprehensive flight dynamics data instead of simulation work.

Keywords: Unmanned Speed Aerial Vehicle; system identification ARX; ARMAX; Box Jenkin's; Output Error; non-linear ARX

Citation: Fatima, S.K.; Abbas, M.; Mir, I.; Gul, F.; Mir, S.; Saeed, N.; Alotaibi, A.A.; Althobaiti, T.; Abualigah, L. Data Driven Model Estimation for Aerial Vehicles: A Perspective Analysis. *Processes* **2022**, *10*, 1236. https://doi.org/10.3390/pr10071236

Academic Editor: Xiong Luo

Received: 17 May 2022
Accepted: 9 June 2022
Published: 21 June 2022

1. Introduction

Over the past few decades, UAVs have become an emerging resource for remote sensing of various precision, agricultural, military, civil [1], and industrial applications [2,3]. The rapidly increasing fleet of UAVs, along with the widening sphere of their utility, therefore presents a serious challenge for the designers to formulate unique optimal control strategies. However, technological advancements in the aviation sector [4–7] and ground control vehicles [8–19] paved the way for the development of hi-fidelity systems. These UAVs help researchers by providing means to collect multi-spectral information with limited resources and data collection times which is critical for time sensitive dynamic data [20]. The pivotal factor for a successful and safe remote sensing research/mission is a robust and fault tolerant UAV control system. Linear and Non-linear control strategies have been used in diverse ways for solving varying control problems to achieve desired objectives [21–24]. Developing such control systems require well-characterized dynamic system models. Similar work [25,26] has been done in the field of ground robotics [8–13,27] as well. Wind tunnel testing (commonly used method for determining the parameters of these dynamic systems) is time-consuming and costly. System Identification (SI) can be used to overcome the limitation of analytical and wind tunnel testing methods for UAVs. There are several techniques in system identification which can be applied to develop dynamic system models and identify model parameters. These methods have been applied to various aerial vehicles in recent years. The developed dynamic model can further be used to design and verify the autopilot control system of the UAV.

1.1. Related Work

Despite their military importance and abundant usage, owing to the proprietary nature of these UAVs, very little information related to design and development is available in literature. Appreciable research work on design and optimization of quad-copters and UAVs has been done by Mir et al. [22,23]. However, the research work available on model identification and optimization of UAVs usually covers one or two techniques of system identification. Perspective analysis and comparison of various techniques on UAVs are quite sparse. Hopping et al. [28] has used a grey box modeling approach applying the prediction estimation method (PEM) to model longitudinal dynamics of a UAV named Taurus.

Mir et al. [29] has further done a tremendous contribution towards soaring energetics of a bio-inspired UAVs. Design optimization of a variable-span morphing wing UAVs and other design optimization and controllability schemes for UAVs have been discussed in detail by Mestrinho et al. & Mir et al. [30,31]. Later, the lateral dynamics of the same UAV was modeled using the same approach and technique by Ahsan et al. [32]. Longitudinal and lateral dynamics of SmartOne UAV were modeled by Rasheed [33] using grey box modeling approach with prediction error method along with performance of error analysis.

Belge et al. [34] obtains an estimate of UAV lateral dynamic system response by using empirical input-output data sets. The accuracy of parametric model estimation (using ARX, ARMAX and OE model structures) and model degrees are compared for different external disturbance effects. The model of Load Transporting System (LTS) originally designed on UAV has been obtained by linear ARX model structure by Altan et al. [35]. ARX system identification model has also been used to identify Multiple Input Multiple Output (MIMO) model of a helicopter. Various transfer functions have been used to analyze the flight dynamics of helicopter. The system identification of a quad rotor-based aerial manipulator is presented in research carried out by Dube and Pedro [36]. ARX and ARMAX models have been obtained from linear accelerations and yaw angular accelerations.

Cavanini et al. [37] has proposed a novel online estimation technique using LPV-ARX model which is both cost effective and storage effective. His method permits to improve the base of knowledge of the provided LS-SVM by introducing the possibility to learn from on-line data, neglecting to perform the time-expensive training phase, such that the proposed approach is suitable for on-line execution. Cavanini et al further [38] presents a Model Predictive Control (MPC) based autopilot for a fixed-wing Unmanned Aircraft Vehicle (UAV) for meteorological data sampling tasks, named Aerosonde. The LPV model is used to design a MPC to drive the UAV. Two different data driven Linear Parameter-Varying MPC (MPCLPV) algorithms have been proposed by using a subspace identification technique. Belge et al. [39] performs the optimum path planning and tracking using Harris hawk optimization (HHO)–grey wolf optimization (GWO), a hybrid metaheuristic optimization algorithm, to enable the UAV to actualize the payload hold–release mission avoiding obstacles. His novel approach generates a fast and safe optimal path without becoming stuck with local minima, and the quad copter tracks the generated path with minimum energy and time consumption. Weng et al. [40] addresses the robust trajectory tracking control problem of disturbed quadrotor UAVs with disturbances, uncertainties and unmodeled dynamics by devising a novel compound robust tracking control (CRTC) approach via data-driven cascade control technique. Marc et al. [41] presents the online updating of the flight envelope of a UAV. His technique is data-driven and the UAV is subjected to structural degradation for the research.

Yu et al. [42] investigates the problem of neural adaptive distributed formation control for quad rotor multiple UAVs subject to unmodeled dynamics and disturbance. Saengphet et al. [43] uses the input and output data obtained from a flight mission of a tail-less UAV for SISO mathematical model using frequency response. Bnhamdoon et al. [44] uses Box-Jenkins model structure and presents a novel method of identification of a quad-copter autopilot system under noisy circumstances.Real time identification of quad-rotor UAV dynamics using a deep learning techniques for has also been studied by Ayyad et al. [45]. Another online estimation method for UAV, using Extended Kalman Filter(EKF) technique has been presented by Mungguia et al. [46].Puttige and Anavatti [47] uses both online and offline models of nonlinear and complex UAV have been obtained using system identification procedure based on Artificial Neural Network (ANN).

Wu et al. [48] provides an approximated solution of the graph partitioning problem by using a deterministic annealing neural network algorithm. The algorithm is a continuation method that attempts to obtain a high-quality solution by following a path of minimum points of a barrier problem as the barrier parameter is reduced from a sufficiently large positive number to 0. A survey of different methods of system identification techniques and its applications for small low-cost aerial vehicles has been carried out by Mir et al. [6,7] and Hoffer et al. [49] . The different control-oriented models of a quad-rotor UAV have been obtained by applying different identification methods presented by Sierra and Santos [50]. Comparison of ARX method for linear estimation and Hammerstein -Wiener method for non linear estimation for ARF-60 UAV identified models is presented by Khalil and Yesildirek [20].

In addition to above referred literature, there are numerous other contributions made by different researchers. Most of the literature is focused on fixed-winged or multi-rotor UAVs used for research work in the fields of military (target drown, target interceptor, aerial munition practice) and non-military (search and rescue, area surveillance, environmental, agriculture) applications. Even for UAVs, the field of side-by-side comprehensive analysis and comparison of different linear and non-linear system identification techniques still has a vast potential for research.

1.2. Motivating Problems for This Paper

Model prediction and performance analysis using experimental flights for UAVs or other aerial vehicles is not feasible due to the involved cost and damage hazard to the system and the environment in case of any crash. Although wind tunnel testing and CFD analysis for model prediction and performance analysis can be done, however, system identification presents a very cost-effective and user friendly solution towards mathematical modelling of the aerial vehicles. Based on the literature review, it has been observed that in-spite of system identification being widely used for UAVs, very little work related to model prediction of UAV using system identification is available. Even in UAVs, the comparison of predicted models using different linear and non linear methods is yet to be explored. The authors of this paper felt that the researchers must be provided with a platform for comparison of linear and non-linear techniques using actual flight data for model estimation and validation. This lack of literature for UAV model prediction using actual flight data motivated the author to fill in this gap through this paper.

1.3. Main Contributions of This Paper

As evident from the preamble of related work, very little research is available in open literature which is based on elaborate comparison of different techniques of system identification for UAVs. Most of the literature for model prediction of UAVs is based on simulation results rather than utilizing actual flight data. Furthermore, when it comes to UAVs, the research contribution using actual flight data along with comprehensive performance evaluation and comparison of linear and nonlinear system identification techniques is even much scarce.

The authors aim to present a base platform for model prediction of UAV utilizing actual flight data after a comprehensive perspective analysis of linear and nonlinear system identification techniques. This paper aims to provide a consolidated platform for the audience which provides a mechanism for model prediction of UAV/UAVs. Besides providing a comprehensive affect of individual training of actual flight data, the presented approach will also help the readers to carry out analysis of several regression techniques in linear and non-linear domain. Moreover, the performance comparison of linear and nonlinear system identification models for quality parameters like final prediction error, residual analysis, mean squared errors and fit percentages further enhances the effectively of the proposed approach.

1.4. Sequence of This Paper

This paper is organized as follows. Section 2 presents the build up of 6-Degree of Freedom (DOF) aerodynamic model for UAV followed by design parameters of UAV. Then a brief overview of all system identification used in this research is given. The results and analysis part gives first presents the flight sorties design conducted for estimation and validation purposes followed by the response of all linear and nonlinear parametric model along with residue analysis of each. A detailed analysis and comparison is carried out for selection of final model. Then the author has also verified the finally selected model by predicting a second actual flight of UAV. Lastly, the conclusion and limitations of the research are presented.

2. Problem Formulation

2.1. 6 DOF Flight Dynamics Model

A 6-Degrees of Freedom (DOF) Flight Dynamics Model (FDM) has been used for studying the motion of UAV in three dimensions. 6DOF refers to the number of axes that a rigid body may freely move in three-dimensional space. It specifies the number of independent factors that define the configuration of a mechanical system. The body may move in three dimensions, on the X, Y, and Z axes, as well as change orientation between those axes via rotation known as pitch, yaw, and roll. FDM assumes a flat and non-rotating earth approximations and is based on dynamic equations (deduced by Stevens, Lewis and Johnson [51]) in body frame reference. These sets of equations, which govern dynamics of translation (Equation (1)), rotation (Equation (2)), kinematics (Equation (3)) and navigation (4)) respectively, are defined as:

$$\begin{aligned} \dot{U} &= RV - QW - g\sin\theta + \frac{X_A + X_T}{m} \\ \dot{V} &= -RU + PW + g\sin\phi\cos\theta + \frac{Y_A + Y_T}{m} \\ \dot{U} &= QU - PV + g\cos\phi\cos\theta + \frac{Z_A + Z_T}{m} \end{aligned} \tag{1}$$

In Equation (1), $\dot{U}$, V, W are the components of linear velocities along the three body axes respectively. $\phi, \theta \& \psi$ are the Euler angles which define the orientation of body frame with respect to inertial frame, P, Q, R are the angular velocities along body x, y and z axis respectively. X_A, Y_A, Z_A & X_T, Y_T, Z_T are the Force and Thrust components along the three axis.

$$
\begin{aligned}
\Gamma\dot{P} &= J_{XZ}(J_X - J_Y + J_Z)PQ \\
&\quad - [J_Z(J_Z - J_Y) + J_{XZ}^2]QR + J_Z l + J_{XZ} n \\
\Gamma\dot{Q} &= (J_Z - J_X)PR - J_{XZ}(P^2 - R^2) + m \\
\Gamma\dot{P} &= [J_X(J_X - J_Y) + J_{XZ}^2]PQ - J_{XZ}(J_X - J_Y + J_Z)QR \\
&\quad + J_{XZ} l + J_X n
\end{aligned} \tag{2}
$$

where J_X, J_y, J_z, J_{XZ} and Γ are the inertia matrix components. Also l, m, n are the roll, pitch and yaw moments.

$$
\begin{aligned}
\dot{\phi} &= P + \tan\theta(Q\sin\phi + R\cos\phi) \\
\dot{\theta} &= Q\cos\phi - R\sin\phi \\
\dot{\psi} &= \frac{Q\sin\phi + R\cos\phi}{\cos\theta}
\end{aligned} \tag{3}
$$

$$
\begin{aligned}
\dot{P_E} &= U\cos\theta\cos\psi + V(-\cos\phi\sin\psi + \sin\phi\sin\theta\cos\psi) \\
&\quad + W(\sin\phi\sin\psi + \cos\phi\sin\theta\cos\psi) \\
\dot{P_N} &= U\cos\theta\sin\psi + V(\cos\phi\cos\psi + \sin\phi\sin\theta\sin\psi) \\
&\quad + W(-\sin\phi\cos\psi + \cos\phi\sin\theta\sin\psi) \\
\dot{h} &= U\sin\theta - V\sin\phi\cos\theta - W\cos\phi\cos\theta
\end{aligned} \tag{4}
$$

where P_e and P_n are the position coordinates alongside the inertial east and north directions. h is the vehicle altitude, J is the moment of inertia matrix, m is the mass, g is the acceleration due to gravity, l, m, n are the angular velocity components (roll, pitch and yaw moments) in the body axis, α and β are the aerodynamic angles representing angle of attack and side slip angle respectively.

Aerodynamic Parameters

High fidelity numerical techniques of Computational Fluid Dynamics (CFD) and USAF DATCOM were utilized for generating FDM based on 6-DOF simulation environment. Flight conditions define aerodynamic forces and moments acting on high speed and are governed by Equations (5) and (6) respectively.

$$
L = q_\infty S C_L,\ D = q_\infty S C_D,\ Y = q_\infty S C_Y \tag{5}
$$

where L, D and Y represent aerodynamic lift, drag and side force respectively in wind axis. C_L, C_D, C_Y are the dimensionless aerodynamic coefficients for lift, drag and side forces respectively, q_∞ is the dynamic pressure and S is the wing area.

$$
l_w = q_\infty b S C_l,\ m_W = q_\infty c S C_m,\ n_W = q_\infty b S C_n \tag{6}
$$

where n_w, m_w and l_w are the yaw, pitch and roll moments in wind axis, b is the wing span, c is the wing chord and C_n, C_m, C_l are the dimensionless aerodynamic coefficients for yaw, pitch and roll moments respectively. The design of UAV under test was optimized based on CFD analysis and comparison of various design configurations.

3. Model Identification

The objective of this search is to build an accurate model for UAV. MATLAB was used for system identification of the system. The adopted research methodology was divided into following steps:

- Acquiring data for two sorties of experimental UAV.
- Pre-processing and filtering the data for whole flight of the UAV.
- Model identification using flight data from one sortie using ARX, ARMAX, Output Error, Box Jenkin's, Non-linear ARX (with various estimators).
- Training of model for each individual technique

- Selection of best fit model on basis of model quality parameters like Final Prediction Error (FPE), fit percentage to actual flight data and residual analysis.
- Validation of selected model on a different flight data and analysis of the results.

3.1. Model Structures

Various model structures are used in this research to model MIMO dynamics of the UAV. The inputs taken were aileron deflection (δ_a) and Vtail deflection (δ_e) whereas the outputs are taken to be yaw rate (P), pitch rate (Q), and roll rate (R).

3.1.1. Auto-Regressive Exogenous (ARX) Model

The second method used is the estimation of ARX model which as per the literature is assumed to be the most efficient polynomial estimation method as linear regression equations are in analytic form whose solution is also unique. The estimation of the ARX model is the most efficient of the polynomial estimation methods because it is the result of solving linear regression equations in analytic form with a unique solution. when the model order is high, then ARX model is preferred. For input $u(t)$, output $y(t)$ and noise $e(t)$, the ARX model is given by Equation (7).

$$A(q)y(t) = \sum_{i=0}^{nu} B_i(q)u_i(t - nk_i) + e(t) \tag{7}$$

where A and B are polynomials expressed in time shift operator q^{-1}. Although ARX model is suited for most high order dynamic systems, it has a disadvantage as the disturbances are part of the system model. The disadvantage of the ARX model is that disturbances are part of the system dynamics. However, this disadvantage can be curbed with a good signal-to-noise ratio.

3.1.2. Auto Regressive Moving Average eXogenous (ARMAX) Model

For dynamics systems with dominating disturbances that enter the process in the early stages like wind gust in case of aerial systems, ARMAX model comes in handy. ARMAX model has advantage over ARMAX model by providing more flexibility for handling disturbances.For input $u(t)$, output $y(t)$ and noise $e(t)$, the ARMAX model is given by Equation (8).

$$A(q)y(t) = \sum_{i=0}^{nu} B_i(q)u_i(t - nk_i) + C(q)e(t) \tag{8}$$

where A, B and C are polynomials expressed in time shift operator q^{-1}.

3.1.3. Box Jenkin's (BJ) Model

When complete system model dynamics are described by modeling the noise and system dynamics separately, this comes under the category of BJ's Model.Very sparse literature is available related to research carried out on UAVs using BJ model as this model is particularly useful when the disturbances enter towards the end of the process. The disturbance is basically the measurement noise. For input $u(t)$, output $y(t)$ and noise $e(t)$, the BJ model is given by Equation (9).

$$y(t) = \sum_{i=0}^{nu} \frac{B_i(q)}{F_i(q)} u_i(t - nk_i) + \frac{C(q)}{D(q)} e(t) \tag{9}$$

where B, C, D and F are polynomials expressed in time shift operator q^{-1}.

3.1.4. Output Error (OE) Model

Output Error model is usually used when there is only the need to parameterize the system dynamics without estimating the noise model. This model is only suitable for theoretical modelling of the aerial vehicles, however, its use in practical system dynamics may be considered after due consideration. For input $u(t)$, output $y(t)$ and noise $e(t)$, the OE model is given by Equation (10).

$$y(t) = \sum_{i=0}^{nu} \frac{B_i(q)}{F_i(q)} u_i(t - nk_i) + e(t) \tag{10}$$

3.1.5. State Space Model

State Space model structure was also used in this research owing to its less computational time in case of iterative analysis which can be attributed to lower model order of the state space model. Equation (11) describes a state space system.

$$\begin{aligned} x(t+1) &= Ax(t) + Bu(t) + Ke(t) \\ y(t) &= Cx(t) + Du(t) + e(t) \end{aligned} \tag{11}$$

In Equation (11), A, B, C, D and K are system matrices. The previously mentioned parametric methods of system identification have their own advantages. However, those models may lead to higher order models and a large number of parameters which could lead to lack of convergence to global minima and extensive computational times for iterative analysis.

3.1.6. Nonlinear-ARX Model

As depicted by Equation (7), the ARX model predicts the output by using the weighted sum of linear regressors i.e, weighted sum of current inputs and past inputs and outputs. The non-linear ARX model provides additional structural flexibility by having a non-linear function F as a model regressor. Three types of model regressors or mapping functions namely wavelet network non-linearity, tree partition non-linearity and multilayered neural network were used for this research. The former two estimators use a combination of offset, linear weights and non-linearity function for computation of output in which units of the non-linear function operate on radial combination of inputs. in the later estimator i.e., multilayered neural network estimator, These networks consists of three type of layers, one is input, one is output and then we can have multiple hidden layers. This type of technique has the ability to find the relation of very complex nature and can cover a large regime of input and output. However, once the network has been trained and is appropriately selected, it will produce good accuracy for the regime it has been trained, but its results could be quite misleading for the values of input and outputs outside what it has been trained.

4. Results and Analysis

The results of various model structures to describe MIMO dynamics od UAV using aileron deflection (δ_a), Vtail deflection (δ_e), height (h), speed (V_t) and thrust (δ_T) as inputs and yaw rate (P), pitch rate (Q), and roll rate (R) as outputs of the function are presented. The data of first flight used to estimate the model has been divided into two parts: the first part for estimation of the model and second part for validation of the model for the same flight. a number of iterations were performed and training was done for each technique. The best model was picked on the basis of Final Prediction Error (FPE) and the picked model quality is further analysed using the following factors:

- Final Prediction Error (FPE)
- Residual Analysis
- Percentage of fit to validation data
- Mean Squared Error (MSE)

The reader is also presented with a comparative analysis amongst the best model of each technique and final model is selected analysing the model quality using previously stated factors. This selected model is further validated to predict the outputs using a different data of the same flight regime.

4.1. *Flight Designs*

The first and foremost step of mathematical modelling of experimental UAV was data acquisition of inputs and outputs for an actual flight. A flight design was sorted out to cover all aspects of control surfaces under different flight conditions. Once, the flight design was finalized, flight was conducted and data was acquired for offline empirical mathematical modelling of the UAV. After modelling, a second flight was conducted to compare the predicted response using finalized model with actual flight response.

4.1.1. Flight 1 (Estimation)

The experimental UAV was given full throttle and held back with the catapult mechanism. The height profile along with x-axis acceleration and thrust shutoff point at time of parachute deployment of the UAV was used to identify the whole flight regime to be used for modelling. The UAV attains a certain height and then maintains that height while performing maneuvers to capture the complete range of control surface deflections for different flight conditions (Figures 1 and 2).

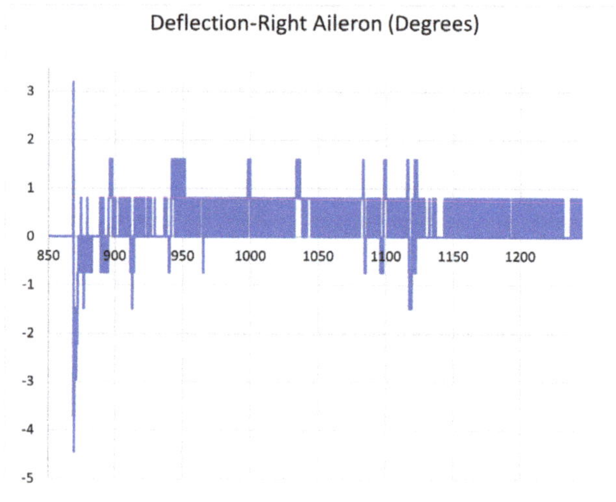

Deflection-Right Aileron (Degrees)

Figure 1. Aileron Deflection.

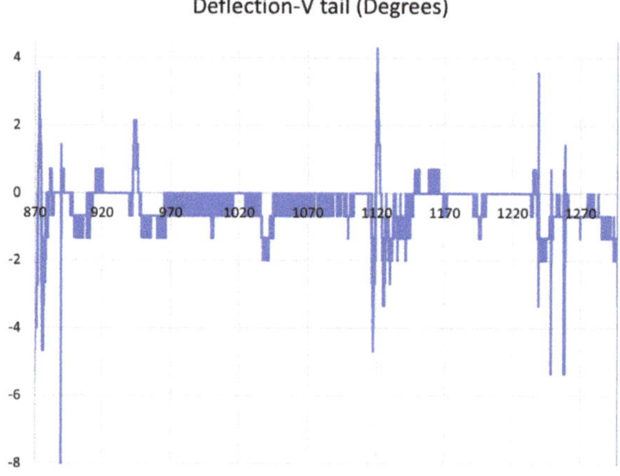

Figure 2. V Tail Deflection.

After capturing the required data, it was trained offline to use it for system modelling. The data was divided in half. The first half comprising the takeoff and part of level flight was used for model estimations and the second half was used used for validation.

4.1.2. Flight 2 (Validation)

Finalizing the the mathematical model using flight 1 data led the authors to conduct validation trial. A second flight was conducted using the same flight design aspects discussed in Section 4.1.1. Different profiles for the validation sortie are shown in Figures 3–8. Complete flight regime (takeoff till parachute deployment) was predicted using the mathematical model and the results are depicted in Section 4.9.

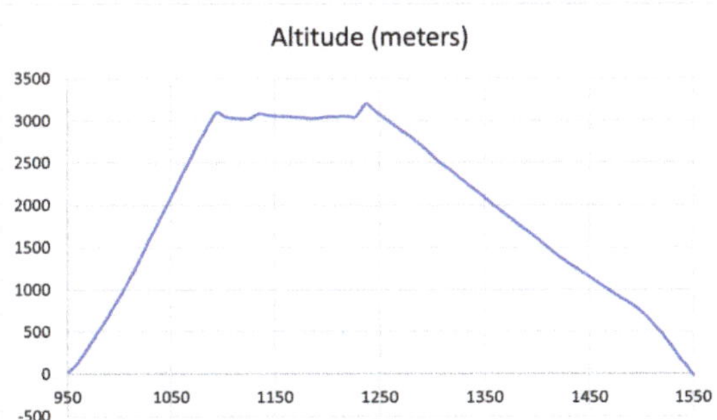

Figure 3. Height: Validation Flight.

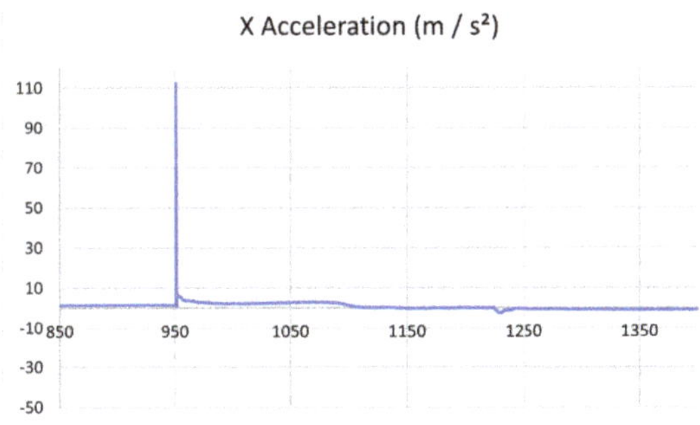

Figure 4. X-acceleration Validation.

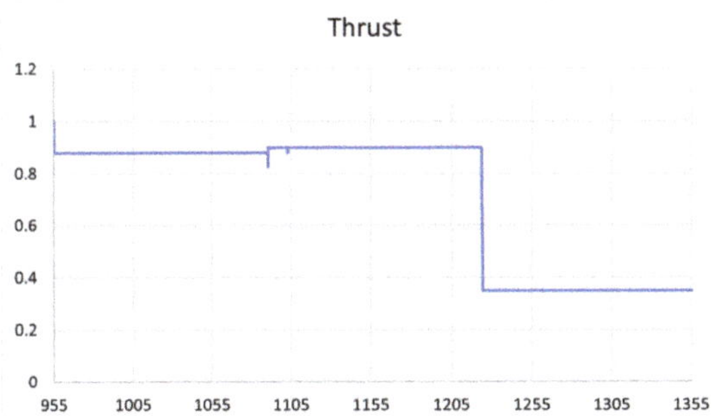

Figure 5. Thrust: Validation Flight.

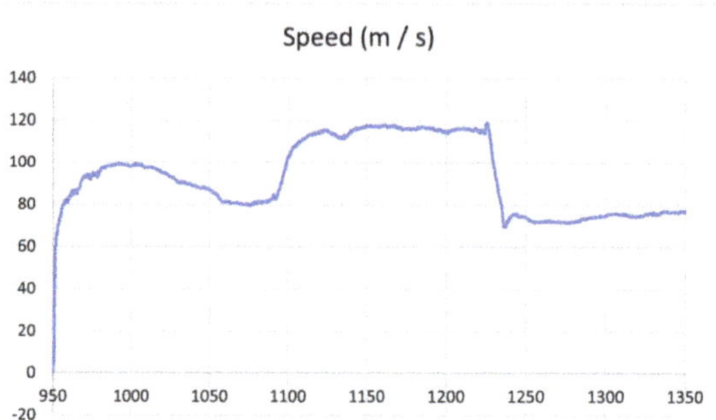

Figure 6. Speed: Validation Flight.

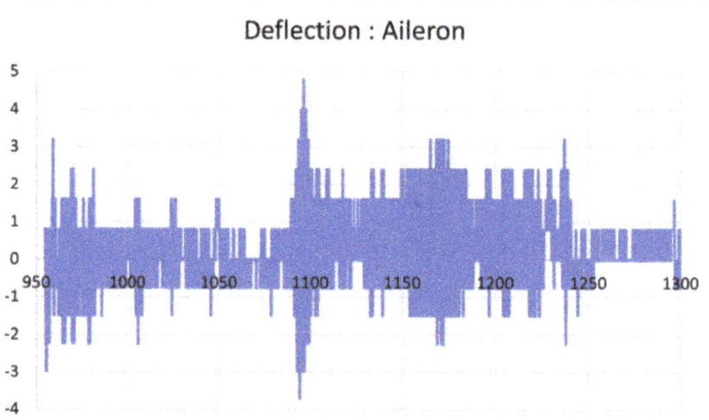

Figure 7. Aileron Deflection: Flight 2.

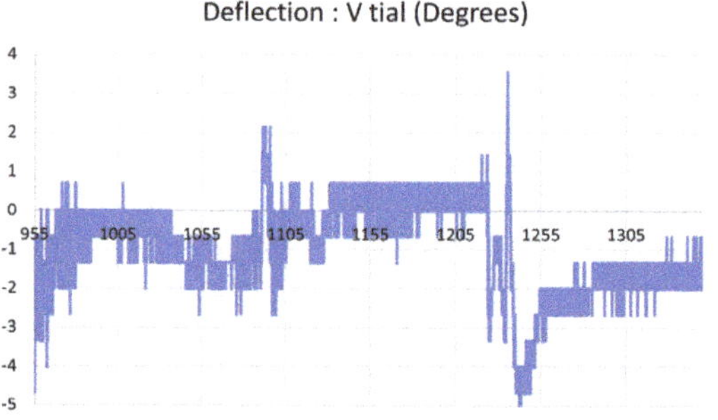

Figure 8. V-Tail Defection: Flight 2.

4.2. *Finite Impulse Response (FIR) Model*

From the FIR model response in Figure 9 the excitation orders for all the inputs are [50 50 50 50 50] and the time delay T_s will be taken zero in further techniques. The FPE and MSE of FIR model is 3.352×10^6 and 2.6×10^4 respectively and fit percentage between modeled output and actual output is for pitch rate, roll rate and yaw rate is −1036%, 11.06%, −9067% respectively. The number of free coefficients of the impulse response model is 1050 which is quite high.

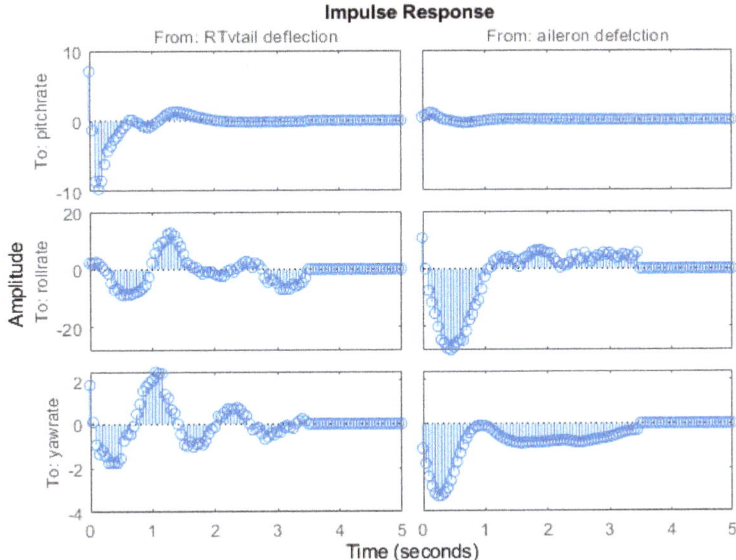

Figure 9. Finite Impulse Response.

4.3. *Auto Regressive Exogenous (ARX) Model*

The model order for the ARX model was selected in an iterative fashion where different combinations of the order of polynomial $A(q)$ (N_a), order of polynomial $B(q)$ (N_b) were chosen. A total of 75 models were derived and training was performed on each of each model before a best ARX model with minimum FPE of 0.00257 and MSE of 0.6356 was finalized. Fit percentages to actual outputs are depicted in 1-step ahead prediction response of ARX model in Figure 10. The auto-correlation and cross-correlation plots of the model response (Figure 11) also shows that the selected model gives good confidence level as the residues are within the range the region marked blue which defines the part of the input response not been able to be predicted by the model. In our research we have set the confidence region to be 98%.

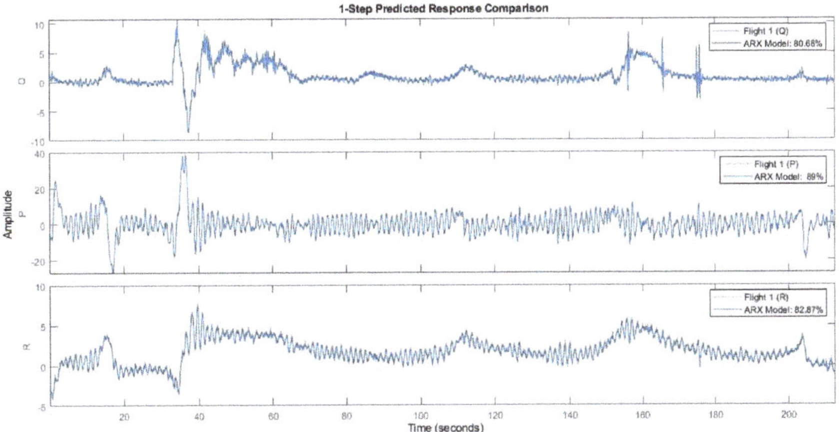

Figure 10. ARX Model Response.

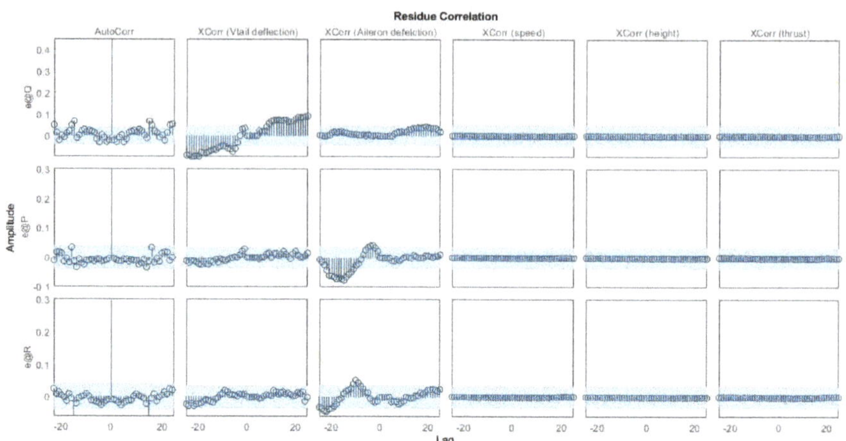

Figure 11. ARX Model Residue Correlation.

4.4. *Autoregressive Moving Average eXogenous (ARMAX) Model*

The same iterative approach was used for model order selection of ARMAX model. Final ARMAX model was selected after deriving and training a total of 125 model with FPE and MSE equal to 0.002208 and 0.6352. Fit percentages to actual outputs are depicted in 1-step ahead prediction response of ARX model in Figure 12. The auto-correlation and cross-correlation plots of the model response (Figure 13) also shows that the selected model gives good confidence level.

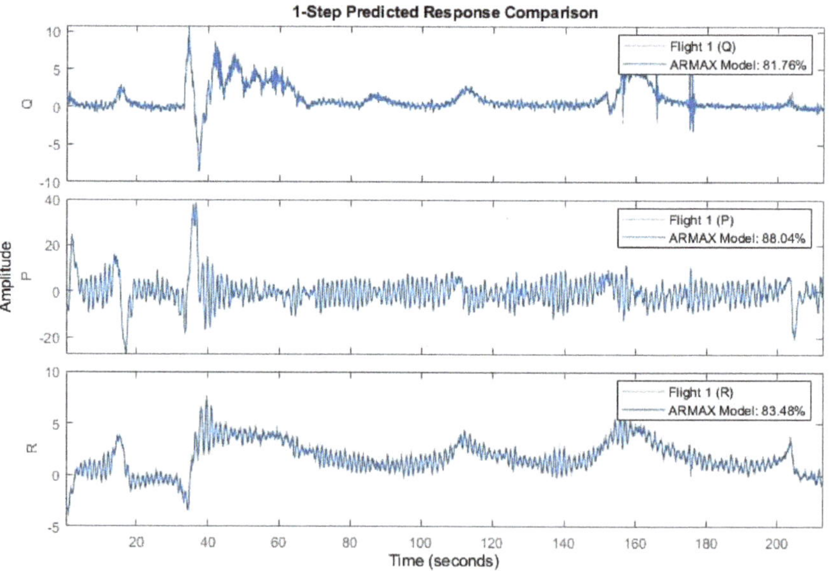

Figure 12. ARMAX Model Response.

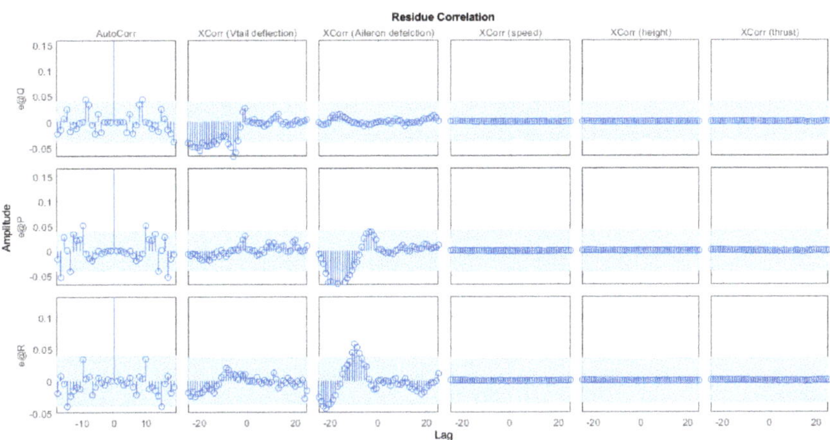

Figure 13. ARMAX Model Residue Correlation.

4.5. Box Jenkin's (BJ Model)

Order selection procedure used in [44] was adopted to select the model order which gives satisfactory residue correlation (Figure 14) and fit percentages (Figure 15) along with an impressive FPE and MSE 0.005407 and 0.8946 respectively.

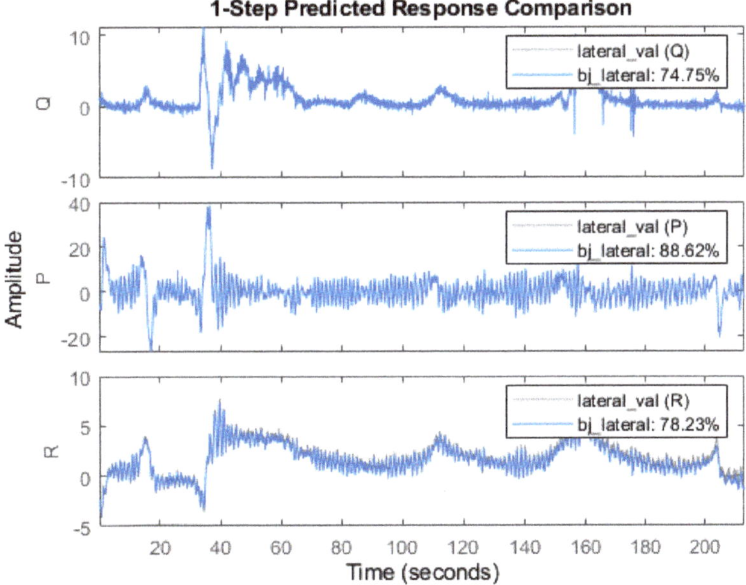

Figure 14. BJ Model Response.

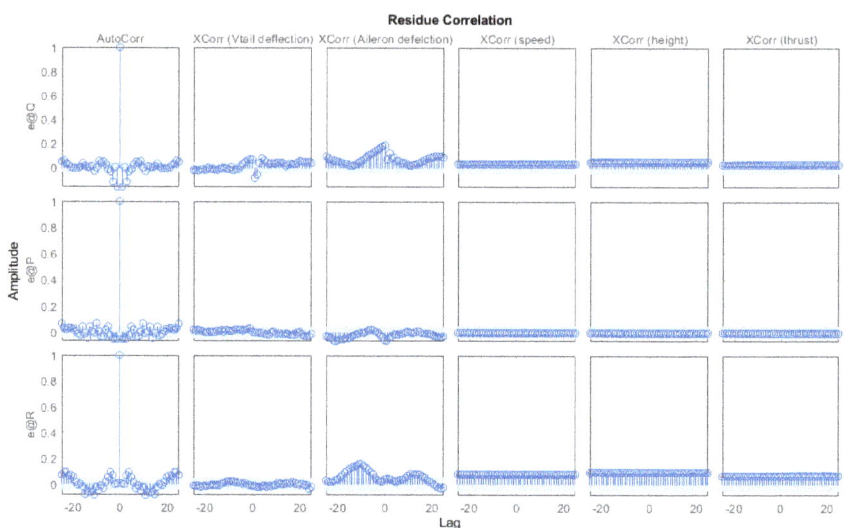

Figure 15. BJ Model Residue Correlation.

4.6. Output Error (OE) Model

The OE model response and residual correlation are shown in Figures 16 and 17 respectively. As shown in the figure, the output error model is not able to predict the 1-step ahead predicted response. This can be attributed to the fact that output error model acts as a similation model in which the model response is computed using input data and initial conditions. Since no past outputs are being used to compute the response the error accumulates and the results deviate from actual response.

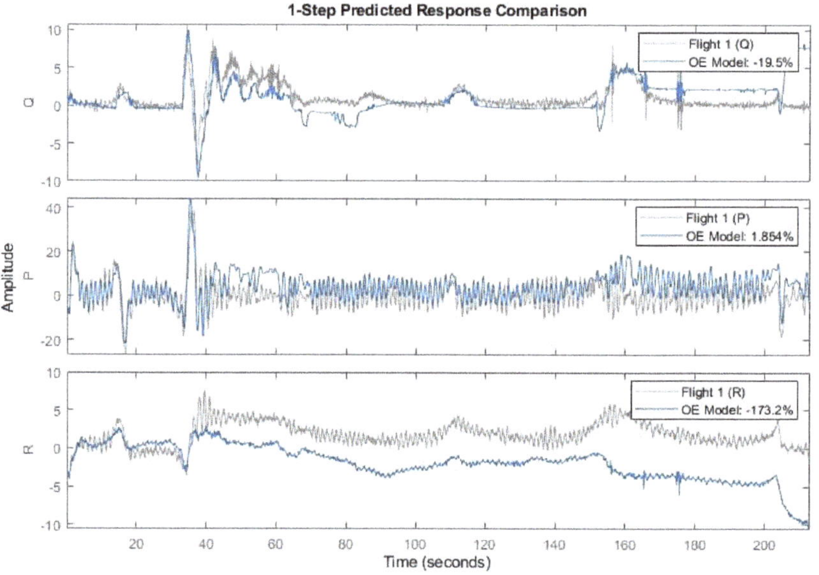

Figure 16. OE Model Response.

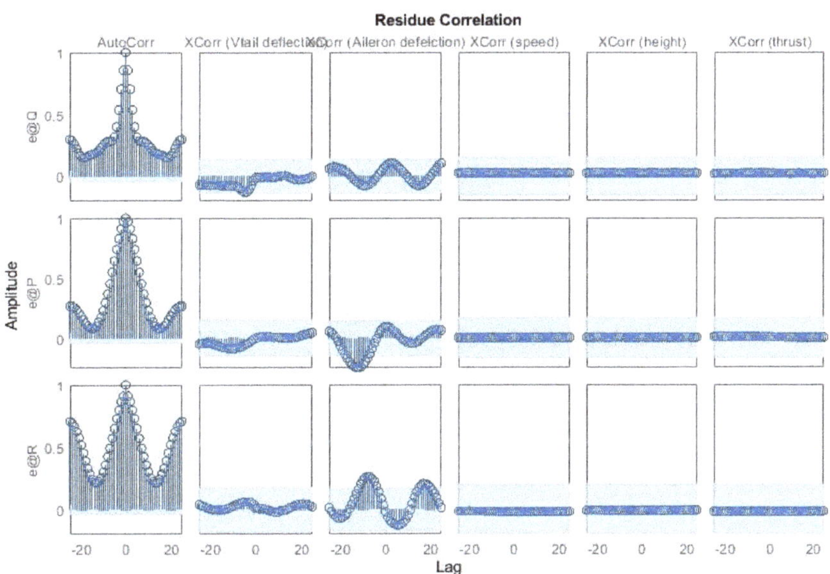

Figure 17. OE Model Residue Correlation.

4.7. State Space Model

Figures 18 and 19 shows the state space model fit percentages and residue correlation respectively. The model order is equal to 6 with FPE and MSE equal to 0.005307 and 0.7822 respectively. The lower model order of the state space model and the residual graphs of the same show that this model provides ease of computation by reducing the order without compromising the quality of the response.

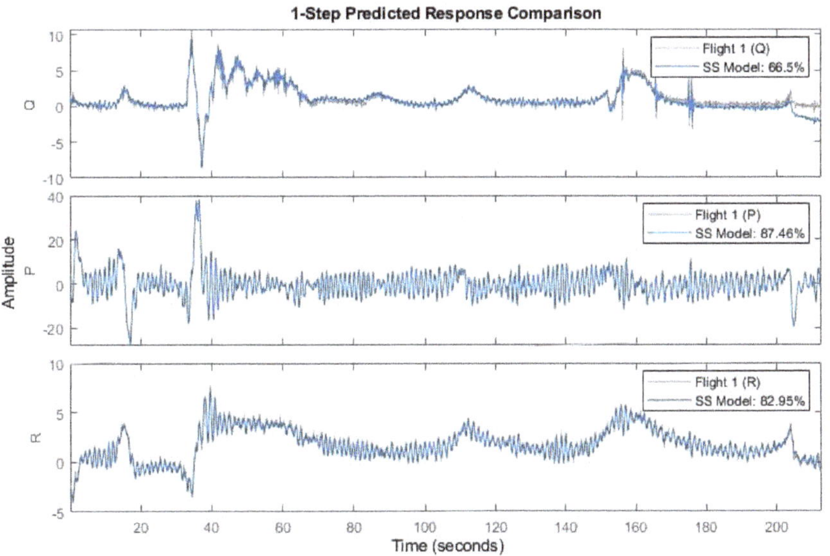

Figure 18. SS Model Response.

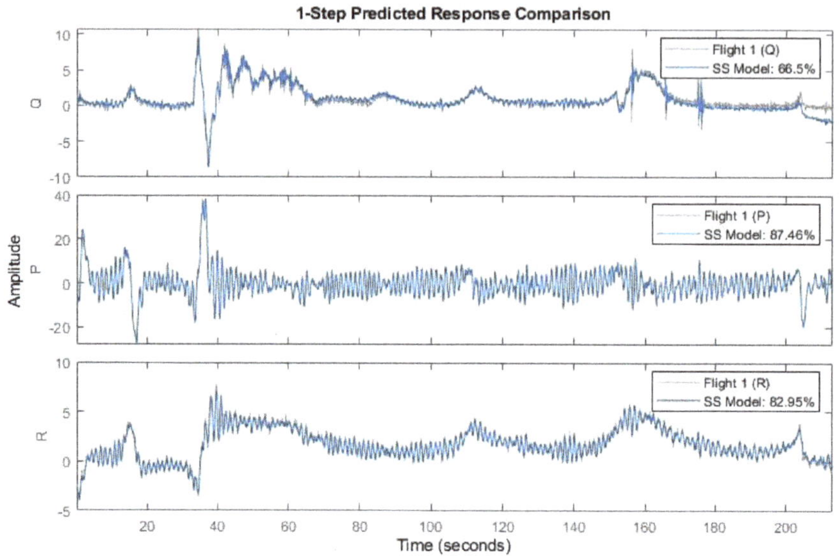

Figure 19. SS Model Residue Correlation.

4.8. Non-Linear ARX Model

The fit percentages and residue correlation plots of non-linear ARX models with tree partition, wavelet network and neural network estimators are presented in Figures 20–22. The Model order is selected to be the same as the best fit ARX model for further comparison. Although the results of wavelet network are acceptable but the added complexity in the model due to non linearity is not favored for time compressed computational environments. the same quality of model response is also provided by linear models.

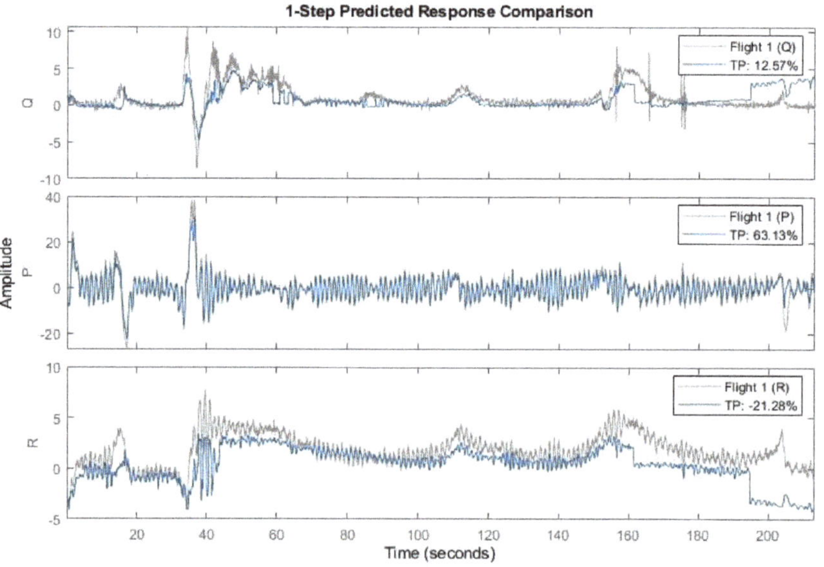

Figure 20. NLARX Tree Partition Model Response.

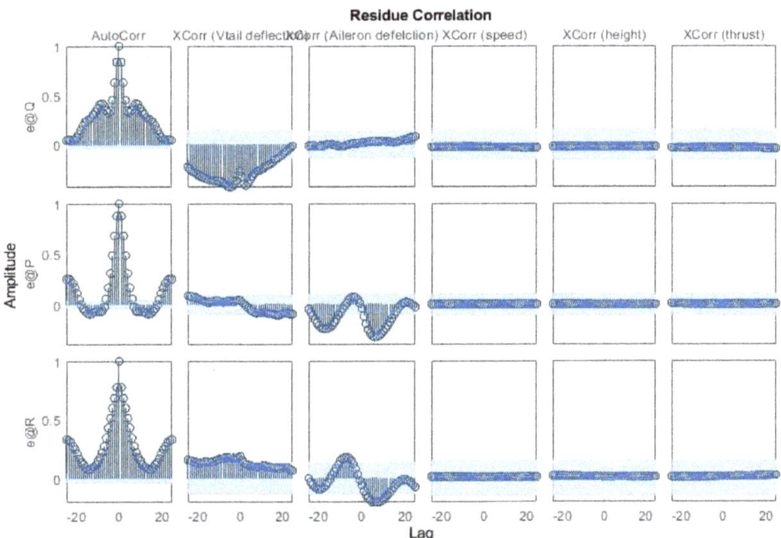

Figure 21. Tree Partition Residue Correlation.

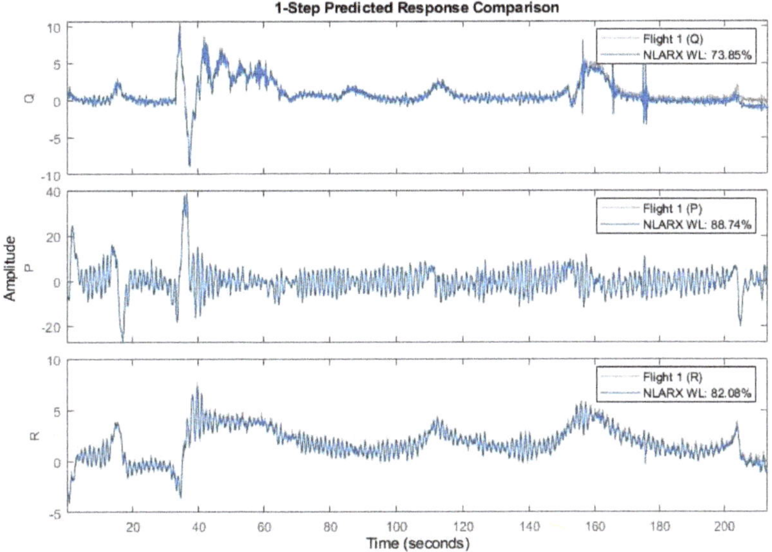

Figure 22. NLARX Wavelet Network Model Response.

4.9. Comparative Analysis

The comparison analysis of results for all linear and non-linear parametric model estimation techniques used in this research are tabulated in Table 1 for selection of best model based on parameters like model order, Final Prediction Error (FPE), Mean Square Error (MSE), fit percentages of roll rate (P), pitch rate(Q) and yaw rate (R), number of free parameters and perspective analysis of residue correlation.

Table 1. Comparison Analysis of Linear and Nonlinear Parametric Model Responses.

Parameter	*ARX*	*ARMAX*	*BJ*	*OE*	*SS*	*NLARX*		
						TP	**WL**	**NN**
FPE	0.00257	0.0022	0.00868	2.611	0.0037	-	0.002688	-
MSE	0.6356	0.635	0.8518	16.29	0.74	3.998	0.643	0.6545
PFit	89%	88.04%	88.62%	1.854%	87.46%	63.16%	88.74%	87.7%
QFit	80.68%	81.76%	74.75%	−19.5%	66.5%	12.57%	73.85%	77.02%
RFit	82.87%	83.48%	78.23%	−173.2%	88.95%	−21.28%	82.08%	81.79%
Coeff.	210	135	72	105	102	-	-	-

ARMAX model and linear ARX model gave best values for FPE, MSE, fit percentages etc. Finally, ARMAX model was selected based on residue analysis. The next step was to validate the selected model on a different flight data covering the same flight regime. The comparison results for predicted output of ARMAX model and actual output data from validation sortie are presented in Figure 23 along with the residue correlation in Figure 24. The fit percentages of the ARMAX model are satisfactory and the residue correlation plot also shows good confidence level.

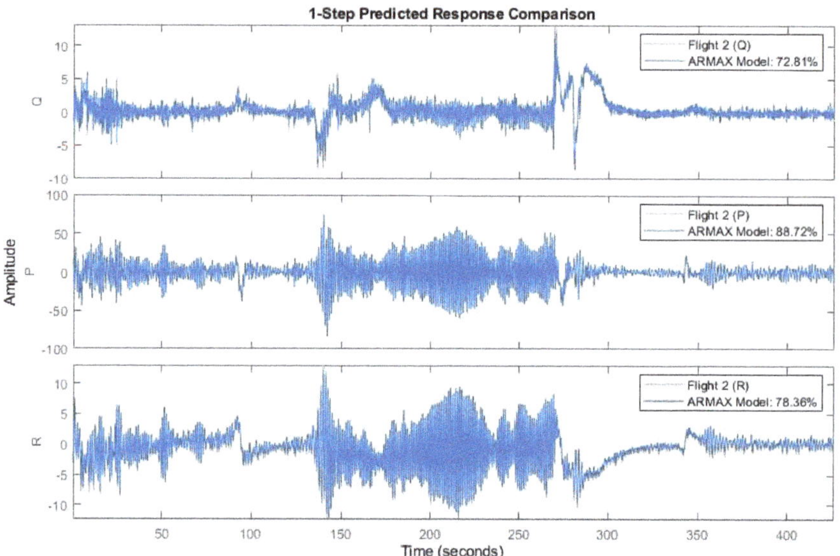

Figure 23. Validation of ARMAX Model with Second Flight Data.

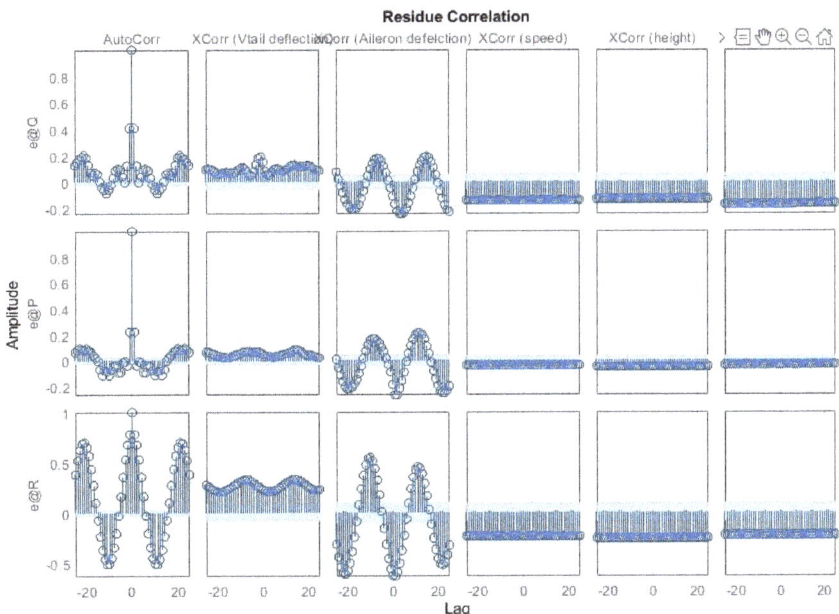

Figure 24. Residue Correlation.

4.10. Discussion and Remarks

As evident from Table 1 the linear and non-linear parametric models give results which are acceptable for use in future design modifications and simulations of the UAV under test. However, the output error model can't be used owing to the high deviations in predicted responses and measured responses. The reason for this can be attributed to the fact that the output error model takes only the previous inputs rather than the previous inputs and outputs for the prediction of the model response. The linear ARX model has higher model order as compared to the linear ARMAX model. However, The ARMAX model provides a better prediction with lower model order and hence, reducing the number of free coefficients in the model, which is usually desired.

Box Jenkins and state space model, although giving even lower model order and lesser number of coefficients in the model, The performance of these models based on fit percentages, mean squared error and final prediction error states render them of less utility in comparison to ARMAX and ARX models. The nonlinear ARX models with tree partition, wavelet network and neural network show acceptable fit percentages but the fact that the model is nonlinear, which adds to the complexity and computational time with high model orders, render them with less utilization. The last factor which has contributed towards the selection of best model amongst all model structures is the residual analysis. ARMAX model gives the best results on this model quality assessment parameter also. Hence, ARMAX model was selected as final model for verification of a different flight data.

4.11. Research Limitations

The research pertains to system identification utilizing various linear/nonlinear techniques. Limited research in this regard is available in literature to make it a benchmark for this research. The flights to be performed for the purpose of data gathering for model prediction clearly presents a financial and administrative challenge. The integrity of model prediction and its accuracy greatly enhances with availability of sufficient flight data. We believe that the technique presented in this research will present even more accurate results with the increase in the available flight data.

5. Conclusions

This paper describes the development of 6 DOF Flight Dynamics Model of the UAV followed by basic parameters of the UAV under test and launch and recovery mechanism of the same. System identification was applied to actual flight data, and various linear and nonlinear parametric models with different model structures were developed. The model structures included the impulse response model, ARX, ARMAX, Box Jenkin's, Output Error, State Space, Nonlinear ARX models with tree partition, wavelet network, and neural network models. Several models were developed for each model structure, and each model was trained before the selection of the final model from that category. A comprehensive analysis was carried out for all the models, and after detailed comparison and analysis best fit model was finalized to be ARMAX model, which has FPE of 0.0022. The model was further used to predict output of a different sortie with same flight regime and the fit percentages of the modeled output to actual output of Roll rate (P) was 88.72%, Pitch Rate (Q) was 72.81% and Yaw Rate (R) was 38.36% which is quite satisfactory. It is imperative to highlight that the proposed framework presented in this study provides a consolidated platform which can be utilized by researchers to perform model estimation for any similar platform. The best fit model structure most suitable for that particular configuration can be selected accordingly as per the proposed benchmarks. The research presented in this paper is purely original and to the best of author's knowledge, such a detailed analysis is presented for the first time in literature.

Author Contributions: Conceptualization, S.K.F., S.M. & F.G.; methodology, S.K.F. & I.M.; software, S.K.F.; validation, I.M. & N.S.; formal analysis, S.K.F., I.M., F.G.; investigation, I.M., F.G. & M.A. resources, N.S., A.A.A. & T.A.; data curation, S.K.F., M.A. & I.M.; writing—original draft preparation, S.K.F., F.G.; writing—review and editing, F.G., I.M., N.S. & L.A.; visualization, M.A., I.M., N.S., L.A.; supervision, M.A., I.M.; project administration, M.A., I.M.; funding acquisition, N.S., A.A.A., T.A. All authors have read and agreed to the published version of the manuscript.

Funding: The authors would like to thank Taif University for funding this research through Taif University Research Supporting, Project number. (TURSP-2020/277), Taif University, Taif, Saudi Arabia. Also, the authors extend their appreciation to the Deputyship for Research & Innovation, Ministry of Education in Saudi Arabia for funding this research through the project number "IF_2020_NBU_432".

Institutional Review Board Statement: Not applicable

Informed Consent Statement: Informed consent was obtained from all individual participants included in the study.

Data Availability Statement: Data is available from the authors upon reasonable request.

Conflicts of Interest: The authors declare that there is no conflict of interest regarding the publication of this paper.

Abbreviations

The following abbreviations are used in this manuscript:

$\bar{c}$	Mean aerodynamic chord (m)
C_D, C_Y, C_L	Coefficients of drag, side force and lift
C_l, C_m, C_n	moment coefficients of Roll, pitch and yaw
D	Drag (N)
g	Acceleration due to gravitational force (ms^{-2})
GCM	Guidance and Control Module
h	Altitude (m)
UAV	Unmanned High Speed Aerial Vehicle
J_x, J_y, J_z	Components of the inertia matrix components in body frame
V_t	Free-stream velocity (m/s)

L	Lift (N)
m	Mass of the vehicle (kg)
n,m,l	(yaw, pitch and roll moments respectively) defined in body frame (Nm)
P_e, P_n	Position coordinates along the inertial east and north directions (m)
p, q, r	Roll, pitch and yaw rates in body frame (deg/s)
q_∞	Free stream dynamic pressure (N/m^2)
S_{Ref}	Reference area (m^2)
X_A, Y_A, Z_A	(axial, side and tangential force respectively) in the body frame (N)
T	Engine thrust (N)
U,V,W	Linear velocity along body x, y and z axis respectively (m/s)
W	Weight (N)
ARX	Automatic Regression eXogenous
ARMAX	Automatic Regression Moving Average eXogenous
BJ	Box Jenkin's
OE	Output Error
SS	State Space
TP	Tree Partition
WL	WaveLet Network
NN	Neural Network
FPE	Final Prediction Error
MSE	Mean Squared Error
n	Model order for state space model
N_b	Order of Polynomial $B(q)$
N_c	Order of Polynomial $C(q)$
N_d	Order of Polynomial $D(q)$
N_f	Order of Polynomial $F(q)$

Greek Symbols

The following greek symbols are used in this manuscript:

ρ	Air density (kg/m^3)
β	Side slip angle (deg)
α	Aerodynamic angle of attack (deg)
ϕ, θ, ψ	Roll, pitch and azimuth angles describing body frame w.r.t inertial frame (deg)
γ	Flight path angle (deg)
δ_a,δ_e,δ_f	aileron, elevator and flap controls respectively

References

1. Gul, F.; Mir, I.; Abualigah, L.; Mir, S.; Altalhi, M. Cooperative multi-function approach: A new strategy for autonomous ground robotics. *Future Gener. Comput. Syst.* **2022**, *134*, 361–373. [CrossRef]
2. Din, A.F.U.; Mir, I.; Gul, F.; Nasar, A.; Rustom, M.; Abualigah, L. Reinforced Learning-Based Robust Control Design for Unmanned Aerial Vehicle. *Arab. J. Sci. Eng.* **2022**, 1–16. [CrossRef]
3. Din, A.F.U.; Akhtar, S.; Maqsood, A.; Habib, M.; Mir, I. Modified model free dynamic programming: An augmented approach for unmanned aerial vehicle. *Appl. Intell.* **2022**, 1–21. [CrossRef]
4. Mir, I.; Eisa, S.A.; Maqsood, A. Review of dynamic soaring: Technical aspects, nonlinear modeling perspectives and future directions. *Nonlinear Dyn.* **2018**, *94*, 3117–3144. [CrossRef]
5. Mir, I.; Maqsood, A.; Akhtar, S. Biologically inspired dynamic soaring maneuvers for an unmanned air vehicle capable of sweep morphing. *Int. J. Aeronaut. Space Sci.* **2018**, *19*, 1006–1016. [CrossRef]
6. Mir, I.; Maqsood, A.; Akhtar, S. Dynamic modeling & stability analysis of a generic UAV in glide phase. In Proceedings of the MATEC Web of Conferences, EDP Sciences, Ulis, France, 10 July 2017; Volume 114, p. 01007.
7. Mir, I.; Eisa, S.A.; Taha, H.; Maqsood, A.; Akhtar, S.; Islam, T.U. A stability perspective of bioinspired unmanned aerial vehicles performing optimal dynamic soaring. *Bioinspiration Biomim.* **2021**, *16*, 066010. [CrossRef]
8. Gul, F.; Mir, S.; Mir, I. Coordinated Multi-Robot Exploration: Hybrid Stochastic Optimization Approach. In Proceedings of the AIAA SCITECH 2022 Forum, San Diego, CA, USA, 3–7 January 2022; p. 1414.
9. Gul, F.; Mir, S.; Mir, I. Multi Robot Space Exploration: A Modified Frequency Whale Optimization Approach. In Proceedings of the AIAA SCITECH 2022 Forum, San Diego, CA, USA, 3–7 January 2022; p. 1416.
10. Gul, F.; Mir, I.; Abualigah, L.; Sumari, P. Multi-Robot Space Exploration: An Augmented Arithmetic Approach. *IEEE Access* **2021**, *9*, 107738–107750. [CrossRef]

11. Gul, F.; Rahiman, W.; Alhady, S.N.; Ali, A.; Mir, I.; Jalil, A. Meta-heuristic approach for solving multi-objective path planning for autonomous guided robot using PSO–GWO optimization algorithm with evolutionary programming. *J. Ambient Intell. Humaniz. Comput.* **2021**, *12*, 7873–7890. [CrossRef]
12. Gul, F.; Mir, I.; Rahiman, W.; Islam, T.U. Novel Implementation of Multi-Robot Space Exploration Utilizing Coordinated Multi-Robot Exploration and Frequency Modified Whale Optimization Algorithm. *IEEE Access* **2021**, *9*, 22774–22787. [CrossRef]
13. Gul, F.; Mir, I.; Abualigah, L.; Sumari, P.; Forestiero, A. A Consolidated Review of Path Planning and Optimization Techniques: Technical Perspectives and Future Directions. *Electronics* **2021**, *10*, 2250. [CrossRef]
14. Gul, F.; Alhady, S.S.N.; Rahiman, W. A review of controller approach for autonomous guided vehicle system. *Indones. J. Electr. Eng. Comput. Sci.* **2020**, *20*, 552–562. [CrossRef]
15. Gul, F.; Rahiman, W. An Integrated approach for Path Planning for Mobile Robot Using Bi-RRT. In Proceedings of the IOP Conference Series: Materials Science and Engineering, Kuala Terengganu, Malaysia, 27–28 August 2019; IOP Publishing: Kuala Terengganu, Malaysia, 2019; Volume 697, p. 012022.
16. Gul, F.; Rahiman, W.; Nazli Alhady, S.S. A comprehensive study for robot navigation techniques. *Cogent Eng.* **2019**, *6*, 1632046. [CrossRef]
17. Szczepanski, R.; Tarczewski, T.; Grzesiak, L.M. Adaptive state feedback speed controller for PMSM based on Artificial Bee Colony algorithm. *Appl. Soft Comput.* **2019**, *83*, 105644. [CrossRef]
18. Szczepanski, R.; Bereit, A.; Tarczewski, T. Efficient Local Path Planning Algorithm Using Artificial Potential Field Supported by Augmented Reality. *Energies* **2021**, *14*, 6642. [CrossRef]
19. Szczepanski, R.; Tarczewski, T. Global path planning for mobile robot based on Artificial Bee Colony and Dijkstra's algorithms. In Proceedings of the 2021 IEEE 19th International Power Electronics and Motion Control Conference (PEMC), Gliwice, Poland, 25–29 April 2021; pp. 724–730.
20. Khalil, B.; Yesildirek, A. System identification of UAV under an autopilot trajectory using ARX and Hammerstein-Wiener methods. In Proceedings of the 7th International Symposium on Mechatronics and Its Applications, Sharjah, United Arab Emirates, 20–22 April 2010; pp. 1–5.
21. Mir, I.; Akhtar, S.; Eisa, S.; Maqsood, A. Guidance and control of standoff air-to-surface carrier vehicle. *Aeronaut. J.* **2019**, *123*, 283–309. [CrossRef]
22. Mir, I.; Taha, H.; Eisa, S.A.; Maqsood, A. A controllability perspective of dynamic soaring. *Nonlinear Dyn.* **2018**, *94*, 2347–2362. [CrossRef]
23. Mir, I.; Maqsood, A.; Eisa, S.A.; Taha, H.; Akhtar, S. Optimal morphing–augmented dynamic soaring maneuvers for unmanned air vehicle capable of span and sweep morphologies. *Aerosp. Sci. Technol.* **2018**, *79*, 17–36. [CrossRef]
24. Mir, I.; Maqsood, A.; Taha, H.E.; Eisa, S.A. Soaring Energetics for a Nature Inspired Unmanned Aerial Vehicle. In Proceedings of the AIAA Scitech 2019 Forum, San Diego, CA, USA, 7–11 January 2019; p. 1622.
25. Hussain, A.; Hussain, I.; Mir, I.; Afzal, W.; Anjum, U.; Channa, B.A. Target Parameter Estimation in Reduced Dimension STAP for Airborne Phased Array Radar. In Proceedings of the 2020 IEEE 23rd International Multitopic Conference (INMIC), Bahawalpur, Pakistan, 5–7 November 2020; pp. 1–6.
26. Hussain, A.; Anjum, U.; Channa, B.A.; Afzal, W.; Hussain, I.; Mir, I. Displaced Phase Center Antenna Processing For Airborne Phased Array Radar. In Proceedings of the 2021 International Bhurban Conference on Applied Sciences and Technologies (IBCAST), Islamabad, Pakistan, 12–16 January 2021; pp. 988–992.
27. Gul, F.; Rahiman, W. Mathematical Modeling of Self Balancing Robot and Hardware Implementation. In Proceedings of the 11th International Conference on Robotics, Vision, Signal Processing and Power Applications: Universiti Sains Malaysia, Penang, Malaysia, 5–6 April 2021; Springer: Penang, Malaysia, 2022; pp. 20–26.
28. Hopping, B.M.; Garrett, T.M. Low Speed Airfoil Design for Aerodynamic Improved Performance of UAVs. U.S. Patent 9,868,525, 29 June 2018.
29. Mir, I.; Maqsood, A.; Akhtar, S. Optimization of dynamic soaring maneuvers to enhance endurance of a versatile UAV. In *Proceedings of the IOP Conference Series: Materials Science and Engineering*; IOP Publishing: Bangkok, Thailand, 2017; Volume 211, p. 012010.
30. Mestrinho, J.; Gamboa, P.; Santos, P. Design optimization of a variable-span morphing wing for a small UAV. In Proceedings of the 52nd AIAA/ASME/ASCE/AHS/ASC Structures, Structural Dynamics and Materials Conference, Denver, CO, USA, 4–7 April 2011; Volume 47.
31. Mir, I.; Eisa, S.A.; Taha, H.; Maqsood, A.; Akhtar, S.; Islam, T.U. A stability perspective of bio-inspired UAVs performing dynamic soaring optimally. *Bioinspir. Biomim.* **2021**. [CrossRef]
32. Ahsan, J.; Ahsan, M.; Jamil, A.; Ali, A. Grey Box Modeling of Lateral-Directional Dynamics of a UAV through System Identification. In Proceedings of the 2016 International Conference on Frontiers of Information Technology (FIT), Islamabad, Pakistan, 19–21 December 2016; pp. 324–329.
33. Rasheed, A. Grey box identification approach for longitudinal and lateral dynamics of UAV. In Proceedings of the 2017 International Conference on Open Source Systems & Technologies (ICOSST), Lahore, Pakistan, 18–20 December 2017; pp. 10–14.
34. BELGE, E.; Hızır, K.; PARLAK, A.; ALTAN, A.; HACIOĞLU, R. Estimation of small unmanned aerial vehicle lateral dynamic model with system identification approaches. *Balk. J. Electr. Comput. Eng.* **2020**, *8*, 121–126. [CrossRef]

35. Altan, A.; Aslan, Ö.; Hacıoğlu, R. Model predictive control of load transporting system on unmanned aerial vehicle (UAV). In Proceedings of the Fifth International Conference on Advances in Mechanical and Robotics Engineering, Rome, Italy, 1 May 2017; Institute of Research Engineers and Doctors: Rome, Italy, 2017; pp. 1–4.
36. Dube, C.; Pedro, J.O. Modelling and closed-loop system identification of a quadrotor-based aerial manipulator. *J. Phys. Conf. Ser.* **2018**, *1016*, 012007.
37. Cavanini, L.; Ferracuti, F.; Longhi, S.; Monteriù, A. Ls-svm for lpv-arx identification: Efficient online update by low-rank matrix approximation. In Proceedings of the 2020 International Conference on Unmanned Aircraft Systems (ICUAS), Athens, Greece, 1–4 September 2020; pp. 1590–1595.
38. Cavanini, L.; Ippoliti, G.; Camacho, E.F. Model predictive control for a linear parameter varying model of an UAV. *J. Intell. Robot. Syst.* **2021**, *101*, 1–18.
39. Belge, E.; Altan, A.; Hacıoğlu, R. Metaheuristic Optimization-Based Path Planning and Tracking of Quadcopter for Payload Hold-Release Mission. *Electronics* **2022**, *11*, 1208. [CrossRef]
40. Weng, Y.; Nan, D.; Wang, N.; Liu, Z.; Guan, Z. Compound robust tracking control of disturbed quadrotor unmanned aerial vehicles: A data-driven cascade control approach. *Trans. Inst. Meas. Control* **2022**, *44*, 941–951. [CrossRef]
41. Lecerf, M.; Allaire, D.; Willcox, K. Methodology for dynamic data-driven online flight capability estimation. *AIAA J.* **2015**, *53*, 3073–3087. [CrossRef]
42. Yu, Y.; Guo, J.; Ahn, C.K.; Xiang, Z. Neural adaptive distributed formation control of nonlinear multi-uavs with unmodeled dynamics. *IEEE Trans. Neural Netw. Learn. Syst.* **2022**. [CrossRef]
43. Saengphet, W.; Tantrairatn, S.; Thumtae, C.; Srisertpol, J. Implementation of system identification and flight control system for UAV. In Proceedings of the 2017 3rd International Conference on Control, Automation and Robotics (ICCAR), Nagoya, Japan, 24–26 April 2017; pp. 678–683.
44. Bnhamdoon, O.A.A.; Mohamad Hanif, N.H.H.; Akmeliawati, R. Identification of a quadcopter autopilot system via Box–Jenkins structure. *Int. J. Dyn. Control* **2020**, *8*, 835–850. [CrossRef]
45. Ayyad, A.; Chehadeh, M.; Awad, M.I.; Zweiri, Y. Real-time system identification using deep learning for linear processes with application to unmanned aerial vehicles. *IEEE Access* **2020**, *8*, 122539–122553. [CrossRef]
46. Munguía, R.; Urzua, S.; Grau, A. EKF-based parameter identification of multi-rotor unmanned aerial vehiclesmodels. *Sensors* **2019**, *19*, 4174. [CrossRef]
47. Puttige, V.R.; Anavatti, S.G. Real-time system identification of unmanned aerial vehicles: A multi-network approach. *J. Comput.* **2008**, *3*, 31–38. [CrossRef]
48. Wu, Z.; Karimi, H.R.; Dang, C. An approximation algorithm for graph partitioning via deterministic annealing neural network. *Neural Netw.* **2019**, *117*, 191–200. [CrossRef]
49. Hoffer, N.V.; Coopmans, C.; Jensen, A.M.; Chen, Y. A survey and categorization of small low-cost unmanned aerial vehicle system identification. *J. Intell. Robot. Syst.* **2014**, *74*, 129–145. [CrossRef]
50. Sierra, J.E.; Santos, M. Modelling engineering systems using analytical and neural techniques: Hybridization. *Neurocomputing* **2018**, *271*, 70–83. [CrossRef]
51. Stevens, B.L.; Lewis, F.L.; Johnson, E.N. *Aircraft Control and Simulation: Dynamics, Controls Design, and Autonomous Systems*; John Wiley & Sons: Hoboken, NJ, USA, 2015.

MDPI
St. Alban-Anlage 66
4052 Basel
Switzerland
Tel. +41 61 683 77 34
Fax +41 61 302 89 18
www.mdpi.com

Processes Editorial Office
E-mail: processes@mdpi.com
www.mdpi.com/journal/processes

www.ingramcontent.com/pod-product-compliance
Lightning Source LLC
LaVergne TN
LVHW070159170726
843515LV00007B/1949

* 9 7 8 3 0 3 6 5 4 7 7 1 8 *